Schonland
Scientist and Soldier

Schonland
Scientist and Soldier

Brian Austin

Department of Electrical Engineering and Electronics
University of Liverpool, UK

Institute of Physics Publishing
Bristol and Philadelphia

British Library Cataloguing-in-Publication Data
A catalogue record for this book is available from the British Library.

ISBN 0 7503 0501 0

Library of Congress Cataloging-in-Publication Data are available

Commissioning Editor: James Revill
Production Editor: Simon Laurenson
Production Control: Sarah Plenty
Cover Design: Frédérique Swist
Marketing Executive: Colin Fenton

Published by Institute of Physics Publishing, wholly owned by The Institute of Physics, London

Institute of Physics Publishing, Dirac House, Temple Back, Bristol BS1 6BE, UK

US Office: Institute of Physics Publishing, Suite 1035, The Public Ledger Building, 150 South Independence Mall West, Philadelphia, PA 19106, USA

Typeset by Academic + Technical Typesetting, Bristol
Printed in the UK by MPG Books Ltd, Bodmin, Cornwall

For Kath

CONTENTS

FOREWORD

by Sir Maurice Wilkes FREng FRS

Basil Schonland, the subject of this biography by Dr Brian Austin, was a distinguished South African who, during and after the Second World War, occupied important positions both in South Africa and in the United Kingdom. He was scientific adviser to Field Marshal Montgomery during the invasion of Europe. He then returned to South Africa to establish a modern scientific and industrial research organization in that country. His final appointment was that of Director of the UK Atomic Energy Research Establishment at Harwell.

It was my great good fortune, during my wartime career, to serve under Schonland for very nearly two years. My close association with him began in August 1941 when he took charge of an Army operational research group, based on Petersham in Surrey, in which I worked. It was at once clear that he was just the sort of immediate chief that I could most desire, and I came to admire him greatly. Like me, he was a Cavendish man, although of an older generation. He had worked under Rutherford, first as a research student and later as a visitor to the Cavendish Laboratory, in the golden age of Cambridge physics. He was one of an illustrious group that included Blackett, Cockcroft, Kapitza and others. The close bond he formed at that time with Cockcroft, later Sir John Cockcroft, had an important influence on his career and forms one of the main themes of Dr Austin's book.

When he returned to his native South Africa, Schonland found it difficult to continue to work in atomic physics so far from the main centres, and took up the study of thunderstorms and lightning, a field for which South Africa provided exceptional opportunities. Before long, he became a world authority in this subject and in 1938 he was elected a Fellow of the Royal Society. South Africa was then rather isolated scientifically, and Schonland told me that he saw it as his role to fly the flag of high quality research.

When I first worked under him, Schonland was a Lieutenant Colonel in the South African Army on loan to the British Army. He had a natural military bearing and it was clear that he found the military life congenial; his campaign medals from the First World War and a military OBE proclaimed him to be a seasoned soldier. He had intense feelings of patriotism both for South Africa and, in the wider sense, for the British Commonwealth. By the Statute of Westminster, South Africa was, as he emphasized to me, an independent country, which happened to have the same king as the United Kingdom.

In Petersham, Schonland found himself in a world that was new to him, namely, that of the British Service ministries and their dependent establishments, where soldiers and civil servants worked side by side.

As a newcomer, Schonland had to get up to speed. He was ready to pick up information where he could, if necessary from his subordinates; for example, I remember explaining D.O. (demi-official) letters to him.

We both lived in the Star and Garter Hotel in Richmond. Relations when off duty tended to be relaxed in Army circles, and I got to know him well. He could at times be almost boyish in his enthusiasm. He had a particular liking for the plays of J M Barrie, which I shared, although they were then unfashionable among superior people. He was never unkind in his judgements of other people; however, he gave vent to a certain mistrust of eminent scientists who made pronouncements on subjects outside their scientific specialities. He used to call them 'public scientists' and I detected a slightly malicious pleasure in the way he said it.

When Schonland took over the Operational Research Group, we were mostly concerned with radar in Anti-Aircraft Command. However, the Army began to appreciate the potentialities of operational research in various aspects of land warfare and it became Schonland's task to develop sections dealing with signals, tanks, field artillery, infantry weapons and tactics, and so on. My own interests did not develop in any of these directions and it was natural that, in the summer of 1943, I should leave the Group to return to mainstream radar. Very soon, I received a posting that took me right across ministerial boundaries, and I found myself at TRE working for the RAF on airborne radar. I thus passed completely out of Schonland's orbit and only heard from time to time of his activities. It was with the greatest interest, therefore, that I read Dr Austin's account of Schonland's adventures in the challenging and somewhat improbable role of Scientific Advisor to Field Marshal Montgomery at 21st Army Group Main Headquarters.

During his later period at the Army Operational Research Group, Schonland saw a great deal of General Smuts, who was then Prime Minister of the Union of South Africa and often in England. Smuts's mind was turning to post-war reconstruction in South Africa, and it was at his personal request that Schonland undertook to set up and preside over a scientific and industrial research organization along the lines of those to be found in other Dominions. Accordingly, he left 21st

Army Group at the end of October 1944 and, after a brief period in London, went back to South Africa.

The Council for Scientific and Industrial Research, as the organization was called, was very successful, but as time went on political developments occurred in South Africa that were far from being to the liking of either Schonland or his wife. He began, therefore, to show interest in approaches that were made to him from various parts of the British scientific establishment. Eventually he accepted an invitation from Sir John Cockcroft, then Director of the UK Atomic Energy Establishment at Harwell, to become his deputy. Later he succeeded Cockcroft as Director and remained in that position until his retirement.

Dr Austin captures well the character and personality of Schonland as I knew him. As well as being a biography, the book will be of interest to many readers for the contribution it makes to the historical record of events, both scientific and military, in Britain and South Africa.

Maurice V Wilkes

Cambridge

December 2000

PREFACE

It was while I was on the staff of the Department of Electrical Engineering at the University of the Witwatersrand (known by staff and students alike as 'Wits') in Johannesburg during the 1980s that a remarkable piece of local history fired my imagination. Though vaguely aware that some most secret wartime work had been done at Wits it was only then that I discovered that an elementary radar set, designed and built just a stone's throw from my office, had produced the first radar echo seen in South Africa on 16 December 1939. The man behind that achievement, remarkable considering the circumstances of the time and the very infancy of radar itself, was Basil Schonland, Professor of Geophysics and Director of the University's Bernard Price Institute of Geophysical Research. Few South Africans knew anything about that event and the remarkable wartime developments that followed from it; others farther afield had no knowledge at all that radar ever existed in the southern reaches of Africa. In 1982, when the University celebrated its Diamond Jubilee, the Department's contribution to the occasion took the form of a series of lectures on radar — one of which, the South African story, was given by Professor Emeritus G R Bozzoli, a former Vice-Chancellor of the University and a significant figure in the development of radar in South Africa. Subsequently, word of South Africa's achievements in this field filtered through to the wider world at an international conference on radar in London in 1985 but was never published. In 1990, by which time I was living in England, the opportunity presented itself to tell the story again and this I did at a meeting in Oxford. But now the focus in my mind, at least, was already beginning to shift from Schonland as the father of South African radar to Schonland the man, but it was now apparent that his name had faded almost completely from the collective memory of his fellow countrymen. Elsewhere he was known to but a few. In 1995 I took the bold decision for an engineering academic to attempt to write a biography of Basil Schonland and I approached his eldest daughter, Dr Mary Davidson, then living in the Schonland family home near Winchester, for permission to do so. Mary graciously and bravely agreed and the project was born, but very much as a part-time activity, since my duties in the Department of Electrical Engineering and Electronics at the University of Liverpool took first call on my time. For the better part of the next five years I led the life of a hermit at home and must therefore thank my wife Kath for all her forbearance and understanding while I attempted to learn the trades of historian and biographer.

No task of this magnitude can be undertaken without the considerable assistance of a vast number of people. One of the many pleasures that the research and writing of the biography has brought me has been the friendships that I have made with a large number of people, both in South Africa and in England. Not only have I received considerable help and encouragement from those who either knew Schonland, worked with him or were just associated with him at various stages of his multi-faceted career, but I found myself almost overwhelmed by the kindness of those of another generation who were just so interested in the project itself and did all they could to assist me. It is to all these individuals and in many cases to the organizations they represent that I owe an enormous debt of gratitude. I shall attempt here to name them all; should I omit anyone, that is a mark of my own fallibility and inadequate record-keeping and I ask you to accept both my apologies for the omission and my heartfelt thanks for your help.

First and foremost I must thank Mary and Freddie Davidson for their hospitality at The Down House and for the remarkable collection of Basil and Ismay Schonland's letters and photographs which they entrusted to me. In addition, Ismay's recollections of Basil written in the hand of the professional historian after his death are priceless. Without all of these the story would certainly have lacked a soul. Mary has also been the recipient of numerous letters and phone-calls from me, often involving convoluted queries about events of long ago, but she never failed to dig deep into her memory to provide an answer. I also thank Schonland's grandson, Dr Robert Davidson, for sending me Basil Schonland's diary kept during his training at the Royal Engineers Signal Depot in Bletchley in 1915, as well as his notebooks written on the field of battle. They are a unique record. My interviews in Johannesburg with Ann Oosthuizen, Schonland's younger daughter, and with her daughter Susan in Cambridge revealed family insights and personal recollections that were invaluable. These laid the bedrock on which to build a story of quite considerable proportions.

To Professor T E Allibone must go a special mention for it was he who wrote the Biographical Memoir of Schonland that appeared the year after his death as part of that invaluable series commemorating the Fellows of The Royal Society in London. Ted Allibone had been associated with Schonland longer than any of his scientific colleagues and so was able to write a most readable and comprehensive account of Schonland's life and achievements. It was the spur that got me going.

Our subsequent correspondence, especially about life at Cambridge and the Cavendish, yielded fascinating cameos from an obviously happy time. To Lorna Arnold I owe a special debt of gratitude. As an employee of the United Kingdom Atomic Energy Agency since 1959, and an historian of considerable note in all matters nuclear, her assistance and advice on Harwell, its personalities and its history were simply beyond comparison. Her own work on the accident at Windscale and her collaboration with Professor Margaret Gowing in writing the definitive account of atomic energy in Britain have been of inestimable value to me in understanding Schonland's part in this massive enterprise. She kindly agreed to read early drafts of the Harwell chapters and I acted immediately on all her comments and suggestions. In addition, my ultimate access to the Schonland files at Aldermaston and Harwell was due entirely to her initiative in setting up the right contacts.

Sadly, the passage of time between commencing my research and actually writing the biography has seen a number of my ever-willing collaborators pass away. None was more poignant to me than the death of Professor G R Bozzoli, simply 'Boz' to every student and colleague privileged enough to know him. A most eminent man himself with an illustrious career in academic life, Boz played a key role in the wartime radar story in South Africa because he was Schonland's chief technical officer responsible for the design of much of the radar equipment and for the running of the Special Signals Services after Schonland left for England in 1941. The interview I conducted with him at his home in Johannesburg in 1997, and our subsequent correspondence until just before his death the following year, were sources of invaluable information and insight. My great regret is that he did not live to read the final account. Others, too, who have since died were Professor R V Jones, Dr J S Hey and Sir Arthur Vick. All had known Schonland in various capacities and at different periods of his career and they responded very willingly to my requests for information about the Bruneval Raid, the jamming of British army radars, and Harwell, respectively.

The wealth of published material on the Second World War contains little about the part played by Operational Research in the various campaigns and engagements. I was, however, most fortunate to discover that the reports produced by Schonland's Army Operational Research Group had been very carefully analysed and catalogued in recent years and they were made available to me by Lt Col Richard Dixon RM of the Centre for Defence Analysis, and by the Royal Military College of Science. Most recently, Professor Terry Copp of the Laurier Centre for Military Strategic and Disarmament Studies at the Wilfrid Laurier University in Canada, has produced the definitive account of the part played by Operational Research within 21 Army Group and he most kindly provided me with much valuable correspondence between

Schonland and his second-in-command at the AORG, Omond Solandt, a Canadian, and I am indebted to him for allowing me to quote from it here.

Numerous former colleagues and associates of Schonland responded to my requests for information and even anecdotes. My thanks go to Dr J V Dunworth for his hospitality at his home on the Isle of Man, and to R M Fishenden, Lord Flowers, R F Jackson, Dr G H Stafford and Professor P C Thonemann for their recollections of Harwell under Schonland. Professor Anton Hales in Australia wrote detailed letters of events long ago at UCT and the BPI with an accuracy that was astounding, while Dr Stanley Hey gave me his first-hand account of the discovery of radio emissions from the sun while at the AORG in 1942, and Schonland's reaction to a discovery of huge significance as one of the forerunners of radio astronomy. His namesake, Professor J D Hey at the University of Natal, provided me with a number of photographs of the Boys camera, undoubtedly the most important scientific tool in Schonland's career. Professor Frank Brooks, Dr John Juritz and Lesley Jennings in the Department of Physics at UCT turned up fascinating historical gems whilst I was there and continue to do so. The part played by Dr Frank Hewitt, formerly of the SSS, the TRL and then finally Deputy President of the CSIR in clarifying innumerable issues has been invaluable. His reading of my chapters on the SSS has ensured their accuracy in any matters in which he was involved. Geoffrey Mangin, now custodian of SSS affairs in South Africa, has been a regular source of valuable information. Professor David Hill, Schonland's Staff Officer in 21 Army Group, was very helpful in recalling events from that period in both his correspondence and at his home near Ripon. My search for information in relation to Schonland's Great War service led me to Sir John Keegan, Brigadier Richard Holmes and Mrs Ann Clayton. Sir Bernard Lovell has been especially helpful in a host of areas, particularly in relation to Schonland's part in the planning of the Bruneval Raid and his work on 'the penetrating radiation'. He put me in touch with Don Preist in the United States whose recollections of that famous raid in 1942 were remarkable. Sir Arnold Wolfendale, the former Astronomer Royal, at Lovell's request provided useful comments on cosmic rays; as did Sir John Mason, whose subsequent research on the charging process within thunderclouds was stimulated by Schonland's work before the war.

Dr R S Pease provided me with much detail about the fusion research at Harwell and assisted in my understanding of the underlying issues that so plagued both the ZETA and ICSE programmes. Dr J E Johnston, also from Harwell, saw the interplay between Schonland and Cockcroft and also introduced me to Mr Nick Hance to whom I am greatly indebted for all his help in tracing photographs from the Harwell period. Dr David Proctor, formerly of the NITR in Johannesburg, helped me place Schonland's work on lightning in perspective, while the events leading to the founding of the CSIR and subsequent developments there were described to me by Dr Bill Rapson in Johannesburg in 1997. Dr John Stewart of the Chamber of Mines has also been most helpful in tracing the Schonland report on research in the mining industry in South Africa. I owe a special debt of gratitude to three Professors Emeriti at the University of the Witwatersrand: Frank Nabarro, Friedel Sellschop and Phillip Tobias. Frank and his late wife Margaret were most patient when I interviewed them at length about their association with Schonland in the AORG and about numerous subsequent events. Frank also read and commented on the complete manuscript. The saga of attracting Sellschop to Wits and his own recollections of those and subsequent activities flesh out the story as told by Schonland's correspondence with Nabarro. Phillip Tobias put into perspective the discoveries of Robert Broom and particularly the antagonistic views held by Solly Zuckerman in relation to South African palaeontology and South Africa in general.

No research of this nature would be possible without the massive assistance of numerous librarians and archivists. I have been exceptionally well-served by Di Arnott and Zofia Sulej at the University of the Witwatersrand, Nicolene Basson and Janie van Zyl at the CSIR, Ellie Clewlow at Gonville and Caius College, Cambridge, Joanna Corden at The Royal Society, John Drew and Lesley Hart at UCT, Valerie Phillips of the Royal Commission for the Exhibition of 1851, Kate Pyne at AWE Aldermaston, Lynne Seddon at the Royal Military College of Science, Godfrey Waller at Cambridge, Christina Tyree at the British Association for the Advancement of Science and her counterpart in South Africa, Shirley Korsman. In addition, the staff of the Public Record Office in Kew, the Churchill Archive, Cambridge, the Imperial War Museum, London, the National Archive of South Africa and the South African Museum of Military History have been most helpful in responding to my numerous requests for obscure information. My special thanks though must go to Sandy Rowoldt, the Cory Librarian at Rhodes University, Grahamstown for her wonderful support that went well beyond the call of duty. The Cory collection of Schonland's letters to his brother Felix were a veritable gold mine of priceless information. For permission to open some that Felix had sealed not quite in perpetuity I must personally thank the Vice-Chancellor of Rhodes, Dr David Woods. I must also thank three professional historians: Philip Bell, formerly at Liverpool, Deon Fourie at Unisa and Bruce Murray at Wits whose capacity to encourage a novice knew no bounds. My thanks go as well to Antony Clark, Headmaster of St Andrew's College, Grahamstown, for introducing me to Doug and Lettie Rivett and for allowing me access to the College archives. Jim Revill, Senior Academic Publisher at the Institute of Physics, has been remarkably patient and most helpful throughout the gestation period of this work.

The sabbatical leave that I was able to spend in South Africa early in 1997 in order to carry out the research for this biography was made possible by the good offices of the University of Liverpool. In particular I wish to thank Professor David Parsons, my Head of Department at the time, for supporting me in this venture and for recognizing the value of such research. Professor Charles Landy, Head of the Department of Electrical Engineering at the University of the Witwatersrand, provided me with an office in his department and supported my wife and me in so many ways while we were in Johannesburg. His friendship is valued enormously. Generous financial support made it all possible and for this I must thank Dr Geoff Garrett, President of the CSIR, the South African Foundation for Research Development, the Ernest Oppenheimer Trust and the South African Institute of Electrical Engineers.

Kind permission to reproduce material has been obtained from Dr Mary Davidson, the Office of the Prime Minister, The British Library, The Royal Society, the publishers of *Nature*, the Royal Commission for the Exhibition of 1851, the UKAEA, the Imperial War Museum, Churchill Archives Centre, the National Archives of South Africa, Oxford University Press, the Universities of Cambridge, Cape Town, and of the Witwatersrand, Rhodes University, and the South African National Defence Force.

Every effort has been made to trace the relevant copyright holders of material used in the biography. Any omissions or lack of appropriate acknowledgement are an unfortunate oversight on the part of the author and are sincerely regretted.

One man whose wise counsel, encyclopaedic knowledge and tactful criticism have been of inestimable value to me throughout my research and writing of this Schonland biography has been Sir Maurice Wilkes FREng FRS. Our discussions in the most wonderful of settings in Cambridge and in London were both memorable and fascinating. They, and our frequent electronic correspondence, touched on far more than just the period when he served under Schonland at Petersham. I owe whatever understanding I have of the structures and personalities of the British scientific establishment of the time entirely to his deep and perceptive grasp of all the issues and, indeed, of the machinations that in some cases lay behind them. I am therefore most appreciative of his kindness and willingness to assist me in writing the biography of the man for whom I know he had the very highest regard. By agreeing to write the foreword to this biography of Sir Basil Schonland CBE FRS he has done me the ultimate honour.

B A Austin

West Kirby

January 2001

CHAPTER 1

EARLY YEARS

In many ways the future of modern South Africa was determined by events in 1896, the year that Basil Schönland was born. The two British colonies, the Cape and Natal, were part of the Victorian empire, secure yet undistinguished. The Boer republics of the Orange Free State and the Transvaal—the homes to a people who had trekked away from the colonial yoke—were about to be dragged back by events which were driven by the discovery of gold just ten years before.

Basil Schönland was born in the Eastern Cape on 5 February 1896 in the town of Grahamstown that lay on the frontier of what, to some, was still darkest Africa and another people: black people whose ancestors had moved south and had encountered the first white men to inhabit this southern tip of Africa just a few generations before. The Eastern Cape was settler country, populated by hardy folk, predominantly from England, who had arrived in 1820, ostensibly to a new Eden but in reality to a rather inhospitable place with restless neighbours. Such, though, was their dogged determination that Grahamstown and the surrounding territory soon flourished. What began as a military encampment rapidly became a town well-served by both churches and schools modelled on the public schools of England. Amongst them was St Andrew's College, founded in 1855 and staffed mainly by men of the cloth whose purpose was undeniably to educate the sons of gentlemen.

But there were other settlers too. Selmar Schönland and his brother Max arrived in the Eastern Cape in 1889. They were German Jews: Selmar, a scientist; Max, the trader. Selmar Schönland was born in Frankenhausen in 1860 and studied botany at the universities of Berlin and Kiel where he obtained a doctorate. He then moved to England and to Oxford where, over a period of four years, he did excellent work and had conferred on him the degree of Master of Arts. However, problems with his eyesight made laboratory work difficult and it was suggested that he should go to the colonies and concentrate on fieldwork [1]. He and his brother, attracted by the commercial prospects that had been bolstered by the recent discovery of diamonds at Kimberley in 1870, set sail together. Selmar soon found himself a position as an assistant to Peter MacOwan,

1

rector and head of Natural Sciences at Gill College in Somerset East, a small town some 100 km north-west of Grahamstown. MacOwan, said to be a 'peppery old Irishman', though born in Yorkshire, was a man of sturdy Church of England stock and was the pre-eminent botanist in the country, who was soon to become the President of the South African Philosophical Society, the forerunner of the Royal Society of South Africa [2].

MacOwan had a daughter, Flora, who was blessed or possibly cursed with her father's temperament. Her fiery personality and fine features offset her lack of height, for though under five feet tall she was a formidable person with a temper to match. She and Selmar were married and soon after left Somerset East for Grahamstown, where he had been appointed curator of the Albany Museum, the second oldest in the country with a fine collection of botanical, palaeontological and geological specimens dating from the early years of the century. There, Selmar pursued his botanical research with enthusiasm and began a collection of dried plants which, over a period of forty years, was to increase to 100 000 specimens. However, there was more to the man than the mere collector, for he was a scientist at heart and he sorely missed the worlds of Oxford and Kiel and their communities of scholars. Grahamstown must have its own seat of learning — a university — and he devoted himself to the task of establishing one in his adopted home. What is now Rhodes University owes its very existence to this gentle, sharp-eyed botanist, Selmar Schönland [3].

1896 was a climactic year in South Africa's history. It was the year of the fiasco of the Jameson Raid. This ill-timed and equally ill-thought-out venture was the springboard from which the Boer War was to be launched some three years later. Dr Leander Starr Jameson was the confidant and close colleague of Cecil Rhodes, then Prime Minister of the Cape Colony. In every way Jameson was Rhodes's right-hand man [4]. Frustrated by the duplicity of President Kruger of the Transvaal Republic, and urged on by champions of Empire, Jameson led a body of irregular horse across the border from Pitsani in British-controlled Bechuanaland with a view to ousting the old Boer and claiming the place for his Queen. But calamity soon struck. On 2 January 1896, Jameson's raiders were surprised, surrounded and out-gunned. After a brief skirmish they surrendered to the Boers when almost on the outskirts of Johannesburg, the seat of *Uitlander* (lit. 'Outlander') power in the Transvaal and Jameson's ultimate destination. Of the six hundred men who had ridden with him from their camp near Mafeking, 65 were dead or wounded and Jameson himself was led away, in tears, to Pretoria and jail. The plan, such as it was, was intended to foment a revolution in Johannesburg, to be supported by the *Uitlanders*, that would unseat Paul Kruger and take over the Transvaal for the British Empire. It failed and Rhodes resigned as Prime Minister of the Cape [5].

War was now on the horizon, but in Grahamstown, almost a thousand kilometres away, life continued with an air of what one rather acerbic commentator of the time described as 'old world ecclesiasticism', and into that world Basil Schönland was born.

*   *   *

He was an extremely tiny baby, so tiny, almost bird-like, in fact, that his first cot was a shoe box and he was fed with a pipette from a fountain pen [6]. Such an unusual method of infusion seemingly had no ill effects because Basil soon grew into an active, alert little boy with qualities of independence and inventiveness much in evidence. The Schönlands first lived in a house in Francis Street but soon moved to 'Bellevue' in Oatlands Road. Two brothers followed. Felix was born in 1899 and Richard, to be known as Dickey and later Dick, six years later. Basil and Felix, naturally closer, shared similar interests, with Basil clearly the leader and much admired by his younger brother. Walking in the hills surrounding Grahamstown was a regular weekend pastime, usually with groups of friends, but frequently the elder Schönland would go off alone to enjoy the challenge and the isolation of the veld. At Howieson's Poort, a favourite picnic spot, he kept in a cave on the hillside a cache of tinned sardines, coffee, sugar and condensed milk; his 'secret store', he called it [7]. As well as an air of independence, soon to emerge were an artistic flair and a hint of organizational ability, for Basil became the organizer of regular circus performances staged in the stable and coach house of the Schönland home. The price of admission was a button, but as the frequency of these events increased the mothers of the audience soon complained to Mrs Schönland that their children's clothing was rapidly falling into a state of some disrepair. Some other currency or at least a diversionary activity was required.

Basil Schönland's first formal exposure to the educational process began at the Victoria Infants school and continued, at the age of six, when he was enrolled at St Andrew's Preparatory School. He was immediately seen to be a good scholar. His school report for that year had him in fourth place in the form but first in Arithmetic and History and second in both English and Scripture. The teacher's comments alongside these marks were 'good' throughout, with English earning extra praise, being both 'very good and improving fast'. However, one detects signs of an intellect not fully stimulated by the curriculum on offer because his conduct was described as 'Good, but rather too fidgetty [sic] & talkative' [8].

St Andrew's was renowned for the quality of its stage productions and they were much appreciated, since the performance of original plays in Grahamstown was something of a novelty according to *Grocott's Daily Mail*, the local newspaper. It described Miss Mullins's production in

Figure 1. Basil Schönland in Grahamstown at the age of eight in 1904. (Reproduced by kind permission of Dr Mary Davidson.)

June 1903 of the *Shaming of the Two* in the typically expansive prose of the day by suggesting that if she 'continues to provide such capital fun from her pupils' acting, in future a larger hall must be procured.' Basil Schönland's role in this epic, which involved 'such exciting material' as robbers, dwarfs, a cave, magic, a prince and much other royalty to boot, was that of a Page. His friend Rupert White-Cooper, who would emerge again in other rather different roles in later years, played the not-unusual part, in an all-boys school, of a princess. At the conclusion of the entertainment, the 'table containing the prizes was put ready' for distribution to the worthy winners by the Bishop of Grahamstown who was accompanied by the Rev Dr McGowan, the Principal of 'the College'—St Andrew's itself. That year Basil Schönland won the prize for Arithmetic.

By the following year Basil had himself been elevated to playing the part of a lady. In this case it was Lady Olga in the school's production of *The borrowed clothes of Princess Rose*, again written and produced by the redoubtable Mullins sisters. Rupert White-Cooper now found himself cast as a shepherdess. Again, the report in the local newspaper was effusive in its praise and, once again, the worthy bishop, shadowed by

4

the headmaster of the College, handed out the prizes. Schönland won those for Scripture and Latin, while emerging was evidence that he was also something of an all-rounder. His lack of physical size was naturally a handicap on the sports field but he more than made up for it by a steely determination that saw him coming in third over 56 yards, a distance no doubt thought appropriate for boys of his age [9].

While Basil was so engaged, Dr Selmar Schönland, in his quiet though very determined way, was applying himself to the task of raising the money for the building of a university in Grahamstown. He turned, remarkable as it may have seemed to some, to the man who until recently had been languishing in jail in England for his misguided adventure up in the Transvaal seven years before. Now, Dr Jameson, having been released early from prison for health reasons, was back in South Africa and very much back in politics.

Cecil Rhodes, though, was dead. The events of the closing years of the century had finally taken a heavy toll on a man who, forty years before, was thought to be unfit for the rigours of the English climate. The African sun was recommended and so the young Rhodes arrived in the country where he was to leave an indelible mark. However, in death, just as during all those years when he dominated both the political and financial worlds in Cape Town and Kimberley, he left another legacy which, to this day, bears his name: the Rhodes Trust and the scholarships to Oxford it has provided to the scholars of selected nations around the world. Jameson was now a Rhodes Trustee and also the recently elected Prime Minister of the Cape Colony, but of much more importance to the canny Selmar Schönland was the fact that Jameson's parliamentary seat was in Grahamstown. Schönland therefore approached his MP for a substantial grant from the Rhodes Trustees for the building of a university to serve his constituents and Jameson readily agreed. Thus, £50 000 would be made available and Selmar Schönland had his university, at least on paper. However, the word of politicians is so often tempered by factors beyond the grasp of mere mammon and, almost inevitably, the promised grant was not forthcoming. Evidently, Jameson had taken his magnanimous decision without consulting his fellow trustees, his continued support by the good citizens of Grahamstown being the more pressing need at that time. However, he soon found himself under severe restraining pressure from a member of his own Cabinet, the redoubtable and parsimonious Sir Lewis Michell, who had been banker to Cecil Rhodes and was himself a Rhodes Trustee. Progress immediately stalled but Selmar Schönland was not to be thwarted and travelled to Cape Town to see Michell himself. There he patiently explained, for one and a half hours, just why the money was needed. Whether it was through sheer exhaustion or the force of argument from this dogged German botanist is not recorded, but Sir Lewis capitulated and Grahamstown was to get its money and thus its university, soon to be known as Rhodes University College [10].

*   *   *

Basil Schönland entered St Andrew's College in 1907 having been awarded, at the age of ten, the first prize in the Junior Division, an examination taken by pupils throughout the Cape, most of whom were some years his senior. His talents apparently knew few bounds and even extended to poetry. Though Kipling's verse and the embers of Empire may have been fading elsewhere, they still smouldered in Grahamstown and certainly at St Andrew's. On 9 March 1906, a memorial to the British soldiers who had fallen during the Boer War was unveiled in the city and this colourful event prompted Basil Schönland to write, with much gusto, five verses of his own [11].

> *Blow sweet and cool O winds today*
> *O southern Sunshine bright*
> *For we are gathered here to pay*
> *A last and solemn rite*
>
> *We see the boyish figure*
> *We hear the pleasant laugh*
> *Gaily in manhood's vigor* [sic]
> *Stepping out on glory's path*
>
> *They fought for king and native land*
> *Our dearest and our best*
> *But God put forth his mighty hand*
> *And took them to their rest*
>
> *The last post sounds in every ear*
> *The list of names is read*
> *The volleys three each one does hear*
> *In honour of the dead*
>
> *O bright and burnished statue*
> *Stand out against the sky*
> *Telling to all who see you*
> *How brave men went to die*

Schönland's arrival at College was somewhat daunting, as well it might have been for a boy so much younger than his peers. He entered form III in 1907 with the school number of 2163. One must presume that his skipping of forms I and II, though nowhere recorded, was due to his academic precociousness displayed while at the 'Prep'. In a letter written some years later to his brother Felix he admitted that he 'was horribly frightened at first' [12], but with the grit that was always a feature of his character, as man and boy, he soon made his mark. The College Magazine recorded in its December 1907 issue that Basil Schönland, at the recent Cape Town eisteddfod, had been awarded first prize for letter-writing. 'As he is only 11 years old this result is highly creditable', it said. His name was to appear again amongst the prize-winners in the September 1908 issue; this time he was awarded the Beit Scholarship — the first of many to come his way. For all this, Schönland was no book-worm. He was greatly interested in sport, especially rugby football, even though he lacked the physical stature usually associated with its more successful proponents. As a 'day boy' at a boarding school he was a member of the imaginatively titled Day House, those English Public School traditions being much a feature of such schools in South Africa too. Basil represented his House at rugby in the somewhat unusual posi-tion, given his lack of physical size, as a forward and though he did not distinguish himself he exuded great energy on the field and much enthusiasm for the game. His team-mates must similarly have lacked the necessary bulk and talent because the College Magazine recorded their heavy defeats in 1908 to the tune of 24 nil, 21 nil and 16–3 at the hands, or feet, of those more hardened boarders [13].

However, Basil Schönland's academic performance was excellent, as reflected in his school report for 1908, the year in which he took the School Higher Examination. He was first out of 23 pupils in all subjects except Latin and French, in which he appeared second. His mathematics master penned alongside a mark of 73%: 'An excellent worker with plenty of ability — did well in Arithmetic in Exam but his Algebra was poor'. An undoubtedly prophetic remark was that from the Acting Principal of the College, one Theodore Cornish, that Schönland was an 'Excellent worker. A very promising boy' [14]. And all this at the age of twelve with just two years left at school was promise indeed.

1909 was a relatively uneventful year by Schönland's standards. By then he was in the fifth form and in a class where the average age was seventeen. His school report yielded an entirely expected result. First place out of twenty-six with only one mark below 70%, that being 54% for Greek, a subject he had just started the year before [15]. At the annual prize-giving ceremony in September, his hand was shaken repeat-edly as he collected the prizes for French, Latin, Mathematics and Science [16]. His subject-spread is illuminating: Mathematics, Chemistry, Latin, Greek, English and French. Not only is an obvious ability with language evident in one who was soon to distinguish himself as a scientist, but the fact that he seemingly studied no Physics is intriguing. One can only presume that St Andrew's had no Physics master! But the best was yet to come.

Schönland entered the sixth form at St Andrew's in 1910. He was not quite fourteen and was the youngest in the class by almost four years, but

7

his talents and his burgeoning self-confidence belied his age. He spoke in the Debating Society in May opposing a motion which averred that 'The sword has done more to advance civilization than the pen'. Schönland stoutly contended that the evidence against this was overwhelming. The College Magazine recorded that 'in a detailed and well-worded speech [he] drew notice to the achievements of printing, from which point of view the pen had undoubtedly been a more potent factor' [17]. In August, the motion for debate addressed an issue, then very topical at the time, with one B Schönland proposing 'That in the opinion of this house women should be allowed suffrage'. The same source records that he opened the debate by hoping that his audience would not be 'prejudiced by the tales of hat pins and strenuous suffragettes'. But they clearly were and though he and his supporter, R F Currey, struggled manfully, prejudice abounded and Schönland's motion was defeated in a division by a large majority! [18]

South Africa itself, in 1910, was on the brink of parliamentary change. The defeated Boer Republics of the Transvaal and the Orange Free State became, with their former colonial neighbours in the Cape and Natal, members of the Union of South Africa under the premiership of one of the most illustrious Boer generals, Louis Botha. His minister of Mines, Defence and the Interior was J C Smuts, the brilliant Cape- and Cambridge-educated lawyer, who had been State Attorney in the Transvaal at the outbreak of the Boer War and who then took to the field as a Boer general of considerable daring and enterprise [19]. But now, a decade later, Afrikaner and Englishman were united under one flag but, as if to satisfy both, the levers of power were distributed between Parliament in Cape Town and the seat of Government in Pretoria. South Africa, though, had even greater divisions because the mass of its population, those of another hue, were excluded from Parliament altogether. Women, as long as they were white, got the vote in 1930.

The young Schönland will have followed these events closely and undoubtedly knew of Smuts for his name was already part of the St Andrew's College legend. Late in 1901, with the Boer War now a guerrilla struggle, Smuts and his bedraggled commando appeared on the outskirts of Grahamstown. The town was at their mercy since all the garrison troops had long since been sent to the battlefields. However, all was not lost. The College Cadet Corps, along with those of their sister schools in the town, were immediately mobilized to defend the good citizens of Grahamstown against a possible incursion. Smuts, however, declined to invade and moved on to continue harassing the real enemy, the British army in the field.

Now, nearly a decade later, when not engaged in the affairs of State, Jan Smuts found his own relaxation in botany [20]. An encyclopaedic memory and a mind of formidable dimensions allowed him to interact easily with the professionals in the field. Not surprisingly, therefore, his path soon crossed that of Selmar Schönland's. Quite soon he would come across the name of the younger Schönland, too.

In 1910 Basil Schönland sat his final examinations for the matriculation. Challenging him for first place in a class of thirty was Ronald Currey, his debating colleague, who was later to become the chronicler of the histories of both St Andrew's College and Rhodes University. Schönland's progress was now being closely observed by the College Principal, the Rev P W H Kettlewell MA (Oxon), who also took the boys of the unremarkable Day House under his wing, and so he saw the young man from many angles. In a letter to Selmar, Kettlewell stated emphatically: 'I am confident he will get a high place'. He did indeed. When the results were announced in Cape Town there was jubilation at St Andrew's. Basil Schönland, though only fourteen years of age, not only obtained the First Class pass expected of him in this national examination but was also placed first in the 'whole of South Africa' [21]. The College certainly had good reason to be proud because Ronald Currey was placed second. Grahamstown's newspaper carried the results under the headline of 'Latest Telegrams', and in the accompanying article it enthused in its congratulations to St Andrew's, calling the feat 'a remarkable triumph'. In addition, said the editor, the 'friends of both these young men are to be felicitated on the excellent promise thus shown'. The Schönland household in Oatlands Road received its share of felicitations too from the Bishop of Grahamstown, on the one hand, to the Government Entomologist on the other [22].

Once again the annual prize-giving was the major event in the College calendar. In 1910 the occasion was made even more significant by the presence of General Lord Methuen as the guest of honour. Methuen had been the commander of the force of eight thousand soldiers whose objective, some eleven years before, had been to relieve Kimberley and its fabulous diamond mine from the besieging Boers. The fact that he failed in this and then enjoyed the dubious distinction of being the last British general to be captured by the Boers [23] in no way diminished the applause which greeted him on entering the splendidly decorated College Drill Hall on 10 October. After all, had not England been triumphant in 1902 and, therefore, Grahamstown too? In 1908, Lord Methuen returned to South Africa as General Officer Commanding all forces in the country and so was clearly a hero amongst admiring friends who were steadfastly maintaining the English Public School tradition in arguably the most British of South African towns at that time.

Methuen's address to the assembled College and guests followed shorter orations from the Principal and the Bishop of Grahamstown. Whereas they had dealt with matters on a broad front, the General had but a single target in his sights — the Cadet Corps and its role in producing the future leaders of the land. To cheers and acclamation he reminded the boys that they were beginning life at the same time as South Africa was beginning hers and 'although we may have aided in making South Africa what she is going to be, it will remain with you to carry out the work which we have begun'. 'Hear, hear' was the resounding chorus. Though some always derided cadets as merely playing at soldiers, the General was quick to assert that:

> there is nothing a master likes more than a cadet corps in a college. It means discipline, cleanliness and self-esteem; it means the power of obeying, and for the boy the rifle becomes a [sic] second nature. The boy is able to command when he leaves college. That comes of volunteering, discipline and drill. There is no school in England in which a cadet corps is not regarded by master and boy as a good thing.

Since both speakers before him had congratulated the boys on their discipline and 'tone', the General's words certainly fell on receptive ears. The world was at peace and a British soldier was there to maintain it in South Africa. Basil Schönland might even have written another burst of verse. However, such euphoria was not to last. In four short years many in that audience would find themselves in uniform and in just such positions of command in 'the war to end all wars', and Schönland would be amongst them. But happier things lay ahead of them that night and Schönland was much fêted amongst the prize winners. Between himself and Ronald Currey they were to take all the VI form prizes except that awarded for Dutch, a subject Schönland had not studied. They shared those for Classics and English, while Schönland won that for French, and Currey those for Mathematics and Science. Basil Schönland was also declared the Jubilee Scholar for 1910.

The evening was brought to a close by the School Concert: a selection of musical pieces, both instrumental and vocal, performed by the boys, the College Choral Society and by the occasional master. Included amongst renditions of violin solos and the lusty singing of *The Sergeant of the Line* was a pianoforte solo, Chopin's *Nocturne in E flat*, performed by none other than one B F Schönland [24].

CHAPTER 2

RHODES TO CAMBRIDGE

Rhodes University College, or 'Rhodes' as it is always known, was just six years old in 1910 when Basil Schönland completed his schooling at St Andrew's. He, or at least his father, was faced with the decision as to which university this young man of outstanding ability should attend. The choice was somewhat restricted because there were not many options in South Africa at that time. Between 1873 and 1918 there was only one, the University of the Cape of Good Hope, which was simply an examining and degree-awarding body, based on the concept of the University of London, and which operated from offices in Cape Town. Its constituent colleges were the South African College in Cape Town, the Victoria College in Stellenbosch and now Rhodes University College in Grahamstown [1]. But Selmar Schönland was naturally more than happy with Rhodes, almost his own creation, and so there was little doubt that his fourteen-year-old son, who had captured the headlines of the local paper and the attention of the local worthies, would pass through its portals—as inauspicious as they were in 1911. Of course, Basil's tender years and the dominant personality of his mother will almost certainly have ruled out the possibility that he might be sent away from home.

Basil Schönland registered at Rhodes on 13 February 1911 in the Intermediate class for the degree of Bachelor of Arts (Mixed) which implied no bias either way in favour of Arts or Science. The university's Bursary Committee saw in him a good prospect and he was awarded a bursary of £21 for two years [2]. He was just fifteen, and whatever the future held in store for him there was no need to rush. In the normal course of events, students spent one year working towards the Intermediate examination—a strenuous enough test requiring the taking of at least five subjects, or, to do very well, up to seven—but Basil spent two years over it, almost certainly because of his age. In his second year, his registration showed him to be a member of the 'Mixed B' class where his choice of subject was still broad. No firm plan for the future had yet been laid but the portents were certainly there.

The closing years of the nineteenth century and the early years of the twentieth was a wonderful period for science and especially for physics.

Heinrich Hertz in Karlsruhe, Oliver Lodge in Liverpool, as well as many others, had turned the mathematical equations of James Clerk Maxwell into reality; some with conviction and fore-thought, others by chance. Maxwell's bold, almost rash, assertion of 1873 that 'light consists in the transverse undulations of the same medium which is the cause of electric and magnetic phenomena' had been shown within thirty years to be one of the most profound discoveries of the age. Early in 1896 the young Guglielmo Marconi arrived in England from his birthplace in Italy to demonstrate his new system of telegraphy without wires. Being blessed with considerable patronage he soon secured the attention of the Post Office in London, the British Army on Salisbury Plain and, by the summer of 1899, the Royal Navy during the annual naval manoeuvres. The potential of the apparatus he demonstrated was beyond doubt and possible applications abounded, both civil and military.

1899 was also the year which saw the outbreak of the Boer War in South Africa. Initially perceived in England as nothing more than 'a little local difficulty' it soon assumed the proportions of a major conflict involving the despatching to South Africa of a massive army of almost a half a million men [3]. Amongst them was a small detachment of Royal Engineers, and some from the Marconi company, whose task was to take to the heart of the battle a revolutionary method of communications—wireless telegraphy. What has been described as the last of the gentleman's wars was actually to be the war that linked two centuries in time, technology and tactics [4]. Trench warfare soon became part of military doctrine, though its use by the Boer forces was a lesson not wholly learnt by the British Army and it was to be a most painful one when repeated just a few years later. Science, in a variety of forms, had certainly arrived on the battlefield. One who was to play a significant part in wars to come, both as scientist and soldier, was then a little boy in Grahamstown who read avidly, almost voraciously, whatever he could find in the library and amongst his father's collection of books [5].

By 1901, Marconi's wireless communications had spanned the Atlantic; in 1904 Sir John Fleming pioneered the new science of electronics, with his discovery of the two-element thermionic diode; and in 1906 R A Fessenden, a Canadian living in the United States, broadcast music for the first time by what was soon to become known, at least on the western shores of the Atlantic, as 'radio'. Progress was indeed rapid. In that same year, De Forest's three-element triode, known as the 'audion', was developed in the USA. It represented a truly massive leap forward because it made possible, for the first time, the amplification of the incredibly weak signals then typical of this embryonic method of wireless communication. Before the end of the decade the Italians Bellini and Tosi had developed methods of determining the direction of incoming radio waves which not only proved vital to the Royal Navy at the

Figure 2. Basil and Felix in their workshop in the old coach house at 'Bellevue' in Oatlands Road, Grahamstown. (Reproduced by kind permission of the Cory Librarian, Rhodes University, Grahamstown.)

Battle of Jutland in 1916 but were, in years and conflicts to come, to have a multitude of scientific and military applications as yet undreamed of [6].

Basil Schönland had read about these pioneering developments in the world of physics, but he was not just a follower of the work of others. He was also the keenest of inventors and experimenters. The old coach house at 'Bellevue' in Oatlands Road was his laboratory and workshop where he and his brother Felix spent many happy hours 'mucking about ... with a few old telephones' [7]. Stories of Basil's exploits with wireless while still a schoolboy began to circulate in Grahamstown. He is remembered as the boy who used 'to experiment in wireless from the tower of the windmill' [8] that stood in the grounds of St Andrews, while his scientific abilities seemingly surpassed even those of the world's authorities in the science because it was even claimed that he and a friend 'spoke to each other by a wireless they had constructed from one Grahamstown hill to another' [9]. Such recollections, however apocryphal they may be, given the primitive state of wireless in those early days, certainly bear testimony to the young Schönland's flair for science and his great keenness to make experiments.

In the South Africa of the time, it was clear that anyone with any academic aspirations at all would attend university in England. However, admission to either Cambridge or Oxford, the only British universities seemingly acknowledged by colonials who had been nurtured by their far-flung graduates, was no automatic or easy matter. Applicants, and particularly those from the colonies, had to be above average in all manner of things. One man who sought to provide for the education of those who spoke the King's tongue was Cecil Rhodes. Never a man to doubt for a moment the superiority of both the British and their Empire, Rhodes had made provision in his will to assist those with certain strands in their make-up to attend his *alma mater*, Oxford. He was especially generous to four South African schools: the South African College and the Diocesan College in Cape Town, the Stellenbosch College, in that town nestling in the winelands just inland from South Africa's mother city, and St Andrew's College in Grahamstown. Such was Rhodes's debt to South Africa that each year these four were allowed to select their own Rhodes Scholars while all other applicants from elsewhere in the Empire and the United States applied in open and fierce competition for these much sought-after awards. Naturally, the holder of the Rhodes Scholarship was a highly esteemed member of the South African society of the day, as such an honour undoubtedly opened many doors to positions of considerable power and influence. In his will Rhodes had stipulated what he was looking for in the man who was to benefit from his munificence. He wanted 'no mere bookworm, but neither did he want the mere athlete—not even the mere athlete of the soundest moral character. He wanted a combination of brain and sinew and character without putting too much emphasis on any of the three...'. So spoke Dr George Parkin, the first Secretary of the Rhodes Trust when he visited South Africa in 1910 and addressed the assembled scholars of St Andrew's. The College believed that it had been delivering such men but, Parkin continued, '...it is a sad fact that the Scholars from South Africa are definitely inferior to those from other countries in intellect and intellectual attainments'. This revelation came as something of a shock and the masters of the College vowed to raise their academic sights. Henceforth no mere matriculation but 'the passing of the Intermediate examination (administered by Rhodes College) would be a necessary condition for candidature' [10].

Basil Schönland spent his first two years as a student at Rhodes as an 'Ink'—the less than imaginative name for all male candidates for the Intermediate examination [11]. His ultimate target, the BA with Honours, would take a further two years of study. He applied himself with enthusiasm and by early 1913, when the results of the Intermediate Examination in Arts were announced, the name at the top of the list with First Class Honours and, once again, first place in South Africa, was his. The newspaper headlines of 28 January announced 'Basil Schönland's Premier Position', and in enthusiastic tones offered hearty congratulations. This young man had 'once again distinguished himself in the University examinations by securing first place', just as he had done in the Matriculation [12]. Here was surely the candidate that St Andrews had been looking for.

But no Rhodes Scholarship was to follow. Instead, Schönland was awarded the Bartle Frere Exhibition while the most prized award of all for the year 1912 went instead to R F Currey, his closest rival throughout their years together at St Andrew's [13]. One can but speculate as to why Schönland was not adjudged to be St Andrew's most worthy candidate that year. Did he not exhibit quite the sinew, as Dr Parkin had described it, in those 'manly outdoor sports such as cricket, football and the like' that Rhodes required? Was he wanting in 'qualities of manhood, truth, courage; devotion to duty, sympathy for the protection of the weak, kindliness, unselfishness and fellowship'? Had he not the 'moral force of character and of instincts to lead and to take an interest in his schoolmates' which, in later life, would 'guide him to esteem the performance of public duties as his highest aim' ? Surely not; but these were indeed formidable heights to climb and would certainly have been daunting to one so young. But was there not possibly another reason? It might well have been that Basil Schönland did not want to go to Oxford, remarkable as this may appear to some.* What exactly were his views and aspirations that may have decided the matter? To answer these questions we need to consider the state of physics at Oxford and Cambridge at that time.

Oxford had its Clarendon laboratory while Cambridge had the Cavendish. Seen from the distance of Grahamstown, through the eyes of a young man whose mind was set on a career in science, they were the pinnacles, the seats of great learning and the cradles of academic destiny. But of the two, it was the Cavendish, at Cambridge, that had spurted ahead during the dying years of the nineteenth century and into the twentieth under such giants as Maxwell, Rayleigh, J J Thomson and Rutherford. Physics was then the dominant science and nowhere more so than in Cambridge. By 1910 both Maxwell and Rayleigh were revered as the pre-eminent natural philosophers, whose brilliance, which seemingly knew no bounds, had laid the foundations of all that was to follow in modern physics. Their disciples continued to dazzle as if the very walls of the Cavendish radiated ideas and inspiration. J J Thomson discovered the electron in 1897, a feat for which he was awarded the Nobel Prize for

---

* In her recollections of her husband, written some years after his death, Ismay Schonland (see [5]), maintained that this was certainly the case. The following year, 1913, the St Andrew's Rhodes Scholar was B G D'Rudd who won the 400 metres and came second in the 800 metres at the Olympic Games in 1920.

physics in 1906, while just two years later, his New Zealand-born student, Ernest Rutherford, won the Nobel Prize for chemistry for his extraordinary pioneering work, started in Cambridge and then continued in Canada and subsequently in Manchester, on the ultimate structure of matter. Rutherford's discovery of what he called 'α and β particles' had led him further and further into the inner core of the atom until, eventually, he deduced that it contained a fundamental, positive particle with charge equal to that carried by the electron, but considerably more massive. He named it the proton. This was the work of brilliance. But Cambridge had produced even more.

The detection and measurement of these minute constituents of matter was made possible by a variety of ingenious instruments, most of which involved the skill of the expert glass-blower to produce. Amongst them were evacuated glass containers fitted with metal electrodes which, when suitably energized, produced the green glow called cathode rays by their German discoverer and which J J Thomson used in his experiments on the 'electrified-particle', the electron. Other similar devices that soon contributed enormously to the understanding of matter in its most basic forms were brilliant, both in their simplicity and in their effectiveness. Paramount amongst these was the invention by another Cavendish man, one Charles Thomson Rees Wilson, always known as CTR, of what has become known as the Wilson cloud chamber. This deceptively simple apparatus consisted of a glass chamber filled with super-saturated air. The passage of a charged particle through the chamber leaves a vapour trail of condensation behind it. The nature of the track under the influence of applied electric and magnetic fields helped, like no instrument before it, to unravel the mysteries of particle physics. For the very first time the images of collisions, and the paths of the resulting deflections, hung in an almost ethereal space to be captured on photographic plate or film. 'CTR' took the first photographs of these tracks in 1911 and was awarded the Nobel Prize for physics in 1927.

This was the physics that inspired Basil Schönland and these were the men who were leading the world, and they were doing so at Cambridge, not at Oxford. We can only speculate as to why Schönland did not win the St Andrew's Rhodes Scholarship for 1912, for such decisions are not recorded, but we know that whatever disappointment there may have been in the Schönland household, for reasons of pride and prestige alone, it was tempered by his relief that the path to Cambridge was now clear, for that is where he wanted to go [14].

However, all that was still in the future. Schönland was only over the first hurdle, though certainly in style. Now he could concentrate on the relatively flat ground ahead for he had a couple of years before he would take his finals. The Professor of Physics and Applied Mathematics at Rhodes was Alexander Ogg BSc (Aberdeen), MA, PhD (Gottingen), one

16

time lecturer at the Royal Naval Engineering College at Devonport, who was appointed to the chair in 1905, the same year in which Selmar Schönland occupied his in Botany [15]. Ogg was an outstanding teacher whose path was frequently to cross Basil Schönland's again in years to come. With him on the staff was R W (Dickie) Varder BA, who had taken his degree at Rhodes before proceeding to Manchester to work in the field of nuclear physics under Rutherford. He returned to Grahamstown in 1909 as Lecturer and Demonstrator in Ogg's department and to a long and distinguished career as an inspirational teacher and a tireless university administrator [16]. These were the men who would help guide Schönland to a future in physics.

*   *   *

The years 1913 and 1914 were uneventful in Basil Schönland's personal life but the outbreak of war in Europe would soon touch the patriotic nerve that ran through many young men of the Empire. When war was declared, Schönland was already a private in the First Eastern Rifles, an Active Citizen Force unit of the Union Defence Force [17]. Lord Methuen's words a few years before had certainly struck their mark and so the young Schönland had readily turned to part-time soldiering to provide a break from his studies. Drill parades, musketry and the occasional bivouac over weekends and during university holidays introduced him to the military way of life and stood him in good stead for what was to come less than two years hence.

During those years the pace of scientific progress had quickened to a gallop. Rutherford's momentous discovery in 1911 that practically the whole mass of the atom was concentrated in a minute nucleus, with a net positive charge, had been the starting gun. Then W L Bragg, while still a student in 1912, made fundamental discoveries about crystal structure by a series of superb X-ray diffraction experiments. These resulted in his sharing the Nobel Prize in 1915 with his father, W H Bragg, the noted crystallographer at the University of London. But possibly the most momentous of all was the achievement, in 1914, of another young Cambridge researcher, H G-J Moseley, when he found that the X-rays produced by metals under cathode ray bombardment decreased in a regular fashion with their atomic number. This seminal result showed that it was the net positive charge on the nucleus and not its atomic weight that was the fundamental property of the atom. The world of atomic physics had seemingly reached its boiling point at the very moment that Europe erupted in the flames of war, and life, as all had known it, would never be quite the same again.

Then something of a bombshell burst in Grahamstown. In December 1914 the Rhodes University College examination results were published and Basil Schönland was awarded the BA with Second Class Honours in Physics [18]. For someone whose academic achievements throughout his school and early university career had always been remarkable this result was equally so, but for a very different reason. What had caused him to slip from the very pinnacle of performance which had almost seemed his by right? A clue is to be found in the minutes of the Statutory Meeting of the University Senate held on 24 February 1915. Evidently an objection had been lodged (by whom is unknown) at some earlier date 'against certain questions in the Physics paper at the last Examination'. The matter was referred to Cape Town for guidance and Professor Ogg read the reply from the Registrar of the University of the Cape of Good Hope (of which, it will be remembered, Rhodes was a constituent college), but the records do not enlighten us as to what it said! Instead we find Selmar Schönland making a statement to Senate [19]:

> Professor Schonland [sic] reported upon what he had done in the
> matter at Cape Town. After discussion Prof. Lord moved that Profs.
> Ogg & Williams be asked to draft a reply dealing particularly with
> the statement that had been made to the effect that four of the
> questions objected to had been set before.

How Ogg and Williams replied we do not know, for their reply also went unrecorded, but there were surely implications for all concerned— particularly the students who sat that examination and one was Basil Schönland, son of one of the players in this Senatorial curiosity. More was to follow; but not immediately. The next occasion when this matter appeared on the agenda was at the Senate meeting of 12 May 1915. But it had to wait as Senate first addressed itself to the matter of the suitability of South's Boarding House for student accommodation. After discussion it was resolved, with Professor Schönland seconding the motion, 'that it be struck off the list'. Its short-comings, or whatever, were not recorded. Then, alongside the heading 'Physics paper 1914' there appeared the cliff-hanger [20]:

> The Secretary read a reply from the University relating to Physics
> paper 1914 dated 20 April.

But what the reply actually said, and what action flowed therefrom, again went unrecorded. We are therefore left none the wiser, since the details never saw the light of day. What is known is that only one First Class Honours degree was awarded in Physics that year and it did not go to Basil Schönland.[*] Was it all down to an administrative bungle and an eventual compromise that satisfied few, or had Schönland just not measured up on the critical day? We will never know. But Selmar and

---

[*] The recipient was P H Copeman, 'Phlop' as he was known, and a good friend of Schönland's.

Flora Schönland knew, and smarted at the indignity of the University's decision, if nothing else.

Whilst all these matters were occupying the Senate at Rhodes, Basil Schönland had already set sail for England. He left Port Elizabeth on 23 March 1915 and arrived in Plymouth on 14 April. His immediate destination was Cambridge where he was admitted to Gonville and Caius College four days later. There, he was to read for Part 1 of the Mathematical Tripos and take the examinations in June with the intention immediately afterwards of volunteering for service in the British Army. One problem remained and that was the matter of his name, and particularly his father's Germanic roots, which conceivably could prove difficult for both of them at this time of conflict between Britain, her Empire and Germany.

Selmar Schönland, ever diligent and always the assiduous father, was particularly worried. He had just had a foretaste of what his heritage might inflict on himself as a result of South Africa declaring war on Germany on 12 September 1914, just a month after Great Britain herself had done so. Fortunately, he was spared the internment, which was the fate of all German subjects living in South Africa on the outbreak of war, by the intervention of General Smuts, who, as the Minister responsible, had sent a telegram to the Chairman of the University Council informing him that the order to this effect would not apply to Professor Schönland [21]. While his own incarceration was no longer an issue, Selmar was concerned about Basil's position as the son of a German immigrant so he wrote in all haste to the Commissioner for Enemy Subjects in Pretoria [22].

5 March 1915

> Sir,
>
> My son, Basil Ferdinand Jamieson Schönland (19 years of age) is intending to sail to England on the 23rd of March from Port Elizabeth to continue his studies at Cambridge University after passing the B.A. degree of the University of the Cape of Good Hope. He was educated at S. Andrew's College and Rhodes University College, Grahamstown where he was also born. He is a member of the Union Defence force but has permission from the military authorities to proceed to England, where, after passing an examination in June he will probably enter an officer's Training corps and offer his services to the British army.

He continued

> I am a German by birth, have lived nearly 4 years in England and over 25 years in South Africa and was naturalised here as a British subject. I have to-day been informed that aliens who have become naturalised in South Africa have to get a passport from the Department of the Interior or a [sic] permission from you before being allowed to leave.

Figure 3. Professor Selmar Schönland, botanist and 'father' of Rhodes University, Grahamstown. (Reproduced by kind permission of the Cory Librarian, Rhodes University, Grahamstown.)

I do not know whether this regulation applies to my son who was born here...

As so very little time is left to make arrangements for my son's departure on the 23rd and as it would prejudice him if he was delayed, I would esteem it a great favour if this application had your immediate attention. Before despatching this letter I shall ask our Magistrate to endorse it.

> I have the honour to be
> Sir
> Your obedient servant
>
> S. Schönland
>
> Professor of Botany, Rhodes University College
> Member of Council University of the Cape of the Good Hope.

P.S. I may add that my wife was born in this country of English parentage.

To which the Magistrate in Albany added his signature having endorsed the contents 'with pleasure'.

Selmar Schönland's entreaties produced a speedy response from Pretoria. On 8 March he was sent a somewhat terse though encouraging reply which informed him that his son was a British subject and that the Department of the Interior was dealing further with the matter [23]. However, it did not rest there, for Professor Schönland was married to a very strong-willed woman whose views on just about any subject were made known with considerable conviction. Her husband, therefore, was left in no doubt that Basil should not suffer any inconvenience on his arrival in England and so Selmar wrote immediately to an old and very influential family friend, W P Schreiner, who was now the South African High Commissioner in London. Schreiner duly replied on 7 April with his view that Basil, as a British subject, was 'fully entitled everywhere to the privileges of his status' and he added by way of reassurance to his old friend that he would 'be very pleased to see him again. I hope he will call upon me when he is London. I am very proud of him as a young South African and I well remember the impression he made on me in 1901 when he was a small boy at Grahamstown. I wish him all the success which he may reasonably anticipate in his future career' [24].

With such assurances the matter could well have rested there except for the rather eye-catching and possibly unpronounceable Schönland patronym. To advertise one's Germanic origins, however remote, would have been rather ill-advised when stepping ashore as a foreigner in England in 1915. So, presumably acting on fatherly advice, Basil removed the umlaut from the family name. Henceforth it was to be Schonland, at least informally and unofficially. In years to come this change in spelling provided Basil Schonland with a nice line with which to lighten his public speeches. Its success with the audience, though, did depend on their understanding of some finer points of mathematics. The shorthand notation for the process of differentiation with respect to time is a single dot above the function being differentiated; repeating the process involved two such dots. The mathematical operation of integration essentially reverses the procedure and removes one dot each time. Schonland therefore, when remarking on the change in the spelling of his name, referred to himself as having been integrated twice with respect to time! [25].

*   *   *

The mathematical Tripos at Cambridge requires the passing of two examinations within the three-year period of the degree. Basil prepared himself well for Part 1 under the tutorship of his mathematics coach, Mr Munro of Queens' College. Both will have been well-satisfied when the results were published on 15 June. There, amongst the list of name of those appearing in Class 1 was *Schönland, B.F.J.* of Caius, with mathematical double derivative too. He was also one of his college's Prizemen and was elected to an Exhibition for one year from mid-summer 1915 for which he was awarded £30 [26].

With his academic credentials now established at Cambridge, Schonland immediately turned his attention to what he saw as his patriotic duty — the service of King and country. The affinity which he felt for England as a little boy at the southern tip of Africa was undimmed. Then, his patriotic fervour had been expressed in the best traditions of thumping verse, as he wrote in his epic, the 'Mistress of the Seas':

> *But it was the same in both those ages*
> *Here were knights and here were pages*
> *Each striving for one thing*
> *To get honour for country and for king*
> *That England might retain her name*
> *Of 'Mistress of the Seas'*

and so on!

Now, in 1915, he was moved to action and early in July went off to London for an interview at the War Office. Arriving there on the eleventh he booked into accommodation provided by a Miss Crampston and that afternoon presented himself before a Major McLinton who 'was very decent' and promised him 'an RE signals job in a fortnight'. The Corps of Royal Engineers (RE) Signal Service had amongst its many and varied tasks that of providing communications services to the rest of the army and men with appropriate qualifications were much sought after for service in the RE's rapidly expanding officer corps. Schonland left feeling 'very braced at getting into such a fine service as signals RE's'. Just a week after Schonland was offered his 'job' one of Cambridge's most prodigious talents, who had preceded him into the same Corps, was killed in the Gallipoli campaign. The loss of Henry Moseley to a sniper's bullet was a devastating blow to physics. Such front-line service was as much the fate of a signals officer as it was inevitable for an infantryman, and both the physical and the moral courage of all who served there would be sorely tested by the carnage they saw around them.

Schonland began to keep a diary in July 1915. 'Don't like diaries', he wrote, 'but this may be interesting when all this excitement is a thing of the past. Starting with a very short summary of the last week, I shall write it up fairly regularly' [27]. And he did just that until his departure for the front six months later.

Basil Schonland certainly wrote well in a fluent, easy style that reflected the age in which he lived — an age when English public schools bred the leaders of the Empire and what, of course, was St Andrew's if not Eton, Harrow or Winchester in the veld? Lord Methuen had said as much just five years before at St Andrew's speech day and his words had imprinted themselves on Schonland's soul. The diary was therefore much more than a bland catalogue of events. The unabashed enthusiasm of a young man facing a challenging, exciting and, thankfully, unknown future bubbles out from every page. The hand-writing is small, precise and neat and displays a sense of order and discipline. In its pages we see an organized, enthusiastic and very perceptive young man of 19 who spent two weeks enjoying the museums of art, science and history in London before joining the colours.

On 16 July 1915 the *London Gazette* published its regular list of new officers and their appointments. Second Lieutenant Basil Ferdinand Jamieson Schönland was now officially commissioned into the Corps of Royal Engineers. Those two dots were still much in evidence. There too was the name of Rupert White-Cooper, with whom Basil had shared the stage in numerous Prep school frolics, and who was now commissioned into the 'Manchesters'. Together the young subalterns visited the Natural History Museum and saw the director 'who knew Dad and showed us SA fossils'. They were fascinated by the Victoria and Albert Museum, and the National Gallery with its fine pictures by Velasquez, Turner, Murillo and Rembrandt. They attended the theatre and enjoyed the revue *'Business as Usual'* and found *'The Man who Stayed at Home'* full of thrills: 'a secret wireless installation, submarines and German spies and codes'; a taste, no doubt, of things to come. At Madame Tussaud's they saw Napoleonic relics, the original guillotine and the Chamber of Horrors, but most spectacular of all was the Science Museum. Schonland went back again on his own and spent two days in the astronomical and electrical sections. It was, he wrote, 'the perfect paradise to the physicist'. He found the collection of X-ray and cathode ray tubes a delight and noted with particular enthusiasm the very tube used by J J Thomson to find e/m, the ratio of the electronic charge to its mass and v, its velocity within the discharge tube. That experiment had convinced him to read physics at Cambridge. When he also saw 'telegraphic apparatus in abundance' that sealed it.

There were also social calls to be made on friends of his parents who had returned home after a sojourn in the colonies. In addition, he visited Mr Schreiner, South Africa's High Commissioner and then, on 7 July, 'waited an hour for the Queen and Princess Mary to drive past, but had a fine view. They looked splendid'. After visiting Hampton Court and rowing on the river with Rupert and Ronald White-Cooper, they all returned to the London residence of his friend's 'Mater and Pater', Dr and Mrs White-Cooper, where they dined and enjoyed their wonderful collection of curios and oriental works of art.

Ronald was a medical student at St Bartholomew's Hospital ('Barts') and the next day he laid on a visit for Schonland to see the X-ray and electric treatment apparatus actually working. In his diary Basil noted that 'One man who was supposed to have a cancer in alimentary canal

23

was made to swallow bismuth in front of the screen; bismuth being opaque to X-rays the path … as it passed down was clearly marked'. That intrigued him. He was shown, and tested as well, 'the new method of using a high-frequency current to heat a joint or tissue of the body using wool soaked in salt solution as electrodes'. Such wonders of science in action had been but a dream back home. More theatrical diversions followed in the form of *'Potash and Perlmutter'*, one of London's major attractions that had been running for over a year. The diary enthused: 'It was really splendid, taking off American Jews. Policemen all consulting photos of some escaped Germans'. And the war in Flanders seemed but a distant diversion.

Letters from home were always high-points in any day. His diary entry for 16 July records the arrival of a batch of eight, including one from 'Dad and Mater' who were very pleased at his success in the Cambridge Tripos examinations. 'Feel glad I didn't raise their hopes too soon and also that I haven't got a second again.' It still hurt.

And so the captivation of a colonial subject in what must have seemed the capital of the world drew rapidly to a close. On Wednesday 21 July 1915 he visited a tailor to have his uniform fitted and noted that he was feeling 'very braced'. The next day he received a letter from the War Office confirming his appointment and informing him to join the RE Signal Depot at Bletchley, a town near Bedford, on 1 August. Providing himself with all the necessary military wherewithal continued apace as he spent the morning 'ordering boots, riding breeches & paraphernalia' but he still found the time to spend part of the afternoon in the Science Museum 'doing the mathematical instruments, sound and heat rooms'. There he saw a 'Michelson harmonic analyser having 80 levers, Joule's original calorimeter and Boys' radio-micrometer and the telescope used in his famous 'candle at a distance of 2 miles' experiment. Credo'. Credo indeed! Here, by chance almost, we see the name of C V Boys. Two decades later that name would reappear in Schonland's life to very great effect.

The round of social engagements continued with visits to other South African families now ensconced in London. Schonland's own equanimity, he found, was not matched by one of his Eastern Cape colleagues who, he noted, was finding it 'a bit difficult to make his Uitenhage ideas embrace the larger world of London'. Basil had clearly adapted to it much more easily. Together with Rupert White-Cooper and his landlady, Miss Crampton, he attended the Sunday service in the Temple church. The building and its stained glass impressed him as did the singing, 'using the ancient tunes and harmonies' which, he informed his diary, was glorious. He dined with the High Commissioner and Mrs Schreiner after they had all had an 'awfully fine time' at the theatre seeing Miss Laurette Taylor in *'Peg-o'-my-Heart'* and then, the next day, he left for

Cambridge to pay his respects to his various tutors and finally packed up the contents of his rooms. This, he recorded, was 'quite a distressing business as these rooms have been my home for the last few months & I have always enjoyed getting back to them'. From now on he was to be a soldier.

CHAPTER 3

FOR KING AND COUNTRY

On Sunday 1 August 1915 Second Lieutenant Basil Schonland arrived at the Signal Depot in Bletchley and reported to a rather forbidding commanding officer. Throughout the course of the day another eleven men turned up from various academic establishments around the country and from the Commonwealth. Word had it that the military regarded them as 'an infernal nuisance' but Schonland did note that 'one or two have been very agreeable' [1].

First thing next morning they left for the Recruit Training Centre at the military barracks in Birmingham where they were to undergo a 'course of discipline' by being drilled on the parade square in the company of ordinary recruits while being 'sworn at by a very peppery captain'. The twelve of them discovered the intricacies of squad drill and 'turnings by numbers' under the eye of an elderly sergeant who displayed admirable patience as his charges came to terms with a way of life rather foreign to most of them. Such was not the view of the officers who had the job of turning them into soldiers-of-a-sort within just a week. They were therefore subjected to some sarcastic comments from the major, while the peppery captain 'was hardly pleasant to us. Got a voice like a hippo' were Schonland's comments to his diary. Musketry drill and section drill took up every morning, while in the afternoons each of the newly-commissioned officers had command of a squad of men, 'some with grey hair', on whom they practised their newly-learned words of command. The RE 'tommy', Schonland observed, 'is a fairly intelligent & very nice type'. As the week of intensive instruction progressed they visited the firing range and he was satisfied with his score of 20 out of 25. The views of either of his officers on this achievement were not recorded. By Friday, Basil Schonland's training as a signals officer had expanded considerably for he was by now learning the Morse code, having bought himself a 'tapper'. The next morning saw them all being examined by the captain in the art of drilling a squad of men. 'Rather an ordeal shouting from 70 yds distance across a flat parade ground. However I did rather well & instead of something sarcastic the Capt. said "Very good, sir! Very good indeed".'

Figure 4. Second Lieutenant B F J Schonland in the uniform of the Royal Engineers at Haynes Park, Buckinghamshire, 1915. (Reproduced by kind permission of the Cory Librarian, Rhodes University, Grahamstown.)

In the company of two of his colleagues Schonland spent his first weekend in the city of Birmingham but was not impressed by what he saw. It was 'very dirty and ugly with narrow streets and a very bad tramway system. The people are mostly of the poorer squalid classes and having plenty of money from munitions work seem to drink very much'. He and his colleagues did discover more appealing features in the place when they came across the botanical gardens, which he admired, and 'on the way saw some of the better class houses'. On the Sunday evening they attended service in the Cathedral where he and his two officer colleagues, 'the Oxford man Maddox and the naval man Barclay', were solemnly conducted to the pew of honour in the very front. 'What it is to be in the army! And especially in the Engineers'.

The week's intensive introduction to the military way of life, as well as to that of his working-class kinsmen, soon drew to a close. The following Tuesday they were all inspected by the major who passed them out in company drill and they then boarded a 'great long motor transport lorry' for the return journey to Bedford. In fact, their destination turned out to be the Signal Service Training Centre at Haynes Park, some six miles away, where they found themselves in a very large and comfortable army mess, which was to be home for the next many months while they became fully fledged signals officers. We should now pause at this point to examine the status of signalling and communications in general within the British Army of the time.

In 1915 all signalling in the Army was the preserve of the Royal Engineers. The Telegraph Battalion, RE, was formed in 1884 with the figure of Mercury, the messenger of the Gods, as its emblem. The art and science of sending and receiving messages, for it was very much both, depended on the telegraph, the telephone, the heliograph and lamp, and even on the waving of flags. Just a decade before, at the turn of the century, wireless communications were at their most embryonic, but leapt into the public mind with Marconi's magnificent feat of spanning the Atlantic with the three dots of the Morse letter S from Poldhu in Cornwall to St John's, Newfoundland, in December 1901. The equipment he used was, to say the least, experimental and its applications were, as yet, unclear. However, the British Army, represented by Captain J N C Kennedy RE, were interested observers on Salisbury Plain some four years before when Marconi managed to achieve successful wireless communications over a distance of nearly three kilometres with apparatus which could almost be described as portable. It would be that same Captain Kennedy who would take a hastily assembled Marconi wireless system to South Africa just three years later at the outbreak of the Boer War for the purpose of improving the Army's communications in the field. However, as it transpired, its performance there was undistinguished, for a number of reasons, until taken over by the Royal Navy who used it to great effect with the Delagoa Bay Squadron. Successful or not, this was indeed the first occasion on which wireless was used in any form of military action [2].

Subsequent progress was rapid. By 1907, the first two wireless companies in the Telegraph Battalion had been formed. A year later saw a change of name when 'Telegraph' became 'Signal' with the birth of the Royal Engineer Signal Service, and in 1911 the Army Signal School came into being. The advent of the motor cycle added another string to its bow with the introduction, in 1912, of the first motor cycle dispatch riders, or DRs ('Don R's in the phonetic alphabet). All members of the Service were soon to wear the blue and white armlets that distinguished them as signallers from the rest of their RE counterparts. The cable wagon, pulled by six horses with three riders plus another five men alongside, was the backbone of the mobile army providing the means of communication between headquarters and the front. The cables were either buried or slung between poles, to form what was known as an airline, and much training went into developing this into a finely honed skill, practised at speed on horseback. When the British Expeditionary Force, the BEF, deployed to Belgium in August 1914 at the outbreak of 'The Great War' it was accompanied by two headquarters signal companies, six divisional signal companies, one cavalry signal company, with eight cable and five airline sections [3].

At Haynes Park, near the village of Bletchley in Buckinghamshire, Schonland found himself surrounded by some grandeur. In the company of a thousand men and nearly a hundred other officers, he was billeted in what was a very large and fine country house, that had become the head-quarters of a significant part of the signals section of the British Army and its remount depot, with 800 horses in training. Schonland marvelled that the grounds were larger in extent than the whole of Grahamstown and were filled with tents, soldiers, horses and stables. The oak-lined drive carried a never-ending procession of motor lorries and inter-weaving DRs on their motor-cycles. The flag above the building was adorned with the winged Mercury, while the capacious rooms were home to Engineer officers and tobacco smoke. There was an undoubted feeling of some luxury about the place with even a pianola which 'alternately discourses ragtime and Beethoven' [4].

Life, though, was not all spent reading the daily papers which lay around, or in languishing in the sunshine on the manicured lawns. In fact, the working day was very demanding and commenced with reveille at 5 a.m., then forty minutes for each man to attend to his military appearance prior to arriving at the stables. Grooming and saddling-up their horses followed until breakfast at 7.00. The rest of the morning was devoted to the cable-cart parade, lunch at 12.30, then riding-parade until 3 p.m., followed by lectures on horse management and veterinary science until 5.00. Then, the return to the stables with their mounts for more grooming and the close of the day's activities at 5.45 p.m. Military riding presented Schonland with no major problems because life in Grahamstown around the turn of the century would have ground to a halt had it not been for horses, so he had had his fair share of practice in the saddle. However, as with most things, the military rode rather differently and the first lesson at Haynes Park required him to master the art of trotting without stirrups and with folded arms! Then came riding and jumping without a saddle. He coped and enjoyed it all. His admiration for the farrier sergeant majors in charge of this aspect of young officer training knew few bounds. They were, he observed, seemingly omniscient when regaling their charges in the most colourful language on all aspects of equestrianism and even about the complexities of the foot of the horse which, Schonland noted when writing to his mother, was '[a] jolly sight more complicated than I thought'.

Cable-cart drill and lectures on the practicalities of laying cable at speed became their next adventure. Each man took his turn at carrying out the functions of the linesman at the various positions around the cart. Some were more strenuous than others and, in the heat of a late summer afternoon, the physical demands took their toll, so much so that Basil admitted to dropping off to sleep during the occasional after-noon lecture. Poor performance in some task, which usually followed such periods of inattention, inevitably led to castigation from the instruc-tor or an officer on a visit of inspection. By dint either of his competence or good luck Schonland managed to avoid the worst of these while confiding later to his diary that 'The C.O. accompanied our cable-cart and strafed everybody most violently. Luckily I was not a working number'.* In between, they were also given lectures on the various instruments of their trade such as the D Mk II and III portable telephones and buzzer and 'on electricity as explained to sappers'. This, naturally, presented him with few problems.

Occasionally the training was interrupted by a special lecture from officers recently returned from France, who gave the new recruits a first-hand account of signalling experiences at the front and told them 'some very interesting yarns'. By now, late 1915, the British Expeditionary Force under Sir John French had already experienced both triumph and adversity as the Germans advanced through Belgium. The British retreat from Mons in August 1914, followed a month later by an advance and a significant victory at Marne, were but small elements in the campaign compared with what was to come. The famous Battle of Ypres in Novem-ber was the most momentous and terrible experience ever endured by a British army in the field, yet it had the effect of producing, not a turning away from war, but an upsurge in patriotism throughout the country [5]. Men rushed to recruiting offices around the land to join the colours so bravely borne by the nation's little professional army. Kitchener's famous call was certainly being answered.

At the end of August the pace of activities at Haynes Park quickened markedly. During the daily parade on the morning of the 30th it was announced that Course S4, Schonland's, along with others, was to be shortened and that they were to be examined on Friday and Saturday. Word filtered through that news from various fronts was not good. Schonland noted in his diary that 'the Germans were now in complete possession of Poland; [*The*] *Arabic*, 1000 tons, torpedoed.... All very bad. Looks as if there will be enough war for all of us to see more than enough action'. He continued: 'Things on the Russian front going from bad to worse; some progress is taking place in the Dardanelles'. Indeed they were, but certainly not of the progressive kind his emissaries suggested. Already fifty signals officers had left camp 'for the port of embarkation and we were informed that we were badly needed at the

* 'Strafing', as used during the First World War, was a verbal admonishment of some fero-city. It meant something rather different during the next conflict.

front. It appears that E.A.R.* or the fifth army is crossing over and they are rushing us through'. So the military rumour machine ground on and the activity in and around the villages of Buckinghamshire became more intense as the Royal Engineers Signal Depot turned out its qualified men for service wherever the need for them might arise.

Basil Schonland had enjoyed himself enormously at Bletchley. He found the cable work interesting, the horse-riding exhilarating and the life of a soldier much to his liking. The need to function in a structured and formalized environment, though often disagreeable and frustrating to many who were thrown into it by the demands of war, was to him a necessary and entirely logical aspect of military life. He found no difficulty in adapting to the discipline. Indeed he relished it and developed a quiet self-confidence—unusual for a man not yet twenty. It was the confidence of someone who had achieved what he had set out to do in a Corps that inspired in him great pride and satisfaction. 'Braced' is what he always called it.

The instructors clearly saw in Schonland the qualities of intelligence, leadership and initiative required of a signals officer and he was marked as a man to watch. One of those instructors was Lieutenant E V Appleton, just slightly older than Schonland and himself a physics graduate from Cambridge, who had been commissioned into the Royal Engineers just a year before and was now senior instructor in wireless at Haynes Park. Appleton's early academic progress at the Cavendish Laboratory, where he had worked on X-ray diffraction as Lawrence Bragg's first research student, had already caught the eye of Rutherford and Rutherford was seldom wrong. Eventually, Appleton, like Bragg, was to win the Nobel Prize just as Rutherford himself had done in 1908. Schonland and Appleton became quite friendly, even though the younger man was somewhat in awe of his instructor and confided to his diary that Appleton, 'a John's man ... knows all the big men at Cambridge, holds a 1st in Part II & is doing research work'. Their regular encounters certainly inspired Basil to foster his own interests in physics and the Corps library at Haynes Park provided the means to do so in the form of the prestigious journal, *Nature*, which he read 'to keep up with the physics world'.

Soon Schonland was exposed to some of the more mundane practicalities of physics, when he worked for four hours every morning with the array of telegraph instruments that formed part of the RE's telecommunications armoury. Their complexity and the ingeniousness of their

---

*Despite exhaustive attempts the author has been unable to determine what the letters E.A.R. might stand for. 'Expeditionary Army Reserve' and 'Engineer Army Reserve' have been suggested, but the true meaning remains elusive. No record of this abbreviation seems to exist amongst any history of the period nor in any Army List or similar document.

construction appealed to his great interest in scientific gadgets, and his thoughts turned frequently to the old coach house in Oatlands Road and to his brother Felix where, together, they had spent so many hours 'mucking about with old telephones'. Now it was all in earnest. Examinations, both practical and written, followed at the end of the week and Schonland passed the instruments-test quite satisfactorily by finding all the faults. Likewise, the written papers on airlines and instruments presented him with no problems. Undoubtedly the most demanding of all, though, was to be the test of his abilities as a horseman, but that was still to come.

The all-male preserve which was the army in those days occasionally gave way to social functions organized in the YMCA tent at Haynes Park. This edifice became the unlikely home to the Fenny Stratford Templars' choir who performed, 'somewhat ambitiously' he felt, the Hallelujah Chorus and the Soldiers' Chorus, 'but did so very well', he recorded. The same tent doubled as the Anglican church on Sundays and Second Lieutenant Schonland was a regular communicant. On occasions he found himself having to borrow a sword in a hurry in order to march, in the accustomed military manner, a group of non-conformists to their church parade at a nearby Baptist chapel. His ear was finely tuned to the quality of the singing and the sermon on all these occasions, and his diary was the frequent recipient of the odd acerbic comment about one or both. Similarly, he had an eye for the delights of the English countryside and these prompted him frequently to comment about the beauty of the surroundings, the fine old country houses, and especially the pleasures of a weekly bath, usually to be had in the town of Bedford which he reached by bicycle. Sunday afternoons in a near-deserted mess were usually spent writing letters. As well as to his mother and his brothers, Felix and Dick, he wrote occasionally to his former mentor, Professor Ogg at Rhodes, thus maintaining an association that was to be renewed at irregular intervals over many years. He wrote as well as to Mr Schreiner at the High Commission in London, and to 'Phlop'. Philip 'Phlop' Copeman was another long-time friend from Rhodes, who happened now to be in Edinburgh, and this presented Schonland with the ideal opportunity for a visit to that fine city. Early in October, he applied for a week's leave to venture north in order to cast an eye over its many delights, and to see his old friend who had beaten him to the only First Class Honours degree in Physics that Rhodes had awarded in 1914.

The Telegraph Battalion riding test took place on 10 October and was observed by the camp Commandant himself, Colonel R H H Boys: an uncommon name but one that would later play a significant part in Schonland's life. Now, it was his skill as a horseman that was under scrutiny and his mount, an ex-competitive jumper, insisted on galloping over the jumps instead of taking them slowly. However, both horse and rider found the experience exhilarating and Schonland was one of 18 out of 24 who passed this crucial test of a signal officer's equestrian competence. Suitably encouraged, and with his application for leave granted, he left on the overnight sleeper for Edinburgh, having abandoned his usual frugality and paid all of 10 shillings for the pleasure. During the long journey he reflected on a rather curious happening that occurred the morning before the important examination on horse management. The farrier sergeant major, presumably intent on keeping his own reputation intact, suggested to the expectant candidates that they might like to consider a few important questions, all of which subsequently appeared, presumably by sheer coincidence, on that afternoon's examination paper. 'Some *finookery* somewhere' was Schonland's observation to his diary, expressed in its most succinct Afrikaans form!

Edinburgh enchanted him as he and Phlop Copeman visited the castle, attended the theatre and then travelled through the picturesque, thickly-wooded countryside to the Firth of Forth where they marvelled at the sheer immensity of its three-spanned bridge beneath which lay at anchor a number of battleships and cruisers. The machinery of war was everywhere but away from the water it seemed to have missed Edinburgh, and Schonland delighted in everything about the Scottish capital, its history and its people.

Refreshed and invigorated he returned to military life and to almost feverish activity. The carnage of war was far from Haynes Park but the reverses suffered by the British Army in France required immediate action everywhere. General Sir Douglas Haig was about to replace Field Marshal Sir John French as Commander-in-Chief and, lower down the chain of command, things were happening as well. On 18 October, the Commandant's orders delivered on parade stated that 'No. 43 airline section will be formed today under 2nd Lieutenant B.F.J. Schonland at Fenny Stratford.' Though straining to appear calm Basil felt elated and 'simply overflowing with joy'. '43' was to be a new motorized airline section and he himself was to exchange his horse for a Singer motor car, while under his command were to be 52 men, five motor lorries and all the equipment of their trade. He threw himself with great energy into the planning and organization of this, his first command.

His diary recorded events as they unfolded. 'A very busy day spent in interviewing colonels, majors, sergeant majors and men ... some can't be spared from cook-house, some are unfit'. In the time-honoured tradition of the army he soon discovered an 'extraordinary number of men with defective teeth or whose wives are dying or who are munition-workers'. Some he discarded, others he ignored as he brought his establishment table up to strength. He then turned his attention to his own chain of command and to his sergeant, with whom he had rather lost

Figure 5. Schonland (seated centre with cane) and the men of 43 Airline Section at Fenny Stratford in 1915. (Reproduced by kind permission of the Cory Librarian, Rhodes University, Grahamstown.)

patience of late. The man was simply not up to the mark. Acting decisively the young subaltern sought the agreement of the colonel and speedily had the man replaced by the 'admirable Corporal Lane', who was rapidly promoted. Now that the section was to be motorized Schonland himself exchanged his horse for a car. 'I am a lucky man ... some job' he enthused to his diary. Field-work consumed every day. He was introduced to the administrative machinery of the army: pay, documents and the quartermaster's stores, and he revelled in the long hours spent in glorious autumn days out in the countryside learning to command a detachment of men who themselves were 'tip-top'. Lunch was taken at the most convenient wayside inn and Schonland was as much at ease with both his command responsibilities and with his men as they were with him.

A week later he marched them to Bletchley station from where they travelled to the holding and drafting depot at Hitchin to train in their new role, and with their new means of conveyance. Schonland was sorry to leave Haynes Park, his colleagues and particularly 'friend Appleton', with whom he had spent many an hour talking about physics. He discovered that his section was to be attached to the headquarters of E.A.R. (Fifth Army) and therefore, being well behind the front lines, presumed almost with a tinge of regret that they were unlikely to see much of the fighting. Their motor lorries arrived but Schonland's Singer was still to come, so he borrowed a BSA motor-cycle from one of the DRs and quickly learnt to ride it. Thus equipped, he pitched his men into building airlines across country, through the villages of Barton-le-Clay, Pinton, Silsoe and on to Bedford — names like those which, just a few years before, would have reverberated around St Andrew's as masters regaled their pupils with tales of England.

Training exercises or 'schemes' designed to simulate the conditions they were soon to encounter across the Channel continued in earnest. New ideas were tested and many rejected as the army attempted to come to terms with the rapidly expanding array of signalling techniques. Telegraph and visual methods had been standardized; the telephone and wireless were recent innovations, but were not yet widely used. In the earliest days of the war, use was made of the Belgian and French telegraph services to pass messages between army commands, while the dispatch rider service was the major means of inter-communication within the BEF itself. In the rear, airlines and cables were laid. Then the retreat from Mons led to drastic revision of the pre-war concepts of military communications as hundreds of miles of cable had to be abandoned, while that still in service suffered severe damage from artillery bombardment. Only the skilful use of visual signalling with flag and heliograph by day, and the Begbie lamp at night, as well as the outstanding bravery of the dispatch riders, managed to keep communication channels open.

Wireless was used between GHQ and the cavalry divisions, and within the cavalry divisions themselves, but with little success because the sets were very unreliable. Not surprisingly, this inspired little confidence in the technology amongst the staff officers, or even within the signals arm itself. In addition, the wireless equipment was very bulky and required large carrying parties, and these seldom escaped casualties [6].

The start of trench warfare in earnest, late in 1914, during the battle of Ypres, brought about its own serious problems of communications. Visual methods were now impossible, both for lack of visibility between stations and the invitation it gave to the enemy for immediate retaliatory artillery bombardment. Cables were therefore laid *en masse*, initially for telegraph use, but then predominantly for telephones, which were eagerly grasped by the staff for the ease of communication they provided. Such vast cable networks immediately caused their own peculiar problems. Whether laid in trenches for their protection from artillery shells, or just on the ground if needs be, they became a source of intelligence to the enemy. The mutual induction between cables not only caused interference between different circuits but, equally, rendered them insecure whenever they were in close proximity to the enemy's cable systems — and they often were, given the nature of trench warfare on the fields of Flanders. Communications were therefore frequently compromised — something that was actively exploited by both sides — and were made even worse by the poor telephone security of those senior officers who never imagined that their unguarded conversations could possibly be overheard. Heavy casualties resulted all too often from such indiscretion. The signals officer was clearly a key man in every battle, for upon his shoulders often rested its very outcome and the fate of the men who fought in it.

By mid-November the Fifth Army was training in earnest. Schonland's Singer had arrived and with it came the snow. Both provided him with some fascination because he had seen neither snow nor such a fine vehicle before. The Singer was brand-new, 'from the shops', and he applied himself to the art of driving it, evidently without mishap, and he found this new experience both easy and exciting. The snow delighted him too, by confirming the picture-postcard view of England at Christmas time — a vision so at odds with the sun-baked festivities he had known in South Africa. On throwing open his curtains he marvelled at the beauty and serenity of the scene and thought it hardly looked real. But it was and soon it turned to slush and then to mud. Only a foretaste of what was soon to come.

The pace of training quickened noticeably. Schonland drew the necessary stores and equipment for the task ahead: '600 poles, 2000 rounds of ammunition, 53 gas helmets, iodine ampoules and field dressings, bandoleers and mess tins ... and bought himself a Colt revolver from the

Q.M.'. He used the remaining few weekends to bid farewell to friends in Cambridge and London, where Miss Crampton not only provided the accommodation, but also a hot bath, which was a rare delight for a soldier. He called on the High Commissioner, Mr Schreiner, and on the large circle of South African friends resident in the city and, with Rupert White-Cooper who was there for the same purpose, enjoyed the distractions provided by the theatre and the music, particularly a concert at the Queen's Hall with music by Weber and Grieg, conducted by Sir Henry Wood.

London was full of soldiers of the 1st South African Infantry Brigade, distinctive with the Springbok on their caps. These men had volunteered to fight in Europe despite the antagonism of so many of their countrymen back home, whose bitter memories of the Boer War lingered long. Schonland noted in his diary that 'Botha seems to have scored over Hertzog for once': those erstwhile comrades-in-arms against the British now led the government and the opposition respectively, and were now bitter enemies. South Africa may have been divided but it had not forgotten its fighting men abroad. The War Sufferers' Aid Society was very busy sending Christmas cards and pairs of socks to all the soldiers. Schonland replied immediately he received his. 'As an old Grahamstown boy, now in the British Army, I wish to thank your society for its kind card of good wishes sent me at Christmas time, and received when about to set out for Flanders with my section. It is a pleasure to feel that Grahamstown has not forgotten one, though far away, and I thank you for the kind thought that prompted the sending of that card' [7].

Back at Hitchin, where he had mastered his new car, Schonland set out just before Christmas to drive the short distance to Bedford to see a friend of his father, Dr Robert Broom. Broom, a Scot, was a medical doctor, now living in South Africa, whose overriding interest was palaeontology. He had volunteered for service in the Royal Army Medical Corps but his age prevented him from serving at the front, so he found himself instead examining and inoculating recruits. In his spare time, of which there was much, he made forays into the local clay beds in search of fossils, or ventured into London, where he studied and bought old Dutch paintings. The young Schonland found his hour with Broom most enjoyable and stimulating, for he had talked to the man whose discoveries in years to come, in a cave just to the west of Johannesburg, would confound the world of palaeontology. On his way home Schonland threw caution and a little discipline to the winds and piloted the Singer back to Hitchin at speed 'with headlights blazing in spite of the Zeps'. Fortunately, the Kaiser did not retaliate.

Final preparations for embarkation for France were now proceeding apace and the young commanding officer, in the time-honoured army tradition, route-marched his men to keep them both fit and occupied. He lectured them on discipline and paraded them at the clothing stores to ensure that all their kit was complete. Convinced as he was by the order and structure of the military way of life he 'catechized' them on all aspects of soldiering and so prepared them to the best of his ability for what lay ahead. 43 Airline Section, with its commander just two weeks short of his twentieth birthday, sailed on 21 January 1916 from Southampton for France.

CHAPTER 4

IN COMMAND

The Army's 'Field Service Post Card' that arrived at Oatlands Road, Grahamstown, early in 1916 was cryptic and concise: 'I am quite well', it read, 'Letter follows at first opportunity'. These were printed lines among a list of choices provided by the Army. The writer chose those most appropriate to his circumstances and crossed out the rest. It was dated '24/1/1915' and initialled 'BFJS', though Schonland surely meant 1916, for the year had already changed. But then there were many other things on his mind. This was the first indication to Mrs Schonland that her eldest son was now at war [1].

In fact, Basil was very well and was turning all the fruits of his training over the last many months into a state of readiness for what lay ahead. By early March, 43 Airline Section was taking its place in the line, one of three such sections with the Fourth Army, then just forming under General Sir Henry Rawlinson. The Fifth Army was still in reserve and not due to come into being until July, but the fluidity of war required that men moved as situations demanded, and signals officers were much in demand. So Schonland and his men found themselves in new surroundings and under an unexpected commander.

However, mere subalterns saw no generals. Schonland's immediate commanding officer was the recently promoted Colonel R G Earle who was filling the equally new post for the army of that time of Deputy Director of Army Signals [2]. Schonland's own domain, which he described vividly to Felix in a letter written from 'Somewhere in France', formed part of the headquarters signals system and consisted of four telegraph offices 'replete with instruments, telegraphic and telephonic, of all conceivable kinds', plus the operators who manned them and all the paraphernalia in the line of stationery needed to make the business of communications function. His two sergeants looked after their detachments, arranged the line stores and instilled the necessary fear of God into the men by way of discipline. The orderly corporal dealt with the correspondence, pay and the multiplicity of indents for stores and related administrative matters, while two others had charge of the maintenance, fuelling and the regular cleaning of the motor lorries. Schonland himself, of course, had his Singer. Two cooks provided for all their necessary sustenance and managed the rations [3].

The Section's task was that of an independent unit. The colonel simply gave orders as to which particular lines were to be built, labelled, tested and maintained and Schonland managed the details. In his estimation, '43' were far and away the best of the three airline sections; a view presumably supported by the colonel because he had indeed designated them the senior section. Acid proof though, Schonland asserted, was provided by the evidence that their lines withstood the rigours of the weather best of all. He himself was hoping that promotion to full lieutenant, and his 'second star', would soon follow, but Earle was of the old school and not given to haste in such matters.

'Somewhere' was actually near the town of Cambrin, itself on the La Bassée road, and between them lay the front line, where the British and German trenches were but a few hundred metres apart. It was here, in the previous September, that the British Army, in the Battle of Loos, had suffered such appalling losses (over 60 000 men for very meagre gains of territory in return) and it was here too that the Army had used gas for the first time. Six months later the positions of the trenches had hardly changed. Two massive armies were dug in and fighting a war of attrition. Surprisingly, after the artillery pounding it had taken, the Le Bassée road was the best Schonland had encountered in France. Of sound construction with well-laid slabs of solid stone, it withstood the impacts of high explosive shells rather well since they did not penetrate its surface. In addition, because of its strategic importance, it was regularly repaired, which was not characteristic of most of the others he had observed as he piloted his Singer between the pock-marks that were now such a common feature of both the countryside and roads alike. His car, though violently shaken at times, ran well as he covered much ground on line-laying missions. One of these was a telephone line that '43' laid to the nearby Royal Flying Corps squadron, and whilst engaged on this task Basil saw up to a dozen machines 'alighting and going up', one after the other. 'I hope to have a flight sometime' he temptingly informed his brother. This flow of news, though subject to the military censor, kept his family abreast of his personal affairs and those of mutual acquaintances such as Rupert White-Cooper who, he informed them, was now 'Assistant Divisional Instructor in Sniping and Scouting. Hot stuff for a division of 10–20 thousand men'. Blessed with a lucid pen and much good humour, Basil provided a picture of the war which, to his eyes at least, was not yet tarnished by its horrors. These were adventurous times for testing one's mettle and he certainly meant to test his.

In April, another line-laying mission on the orders of a Major Barrie 'that prince of noodles', as he described his senior officer, took '43' right to

the trenches, where they were to build a line for none other than Lieutenant Lawrence Bragg from the Cavendish. Bragg, whose share of the
Nobel Prize with his father for their work on the analysis of crystal structure by means of Röntgen rays was but months away, had been training in
the Royal Horse Artillery when he was selected to go to France to investigate the method the French used for locating enemy guns by the sound
of their firing. With the assistance of one of his corporals by the name of
Tucker, in peace-time a lecturer in physics at the Imperial College in
London, Bragg developed a technique of sound-ranging that was most
effective [4]. The significance of this acoustic predecessor of direction
finding by radio (known as the 'MUM machine') was not lost on
Schonland, and though the military censor prevented him from divulging
its exact purpose to Felix, he confided to his brother that Bragg '(one day
you will hear of him) . . . is ——ing German ——s by means of a very cute
invention' [5].

The Army was at last beginning to realize that warfare in the age
that produced the technology of the machine gun, the aeroplane, high
explosives and poison gas required men of the highest intellectual calibre
to devise and design methods for their use and also to counteract their
effects. Thus, men like Bragg and Appleton were now usefully employed,
in stark contrast to the policy of just a short while before that led to the
almost inevitable death of Henry Moseley, their brilliant Cambridge
colleague, in the Dardanelles. But this was war, and war of the most
violent kind and anyone in uniform, regardless of rank or station, who
ventured near its front lines was vulnerable. Schonland was right on
that line but, by one of those accidents of fate or luck, was himself
destined to survive the carnage.

In May 1916, 43 Airline Section was recalled to Fourth Army HQ,
much to their commander's disappointment and that of his men. Their
performance at the front had earned them the praise of 'the Corps' who
paid them 'some nice compliments', wrote Schonland, but they now
found themselves rather under-employed 'just doing odd jobs'. However,
it did present him with the opportunity to turn his hand to some invent-
ing and what he produced was a piece of equipment for winding up
cables. Until then this process had been particularly laborious, requiring
much human effort. The idea struck him to use the power of the engine of
one of the lorries to drive a cable drum and so he designed a device which
did this almost automatically. It would obviously require some gearing
and this he arranged by means of a belt and two pulleys driven by the
drive-shaft of the lorry. To a parallel counter-shaft he attached a large
steel disc that functioned as a friction drive. At right angles to this he
then mounted the cable drum, the shaft of which was connected to a
small wheel, whose position on the friction drive could easily be changed,
thereby providing control of the take-up speed. His prototype worked

well but, before it could be manufactured in the base workshop, Schonland had left the Airline Section so was unable to test his invention in the field. Later, he was assured by Lieutenant Watson, his successor, that it worked very well (with one modification incorporated, of course, by Watson himself!). In describing it later to his brother, Basil was philosophical about whether he would ever receive much, if any, credit for it: 'However I don't mind, as I know it works, and anyway it will never be of any use to me.... I designed it and got it made.' [6]

The nature of the war was now changing and Basil Schonland's own role in it was soon to change too. The front remained static for long periods as the armies endured the murderous ferocity of artillery bombardments that flattened whole towns and left hardly a tree standing on what soon became a waste-land from Ypres in the north to Arras in the south. Until then, communications between HQs and all subordinate formations had been achieved by a variety of means. These ranged from the few wireless sets supplied to the cavalry, to the hundreds of miles of telegraph and telephone line linked to the airlines and buried cables that laced the countryside. Finally, in the trenches themselves, brave men waved flags and aimed heliographs and lamps, or released carrier pigeons and messenger dogs, and then prayed. All were supplemented by the intrepid DRs on their motor cycles and even suicidal runners on foot. But communications, like life itself in that muddy Hell, were only tenuous and often brief. The shrapnel that fell like hail and the direct hits from the 'Boche 5.9s' soon devastated any cables not buried underground, while the withering curtain of machine-gun fire rapidly put paid to any signalling system which required an operator to raise his head, or hand, above the parapet of a trench. Airlines, other than those well out of range of the artillery, were therefore soon to be assigned to the pages of history; all cables, and there were to be thousands of miles of them by the end of the war, would have to be buried at considerable depth. Only wireless systems could possibly work, but few sets were available. Great effort then went into developing this newest of technologies and, as fast as sets were produced, so they were rushed into service along with the skilled operators to use them. Work was underway in the RE Signals Section in all these departments for what was soon to become known as the Battle of the Somme.

*   *   *

The first wireless equipment taken to France by the BEF consisted of 1.5 kW sets mounted in lorries or in limbered wagons for use with the cavalry. These sets were large and cumbersome and were soon found to be unsuitable once the opposing armies were dug in, and equipment better suited to such an environment was rapidly sought. Wireless telegraphy was still in its infancy and its use in trenches was very much experimental. At this stage of the war only the Royal Flying Corps had equipment remotely small enough for rapid deployment within the confined spaces of a trench. Experiments using it were undertaken in the field and soon a tentative design for an Army set was sent to the newly established Signals Experimental Establishment at Woolwich, where 100 sets were manufactured. This became the famous 'Trench Set, Spark, 50 watts', better known as the BF set. This nomenclature puzzled many, but not its users, who soon assumed that it was meant to be foolproof even when used by the many hastily-trained operators rapidly assembled from the ranks of infantry. When demonstrated to senior officers and other dignitaries, however, it was always referred to as the 'British Field Set', thus preserving some decorum.

The BF set could be carried by three men and was simple to use. In practice, however, the party had to be augmented by many more whose task was to carry the massive accumulators which provided the power to sustain operation for the duration of lengthy engagements. The set operated at wavelengths of 350, 450 and 550 metres, agreed after early experiments had confirmed the need for careful coordination of frequencies with the Royal Flying Corps in order to prevent interference between their respective wireless communication networks. Needless to say, the enemy were developing wireless equipment of their own for very similar purposes and cases of mutual interference between them were also common. Soon, such intentional interference with the enemy's communications became just another military tactic and 'jamming' became another weapon in an expanding armoury of military technology. Not long after, wireless listening devices, and the intelligence they provided, also assumed increasing importance [7].

On 1 July 1916 the Battle of the Somme commenced with an assault by fifteen British and five French divisions across a front of almost 23 miles. It was to last until mid-November as a battle of attrition, which accounted for huge losses on both the Allied and German sides. 43 Airline Section were again in action. In all, the Royal Engineers, with massive muscular support from any labour to hand, laid 43 000 miles of overhead lines and 7000 miles of buried cable to provide the communications for the campaign [8]. Trenches snaked across the countryside with the opposing armies frequently within shouting distance of each other. Conditions within the trenches were awful. When it rained, glutinous mud enveloped everything and above ground the thousands of shell holes were flooded death-traps for the unwary, the exhausted and the wounded. Men lay for days where they fell before some makeshift burial was possible, if at all. In a letter to his mother Schonland described a scene he encountered: 'We had a very unsavoury piece of work to do, to run a line for a station, through a trench full of dead; three weeks old too. It is curious however how different things are in actual fact to what one's imagination pictures. Quite as horrible of course, but so much more "ordinary". I won't say any more.' He was now in the thick of battle, 'right amongst the guns near Combles' where 'the whole earth shook and rocked and the air was full of shrieks and hisses of shells... while all around were the spurts of flame from the guns' [9]. A mere mile away was Delville Wood where the South African troops that he had seen in London just a few months before were now fighting one of their greatest battles as part of the 9th Scottish Division.

The Cavalry were again pressed into service and they, possibly more than any other Corps within the Army, required wireless communications. Schonland requested permission to go forward with a cavalry unit that needed a wireless station 'to keep up communication' once it crossed the line. Permission was granted and he set about preparing his equipment while his batman polished his revolver. To his great disappointment, though, the action was cancelled and the unit returned to its billets. 'By Jove, I was fed up!', he informed his mother, but then reassured her that he was quite well and feeling 'very braced' with the successes of the past few days. 'At Thiepval, for the first time in this war, the Boche threw away his arms and fled', he wrote. What also helped contribute to his positive frame of mind was his first sighting of a tank which, he said, 'the Boche wireless press [called] a disgrace to modern warfare' [9]. This secret weapon of the British Army, until then, known only as 'hush-hushes', made its first appearance on the battlefield on 15 September when thirty-two of the iron-clad monsters lumbered into battle at under 3 mph near Flers-Courcelette. This formidable, though still experimental, weapon was understandably called a 'priceless conception' by Winston Churchill, its foster father, when he criticized Field Marshal Haig for revealing it to the enemy too soon. However, the immediate impact of the tank, though initially terrifying to the enemy, was short-lived because neither the machine itself, nor the tactics for its effective use, had yet been fully developed [10]. When viewed from the trenches, and even by an itinerant inhabitant of them such as Schonland, a leviathan like this must have seemed like a gift from Mars himself.

By 23 August 1916, Schonland had handed over command of 43 Airline Section and was now attached to the Cavalry Corps and its Wireless Squadron or W/T CC. There, in the company of Lieutenant Long, his superior officer, they had under them 100 men, 110 horses and eight wireless stations. His letters home were couched in oblique phrases with no reference to any details, for it was 'all more or less of a secret nature', but his enthusiasm for this most recent wonder of physics was evident in every one. To his youngest brother Dick he enthused in the occasional letter he wrote to him. 'Wireless is a fine job and though I mustn't say much about it, it is very much connected with the fighting. For in these days of trench warfare telephone cables are useless and it is the only thing that will keep up communication. Of course you can't flag-wag in the trenches! So the old "wireless wallahs" have to do the trick for them' [11]. To Felix he called it 'a splendid job', and the censor allowed him to tell his brother about the way the British and French armies communicated with each other by wireless at any time:

> All this requires arrangement of procedure (for they "work" differently to us), codes, wavelengths, etc., etc. It works splendidly though. You've simply got to turn on to the right wavelength, press the key and call — V —, and back it comes — de —, "I am ready to receive"!

Since soldiers fight on their stomachs Basil was also greatly appreciative of the biltong[*] he occasionally received from home and he assured Felix that he kept it for 'special feast days' and also used it for emergency rations 'for it is fine stuff to take with you when you go off into the blue on a job of work'. South African cigarettes too were particularly welcome, for the English variety had not come up to his expectations. He also inquired about the girls he used to know: 'Gertie Rutherford & Co and Grace Bartlett (my old flame). The Jennings girls are the only ones I have heard of and they *have* been a collection of surprise packets' but 'no time for girls now there's a war on' [12]. Dick, now aged 11, was just about to enter the College from Preparatory School and the significance of such a step in a young man's life was not lost on his eldest brother who mentioned that his own recent attempts at growing a moustache had not been too successful. He then turned to St Andrew's. 'What is College 1st XV like? Are they playing Kingswood this year? Nearly everyone one meets this side of the equator plays soccer. Even out here, once a battalion comes out of the line they are hard at it, as always, soccer'. Intriguingly, the address on this letter is 46 Stationary Hospital, and again, just 'somewhere' in France. If Schonland had been wounded or injured he did not let on but said only that 'he had bags of time on (his) hands'.[†]

Felix was desperate to play his part in the war but was short-sighted, and since he was only about to leave school he had no special skills to offer. He enquired of the brother he so admired what he would have to do to be accepted into the Royal Engineers or, failing that, the Administrative Services Corps. Basil wrote back kindly and indicated that he was not optimistic about Felix's chances of getting into the Engineers without technical qualifications or long service as a cadet. He pointed out that he himself did not expect to get in and that even his good

---

[*] Biltong is an Afrikaans word, used universally in South Africa, which refers to dried meat (usually game) that is regarded either as something of a delicacy or even a staple diet, depending on one's predilection.
[†] An account in a South African newspaper in 1960 said that he was wounded at Arras. Schonland never talked of it after the war but might this have been the occasion?

friend, Rupert White-Cooper, could not. He was decidedly forthright in his views of the other Corps, however: 'I hope you won't join the last named. It's an awful corps and rotten work. Dishing out rations or stealing jam from the men in the trenches. I think they deserve the fun that is made of them'. As an alternative he suggested the Royal Field Artillery, but promised anyway to make inquiries about Woolwich and the RE.

*   *   *

Between July and September, at the height of the battle, the front line edged forward by no more than 10 km in the area to the east of the town of Albert. At its eastern end lay the devastated village of Longueval, and nearby was Delville Wood with its quaint, rectangular layout of tracks to which the soldiers had given such inappropriate names as Princes Street, Regent Street, Bond Street and Rotten Row. On 14 July 1916, 3173 of Schonland's compatriots of the 1st South African Infantry Brigade entered the wood in an attempt to take it from well dug in German troops. To its south was Waterlot Farm, which the Germans had fortified with heavy machine guns that subjected the advancing South Africans to devastating enfilading fire. Six days later, after one of the bloodiest battles of the Somme, only 748 men emerged [13]. The wood, an inappropriate term if ever there was one to describe the shattered tree stumps, the bomb-ravaged earth and the bodies, was theirs. But at what cost?

Wireless sets from the W/T CC played their own part in that battle. On a rise in the ground just a few hundred metres from those German fortifications at Waterlot Farm, one of a network of BF sets had been set up to provide vital communications with Maricourt some 5 km farther back. At the height of the battle the intensity of the bombardment was such that it was impossible to keep erect the two 5m masts that supported the usual wire antenna and so 'ground aerials' (insulated cables laid along the ground for a hundred metres or so) were used. 'Time and again these sets got "through" important messages when all other means of communication failed', wrote Schonland in an article after the war. Other BF sets at places like Guillemont, Mametz Wood and Bazentin-le-Grand were similarly communicating with Maricourt and Fricourt as the Fourth Army fought its way forward through the mud and into a wall of lead. The wireless operator, 'Sparks' to his fellow occupants of their hole close to Hell, was alternately humoured or cheered, depending upon the success he achieved with his 'wireless magic'. Frequently, as in the battle of Arras, it was a BF set that called the artillery into the pulverizing reaction that was to save a beleaguered observation post, such as that on the hill at Moncy-le-Preux, from the teeth of the rapidly advancing enemy. Then, even the most hardened of sceptics believed that these 'miracles' really did work.

> The success first obtained on the Somme was maintained in
> succeeding operations and gradually prejudice and opposition were
> won over. More men and better equipment were provided, with
> correspondingly better results from wireless communication. Schools
> were started during the winter of 1916–1917 for Signal Service
> officers, to give them some knowledge of the working conditions,
> possibilities and needs of the sets. The number of sets was increased
> by production in special factories in England and operators were
> trained at depots in England and France. The need being recognised,
> a staff of wireless experts at Woolwich was busy investigating the
> trench wireless problem, and applying to it all that modern wireless
> practice could suggest. In particular, that revolutionary instrument,
> the three-electrode valve, was called in to help.

So wrote Basil Schonland in 1919 when he published possibly the first description of how the wireless telegraphy (W/T) units of the Royal Engineers Signal Section functioned during the war just ended. In a four-part article, aptly titled 'W/T RE', that appeared in the popular journal *The Wireless World* between July and November [14], he described in graphic style how wireless was used and how it performed at the front. Personalities played no part in it, least of all himself, but the content portrays its author as someone not only well-versed in the nuts and bolts of his subject but as an enthusiast for this exciting new science that had just undergone its own baptism of fire. Nuts and bolts indeed, for the wireless sets of that time were as much mechanical contrivances as they were electrical, and their users were as adept at adjusting spark gaps and polishing brass connections as they were at nursing carborundum detectors and at 'tuning up'.

As well as the BF Trench Set the Army was also equipped with two other pieces of wireless apparatus by the end of 1916 and Schonland described them both. They were the Wilson Set and the Loop Set. Like their famous BF counterpart they used spark transmitters while their respective receivers were what subsequently would be known as 'crystal sets', containing no amplification whatsoever. As a result, they were insensitive devices and successful communications required the use of considerable transmitter power. The three-electrode valve, invented in 1907, was not yet used in any British wireless set, though it was in use by this time in the 'IT' set, a French amplifier, which proved particularly useful for 'overhearing' German telephone and telegraph conversations by means of the currents that were always conducted and induced into the ground by the labyrinthine system of cables that criss-crossed the battlefield. What was initially seen as a serious weakness was soon turned to advantage when it was realized that such 'leakage' current could be used to eavesdrop on the enemy. As with so much scientific pro-gress during wartime it was the lead that the Germans themselves had established in the use of such eavesdropping techniques that provided the impetus for its development within the British Army, and the French three valve amplifier, the *Amplificateur* Model 1, was the basis of this pioneering listening device.

By now, Schonland was much involved with all this technology. His own notebook, solidly bound and inscribed 'ARMY BOOK 3', with his name and position as Wireless Officer, Cavalry Corps, written within is a mine of information [15]. In it one finds his carefully drawn circuit diagrams of each and every one of these sets plus those of many more as they appeared in the general signals armoury. The diagrams are accompanied by detailed, hand-written notes that describe the principle of operation of the sets and, of importance for what was to come, his own ideas on improvements that might be introduced. Most important were his views on the use of the new valve sets and his suggestions for generating controlled oscillation and a completely new method of wireless communication.

As the end of 1916 approached, Schonland's own role within the Signal Service was about to change yet again. Soon, he would abandon pliers and soldering iron, the tools of the signal officer's trade, for the chalk of the instructor as he applied himself to teaching 'wireless' to the new officers flooding into the Corps. In December, he was withdrawn from the line and posted first to Abbeville and then to GHQ Central Wireless School at Montreuil, the little walled town just a few kilometres from the seaside resort of Le Touquet, where he was to serve until March 1918. There, he carried the dual responsibility as Wireless Research Officer, attached to the Wireless Experimental Section BEF, and instructor-in-charge of Officers' courses of instruction [16]. On 15 February 1917 his promotion to Acting Lieutenant was announced and he could put up the long-awaited second 'pip'. His duties involved much lecturing to large classes as well as practical demonstrations of the whole array of wireless equipment then in service with the Army. It was soon evident that Schonland possessed a natural bent for teaching and the Chief Wireless Instructor, Major Rupert Stanley, was lavish in his praise of the courses that Schonland put on. Of significance too was the relish with which he undertook 'important researches on Radio Signalling' leading to his being regarded as one of the best-equipped officers on the School's technical staff. Major Stanley was most impressed, too, with Schonland's 'fine knowledge of Mathematics and Science', and with the fact that he was 'an enthusiast in their application' [17].

*   *   *

Letters to his family continued to and fro but their content lacked any whiff of the military actions of the previous two years, while the censor ensured that no details about wireless was breathed to anyone because, as Basil told Dick in October 1917: '[it] is doing great things in the war now. Of course, we can't talk about it now but after the war I will have some exciting tales to tell you'. His new position had given him an introduction to Colonel Gustave Ferrié, Chief of the French Wireless Services. Ferrié had been responsible for placing France ahead of all other countries in the strategic use of wireless communication in the years immediately preceding the war, while two of his major contributions afterwards were the development of the most successful valve of the period, the 'TM' ('Télégraphie Militaire'), which became famous as the type 'R' in its better-known civilian guise later on, and the directional frame aerial that was introduced into the battle zone in the final year of the war. Schonland would himself become well acquainted with both in years to come; now he hoped to visit Ferrié's famous wireless station in the Eiffel Tower from where daily time signals had been transmitted since 1910, making Paris the 'Time Centre' of the world [18].

It was while he was based at the Central Wireless School that Schonland carried out some research, possibly his first, and certainly the first to lead to a publication under his name in the scientific literature [19]. With a fellow officer by the name of Spencer Humby[*] he investigated the characteristics of oscillating valve circuits and it was in that year, 1917, that the first continuous wave or CW sets made their appearance on the battlefield [20]. This was a major leap forward in the technology of wireless transmission for it changed the whole concept from one where a spark discharge set up a damped oscillation in a tuned, or resonant, circuit every time the Morse key was pressed to that of a continuous, almost pure, sinusoidal waveform that was produced by keying an oscillating valve. The advantages of the CW set were two-fold: first, it produced radiation from its aerial or antenna that occupied a very narrow band of frequencies (the bandwidth in modern parlance) and so reduced the interference to other wireless systems operating on nearby frequencies; second, it was considerably more efficient in terms of the conversion of battery-generated power to radio frequency energy. Both were to be of much importance to the further use of wireless communication systems, not just in wartime, but for all subsequent developments that took place in the years following the armistice.

Humby and Schonland made a series of experiments during the winter of 1917–1918 using a triode valve operating as an oscillator. They investigated the conditions under which oscillation occurred

---

[*] Spencer Humby (1892–1959) graduated in Physics from Cambridge in 1913 and was commissioned into the RE in 1915. He won the Military Cross in 1918. During the inter-war years he was a schoolmaster at Winchester, where he was highly regarded as a teacher of Physics and as the author of some very successful texts on the subject. He was to serve under Schonland in the Second World War in the Army Operational Research Group (AORG) and became its deputy superintendent in 1945.

when the coupling and phase relationships were changed between the two resonant circuits at the grid and plate of the valve. The circumstances of war prevented them from publishing their findings in the open literature until almost a year later, by which time many others had addressed themselves to the same problem. Amongst them was none other than Schonland's old colleague from Haynes Park in Bedford, Edward Appleton who, almost a year before, had published a paper in the same journal that also examined the conditions necessary to produce oscillations from the three-electrode valve [21].

In March 1918, Schonland was posted as Wireless Officer to the VI Army Corps [22]. Regular monitoring of German wireless traffic was just one of the Corps Wireless Officer's functions and this task even included listening to 'Boche ships in the Baltic' whilst 'the big stations like Petrogna, Carnavon, Berlin, Rome, Paris, Lyon etc. are mere commonplaces' Basil wrote to Felix. There too he found himself with an array of apparatus under his control. 'Gadgets Limited' is how he described it to his brother. By this stage of the war both sides were equipped with a variety of wireless sets. The 'aether' was subjected to its own barrage — a wireless barrage — and 'was in a terrible state of agitation' according to Schonland. The number of spark transmitters had increased many fold since the outbreak of war. As well as the ubiquitous BF sets, there were the more powerful Wilson sets with their associated Mark III Tuners or receivers, and the much smaller 'loop' set. This equipment was a direct result of the call for a smaller, more portable wireless set and particularly one whose presence was less obvious to the Germans, who so rapidly sought out any elevated wire as a vital target. Destroy it and your enemy was isolated and alone, unable to report his progress or to call for assistance when needed. The loop set achieved this visual obscurity by using, as its antenna, a metre-square brass loop with self-contained spark-gap and tuning condenser, or capacitor, mounted on a bayonet. The operating wavelength was in the range from 65 to 80 metres, by far the shortest employed by any British wireless equipment used in the war and the transmitter produced about 20 watts of power. It was capable of communicating effectively over a distance of two to five miles and so served its purpose admirably between trench and Brigade HQ. The CW set, with its brightly glowing valves, operated between 600 and 2000 metres and was now used almost exclusively between the observation posts, or OPs, of the artillery and their guns to call down a barrage on some selected target. By now, too, wireless sets had even been fitted to some of the tanks that so changed forever the face of trench warfare when they overwhelmed the German defences at Cambrai on 20 November 1917 [23].

The military situation was becoming critical because a major German attack was expected and Schonland found himself busy, day and night, 'on a new stunt' that involved providing effective wireless links between the BEF and the French Army. The attack duly came when Ludendorff unleashed his Storm Troopers from the Hindenburg line. Its aims were to separate the British and French armies, to capture the Channel ports and then ultimately to destroy the encircled British forces. This was the most massive and powerful German offensive of the war and the British Army suffered heavy casualties but it held — just. However, in little more than a fortnight the German onslaught faltered and the tide of war began, for the first time, to turn against them. Then, on 9 April in one desperate, almost dying lunge, the attack switched north to Flanders where it had all begun nearly four years before. The Germans fought like men possessed. Schonland's descriptions of the battle to Felix were graphic. 'Tonight there appears to be a terrific strafe on. The whole sky is lit up with the flashes. Apparently he is striking another blow for Arras and the Vimy ridge. His last attempt was stopped dead with enormous loss to him...'.

In June 1918 Basil Schonland was promoted acting Captain [24]. Soon after, he applied for leave and set off for London where the 'endless taxis, buses, tubes etc.' came as something of a shock after the sites of the rather different world that he had recently been inhabiting. He did enjoy the opera: 'it was magnificent' he said in another letter to his brother who was now reading engineering at the university in Cape Town. 'By the way do you often go to the Cape Town orchestra? You should, old man, for it is a jolly good one and the things they play are an education in themselves'. Schonland's evident *joie de vivre* was reinforced by the recent news that his captaincy had been backdated to February, with over £75 in back pay. Just by way of a footnote he added: 'I hear GHQ are making a wonderful new wireless set as a result of that research Humby and myself undertook' [25].

His leave over, he returned to France to the news that four of his men were to be decorated with the Military Medal, following two others who had already received it. Schonland felt immensely proud and went around congratulating each of them. Soon, his own prowess received further recognition when he was placed in charge of all wireless in the First Army under the command of General Horne, who had distinguished himself those years before when he led the gallant British rearguard action at Mons in 1914. Schonland's responsibilities had now increased enormously. At the age of 22 he had reporting to him approximately 30 officers and 900 men with an equipment complement of 300 wireless sets [26].

Wireless had indeed made its mark. And when it finally came the war was ended by wireless. At 5 a.m. on 11 November 1918 the powerful transmitter in the Eiffel Tower 'rang down the curtain on the world-war' by transmitting the epic message from Marshal Foch, the Allied Supreme

Commander: 'Hostilities will cease at 11 a.m. . . .'. This was soon followed by 'every set in the British Army, French, Belgian and American Armies ... repeating it as an order to the fighting troops'. The coast stations had also taken it up and the aether 'was simply full of it'. So Basil described it to Felix. For the first time in nearly three years he was able to announce where he was, at Valenciennes, and very comfortably accommodated in a château that was serving as the headquarters of the First Army. He expected 'to be turfed out soon' as the civilians flooded back to reclaim their town, or what was left of it, after it had been formally handed back to the Burgomaster by General Horne [27].

CHAPTER 5

FROM 'THE CAVENDISH' TO CAPE TOWN

Captain Basil Schonland RE was discharged from the British Army on 15 January 1919, but not before some effort had been made to retain his services [1]. When the fighting ended he was commended by the Chief Signals Officer, Colonel H T G Moore, for the energy, resourcefulness and ability with which he had controlled the wireless communications throughout the British 1st Army as it advanced from Arras to Mons [2]. This was high praise indeed and must have marked Schonland out for serious attention, for it was soon followed by the offer of the post of Chief Instructor in Wireless in the Army and with it the rank of Major. Others had noticed his all-round ability as well. The Marconi Wireless Telegraph Company offered him a position in their Chelmsford Works. Both were tempting but he declined them. It had always been Schonland's intention to return to Cambridge and to resume the academic career he had left in abeyance four years earlier in order to do his duty as a soldier. Though flattered by the offers made to him he remained resolute in his intention to become a scientist [3].

Of immediate importance, though, was to return to South Africa as soon as possible to see his family. The process of booking a passage on a steamer that would take him from post-war England to Port Elizabeth was not easy. Thousands of demobilized soldiers from the countries of the Empire, who had responded so nobly to Britain's call for fighting-men, were now of one mind and that was to get away from it all and to get home. After much frustration and time-wasting Schonland eventually arrived back in Grahamstown in July 1919, but only after the weighty intervention of none other than the South African Prime Minister. Getting a berth was proving far from easy and Basil communicated his feelings at the interminable delay to his father and Selmar Schönland, who had shown his tenacity and willingness to appeal to authority before, took action by approaching his botanical kindred-spirit, General Smuts himself. Of course, by the end of the war Smuts was a man of some influence in England; the former Boer general was not only a member of the War Cabinet but was a signatory of the Treaty of Versailles and was able to pull the odd string or two [4].

Selmar Schönland later wrote to the Prime Minister to express his thanks for his intervention [5].

> You will understand that I would not have worried you if I had not known that many people got passages who had done nothing to deserve them and that plenty of our boys were waiting to get home. We have had him here now for 4 weeks and he still has about 3 weeks before he returns to Cambridge. I am pleased to say that he got the OBE as a parting gift. He is in the best of health and spirits in spite of the hardships and dangers he has gone through and hopes to take his 2nd Tripos at Cambridge next June. He got a First in the 1st part 4 years ago...

The 'parting gift' of the OBE (Military Division) awarded to Acting Captain Basil Ferdinand Jamieson Schonland 'in connection with military operations in France' was indeed announced in the *London Gazette* of 3 June 1919 and was followed shortly after, on 5 July, by a Mention in Dispatches [6]. If it was possible, given the carnage and slaughter synonymous with places like the Somme, Ypres, Thiepval and dozens more, then Basil Schonland had had a good war. However, it was one that he soon wished to forget.

*   *   *

Schonland was amongst a throng of ex-servicemen who returned to Cambridge for the start of the Michaelmas term of 1919. Amongst them was another South African by the name of William Hofmeyr Craib, always known as Don, having been informally christened 'Donal' by a Scottish relative who claimed that 'no Scotsman is ever William' and that 'he had never even heard of the name Hofmeyr'! Don Craib hailed from the town of Somerset East, near Grahamstown, where his father, James, was head of mathematics at Gill College, an institution of higher learning that prepared students for degree courses. Craib's mother was a Hofmeyr, a name with great resonance in South African history, and a granddaughter of Andrew Murray, the pioneering Scots Calvinist missionary to South Africa.

Don Craib was a year older than Basil and was a doughty character. Academically gifted and also schooled at St Andrew's, he had commenced his studies in civil engineering in 1912 at the South African College (the forerunner of the University of Cape Town). His progress was such that he was awarded the Jameson Scholarship for overseas study but before he could take it up war was declared and Craib found himself conscripted into the South African army and posted as a latrine attendant to South West Africa. Such an inauspicious start to his unplanned military career was not what he had in mind and so, without let or leave, he made his way to England to offer his services in a more demanding capacity. On arrival he was promptly arrested for desertion. However, once again W P Schreiner, the South African High

Commissioner in London, who was similarly to watch over Basil Schonland, intervened and, with Smuts's agreement, Craib was transferred to the British Army on condition that he served for the duration of the war—assuming he was fortunate enough to survive that long. Serve he did indeed and was soon commissioned into the Royal Field Artillery. By the war's end, as Captain Craib, he had been awarded the Military Cross and Bar within the space of one month in 1918, and was twice Mentioned in Dispatches. Along with so many others who had endured the horrors of that war, Don Craib was left with deep scars on his soul if not his body, and so massively was he affected by what he had seen that he abandoned engineering and switched to medicine 'to do something to make some reparation for the war' [7].

Schonland and Craib decided to share accommodation at Caius and were allocated a suite of rooms above the Gate of Virtue in Caius Court that had previously been occupied by E A Wilson, the doctor on the ill-fated South Pole expedition of 1912 under Captain Robert Falcon Scott [8]. There, in such heroic surroundings, Schonland was soon to meet Craib's sister, Isabel Marian, who was always known as Ismay. She was reading history at Newnham College. Born in Somerset East, Ismay was nine months younger than Basil and had matriculated at the Bellevue Seminary there, before obtaining the BA degree at the South African College. She was then awarded a scholarship, which took her to Cambridge in 1919 [9].

During the following academic year Basil Schonland was a College Exhibitioner, the award that had been made to him in 1915 on completion of Part 1 of the Mathematical Tripos. Two choices were now open to him. He could, after two years' residence, submit a thesis on research work he had carried out and, if it was deemed acceptable, would be entitled to a Cambridge University Master of Arts degree. The other option was to complete Part II of the Tripos before concentrating on research [10]. Schonland chose the latter route by opting to follow a course of advanced physics and mathematics in Part II of the (Physics) Natural Sciences Tripos and then to embark upon his research. Three years in uniform had not in any way dimmed the Schonland academic flame. In June 1920 he completed the Tripos with First Class Honours and was awarded the Schuldham Plate by Caius College. This was a prize dating from 1776 and given 'to some scholar as after due examination shall be most deserving'. In addition, he received a George Green studentship for research for a period of two years [11]. Ahead of him now lay the excitement of research in the laboratory that had captivated him all those years before in Grahamstown.

*   *   *

1919 was a significant year at the Cavendish Laboratory, for it was then that Sir Ernest Rutherford returned, at the invitation of Sir J J Thomson, after his wonderfully fruitful years of research, first at McGill University in Montreal and then in the Langworthy Chair of Physics at Manchester. In 1908 Rutherford won the Nobel prize for Chemistry, an unsurprising fact given the proximity between physics and chemistry at that time, and, soon after, found the holy grail of alchemy when he became the first person to effect the artificial transmutation of an element. Rutherford's scientific tools were α particles, the positively charged ions that constituted the nucleus of the helium atom. Virtually all his famous experiments, amongst the greatest examples of practical genius since Faraday, made use of these atomic missiles in his quest to discover the innermost details of the atom itself and all the while his apparatus was essentially so simple. Such frugality was a characteristic that he instilled in some, though not all, of his students and protégés: Schonland was certainly one.

Amongst Rutherford's scientific *tours de force*, that not only anointed him as one of the giants of science but which changed the face of atomic physics forever, was an experiment in 1909 that had greatly inspired the young Basil Schonland in his coach-house workshop in Grahamstown. Thirty years later, having himself lived and worked in Rutherford's shadow, Schonland would meet one of the men who had performed that experiment and their meeting would change profoundly the direction in which Basil Schonland's life would run thereafter. That man was Ernest Marsden.

It was in Manchester, in 1909, that Rutherford, with the assistance of a visiting German physicist, Hans Geiger, and a 20-year-old student from Blackburn by the name of Ernest Marsden, investigated the scattering of α particles by gold foil. The source of the α particles was the so-called 'emanation', the radon of today, which is a decay product of radium. The radium source was placed within an evacuated, thin-walled glass bulb which also contained a thin gold foil some distance away. The inner wall of the glass was coated with zinc sulphide that fluoresced or scintillated when struck by charged particles and so served as an easy method of indicating their presence. An observer, whose eyes had been adapted to the dark by spending at least half-an-hour in a darkened room, counted the individual scintillations by squinting through a microscope. This uncomfortable yet simple technique had much to commend it, but it was more successful when practised by young eyes, and so Rutherford always had young researchers working with him, who performed this task that required considerable visual acuity.

What had been conceived initially as a useful exercise in teaching experimental technique, now produced a result of fundamental importance in the understanding of the nucleus of the atom and it was the young Ernest Marsden who discovered it. After some initial hesitation followed by very careful checking, in case his technique was flawed,

Marsden reported to Rutherford that some of the α particles were being scattered through angles greater than 90 degrees. Rutherford was profoundly sceptical when he heard this and set out to verify it for himself. He found that Marsden was indeed correct and immediately exclaimed in characteristic Rutherford style that it 'was about as credible as if you had fired a 15-inch shell at a piece of tissue paper and it came back and hit you'! But it was truly remarkable and Rutherford was soon convinced that at the centre of the atom was a small nucleus which contained both the major part of its mass as well as a positive charge. For many months he worked on the problem, performing a series of meticulous tests, before announcing almost informally that he knew what the atom looked like. The accompanying mathematical theory was based on some elegant, though essentially straight-forward, Newtonian mechanics and the physical laws of Coulomb forces, yet its implications were profound [12].

Rutherford's announcement overturned the prevailing theory of the time that the nucleus consisted of an ethereal sphere of positive charge within which were embedded the electrons: the 'plum pudding' model first propounded by his immediate predecessor in the Cavendish Chair, J J Thomson. The way was now clear for Rutherford and the Danish physicist, Niels Bohr, who was soon to arrive in Manchester, to work together in producing one of the most dramatic advances yet seen in physics, in the form of the quantum theory of the atom [13].

Ernest Marsden's subsequent career became multi-faceted and its almost uncanny similarity to that of Schonland's is one of the themes that ran through their occasionally intertwining lives. In 1914, Marsden was appointed on Rutherford's recommendation as Professor of Physics at Victoria University College in Wellington, New Zealand, but soon returned to Europe, in uniform, and served with distinction as a signals officer, first in the New Zealand Expeditionary Force and then on secondment to the Royal Engineers, where he did important work on sound-ranging. In the rank of Major when the war ended he had been Mentioned in Dispatches and was awarded the Military Cross in 1919. He then returned to New Zealand to resume his research career [14], just as Schonland was about to commence his at the Cavendish. And it was to the Cavendish that Rutherford himself was soon to return as Professor of Experimental Physics in succession to J J Thomson, who had graciously stood aside for his young protégé, now at the height of his powers.

Once there, with his precious 250 mg of radium that he had brought with him from Manchester, Rutherford set about rebuilding the laboratory after the years of torpor into which it had fallen during the war. His task was eased considerably by the influx of so many returning students of considerable talent, plus others who were coming up for the first time. E V Appleton, F W Aston and James Chadwick were amongst the former, while P M S Blackett was one of many serving officers sent in 1919 by the Royal Navy on what were called 'holiday courses' to compensate them for time served at sea. Inspired by what he saw there, Blackett resigned his commission and stayed [15]. All were to become Nobel laureates. If the previous twenty years at the Cavendish under Thomson had seen the laboratory ascendant, the next twenty were to be its apotheosis, its golden age, as Rutherford inspired and guided 'his boys' as they charged through the gates of modern physics that he had opened for them.

Basil Schonland was admitted to the Cavendish by Rutherford in June 1920 to pursue research for the degree of Doctor of Philosophy and, as was normal practice at the time, was expected to take full responsibility for a research topic of limited, though well-defined, scope. The task he was given was one of considerable importance to the understanding of atomic physics since it involved the β particle, which had not proved to be nearly as amenable to lucid argument and rigorous experiment as Rutherford's α particle had been. Not only was the scattering of β particles more complex and therefore less tractable to analysis, but some experimental results had even cast doubt on the very model of atomic structure that was Rutherford's showpiece. It was clearly of great importance to him that the problem should be solved and the conflict resolved once and for all. Then the war intervened. Now, with four years of devastation behind them and a new professor installed, the resuscitated Cavendish laboratory was about to bring its considerable intellectual resources to bear on the problem once more. These resources, most effectively marshalled by Chadwick who soon established himself as Rutherford's right-hand man, essentially confirmed the validity of the nuclear model of the atom—the Rutherford model—yet there still remained a serious discrepancy between the theory and experiments on the scattering of β particles. Even though Rutherford's nuclear atom was now unassailable, these other experimental results were puzzling. Rutherford's elegant analysis of 1911 had assumed that all the forces associated with the atom simply followed the inverse square law with distance between the incoming particle and the atomic nucleus. Could it possibly be inferred that there were unexpected variations in the forces within the very core of the atom that affected the light β particle in some way differently from the α particle? A rigorous investigation was necessary and the man Rutherford selected to undertake it was Schonland.

In terms of Rutherford's philosophy, research was best done by people working individually, rather than each just having a small role as a member of a large team, and so he set Schonland to work on his own, but under the supervision of Dr J A Crowther. James Crowther,

Figure 6. Staff and students of the Cavendish Laboratory, Cambridge in 1920. Back row (L to R): A L McAulay, C J Power, G Shearer, Miss Slater, Miss Craies, P J Nolan, F P Slater, G H Henderson, C D Ellis. Middle row (L to R): J Chadwick, G P Thomson, G Stead, Prof Sir J J Thomson, Prof Sir E Rutherford, J A Crowther, A H Compton, E V Appleton. Front row (L to R): A Muller, Y Ishida, A R McLeod, P Burbridge, T Shimizu, B F J Schonland. (Reproduced by kind permission of the University of Cambridge, Cavendish Laboratory.)

university lecturer in physics as applied to medical radiology, was already working on the β particle, essentially an electron, and it was his findings that had intrigued Rutherford. From as early as 1906, Crowther had been measuring the diffusion or scattering of a beam of electrons as they passed through thin sheets of matter and in 1910 had published a paper that seemingly confirmed the validity of the Thomson 'plum-pudding' model of the atom. This caused Rutherford some understandable concern and caused him to examine Crowther's results very closely. He concluded that Crowther's interpretation of his data was the source of the confusion, with the crux of the problem being the nature of the scattering itself. Was the scattering the result of just a single encounter between the β particle and an atom, or were there many such encounters leading to multiple scattering? The matter required careful and immediate investigation. Armed with this information, plus a supply of radium 'emanation' kindly provided by Rutherford, Schonland began his research.

* * *

The life of a research student within a university always involves more than the immediate problem in hand. In Schonland's case he was assigned as a teaching assistant to the formidable Dr G F C Searle, lecturer in experimental physics and one of very few people ever to gain the trust of that eccentric genius, Oliver Heaviside, whose theories had recently turned elements of electrical science on its head. Searle had been at Cambridge since 1886 and was to remain there for forty-five years. Every student reading physics encountered him, for he ran what all knew as 'Searle's Class'—the practical class during the first year of the Tripos that met on three days a week. There, under Searle's stern gaze, and as the recipients of his occasional acerbic comments, they learnt the art of scientific experimentation, for he had devised a series of experiments that were legendary in their rigour and effectiveness in demonstrating fundamental principles. Searle was regarded by his students with a mixture of awe, amusement and affection and he frequently boasted that he had taught more Nobel prize winners than any other man [16]. He already knew Schonland's ability from Part 1 of the Tripos and was therefore very glad to secure his help in teaching his class. He was later to assert that Schonland was a particular asset because 'it was refreshing to have as an assistant someone who knows some mathematics, a subject which plays a much more important part in the study of Physics than many in the "experimental" school suppose' [17].

In his third year of the Tripos, Schonland had attended the lectures and laboratory classes of another of the legends of the Cavendish: C T R Wilson, who possessed a personality as different from Searle's as was humanly possible. 'CTR' (for he was always known simply by his initials) was a Scot of the gentlest disposition and had a manner of speaking which, to many, appeared as an appalling stutter but which was much more related to the depth of thought that lay behind everything he ever said and, indeed, ever did. CTR was no orator. Those who attended his lectures on physical optics for the first time encountered an apparently timid man whose hand-writing verged on the indecipherable and whose delivery was tortuous in the extreme. Before saying anything he would contort himself into a position with one leg almost wound around the other, and would then proceed to unwind or unravel himself as he disgorged a sentence in a torrent and flurry of words [18]. To those in the audience for the first time this was both startling and hilarious and inevitably there were some who attended CTR's lectures for the entertainment alone, but they soon melted away once their content went beyond them. Those who remained, no doubt the more perceptive, recognized the quality and depth of scientific insight that was buried beneath the staccato barrage. Lawrence Bragg, the Nobel laureate crystallographer, wrote that CTR's lectures on physical optics combined the best content with the worst delivery he had ever experienced! [19]. CTR also ran the advanced 'Practical Class' at the Cavendish and had designed a set of experiments that covered the whole of physics. In his hesitant, gentle way he encouraged every student to regard each experiment as 'a small research' and inspired them to look for unexplained effects, never letting them give up until they had got everything possible out of it. This was a marvellous training ground for anyone about to embark upon their own experimental exploits. Schonland had been through it and was soon striking out on his own.

C T R Wilson's most famous contribution to physics was made in 1911 and it won him the Nobel Prize sixteen years later. Ever since it has been known as the Wilson cloud chamber. As was so customary of the Cavendish of those days, the sheer simplicity of this device was almost as remarkable as its significance and this caused Rutherford to describe it as 'the most original apparatus in the whole history of physics' [20]. The equipment consisted of nothing more than a glass cylinder filled with air, saturated with water vapour and kept at constant pressure by a ground-glass piston. By means of an ingenious system of valves, taps and subsidiary vessels, the piston could suddenly be dropped, so bringing about the rapid expansion of the air. This naturally caused it to cool and become supersaturated with water vapour. If a charged particle should pass through it, the air would ionize and the trail of ionization became the nuclei for the formation of water droplets. These fine, spider-web-like traces were then the visible evidence of the path followed by a particle, its subsequent collisions and its ultimate demise. This remarkable creation provided the first means by which the elementary atomic

61

particles that were the daily bread of the Cavendish could actually be seen, or at least their ethereal presence detected.

And so Basil Schonland found himself at the very heart of physics as it was shaping itself for a new century with all the excitement of fundamental discovery and the application of a host of seemingly wild and new theories. He was surrounded by intellects of a calibre that he had never encountered before and rubbed shoulders with men who were engaged in overturning much of the accepted dogma of the last century and in replacing it by new models of the most fundamental components of matter — and he was part of it. A feature of the Cavendish was the broad international base from which it drew its research students. What had paved the way for this was Cambridge's decision in 1894 to admit graduates from other universities for the purpose of conducting research. It was actually made possible by the munificence of the Commissioners who administered the proceeds from the famous *Exhibition of 1851* at which 'the Works of Industry of all Nations' were put on display in grand surroundings in London. They were able to set aside a large sum of money for the encouragement of scientific research and so the Cavendish became a melting pot where young men, as well as a few women, with diverse backgrounds, cultures and even customs came together in the most stimulating of environments. A number of Research Exhibitions, as they were called, to the value of £150 per year were first awarded in 1890 to selected students from British and Dominion universities. The Cavendish attracted two out of 29 in the years between 1891 and 1895, but received 60 of the 103 awarded in the 25 years from 1896 to 1921. The first, a portent if ever there was one, went to a New Zealander, Ernest Rutherford [21]. But it was not just British students and those from the Dominions who came to the Cavendish. Rutherford's reputation and his almost magnetic attraction brought in research students from all parts of the world. Ties were naturally strong with Australia, Canada, India, New Zealand and South Africa, and all were represented but so, over the years, were China, Bulgaria, Denmark, Germany, Italy, Japan, Poland, Switzerland, the USA and Russia [22].

The international character of the Cavendish also extended to the colleges of the university. At Caius, where Schonland was now comfortably settled, it had been felt for some time that a club should be formed consisting of men who had been in some continent other than Europe and so in 1907 the Raleigh Club, named after Sir Walter Raleigh, came into being. It held regular meetings at which the highlights were the informal talks by members or their guests about their experiences in other lands. These were then followed by a discussion and the inevitable 'social'. The event of the year was always the Annual Dinner where the guest of honour was some notable person whose connections were with the world at large. Basil Schonland wasted no time in applying for membership and was duly elected in 1919. By May of the following year he had made a sufficient impression on his fellow members that he became the Club's vice-president [23].

Schonland's research began in earnest in the late Summer of 1920. He started by reading Rutherford's own masterwork, *Radioactive Substances and their Radiations*, and then considered James Crowther's paper of 1910 that seemed to have set the old Thomson cat amongst the young Rutherford pigeons. Sufficient fluency in German, thanks to his father, and schoolboy French, enhanced no doubt by his sojourn amongst the local population during the war, allowed him access to the work then being done by physicists on the Continent. Two sources of inspiration that Schonland thought excellent and which tested both his mathematical ability and his German were the standard texts of the day. Arnold Sommerfeld's *Atombau und Spektrallinien* provided stimulus on both fronts while Einstein's *Relativity*, in English, he thought was 'simply splendid'. 'I really know what they are talking about now' he wrote to a former Rhodes colleague, Sherwood Watson in Grahamstown, who had been at the Cavendish in 1915 [24].

The practicalities of atomic research meant that Schonland had also to make the acquaintance of Mr Ebenezer Everett and Mr G R Crowe in the workshop at the Cavendish. Everett had been there since 1886 and was a highly skilled, self-taught glass-blower who also set up apparatus and, as he said, 'got things ready'. He saw his role in the unravelling of the mysteries of the atom as simply providing the necessary hardware and defended stoutly his assertion that he did not wish to have the underlying science explained to him. George Crowe was Rutherford's personal assistant throughout the period of his directorship of the Cavendish. He had joined the workshop staff in 1907, left during the war and was re-appointed as soon as Rutherford took over from Sir J J Thomson in 1919. Partially deaf and therefore seemingly oblivious to moments of vented frustration around him when experiments went wrong, Crowe was the most competent of technicians. He was a skilled craftsman in metal, glass and photography and had developed considerable prowess in the preparation of radioactive sources for whichever experiment demanded them. Such men were the salt of the Cavendish earth and made the science possible. Rutherford had decreed that no paper would be submitted for publication unless it carried the authors' acknowledgements to Everett and Crowe for their assistance in constructing the equipment and for filling the 'emanation' tubes. This latter task was recognized as being dangerous and so required special precautions but Crowe, particularly, often neglected to wear the heavy gloves provided and so eventually lost a finger due to radiation damage [25].

The apparatus Schonland used for the investigation of the scattering of rays of β particles was based on that developed originally by Crowther, but it included many modifications suggested by subsequent experience and made possible by the greater intensity of the radioactive source now provided by Rutherford. At first sight the small brass cylinders, blocks of lead and iron, an electromagnet, a vacuum pump and some means of measuring the ionization might have appeared unsophisticated, even crude. However, they represented the type of apparatus beloved of Rutherford — simple, yet ingeniously designed and constructed by his technicians, with meticulous care and attention to detail. Of the three radiation components produced spontaneously by the radium emanation, the β particles are very easily deflected by a magnetic field, the α particle much less so, and in the opposite direction, while the more recently discovered γ rays were not affected at all. Since Schonland's work was intended to examine the interaction between various substances and β particles alone, a method had to be devised to separate them from the other two radiation components. Crowther had used a magnetic field to rotate the β particles through 90 degrees and thus away from the direction of the other two. In his version of the equipment Schonland increased the intensity of the field to produce a full 180 degree rotation to reduce even further any unwanted interaction with the γ rays, particularly. Schonland also incorporated more lead and soft iron shielding around his apparatus to try and prevent any interference with the β particles once they had entered the evacuated chamber containing the sliver of scattering material of either gold, silver, copper or aluminium. The equipment was delicate, its operation required skill and considerable patience and the interpretation of the results was both challenging and difficult.

In his letter written to Sherwood Watson, Schonland confided that the 'research is hard and full of snags but going strong and I hope to get some good results this term. In fact, with luck, we can make a preliminary announcement soon. The prof. doesn't want it shouted about for fear of someone pinching the idea.' [26]. After asking Watson to keep it under his hat, except for a quiet word to their old Rhodes mentor, Dickie Varder, who had himself been part of Rutherford's team in Manchester, he then elaborated:

> Briefly, I've measured the scattering of β rays by thin foils of Al
> (from 7.10^{-5} cm up) using rays of vel 400,000–1,400,000 volts.* It was
> expected to be a case of "single" scattering but it is 10–100 greater
> than that and greater even than J.J's old idea.

What had surprised Schonland, but particularly Rutherford, when he viewed the results, were the peculiar paths that the scattered β particles

---

* At the time the energy of the bombarding particles were quoted in volts. Subsequently, this became electron-volts (eV), the energy acquired by a particle with a charge equal to that of an electron when it falls through a potential difference of one volt.

followed after encountering the target aluminium atoms. These were revealed beautifully in a Wilson cloud chamber that Schonland had set up in order to be able to photograph them. Rutherford had expected the tracks to be completely random with no special features; instead the trails of droplets traced out spiral configurations with definite, well-defined characteristics. This set pupil and master thinking and, collectively, according to Schonland, they arrived at an explanation. He continued, in his letter to Watson:

> We think the solution is that the electron has definite polarity, magnetic no doubt. And speculation as to a possible spin (with a quantum of spin energy) due to emission from an atom.... But this looks like a first class demonstration of a new property of an electron or atom or both. Why we want it kept dark is that we want to go on to cathode rays soon (fast 'uns) and see if they have the same scattering law.

It had been expected from Rutherford's theory that a β particle, on encountering the intense electric fields in the vicinity of the atomic nucleus, would be swept around and out again along a hyperbolic path, just as the α particles had been in many previous experiments conducted both by Rutherford and others at the Cavendish. The form of a graph showing the variation between the relative intensity of the particles and the thickness of the foil would then indicate the type of encounter that had occurred within the atom. If just a single encounter between nucleus and particle took place, or 'single scattering' as it was known, then the graph would be a straight line of negative slope. On the other hand, if multiple scattering occurred, the line would be curved and of complex shape. Much therefore rested on this graphical evidence.

Crowther and Schonland's results, obtained with subtly redesigned equipment intended to generate rays of uniform velocity and minimal divergence, produced a straight line, unlike that in Crowther's earlier experiments, and so single scattering was naturally inferred. But the amount of scattering was much greater than expected, by as much as ten to a hundred times, as Schonland told Watson in his letter, and the effect was even more pronounced with the heavier metals such as silver and gold. This intrigued Schonland and undoubtedly concerned 'the prof.' for his theory was now being questioned.

It was fully appreciated at the time that the scattering of the light β particles was considerably more complex than that undergone by their heavier α counterparts, but one factor not taken into account by Crowther, when he performed his experiments a decade earlier, was the need to introduce a correction for the relativistic effects, which would undoubtedly influence the electron as it homed in on its nuclear target. Since it was travelling at speeds that were a significant fraction of that of light, such effects could not be ignored, as had been pointed out in 1913 by C G Darwin, a mathematical physicist at the Cavendish and a grandson of the father of evolutionary theory whose first name he shared. At Schonland's request Charles Darwin tackled this problem and developed the mathematical equations that showed how the interaction between the nucleus and the β particle had itself evolved from Newtonian mechanics and Coulomb's laws of electrical interaction to include Einstein's laws of relativity.

But even this extra sophistication failed to provide a satisfactory explanation and it was at this point that the Wilson cloud chamber was pressed into service to examine the paths of the scattered particles and to help in deducing the scattering process. The ionised droplets in the saturated water vapour traced out gentle spiral curves 'as if they had a bias', when all expectations were that they should follow an entirely random path as they traversed across the chamber. This was a surprising result but to talk of some new property of an electron, or even of the atom itself, was all very well in Schonland's letter to his friend but Rutherford would never allow anyone at the Cavendish to venture into print without cast-iron evidence, and so far that had eluded them.

In November 1921, Crowther and Schonland submitted a paper entitled 'On the scattering of β-rays' reporting their findings and conclusions in the *Proceedings of the Royal Society*. As was the custom, papers written by those who were not themselves Fellows had to be communicated to the Society by a Fellow and Rutherford readily agreed to do this. The paper duly appeared the following year [27]. It reported in great detail on the experiments that Schonland had carried out and on the various ideas that he, Crowther and Rutherford had advanced to explain what were clearly rather surprising results. The fact that discrepancies remained between the experiments and Darwin's increasingly sophisticated mathematics indicated that the motion of a β particle in close proximity to an atomic nucleus was more complex than any of them had imagined and would require further study. As insisted upon by Rutherford the paper was appropriately circumspect about the interesting pictures provided by the Wilson cloud chamber. They cautiously offered an explanation for the far from random deflections they had observed and suggested that they might be due to the effect reported just a while before by the American physicist, A H Compton, who had shown how the track of β particles would not follow a random path but would be biased in a particular direction 'due to induced magnetism, produced in the scattering substance during the passage of the particle through it'. However, Compton's postulates were by no means applicable to the microscopic scale of the Cavendish experiment and though Compton was himself working on such a problem, his findings were not yet to hand. Much therefore still remained to be discovered. As instructed, Schonland and Crowther duly concluded their paper in the accustomed style:

> We wish to express our warmest thanks to Prof. Sir Ernest
> Rutherford for his continued interest in these experiments and for
> his kindness in supplying the considerable quantities of emanation
> employed in this work. We are also indebted to Mr. Crowe for
> constructing and filling the emanation tubes.

Schonland continued to work on the problem on his own and within six months had a second paper ready to submit to the *Proceedings*. Its slight change of title to 'On the scattering of β-particles' was not only more appropriate to the present investigation but also recognized the wave–particle duality that had so transformed all thinking in physics when first postulated in the earlier years of the century. The experimental apparatus was modified slightly to allow for larger angles of deflection as the β particles approached the nucleus. In addition, possible sources of error in the previous results due to the presence of stray magnetic fields, or secondary particles scattered from the walls of the apparatus, were eliminated leaving only one significant source of error—the divergence of the beam itself. Schonland's treatment of the problem represented an elegant use of Darwin's theory, modified now to include a further proposal that Schonland himself had made. He argued that the degree of scattering to which the β particles were subjected depended upon the thickness of the metal and on its atomic number. Consequently, aluminium would behave very differently from gold, and indeed it did, but not in the way predicted by his and Crowther's previous theory.

After carefully examining the effects of beam divergence, Schonland used the current theory due to Niels Bohr, which described the distribution of the electrons around the nucleus, to proffer an explanation. He proposed that there was a screening action due to the electrons within their various orbital positions and this had the effect of changing the scattering angle as the atomic weight of the metal increased. In other words, the particles approaching an aluminium nucleus penetrate within Bohr's K 'ring' of electrons (those closest to the nucleus) and so come solely within the influence of the positive nucleus itself. By contrast, particles approaching within a similar distance to the nucleus of gold would only penetrate the L ring, farther out, and so would be somewhat shielded from the nucleus by the electrons in the K ring. The resulting partial screening of the nucleus in this way, and the mutual repulsion between the electrons themselves would account, he believed, for the differences between the previous theory and the experiments. Schonland readily conceded that his proposal was 'to some extent artificial' in view of Bohr's later views that the electrons surrounding the nucleus were not simply distributed in 'a discontinuous set of rings'. However, with Darwin's assistance once more, a theoretical description of the mechanism was produced that took account of this shielding effect when calculating the new relativistic corrections required. The results appeared credible, Rutherford's model stood unchallenged, the nuclear forces obeyed accepted laws and the scattering still appeared singular. It was a most satisfying conclusion and Schonland's paper appeared in the *Proceedings of the Royal Society* late in 1922 [28].

*   *   *

Scientific progress had been most satisfactory; the Cavendish, as always, was exciting and stimulating, while the combination of Caius and Don Craib had provided the other fillip in Schonland's life. By now Basil and Ismay Craib had become particularly attached to one another. They were intellectually well-matched and were imbued with the same pioneering spirit from the Eastern Cape. Ismay had strong views on society that embraced a socialist ideal and, though Basil was far less forthright in his, they found that they had much in common. Of course, Cambridge accommodated the complete spectrum of political and other views and its Fabian Society provided the ideal forum for debate and discussion across the broadest front. In addition, reading parties were a feature of College life and Basil and Ismay entered into the spirit of these together. Echoes of St Andrew's of a decade before no doubt drifted through the quads and corridors of Cambridge, for this was the England upon which Grahamstown, its cathedral, colleges and university had been built. Life here, it seemed, was close to perfection.

Soon Basil and Ismay were to become engaged, although she was about to return home to a lecturing post at Rhodes. Schonland now had to consider his own future and it clearly lay in South Africa. Within days of submitting his paper to the Royal Society there appeared in *The Times* a notice regarding a vacancy for a senior lecturer in physics at the University of Cape Town (UCT). The department there had as its head none other than Alexander Ogg, who had held the same position at Rhodes when Schonland was there as a student with such potential. Ogg had, in the meantime, briefly occupied the chair of physics at the South African School of Mines and Technology in Johannesburg, which was soon to become the University of the Witwatersrand, and had even served as its Acting Principal for a while before accepting the chair in Cape Town in 1919. In the Cape he soon built up a small but vibrant research group in his Department, working in the area of geomagnetism, and he had a colleague there who was using the Wilson cloud chamber to determine the speed of β particles emitted from elements exposed to X-rays [29]. The expansion of student numbers and the demands for other courses made it necessary to appoint another man to the staff and so advertisements were placed in the British press. The salary attached to the post was £450 per annum and it was hoped that the new lecturer would be able to commence his duties on 1 March 1922. Applicants were instructed to submit their details with the accompanying testimonials to the Office of the South African High Commissioner in London with a deadline of 21 December 1921.

Schonland immediately set about obtaining the necessary testimonials in support of his application and soon assembled an impressive list, headed by Rutherford, of course, with another from Dr H K Anderson (the Master of Gonville and Caius College). He also obtained one from his old wartime friend Edward Appleton and from Dr Searle. The British Army too was called on to provide supporting fire in the persons of Major H Thirkill (late of the Royal Engineers and now University Demonstrator in Experimental Physics at Cambridge), Major Rupert Stanley in his capacity as Chief Wireless Instructor in France and from no less a figure than the Chief Signals Officer, Colonel H T G Moore himself.

Rutherford wrote of Schonland's important research in a difficult area, of his ability and originality in attacking the problem and expressed the view that he was likely to do good work in the future. Appleton commended his colleague's exceptional qualities for research and praised him for reinforcing his basic mathematical and physics skills before embarking on research. Searle was not only impressed by Schonland's mathematical ability but noted that as a teacher he had earned the respect of the 18 to 20 students he had supervised in the laboratory throughout 1921. Candid as ever, and with this first-hand knowledge to draw on, he commented that 'students are probably much better judges of a teacher's capability as a teacher than those of much more exalted positions', and he felt therefore that Schonland would have no trouble in managing a class. Of course, Searle would now lose Schonland's services in his laboratory but he was most generous in recommending his former assistant and he did so with 'much pleasure tinged with regret'. Thus armed, Schonland completed the Form of Application provided by the High Commissioner's Office, with the dots of double differentiation again evident in his printed name, and submitted it on 29 December 1921.

As always, little happened in the life of Basil Schonland that escaped the attention of his father and this lectureship at UCT was certainly no exception. Though Basil's application had missed the prescribed deadline by a few days this in no way jeopardized his chances because the Registrar of the University was fully aware that it was on its way. In fact, he had already received an application, dated 19 December, completed on Basil's behalf by Selmar Schönland himself! While not appending a detailed summary of his son's career, Schönland senior listed, in broad outline, what Basil's achievements had been. Included was the dreaded '2nd class Hon. in Physics' from Rhodes, alongside of which Selmar wrote that 'Professor Ogg was acquainted with the circumstances over which his son had had no control and through which he had failed to obtain a 1st class'. But the ignominy clearly still stung!

Then followed an intriguing piece of cryptography. Selmar indicated that he was expecting a cable from Basil in which there would be, in the absolute minimum of words, a suitably coded response to indicate his son's intentions. For the Registrar's edification in such abstruse matters he appended the Schonland code:

CABLE CODE:

YES means: I wish to apply and can enter upon the duties in March.

NO   means: I do not wish to apply.

YES with the name of the month added means: I wish to apply, but cannot be in Cape Town until the month named. Thus YES APRIL would mean: I wish to apply but could not manage to enter upon my duties before April.

In the middle of January another missive arrived on the Registrar's desk from Selmar Schönland advising him of the contents of the cable he had received that day from Basil: 'Yes, April first applied asking July', it said.

Selmar then attempted to decode for the Registrar's benefit this intriguingly ambiguous sentence but managed only to contort it still further. In the meantime, Professor Ogg had prepared, from amongst the applications received, a list of the eight most worthy candidates for the post and in his pencilled footnote left no doubt which he most favoured: 'The appointment of Mr Schonland is recommended to commence duties on 1 April 1922' [30].

Basil Schonland bade farewell to Cambridge, at least for the time being, and sailed for South Africa early in March. His journey was punctuated by occasional sea-sickness but that was easily offset by the attractions and challenges that lay ahead. Immediately on arrival he hurried to Grahamstown and then on to Gardiol, the Craib home just outside Somerset East, where Ismay, who had left England ahead of him, was waiting. On 4 April he formally signed a Memorandum of Agreement with the University and so took up his appointment as senior lecturer in the Department of Physics at the University of Cape Town.

CHAPTER 6

AND THEN THERE WAS LIGHTNING

The University of Cape Town, always UCT to all who know it, nestles on the slopes of Table Mountain with panoramic views of the city below. Its location is breath-taking, in what was once the Groote Schuur Estate, literally the 'Great Barn', that provided sustenance for the earliest settlers in the 17th century. In later years it was to be permanently reserved for the South African nation by Cecil Rhodes and the university came to occupy this most majestic of sites by a number of rather circuitous routes.

Its earliest progenitor was the South African College, which opened its academic doors in October 1829, with the Professor of Mathematics rather stretching his abilities in order to teach 'the whole of science' [1]. In 1873 the University of the Cape of Good Hope came into being, but it was only a degree-awarding body to which various university colleges elsewhere in the country were affiliated. It actually operated from offices in the centre of Cape Town, rather than from the side of the mountain, and ceased to exist under that name in 1916. Individual brilliance or, more likely, the scarcity of suitably qualified individuals saw the novel custom continue of professors spanning vast oceans of knowledge. In 1876, for example, the same man filled both the Chairs of English and Physical Science, but soon, as the South African College began to attract good men from overseas, specialists occupied their rightful posts and in 1903 physics was separated from applied mathematics, and both engineering and zoology were added to the array of science courses now on offer [2].

But 1916 was really the important year in the history of university education and research in South Africa, for it was then that no fewer than four universities came into being by special Act of Parliament, and the South African College was renamed the University of Cape Town [3]. Its physics pedigree had already been set, as it were, in stone as a result of the magnetic surveys of the continent that had been carried out between 1902 and 1906 by its then head, Professor J C (later Sir Carruthers) Beattie, an 1851 Exhibitioner who arrived in Cape Town three years before the turn of the century to take charge of the university's fledgling activities in physics [4]. When Beattie became the new university's first Principal in 1919 he left to Alexander Ogg, his successor, an embryonic culture of research resting on a sound foundation, much as the University did on Table Mountain with Devil's Peak above it and the two oceans at its feet.

Amongst the first tasks Professor Ogg set himself was to attract to his Department good staff and that meant attracting applicants from overseas; the local pool of talent being very meagre. When Basil Schonland's application arrived from Cambridge, suitably reinforced by all his father's correspondence with the Registrar, Ogg was well pleased, for not only had he attracted the interest of a man who was a product of the Cavendish, but he was a South African to boot and one with quite some reputation. Ogg's Physics Department in Orange Street was housed in a red-brick building that contained a fairly large laboratory, a lecture room, workshop, dark room, a small room for research, plus an office that was shared by the professor and his senior lecturer. Schonland, on his arrival, noted that the Department was well-equipped for research and possessed some useful apparatus. He noted too the Wilson Cloud Chamber much in evidence and so was on familiar ground and happy to be there. As was the custom, the professor took it upon himself to teach physics to those who were fresh out of school. There were large classes of aspirant scientists as well as those from other faculties, such as engineering and medicine, who needed this sound under-pinning of their own disciplines. Schonland's first teaching responsibility was for those who had progressed beyond this preliminary stage and, as well as lecturing to them, he conducted the practical classes in the laboratory in the manner of 'Searle's Class' to which he was now well-accustomed. Postgraduate courses, though few, were given in some classical aspects of physics as well as in areas at the forefront of research abroad. The textbook that provided much of this material was J A Crowther's *Ions, Electrons and Ionizing Radiations*, first published in 1919 and already into its third edition, so vibrant was the subject area. The influence of Cambridge and the Cavendish was almost all-pervasive [5].

The challenge that now faced Schonland was to maintain his research activity at the level he had established over the past two years at the Cavendish. There he had shown great energy and determination in pursuing the research that would soon earn him his doctorate. But at Cambridge he had been in the company of those who were collectively leading the world in unravelling the mysteries of the atomic nucleus; now, 10 000 km away, he was almost isolated. The young man who had sat at the feet of many masters had rapidly to assume the mantle of the master himself.

But Schonland's thoughts were not just on physics for he and Ismay were also contemplating their forthcoming marriage, their home and their future together in South Africa. During the first university vacation of the year she was able to leave Rhodes and travel to Kalk Bay, a pleasant fishing village near Cape Town, where Basil had moved into temporary accommodation nearby. Whilst there he used the opportunity to reply to a letter from Sherwood Watson who was now the mainstay in the physics department at Rhodes under Dick Varder. Watson had not yet found an area of research for himself but was dabbling in wireless and he sought Schonland's advice on the use of thermionic valves in high frequency receivers. Much, of course, had happened since Schonland last found himself working with such devices and he informed Watson that he was now 'awfully bad on wireless' but proceeded to offer some suggestions of the way he would interconnect them to make such a receiver. Then, turning to UCT, he mentioned that Professor Ogg was as nice as ever and was also most efficient and well looked-up-to by all in the Department. He also conveyed to Watson the good wishes of C T R Wilson, whose intellectual powers had enthralled them both, just as they had all who had really taken the trouble to listen to him [6].

> I've proposed to CTR that one day you and I should write up and publish his lectures. For the present generations of Cambridge know not Joseph − nor $d\theta = \lambda/A$! As a matter of fact it would be well worth doing.

Indeed it would have been, but sadly Schonland's suggestion never progressed, as other matters intervened and so the collected lectures of C T R Wilson, so magnificent in content yet so contorted in delivery, were never written up by two of his most enthusiastic students.

While Watson may have found ideas for research hard to come by, Schonland was faced with just the opposite problem. 'I wish I had time to do all the things I should like to. Do you want to photograph the night sky (auroral) spectrum?', he asked and then proceeded to tell Watson of efforts in England to photograph a bright green line seen whenever there was an aurora but which had, so far, defied explanation. Schonland's enthusiasm for a novel piece of spectroscopic observation seemed not to move Watson to do anything about it even though he was now in a part of the world where night skies were more often than not without the sort of cloud-cover that made observational astronomy much more of a lottery in England. As it turned out, the wavelength of that green line was measured the following year but its chemical origin was not identified until 1925. It would be a further 25 years before it was shown beyond doubt that at least part of the luminous auroral phenomenon within which it occurred is caused by the arrival of energetic particles from some higher region in the atmosphere [7]. Soon, Schonland would turn his attention skywards but at the moment his efforts were focused entirely on the β particles generated in the laboratory.

It was as a lecturer that Schonland first made his mark in Cape Town. His enthusiasm and all-round competence was the source of much inspiration to his students, and in years to come many notable South African physicists recalled with pleasure the lectures he gave. Amongst them was E C Halliday, who arrived at UCT in 1922 with only vague ideas about where his future lay. Then he encountered Schonland who made such a deep impression on him. Near the end of the year Schonland took his class on a short tour around the research section within the Department. It was really nothing more than a single large room, but its contents were inspirational to Eric Halliday, at least. There he saw the apparatus that Schonland was constructing to continue his work on the passage of β particles through matter, and he saw too the remarkable Wilson cloud chamber that was the very embodiment of brilliance. For Halliday, physics now assumed a new dimension: no longer was it just an intellectual exercise but it became a pursuit in which one might actually discover something entirely new. For that vision he had to thank Schonland [8]. One year behind Halliday was a young man, D J Malan, whose ancestry was firmly rooted amongst the French Huguenots, whose arrival at the Cape in 1688 had introduced a rich vein of culture and talent into Africa. Dawie Malan, too, was capti- vated by Schonland's lectures and by his enthusiasm for his subject. Years later, after Malan had completed a doctorate at the Sorbonne, he and Schonland would become powerful collaborators in their quest to understand the mechanisms of lightning. Another student of Schonland's was the youthful Richard van der Riet Woolley with his precocious ability as a mathematician. He too had been stimulated by Schonland and Ogg's lectures and, after completing an MSc at UCT, went to Cambridge where he switched to astronomy. In 1956 he became Britain's Astronomer Royal [9].

Basil Schonland's own research depended upon a ready source of β rays, but the radioactive material, such as radium, that produced them naturally was a rare and therefore a highly-prized commodity in any physics laboratory in the second decade of the century. It was certainly beyond the resources of a small group of scientists at the southern tip of the darkest continent, and so a different method was needed to generate them. A source was to hand in the form of the 'cathode rays' that had been discovered in Germany in 1876. The release of the β rays by this process depended not on any exotic material but on the curious emissions that occurred within a 'discharge tube', an evacuated glass envelope containing two electrodes, the anode and cathode, connected to a very high voltage. The design of such a device lay well within Schonland's grasp and its construction was within the area of competence of Mr J A Linton, the technician of long-standing in the UCT Physics Department. It was J J Thomson at the Cavendish who had shown in 1897 that the rays emitted from the cathode were the fundamental particles now known as electrons and, of course, these were the same β-rays produced by the radium emanation that had so exercised Schonland's mind and his experimental talents over the previous few years. Then, he had examined how they were scattered by thin metallic foils. Now he set out to study their absorption by those materials. While the natural radiators of β-rays, such as radium, produced beams of considerable velocity, their intensity was low. By contrast, cathode rays, though slower, were much more intense and had one significant advantage over the naturally radiated kind in that they were generated without any of the accompanying γ-rays that so dogged all previous scattering experiments [10].

Schonland took considerable care over the design of his experimental apparatus. Rutherford's emphasis on simplicity had rubbed off well, which was fortunate because there was certainly little scope for elaborate technology in South Africa in 1922. Between them, he and Mr Linton made full use of whatever was readily to hand in their own laboratory. The cathode rays were to be produced within a glass lamp-chimney, of the type found on gas lamps, within which Linton had fitted a movable cathode and a large, hollow, water-cooled anode that operated at voltages between 10 and 50 kV. The cathode rays passed through a hole in the anode and thence into a magnetic focusing system beyond. Primary power for the generator came from an induction coil connected to a bank of batteries operated through a slow-speed mercury break switch. As in the Cavendish experiments, the beam of cathode rays was focused, and its speed controlled, by being bent through 90 degrees by an electromagnet before entering what was called the firing tube. This cylindrical device contained a number of metal discs, each with a small, concentric hole through which the beam passed, while preventing any stray or reflected rays from entering the chamber containing the metal film under examination. John Linton's craftsmanship came fully into play in the construction of this piece of equipment and rivalled that of Messrs Everett and Crowe at the Cavendish. To complete it he mounted two brass cylinders, about 4 cm in diameter and 10 cm long, one above the other and directly above the firing tube. They were separated by an ebonite arrangement that held the thinnest of metal foils, of aluminium, copper, silver or gold, whose absorption properties were to be examined. Both the upper and lower cylinders, as well as the metal foil, were connected by switches to a pair of sensitive galvanometers that could be selected to display the transmitted, reflected and absorbed portions of the beam that struck the metal foil. Since the complete system had to be evacuated, the quality of the vacuum depended on the ebonite seal between the cylinders and this was carefully ground and greased before all the pieces were assembled. It was no wonder that Eric Halliday, when he saw this marvel of engineering, was so captivated by the possibilities of making some scientific contribution of his own.

Having well designed and constructed apparatus was one thing; quite another, and just as important, was its accurate calibration and the elimination of any errors of measurement that were an inherent part of any experimental system. When performing preliminary experiments Schonland was aware that the main beam of cathode rays, on striking the foil, were dislodging what had come to be known as secondary or δ-rays. These recoil electrons were of much lower energy than the reflected rays he wished to measure but they would also be detected and the results would therefore be in error. To eliminate them Schonland devised an elegant, yet simple technique that later, and quite independently, would find application in the thermionic valve, then in its earliest stages of development. He positioned a fine grid of wire mesh on either side of the metal foil and just 2 mm away from it. By bringing out a wire from this mesh or grid and connecting it to a negative potential he was able to repel these secondary electrons and so prevent them from interfering with the reflected rays. Further experiments showed that this grid did not adversely affect the primary rays in their passage from cathode to foil and so it turned out to be a remarkably simple, yet effective, solution.*

The use of this equipment to measure the absorption of the cathode rays as they passed through slivers of metal foil at speeds between 20 and 40 per cent of that of light took many weeks of painstaking work. The data, when processed, yielded an interesting result. Each of the metals had its own characteristic depth of penetration, called its range, and the values measured by Schonland with his cathode rays were of similar character to those obtained a few years before by Dick Varder while working under Rutherford in Manchester with a radium source. Of much more significance, though, was the excellent agreement obtained with the results predicted by Niels Bohr, whose collaboration with Rutherford in Manchester had set in train the whole new world of modern physics. Bohr's theory had taken full account of the need to include the change in mass of the β particle which Einstein's theory of relativity made necessary. Schonland's immediate aim was to increase the speed of his cathode rays to examine this effect in more detail, but to do so required a re-design of the equipment. Before that could be tackled he had to write up the results of the current experiment for publication. Early in 1923 the paper was ready and was sent to Rutherford who communicated it to the Royal Society in April [11]. By then the new apparatus was also taking shape and Schonland's activities in the Cape had brought him to

---

* The same approach was followed in the pentode valve, patented by the Philips company in Holland in 1926. It contained a third grid, known as the suppresser grid, which was intended specifically to suppress the secondary electrons by connecting it to a suitable potential with respect to that on the anode. Apparently it was conceived without any knowledge of Schonland's earlier use of the idea.

the attention of the Government's Research Grant Board that had been set up soon after the First World War to foster research at the universities [12]. His application for a grant in aid was viewed favourably and duly approved and this helped in some way towards meeting the costs in the laboratory [13].

But there were two other very pressing matters that also required his attention. The first was his marriage to Ismay; the second, the submission to Cambridge of his PhD thesis based on the work he had done whilst there, as well as this most recent research at UCT.

*   *   *

Basil and Ismay were married in Cape Town on 21 July 1923 [14] and set up home in a rented house in Milner Road, Tamboerskloof, a suburb of Cape Town [15]. This was to be the first of many rented homes in the Cape as their family circumstances altered, first with the arrival of a son, David, in 1924, followed the next year by a daughter, Mary. Their third child, Ann, was born in 1931. It was to be the happiest of marriages. Ismay was a pillar of strength on whom Basil was able frequently to lean when matters to do with affairs much larger than either of them required an attentive listener and a wise counsellor. She was both, in good measure. Though they shared no common scientific interest, Schonland was no Philistine. He was, when time allowed, an avid reader with catholic tastes and an amiable man with a great love of music. To his children he was an excellent listener, and an amusing and devoted father. As the Schonland saga began to unfold it soon became clear that the major decisions of life were seldom ones that he reached alone. Ismay was his closest confidante.

The matter of Schonland's PhD required Cambridge to change its usual practice to some extent. In the normal course of events a candidate having submitted a thesis for examination would appear for a *viva voce* — an oral examination or interview at which he might be required to expound upon ideas in his thesis or digress into related areas to assess his grasp of the general subject area. In Schonland's case this was hardly feasible now that he was living in Cape Town, so the procedure adopted was to have him sit a written examination, set by the Cavendish, to establish his fitness to hold the Cambridge doctorate. So, as well as having to function as husband, teacher and researcher Basil Schonland now found himself very much the student as well. His spare time at home was taken up by a certain amount of study of all aspects of modern atomic physics that he considered likely topics in his forthcoming examination. This he duly sat in January 1924 under the personal supervision of the Registrar of UCT and, as he later recounted to Sherwood Watson [16], he was rather aghast at the questions that were asked. Instead of pitting himself against the theories of Bohr, of which he was 'chock full', they asked him to discuss the theory of the ultramicroscope and the resolving power of a prism and telescope. Whereas he claimed to have forgotten all about 'those wonderful $d\lambda/\lambda$s of C T R Wilson' he clearly had not, for on 17 March the Degree Committee voted unanimously in favour of the dissertation 'being of distinction appropriate to the degree of PhD' and their decision was duly endorsed by the Board of Research Studies. The degree of Doctor of Philosophy was awarded to Basil Schonland, by proxy, on 7 May 1924 — only the forty-fifth PhD degree to be awarded by Cambridge since the University had adopted it from American practice a matter of just five years before [17]. Given his single-minded pursuit of the topic Schonland's thesis carried the almost predictable title of 'Scattering of β particles' and it reflected the state of knowledge at that time in an area that was readily agreed to be beset by both theoretical and practical problems.

With the formalities of the PhD now behind him, Schonland could devote almost all his energy to further research. His teaching duties, though still demanding, were somewhat eased by the appointment of another lecturer to the Department, E H Grindley, an enthusiastic and industrious young graduate from Bristol [18]. Schonland's immediate intention was to produce cathode rays of even higher velocity, but this required a radical change in the construction of the cathode assembly because at voltages much above 40 kV it had proved troublesome. The first alternative tried, a heated tungsten element with water cooling, though capable of operating up to 57 kV, needed several hours of pumping to remove occluded gas whenever the assembly was opened to change the metal foil. This was clearly unacceptable, so an idea based upon an X-ray tube then being used at the Royal Institution provided a possible solution.

The cathode now consisted of an aluminium rod cast inside a brass tube, with the face of the rod made concave and highly polished so as to be an effective emitter of cathode rays when the voltage between it and the nearby anode, both of which were water-cooled, was high enough. Changing the foil now only took about fifteen minutes. With this new apparatus, operating at high vacuum, Schonland was able to increase the voltage to 100 kV, being limited only by the onset of voltage breakdown within the assembly. At the time this was reputed to be the highest voltage yet used to generate an electron beam by this method. With a beam velocity of 0.55 times that of light, he was now able to achieve conditions similar to those using radioactive sources [19].

Once again the skills of John Linton were called upon to construct the equipment which, though similar in concept to the original version, contained many modifications based on Schonland's and Linton's collective experience. The higher velocity cathode rays required a more intense magnetic field to rotate them through the 90 degrees so that they would fall within the aperture of the firing tube and this, in turn, meant that the electromagnet assembly had itself to be provided with its own water-cooling system. Considerable ingenuity also went into devising and using a simple method to measure the intensity of that field so that the beam velocity could be determined with the highest degree of accuracy. No effort was spared in trying to reduce as much as possible all sources of error within the system and the process of gathering results was both long and tedious.

But there were also moments of light relief. The University's lively magazine, the *University of Cape Town Quarterly*, which had reported the wedding of Basil and Ismay the previous year, and had congratulated him on the award of his doctorate from Cambridge, now carried another item of interest in its very next issue [20]. It came from the Honorary Secretary of the Emergency and Civics Corps. Apparently, Schonland had been prevailed upon to deliver a lecture to this band of enthusiastic volunteers and he chose the intriguing title of 'Radio for Rabbits'. The venue was the Physics Lecture Theatre and in front of him he had an audience of 230. Whatever he told them met with a most enthusiastic response and this led the scribe to suggest that there would undoubtedly be 'requests for a repetition of this lecture next year'. The old wireless flame clearly still flickered despite the lecturer's recent claims to the contrary. In fact, Schonland and Watson were soon in contact again on the subject. This time Basil offered his Grahamstown colleague the chance to make a name for himself, or possibly even for both of them: 'If you can make a short wave wireless transmitter with reflector I will tell you a way to make our fortunes', Schonland wrote in June 1924 [21]. But he did not elaborate on what this tantalizing application might be and so poor Watson was none the wiser, as are we.

One matter that was very much on Schonland's mind about this time was South Africa's isolation from the centres of physics in Europe, and especially at the Cavendish. He missed very much the opportunities of discussing problems with colleagues who were working in similar research areas to his and he missed, as well, the frequent seminars and informal meetings that were such a feature of life in Rutherford's laboratory. In Cape Town he toiled alone. The only means of ensuring that he kept in touch with the developments taking place at an astounding pace elsewhere were through the scientific journals that he read avidly, and pre-eminent amongst them were the *Proceedings of the Royal Society*, the *Physical Review* and *Nature*, to which he had turned for inspiration almost a decade before when he was training for war. But by the time these publications reached South Africa, after a journey of some weeks by sea, the scientific progress they reported had more than likely been superseded by more recent discoveries. Schonland was painfully aware that his laboratory was a very long way from the centres of scientific excellence. However, for all that isolation, he was confident that his research would stand up to scrutiny from any quarter, and so it was that he mounted a robust defence of it in the columns of the April 1925 issue of *Nature* [22].

His *Proceedings* paper of 1923, in which he first reported on the passage of cathode rays through matter and which also made reference to his use of the 'suppresser grid', had been criticized by an American scientist at Columbia University who suggested[*] that Schonland's measurement of the velocity of his β-rays was inaccurate when compared with those of other workers in the field. The Schonland riposte was immediate:

> This is scarcely correct. A variety of causes rendered the work of earlier writers unsatisfactory from a quantitative point of view, while the experiments of Whiddington[†] suffered from an important defect, for no precautions were taken against the disturbing effect of the emission of secondary rays from the foil. My apparatus was designed to remove this source of error, and when allowance is made for the secondary emission, Whiddington's results are in satisfactory agreement with my own.

He then turned his attention to the man who had levelled the charge. Whereas it had been suggested that Schonland's quoted value of an important parameter was 20 per cent in error he demonstrated convincingly that, as before, the unsuppressed emission of secondary rays from the metal foil would lead to just such an erroneous supposition. Considerable attention to experimental detail, he emphasized, was absolutely vital if any reliance at all was to be placed upon the results of such an experiment. Any stray electrons, from whatever source, would simply destroy the credibility of the method used. His critic, he charged, had fallen into that very trap himself.

Schonland was not only forthright but was convincingly so. His experimental technique had been honed in the Cavendish where slapdash procedures or sloppy thinking were as likely to incur blistering verbal fire from Rutherford as they were to encounter the withering stare from the severe countenance of James Chadwick. But not only were his arguments watertight and well-constructed but so was the prose that presented them. Schonland wrote well and this meant that a strong case was delivered in style.

*   *   *

Early in 1925 Basil Schonland's feeling of creeping isolation was brought to a sudden halt by some heart-warming news. Sir Ernest Rutherford would soon be arriving in Cape Town. In 1925 Rutherford was elected President of the Royal Society and one his duties was to visit the

---

[*] H M Terrill *Physical Review* Dec. 1924.
[†] Another worker in the field.

Dominions to deliver his Presidential lecture. In July, he and his wife set sail for Australia and New Zealand via the Cape where they were the guests for a few days of Sir Carruthers Beattie, now the Principal of UCT. Whilst there Rutherford made a special point of visiting Schonland's laboratory to see, at first hand, what his former student was doing [23]. He was, of course, entirely familiar with the papers that had reached him from time to time from Cape Town for onward communication to the Royal Society and would later express himself much impressed with Schonland's 'strong experimental ability' and 'his capacity to bring a difficult investigation to a successful conclusion' [24]. It was about this time that Rutherford was giving serious thought to updating his textbook on the nature of atomic particles and their characteristics. The first edition had appeared in 1904; the second, seven years later. It was now time to bring together all the accumulated knowledge that had appeared in the scientific literature since then. Soon, with the collaboration of two of his senior colleagues, James Chadwick and Charles Ellis, Rutherford would write what was to become, for many years, the standard reference on virtually anything to do with atomic radiation and radioactive materials. Entitled *Radiations from Radioactive Substances* [25] it was published in 1930 and it contained details of Schonland's precise measurements of the scattering and absorption of cathode rays by various metals. His results provided much of the data from which the variation in the range of electrons with different velocities was determined and went some considerable way towards confirming the theories of Bohr, which were now crucial to a full understanding of the structure of the atom and of its properties.

The experiment that Schonland would undoubtedly have demonstrated to Rutherford was yet to appear in print. It is arguably the most important scientific paper at that stage of Schonland's career and reflected his return to the most challenging of problems — the scattering of fast-moving electrons by an atomic nucleus [26]. By now he was a seasoned experimenter, well-versed in both the details of the equipment and in the interpretation of complex data. Of particular importance was his use of cathode rays instead of β-particles. Not only were they relatively easy to generate but the magnetic deflection technique allowed them to be formed into a homogeneous beam with a velocity that was readily determined. Because their intensity was so much greater than naturally radiated β-particles, they were also easier to measure and they were not contaminated by γ-rays. The only uncertainty was whether they contained sufficient energy, at their lower velocity, to approach close enough to the nucleus of the target to experience the single scattering he hoped to measure. Since he had last done any work on scattering, others had looked at these problems, both in the laboratory and in theory. A particularly useful criterion of what constituted single scattering had been

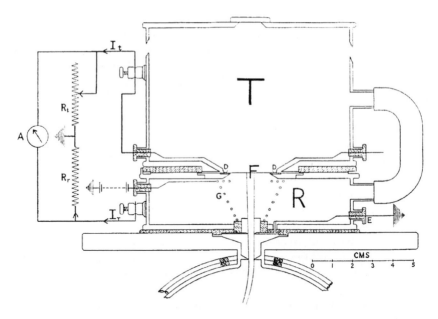

Figure 7. Schonland's apparatus of 1926 devised to measure the scattering of cathode rays. To prevent the emission of secondary or δ rays from the target foil at F, he included a negatively biased conical grid of wires, G, which was based on a technique he had first used in 1923. This was apparently the first use of what subsequently became known as a suppresser grid in the thermionic valves then being developed. ([26, figure 1] reproduced by kind permission of the Royal Society.)

determined in 1923 by Wentzel in Germany and Schonland found that his cathode ray technique satisfied it as well as, if not better than, any previous β-ray experiments. Again he employed his suppresser grid, now in an improved conical configuration, to eliminate secondary electrons.

The measured results that he reported were in excellent agreement with the scattering theory of Rutherford, which included the necessary correction terms due to relativity effects. When plotted graphically they showed the features which readily identified the regions of single and plural scattering and then, almost as a *pièce de résistance*, Schonland used his data to calculate the atomic numbers of the four scattering materials. Good agreement here would have provided final proof of the validity of his results, and indeed it did in the cases of aluminium, copper and silver but not for gold. He offered an explanation for this discrepancy that was based on the binding energy between the various shells of electrons surrounding the gold nucleus. The energy imparted to his cathode rays was very close to the binding energy of the K and L shells of electrons and this would lead to the abnormal emission of secondary rays from the gold foil. Since gold exhibited the least amount of single scattering anyway, even the smallest secondary emission would upset the results. The only solution would have been to increase the energy of the incident rays and to do this would require running his equipment at an even higher voltage. But that was beyond its capabilities and would require the design of a radically different and, no doubt, more complex piece of apparatus of the type now taking shape at the Cavendish.

Schonland's conclusions in the paper he published were buoyant: he had achieved single scattering—for so long a most contentious point. The forces that prevailed between the moving particle and the atomic nucleus followed the inverse square variation with distance, thus putting paid to another equally contentious piece of speculation that had suggested otherwise; and finally, he had shown that Rutherford's scattering equations were just as valid for electrons as they were for protons, as long as they included the appropriate relativity correction.

With the publication of these results Schonland had reached a most important point in his scientific career and he paused to collect his thoughts. It was as if a pendulum had swung to its maximum extent and paused ever so momentarily, before commencing its cycle again. He began to think deeply about the direction in which his research should go. He had been working on problems that he had transplanted from the Cavendish, but Cape Town was too far from the scientific centre of gravity for him to seriously believe that he could make any further contribution in the field of atomic physics. He discussed his predicament with Ismay and later she recalled how he believed that he should find a project more suited to the circumstances of South Africa, since the atomic work at the Cavendish was beginning to require engineering and technical resources that were just not available in South Africa [27].

Schonland's determination was beyond doubt, as Rutherford had readily acknowledged. So too was his capacity to tackle work of an intricate nature. He was not interested in large and complex pieces of equipment, yet his research was now leading him in that direction. His aversion to large apparatus may well have been due to the influence, yet again, of Rutherford, for it was the very simplicity of the Cavendish, with its string and sealing wax approach, that had so struck the young man from Grahamstown when he worked in those most hallowed of buildings in Free School Lane. That some of the greatest ever scientific discoveries were made there using apparatus that looked positively antediluvian seemed astounding, yet it was true.

*   *   *

The change of direction in Basil Schonland's career occurred during the long summer holidays over Christmas 1925 when Basil, Ismay and their two children left Cape Town for the Craib family farm, Gardiol, near Somerset East, not far from Grahamstown. Unlike Cape Town in the western Cape, that part of the country experiences the thunderstorms and accompanying lightning that are very much a spectacular feature of many parts of South Africa during the summer season. At an altitude of 750 m above sea-level, and in the path of the storms as they move from west to east, Gardiol was the ideal place from which to observe the lightning as it ripped through the skies above the East Central Karroo.

Whether it was a sudden realization on his part that here was a little-investigated natural phenomenon that he could make his own, or whether his thoughts had drifted in that direction because of his reading of the recent scientific literature is unknown. What is certain is that Schonland would have been well aware of the work that his old Cambridge mentor, C T R Wilson, had done from as early as 1914 on lightning and its electrical characteristics. CTR had published his theories and experimental results in two important papers that appeared in 1916 and 1921. Then, more recently in 1924, there appeared another within the *Proceedings* of a joint meeting in London of the Physical Society and the Royal Meteorological Society [28]. It dealt with the ionization in the atmosphere and its influence on the propagation of wireless signals, and amongst the participants were some legendary figures from the world of wireless communications including two whose impact on the subject within just a few short years would be monumental. They were Edward Appleton, Schonland's former colleague from their days in uniform, who was soon to leave the Cavendish to take the Wheatstone Chair of Physics at King's College, and R A W Watt, from the Radio Research Station of the Meteorological Office.

Appleton's famous experiment in which he established the existence of the ionosphere was just a month away, while Watt[*] was destined to reappear in another guise, and to massive effect, within a decade. Now, they were continuing a collaboration they had started some few years before on the origin and features of 'atmospherics', those transitory crashes and bangs that punctuate one's listening to the 'wireless'. Also present was C T R Wilson, no doubt almost invisible and certainly almost mute alongside such powerful personalities. However, Wilson's contribution to the discussion on 'The Electric Field of a Thundercloud and some of its Effects' was, as ever, deeply thought-provoking and precise in its discussion of the underlying physics of the problem. CTR drew attention to three effects of the electric field of thunderclouds which, he felt, were worthy of further investigation: the possible ionization of the air at great heights above the earth, and the likely discharge of energy from

[*] R A Watson-Watt only assumed the hyphenated form of his surname when he was knighted in 1942. He was christened Robert Alexander Watson Watt and until 1945 all his published scientific papers bore the name R A W Watt, but during the early part of the Second World War he chose to sign himself R A Watson Watt, without the hyphen [*Biog. Memoir of Fellows of the Royal Society* 1975 **21** 563].

clouds to those ionized regions; the discharge between earthed conductors and the base of thunderclouds; and the possibility that penetrating radiation was produced within a thundercloud. All three would, in time, attract much attention from various scientists and all three ideas, when he read about them, fired Schonland's imagination. But the scientific curiosity that really launched Schonland into this new field of research was a remark of Wilson's that caused a sharp rejoinder from none other than the Director of the Meteorological Office of the Air Ministry, one G C Simpson.

CTR had stated that the charge in the upper regions of a thundercloud may be either positive or negative 'but ... it is more frequently positive than negative' [29]. This brought Simpson smartly to his feet. There was, he said, 'no experimental or observational evidence [that] has yet been produced to show that thunderstorms ever have negative electricity under positive electricity and until that was done he was unable to accept Mr Wilson's explanation of the normal negative charge found on the earth's surface' [30]. Wilson either did not or more likely could not reply, for no reply was recorded, but he was not in the least deflected in his view, a view soon to be proved correct, and the man who was to provide much of the evidence was Schonland.

The speed and urgency with which Basil Schonland launched into this new field of research is quite remarkable. His plan to investigate one of nature's great phenomena, with which South Africa was blessed or maybe cursed in abundance, was hatched late in 1925 after corresponding with Wilson [31]. Schonland began by repeating the experiments made by CTR in England some time before and this required the construction of precise duplicates of Wilson's apparatus, a task given to Mr Linton. The apparatus, in the true Cavendish tradition, consisted of very simple field-measuring devices involving an ingenious electrometer for indicating the changes in electric field strength, and an automatic method of photographing the records. Such was its simplicity, yet so important were the results that it produced that they bear describing in some detail.

A 30 cm copper sphere mounted atop a 5 m hinged conducting pole was used to measure the fields from distant electrical storms. For nearer storms the technique used what can only be described as pure Wilsonian artistry underpinned by the clearest understanding of the physical processes involved in the lightning discharge. A circular metal sieve, 55 cm in diameter and 14 cm deep, was filled with soil and then placed within a hole on sulphur-ebonite insulators so that its top surface was flush with the ground. Immediately above it, on a pivoted swinging arm, was a circular metallic cover, 92 cm in diameter, that could be positioned directly above the buried sieve and then swung away by means of a loop of rope operated from a little wooden hut carefully sited nearby. The principle was simple. An electric field from a lightning stroke would deposit charge on the earth in the sieve whenever the cover was removed but would not do so when the cover was in place. An insulated wire from the sieve, or from the pole and ball (depending on whether the storm was nearby or distant), carried these collected charges to the hut where they registered on a Lippmann capillary electrometer, another device of remarkable simplicity yet one that was ideally suited to this particular task. It consisted of a glass tube, of very fine bore, linked to two mercury-filled chambers, one electrically connected to earth and the other to the buried sieve or pole. The tube also contained a small amount of dilute sulphuric acid which formed a film on the mercury and a bubble between two thin mercury filaments. An electrical charge applied to one side of the electrometer would cause the bubble to move, with its displacement being proportional to the magnitude of the charge, and its direction of movement corresponding to its sign — either positive or negative. The movement was almost instantaneous and the restoring action, on removal of the charge, equally so. Visual observation, or photographic recording of the position of the bubble, would thus indicate both the polarity and the magnitude of the deposited charge. And so the electrometer, a carefully focused microscope and the recording camera were housed in the wooden hut, so constructed and sited, *à la* CTR, to reduce, as much as possible, any interference to the fields being measured.

This delightfully simple yet quite intricate apparatus was set up on the farm of James Craib MA, mathematician and sometime farmer, who was Basil Schonland's father-in-law. Craib himself thus became an enthusiastic participant in these investigations of the electric field of the thunder cloud.

The underlying theory which they set out to test was based on Wilson's simple bipolar model of a thundercloud. He had postulated that the cloud contained two centres of charge, with the upper one being positive; what he then termed 'a positive cloud'. This was the major point of contention with Simpson's view in which the lower regions of the cloud were supposed to be dominated by positive charge. Wilson's mathematical model, which described the changes in electric field that accompanied a lightning flash, was as simple as his apparatus. All hinged on the polarity of the measured field-change registered by the movement, either to left or right, of that little bubble of sulphuric acid in Lippmann's electrometer. In their published paper describing the results they obtained, Schonland and Craib showed how the polarity of the discharges from a distant lightning flash were indeed readily identifiable from the deflection of the bubble. By using CTR's theory it was possible to show that a flash totally within a positive cloud, from positive to negative, would produce a negative field change; that a flash from the underside of the cloud to ground would cause a positive field change; while a flash from the upper positive pole to the ground would cause a negative change.

By contrast, if the lightning storm were nearby then things would change, but again the mathematics was clear and unambiguous. The first two events, described above, would now produce positive deflections of the bubble, while the last would still be in the negative direction. The only issue to be resolved was what constituted a distant storm or one that was nearby and here, yet again, simple observation aided by CTR's analysis provided the answer. Thus, suitably armed, Schonland and Craib commenced their campaign of observations until Schonland had to return to his academic duties in Cape Town in February, when he left the experiment in Craib's care.

At the end of the lightning season they analysed their results. In the six months between January and June 1926, 23 thunderstorms were observed in a year of considerable drought in the Eastern Cape. The great majority produced lightning discharges that passed from the upper to the lower charged regions within the clouds without striking the ground: a common observation in South Africa. Discharges from the lower regions of the cloud to ground occurred roughly 10 per cent of the time while those from the upper regions to the ground were comparatively rare. Of the 18 distant storms they observed, 83 per cent produced negative shifts of the mercury meniscus while 81 per cent of those nearby yielded a positive shift. These results provided conclusive proof that the cloud-base was negative, and Schonland announced them at a meeting of the Royal Society of South Africa in Cape Town on 16 June 1926. After consulting Wilson on the further interpretation of the observations, he and Craib wrote a detailed scientific paper and sent it to him in January 1927 for communication to the Royal Society in London.

In the paper Schonland was forthright and quite unequivocal in his observations. The experiment had confirmed the validity of Wilson's bipolar model of the cloud and showed that there was a 'strong predominance amongst such clouds of a type in which the upper pole is positive and the lower negative', the positive thundercloud. Simpson's theory, he stated, 'must either be rejected or radically altered' [32].

Simpson's response was not long in coming, but before we consider it and Schonland's reaction, it is important to dwell briefly on another observation of Schonland's that deserves some attention in its own right. Not only did it confirm one of the three effects first described by Wilson at the meeting at Imperial College in 1924 but it also provided visual confirmation of a most peculiar phenomenon. Schonland reported that during their observations of thunderstorms at night he and Craib had often seen an intermittent upward glow from the top of a distant cloud. This was the discharge from the top of a cloud and the upper regions of the atmosphere that CTR had postulated would exist and which today is known as a 'sprite'. The first recorded sighting of such upward lightning strokes was made by C V Boys in about 1876 near the village of Wing in Rutland, but he only published an account of it a matter of a few months before Schonland and Craib published theirs [33]. Schonland noted this phenomenon almost in passing; only many years later would it receive considerable attention in the scientific literature [34]. Soon Schonland would be linked with another pioneer in the field when the names of Boys and Schonland became almost synonymous with the photographic observation of lightning.

The subsequent exchange between Schonland and Simpson took place in the pages of the *Proceedings of the Royal Society*, that most august journal of the day. Simpson's stout defence of his assertions appeared late in 1927 under the title 'The mechanism of a thunderstorm' [35]. What followed was scientific debate at its finest. In measured tones, each word having been weighed and every sentence carefully balanced before launch, the proponents of two conflicting theories sought to convince the other of the merits of the greater case. Simpson saw aligned against him not only an unknown duo from the southern reaches of Africa but also others of considerable scientific pedigree much closer to home in the persons of Wilson, Appleton and Watt. Undeterred, however, he chose no head-on rebuttal but instead set out to show that the mass of experimental evidence stacked against him actually confirmed the validity of his own hypothesis!

Schonland countered early in 1928. Armed with even more results from another season of storms during January and February 1927 — gathered this time by himself and Mr Linton, with some assistance from Ismay — he presented a tightly-argued case supported by many photographic images of the rapid displacement and recovery of the bubble in Lippmann's electrometer. In addition, he analysed a new situation, suggested by Simpson, where the lightning flash was not vertical at all, but inclined towards the observer. Evidently such occurrences, though uncommon, would produce an apparently anomalous result in an inter-cloud discharge because the direction of the instantaneous measured field would be reversed, which could lead to the conclusion that the cloud was positively charged at its base: Simpson's own claim, when, in fact, this was not so. Schonland was forthright. No evidence had been found in his extensive measurement campaign, he asserted, to support the contention that discharges within thunderclouds 'ever involve the upward movement of positive charge such as must frequently occur within a cloud of negative polarity'; in fact, the contrary situation predominated, with the discharges being invariably downwards as if they took place between the poles of a cloud of positive polarity [36]. And there the matter rested; at least for the time being.

These summertime expeditions to Gardiol continued to yield a rich harvest of scientific information and allowed Schonland to test yet another of Wilson's theories. Evidence had been found that explained how it was that the natural, fair-weather, electrical charge on the surface of the earth was negative. If ever there was an experiment in physics that used nature's own apparatus with just minimal intervention from man this was it.

The countryside around Somerset East is covered by small thorn-trees, known as the Karroo *Acacia*, which boasts a formidable array of long, sharp thorns. Schonland cut down a thriving specimen and mounted it firmly above the ground on ebonite-sulphur insulators. To the trunk he attached a wire that he connected through a galvanometer to earth. The principle behind the experiment was simple. Pointed or sharp conducting objects distort an electric field in their immediate vicinity and cause it to be concentrated, thereby increasing its intensity and so producing ionization of the air. If the natural electric field should increase, as would happen whenever a thundercloud approached, then the sharply-tipped, moisture-laden thorns would cause an enhancement of the field and the ionization of the air in their vicinity. The negatively charged cloud-base would then attract the positive ions and so leave the surface of the earth with a net negative charge. In addition, every lightning stroke from that cloud would deposit additional negative charge on the earth beneath it and so increase it still further. By contrast, the charge deposited by rain, as Wilson had already shown, was almost always positive. This happened as the raindrops encountered the rising positive ions and accumulated some of their charge as they fell to earth. A preponderance of negative charge on the earth means that the point-discharge effects of all sharp and pointed objects, even blades of grass, and the occasional extra doses of negative charge from thunderclouds would outweigh the effects of rain. Would the experiments with the thorn bush substantiate the theory? Indeed they did. The steadily increasing deflection of the galvanometer indicated that Schonland's thorns beneath the thunderclouds were prolific sources of positive ions that were rapidly wafted skywards, leaving their negative counterparts on the earth. Every lighting stroke to ground increased that negative charge still further, while the positively-charged rain, though causing the galvanometer to fluctuate, never offset the dominant negative charge. The results soon appeared in print and Schonland and Wilson began a correspondence that would change the direction of Schonland's research and make South Africa the lightning laboratory of the world [37].

CHAPTER 7

ENTER COCKCROFT

Science in isolation is sterile.

To thrive and, indeed, to prosper it needs the catalytic action provided by contact between like-minded individuals. When the scientific community was as small as it was in South Africa in the earliest years of this century, its very survival depended crucially upon two learned institutions for support, encouragement and occasional strong representations to government. The bodies concerned were the Royal Society of South Africa and the South African Association for the Advancement of Science, later to be known, rather appropriately, as S_2A_3. While the former bore the much revered title of its illustrious British counterpart, it is not to be confused as a mere local branch or agency of the London institution, for it was awarded the Royal Charter in its own right in 1908 after first coming into existence in 1877 as the South African Philosophical Society. In 1902 the South African Association for the Advancement of Science was formed, not in competition with its philosophical counterpart but in close collaboration with it, along the same lines as the links that exist between the Royal Society of London and the British Association [1]. The bonds of Empire were indeed strong.

When Basil Schonland returned to South Africa from Cambridge in 1922 his natural inclination was to become involved not just in the practice of scientific research but to participate in its organization, and so within little more than a year he was a member of S_2A_3 and by 1927 had become Secretary of the Royal Society of South Africa, whose President was Professor Alexander Ogg. The fact that the Head of the Physics Department and his Senior Lecturer shared an office at UCT presumably contributed greatly to the smooth running of the country's leading scientific society during the years when Ogg and Schonland were in harness together. Such grounding in the mechanics of scientific management will have also augured well for the future of science in South Africa, while the close association between Schonland and the man who had known him as a brilliant student at Rhodes would soon play a crucially important part in the next phase of the younger man's career.

In April 1927, the month Basil Schonland was elected a Fellow of the University of Cape Town [2], Ogg wrote to the secretary of the *Royal Commission for the Exhibition of 1851*. This grand title described the body in London that was charged with mounting the great exhibition held that year in the Crystal Palace situated in Hyde Park. Such was its success that the Commissioners were able to use the considerable financial surplus it generated to establish an educational institution of world renown in South Kensington dedicated to technological, scientific and artistic education—the Imperial College. In addition they launched, in 1890, their scheme of postgraduate awards that were to provide some of the most talented young scientists in Britain, her fading Empire, the Dominions and then the Commonwealth with an unrivalled opportunity to pursue their research at the leading universities of the time. In his letter Ogg recommended Schonland for the award of a Scholarship and set out the rather special circumstances that applied in his case, for Schonland was 31 years of age and a married man with two children; not the type of candidate usually considered by the Commissioners for its highly prized scholarships. However, as an Exhibitioner himself at the University of Göttingen from 1896 to 1898, Ogg knew that some special pleading would not go amiss and he also knew that he had a mighty ally.

Ernest Rutherford had arrived in Cambridge from New Zealand in 1895 on a research scholarship from the 1851 Exhibition. By 1908 he had won a Nobel Prize and was now regarded as the greatest experimentalist since Faraday. His too was the commanding presence at the Cavendish and, of course, it was under his seemingly omniscient gaze that Basil Schonland had carried out his work on β scattering. Then, when the time came to publish the results, it was Rutherford who had always communicated them to the Royal Society, while later he had enthusiastically supported Schonland's application for the lectureship at UCT. Naturally Ogg was quick to quote him in support of his nomination of Schonland. In the course of correspondence between them Rutherford had written [3]:

> As you know, I have been fairly closely in touch with the work of Schonland since he left Cambridge which has been published in the Proc. Roy. Soc. Considering the isolation under which he works I think he has done admirable work and shown that he has not only strong experimental ability but a capacity to bring a difficult investigation to a successful conclusion. He has shown also a good deal of versatility as seen in his work on the electrical effect of Thunderstorms.

Recognising so well how isolated he himself would have been had he remained in New Zealand, Rutherford continued:

> I consider it would be of great advantage to him if he were to renew his contact with European Physics. It is difficult for one at a

91

> distance to follow the rapid changes in theories and experimental
> methods that are taking place. To obtain a reasonable perspective
> on these movements it is important to be in a position to speak with
> those who are in the forefront of the advance. The amount of
> published matter is so great that it is impossible to read more than
> a fraction of it but a general knowledge can be obtained by talks
> with those who have special knowledge in the various directions. I
> have always been of [the] opinion that a teacher in a university of
> our distant Dominions should be given frequent opportunity to visit
> Europe or America to keep in touch with his subject.

Ogg then provided the Commissioners with the broad details of
Schonland's career and academic achievements to date, including the
fact that he had already proved himself to be 'a most excellent teacher
and researcher' with several papers in the Proceedings of the Royal
Society communicated initially by Rutherford and, more recently, by
Professor C T R Wilson FRS, who now held the Jacksonian Chair of
Natural Philosophy at the Cavendish. When approached, CTR readily
gave his support to Schonland in a way that, by his standards, was
positively effusive [4]:

> I consider the results of this preliminary work are themselves of
> great importance and that we may expect the future work at
> Somerset East to add greatly to our knowledge of Thunderstorm
> electricity. I am very glad that Schonland has taken up this work. I
> think that it is very fortunate that the great opportunities which S.
> Africa affords for the study of Thunderstorms should be utilised by
> one who is so competent to do so.

The quality of Schonland's scientific work seemed beyond doubt, but his
age could well have counted against him had Ogg not pressed his claim
with great fervour. He explained that the University was prepared to
grant Schonland six months' leave, three on full pay and three on half
pay, but any additional leave beyond that would be without pay. Some
financial sacrifice was on the cards but the benefits would be great. Ogg
continued: 'A scholarship for a year would be well spent on one who
has proved himself so capable'. And then he played his final, patriotic
card by informing the Commissioners that 'Dr Schonland had spent
nearly four years on Active Service in France'. If Rutherford and Wilson
had not the influence to bring to bear then King and Country surely did.

Schonland's application for a scholarship, and the letter of support
from Ogg, duly appeared before the 1851 Commissioners, who referred
them in the customary way to academic referees of standing and
repute. Schonland's was seen by Sir William Bragg at the Royal Institution, and by Rutherford whose views on the matter were already pretty
clear-cut. Bragg's letter of recommendation to Mr Evelyn Shaw, secretary
to the Commission, is worthy of attention both for its recognition of

Schonland's standing as a scientist and also for its balance [5]:

17 June 1927

Dear Shaw,

    The Cape Town candidate, Schonland, whose papers you sent me and I now return has a very good record indeed. The subject of most of his work, namely the Scattering of B rays is a very confusing and complicated question but its importance has attracted a number of investigators. I have looked at one or two of his papers and have the opinion, confirmed by what Rutherford and others have said, that Schonland's work is a very real contribution to the subject. I feel that he is a safe man for the award of a scholarship.

    His age is rather against him, but war must always be accepted as an explanation of delay. He has in the time possible to him already carried out important investigations and I do not think we can ask for more.

Yours sincerely,

W.H. Bragg

The Commissioners were satisfied and so, in December 1927, Schonland travelled to England as the recipient of a £250 scholarship. He took up residence at 4 Jesus Terrace, Cambridge, and prepared to spend a year at the Cavendish working under the guidance of the none other than C T R Wilson, whose theories on thunderclouds and lightning Schonland had recently subjected to such rigorous experiment. It had been agreed in correspondence between Schonland and CTR that he would work on the 'penetrating radiation' and its origin; the passage of cathode and β rays through matter and on thunderstorms'. Rather a tall order it might seem for anyone to tackle in a year, but there was even more to come! CTR had, in the meantime, come up with a new idea altogether. It was, he said, 'a wee gadget' and its construction required manipulative dexterity of the highest order and eyesight to match. What he had conceived of was a new type of electroscope, an instrument in general use in those days for measuring electrical charge, but there was need for one sensitive enough to measure the very small charges that might result from the passage through space of what was then called 'the penetrating radiation'. Wilson had thought of a way of doing this but it required the construction of a very small and delicate instrument and he believed Schonland had the ability to do it.

    The stories about CTR's patience are legion, and Schonland needed it too in great measure when he set about the project. He had to cut a concave mirror, just 4 mm square, from a sliver of silvered mica and then affix it by means of two hair-like gold hinges to a larger, central electrode that formed the other plate of the electrometer. The principle of the instrument was well-known and depended upon the repulsion that two like-charges have for one another. Any charged particles alighting on the mirror and the central electrode would produce a repulsive force that would cause the mirror to rotate about its minute hinges. A light source, focused on the mirror and illuminating a calibrated scale some distance away, would enable very sensitive measurements of the electric charge to be made. Some months had passed and many broken slivers of mica littered the floor before Schonland had a mirror which satisfied his requirements. All the while Wilson hovered in the background offering encouragement and advice when needed; now the moment of truth had arrived. Measurements showed that the device produced a deflection of the light beam of more than 4 mm when charged with the equivalent to a single α particle: a remarkable sensitivity indeed and considerably better than any achieved up to that time. In May, Schonland described this new electrometer at a meeting of the Cambridge Philosophical Society and then wrote a short paper for the Society's Proceedings, in which he discussed the underlying theory and its method of construction [6].

Occasionally, by way of diversion or even relaxation, Schonland found the time to work on a theoretical problem he had brought with him from Cape Town. It was to do with his β-particle scattering experiments of a year or so before, and his return to the heart of atomic physics, as the Cavendish most assuredly was, provided the ideal opportunity to resurrect the topic. The problem involved the relativity correction that had to be included in the mathematical formulation as a result of the increase in particle velocity, which occurred as it approached the scattering nucleus. Further thought had convinced him that the existing theory inadequately described the process because no account was taken of the spiral nature of the electron's path, nor had any consideration been given to the resulting loss of energy by radiation that took place. Whereas he could account for the additional geometrical features of that path, he knew of no way of determining this radiation loss. The paper describing Schonland's work on this problem appeared in 1928, having been completed while he was in Cambridge [7]. It was a most elegant piece of theoretical work and merited the attention of Rutherford, Chadwick and Ellis when they published their monumental book on the radiations from radioactive substances in 1930 [8]. But many questions still remained unanswered.

Schonland's had been just one of many attempts to explain the scattering processes involving β-particles. James Chadwick was one who had set out to do so, but he concluded that all previous experiments still left much to be desired. It was, though, he conceded, Schonland's use of cathode rays and the negatively biased grid to suppress any secondary electrons that really provided a new approach to the problem and his results looked promising. However, even that method had its shortcomings. Now, Schonland's latest theoretical treatment during this sojourn in Cambridge was an elegant new attack on the problem, that benefitted greatly by having Charles Darwin, the mathematical physicist, nearby. But however elegant the mathematics, it was the physics that presented the more fundamental problem. For the experiment to have any meaning the encounter between the approaching β-particle and the nucleus had to satisfy the condition of single scattering, but with the apparatus available to him at UCT Schonland had only been able to achieve this under conditions that produced spiral orbits around the nucleus, and these led to loss of energy by radiation. Schonland wrestled with the problem and described in elegant style the dilemma any experimentalist would face: [9]

> To steer between the Scylla of spiral orbits and radiation loss and
> the Charybdis of plural scattering requires the use of fast β-rays and
> moderate angles of scattering...

A complete test of the theory of β-scattering, he wrote, would require very high voltages and exceptionally thin metal foils, both of which would lie well beyond his reach once he returned to South Africa. And so, in the end, he conceded that his cathode ray scattering experiments were not an adequate test of the relativity correction to the orbit of a β-particle deflected by an atomic nucleus. However, he had at least provided an estimate of the amount of scattering to be expected. There was also some consolation to be had from the fact that, whereas Rutherford's early calculations for the scattering of the α-particle had predicted a value only 0.44 of that actually observed, Schonland's predicted value for the scattering of the β-particle was 0.83 of the measured value. To have improved on the work of the master was no mean feat in itself and Schonland's work therefore represented a significant contribution to the state of knowledge at the time. All were now agreed that the remaining fraction could probably be accounted for by the radiation loss, but this was a problem which still defied analytical solution.

Basil Schonland bade farewell to the study of atomic physics with the publication of that paper in 1928. And so he took the decision to navigate a different route. Henceforth his encounters with the wonders of nature would be dominated not by the atoms of Democritus but by the lightning of Zeus.

*   *   *

Life in Cambridge that year was not totally taken up by work, for in the summer Ismay, having left the two children with her mother in Somerset East, sailed for England. She stayed for three months and it was a happy time as the Schonlands soon fell into the way of life they had known as students a few years before. They revisited old haunts and punted on the river with their many friends, old and new.

Amongst the friendships that Basil made and renewed that year were some that would be both long-lasting and of great significance in his future career. At the Cavendish were men from most corners of the world, rubbing shoulders and sharing workbenches with their British colleagues. He enjoyed their company and the natural good humour of the place that was so associated with the massive presence of Rutherford amongst 'his boys'. There were lively discussions with T E Allibone, who was just embarking on work in the field that Schonland was leaving, and with B L Goodlet, an electrical engineer and a frequent visitor from the Metropolitan-Vickers Company, in Manchester, who was assisting Allibone with his experimental apparatus. Peter Kapitza, the flamboyant Russian whose presence so enlivened the Cavendish, entertained some unusual ideas about a phenomenon known as ball-lightning. Schonland listened with interest but could not hide his scepticism [10]. Another of his new acquaintances was M L E Oliphant, recently arrived from Australia as an 1851 Exhibitioner, who was a kindred Colonial spirit, while Basil renewed his contact with Patrick Blackett whose spirit was of a different kind — 'a young Oedipus' was how some described him: certainly an imposing presence and a formidable intellect [11]. Blackett, the former naval officer turned physicist, was destined to win the Nobel Prize for his work on cosmic rays, for which he used that equally remarkable of instruments, the Wilson Cloud Chamber, with which Schonland was certainly familiar. All their paths would cross again many times and their subsequent careers would owe much to those friendships that developed at the Cavendish. None, though, was to be more significant to Basil Schonland than his friendship with J D Cockcroft.

John Cockcroft was a year younger than Basil and was born in Todmorden, just on the Yorkshire side of the border with Lancashire. After attending the local schools he entered the BSc course in mathematics at Manchester University in 1914, right at the outbreak of war. Cockcroft's real interest lay in the exciting new world of atomic physics and he was especially fortunate in having Rutherford take over the first-year physics course when the previous lecturer failed to keep order in the class. His effect on Cockcroft was electric, for Rutherford commanded enormous respect and was clearly the master of his subject. On passing the Intermediate Examination at the end of his first year, Cockcroft volunteered for service and was soon drafted into the signalling branch of the Royal Field Artillery, where he tackled the Morse code, spliced wires and cables, and learnt to ride a horse. He was then dispatched to France. Service in the trenches and in artillery observation posts, under threat of instant annihilation, instilled in John Cockcroft, just as it had in all those who saw carnage close up, a deep hatred of that war, which later manifested itself in his refusal ever to talk about it: a characteristic he shared with so many who had endured and survived the terror. His bravery under fire saw him mentioned in dispatches and in 1918 he was commissioned as a second lieutenant in the RFA. Immediately on demobilization

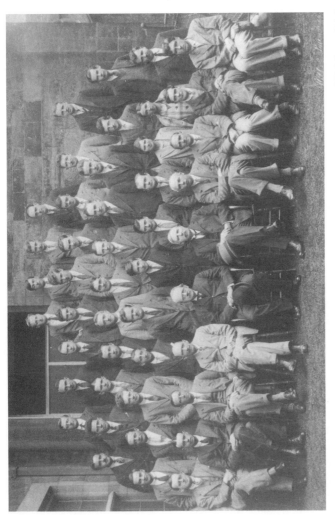

Figure 8. Staff and students of the Cavendish Laboratory, Cambridge in 1928. Fifth row (L to R): G C Lawrence, H M Cave, C A Lea, E A Stewardson. Fourth row: G Millington, C E Eddy, F A Arnot, D S Lees, E E Watson, C E Wynn-Williams, F A B Ward, J D Cockcroft, L H Gray. Third row (L to R): F R Terroux, M L E Oliphant, N Feather, R R Nimmo, G H Aston, N de Bruyne, E T S Walton, Prof E L Harrington, M C Henderson, J Chariton, J L Hamshere. Second row (L to R): C F Sharman, E P Hudson, W R Harper, B F J Schonland, W L Webster, D C Rose, E J Williams, T E Allibone, Miss Mackenzie, Mrs Salamon, H J J Braddick. Front row (L to R): G H Henderson, G Stead, J Chadwick, Prof C T R Wilson, Prof Sir J J Thomson, Prof Sir E Rutherford, E W Aston, Prof G I Taylor, P Kapitza, P M S Blackett. (Reproduced by kind permission of the University of Cambridge, Cavendish Laboratory.)

he returned to university, on a special course for ex-servicemen, but switched from science to electrical engineering. After completing the BSc Tech he became an apprentice at Metropolitan-Vickers, and while there wrote a thesis on the harmonic analysis of voltage and current wave-forms for which he was awarded the MSc by Manchester University. By now it was evident that Cockcroft was much above the average, both as a researcher and as a most competent young engineer, and this brought him to the attention of A P M Fleming, Manager of the Research and Education Department of Metro-Vick, as the company was always known, who suggested that he should go to Cambridge to improve his mathematics, since it 'would be advantageous for future work in the industry'. With generous financial assistance from his former professor at Manchester, and a college Exhibition from Cambridge, Cockcroft left Manchester in June 1922 to take Part II of the mathematical Tripos, having been excused Part I since he was already a graduate. Over the next six years he steadily enhanced his reputation by becoming a B Star Wrangler in Mathematics and then, after catching Rutherford's eye, commenced research for his PhD in the field of atomic physics. For all his evident mathematical ability, Cockcroft always showed a strong bent for practical engineering and was fascinated by large machines. This was soon revealed when he designed and constructed some of the most important, and certainly the largest, experimental apparatus to be used at the Cavendish. Very intense magnetic fields were soon produced by Kapitza using a Metro-Vick alternator, short-circuited by a special copper coil designed for him by Cockcroft. This technology opened up a new field of research at the Cavendish and essentially heralded the end of Rutherford's era of string and sealing wax [12].

John Cockcroft was an enigma. Of rotund build and owlish appear-ance, he was by nature a man of remarkably few words. To some he appeared hard to know but he had a happy knack of befriending new-comers to the Cavendish and making them feel at home. He had done this when Oliphant and his wife arrived from Australia in October 1927 [13] and now he did so again when Schonland returned to his old laboratory just a few months later. Schonland and Cockcroft became firm personal friends during that year at the Cavendish and it was a friendship that was to last for 40 years until Cockcroft's death. Over the next decade they were to meet only occasionally, but when they did the old bonds were still strong. When war darkened the world's horizons for the second time, it would be such bonds and old acquaintanceships that were to be so important in the marshalling of Britain's scientific resources to meet and counter the Nazi threat. But that was still in the future.

Now, with his year at the Cavendish fast approaching its end, Schonland wrote a report on the work he had undertaken and sent it, as was required, to the Commissioners for the Exhibition of 1851, who appointed Sir Richard Glazebrook to examine it. Glazebrook was one of the elder statesmen of the Cavendish who had been there as a student of James Clerk Maxwell in the 1870s and now, as Chairman of the Commission's Science Scholarship Committee, he read Schonland's report and pronounced himself well-satisfied with it. 'Mr Schonland's Scholarship was well deserved', he declared and then drew particular attention to Schonland's own concluding remarks [14]:

> It is possible for an isolated worker to keep in touch with current thought for a time by the study of the journals concerned, but it is difficult to do so for long. The atmosphere and spirit of the Cavendish Laboratory, the opportunity of hearing and conversing with so many workers in the subject both here and elsewhere in Europe, constitute a revitalising influence which I have felt very strongly. I owe my thanks to Sir Ernest Rutherford for the hospitality extended to me in the Laboratory.

That revitalizing influence was all that Basil Schonland needed and he returned to Cape Town in 1929 brim-full of ideas for new research, stimulated by numerous discussions about atmospheric electricity with C T R Wilson. Schonland's own forays into the field, and the almost unique environment that South Africa offered for such work, were now the spur. He immediately set about planning an experiment intended to test Wilson's theory that the electric fields within thunderstorms should exert an important accelerating effect upon β-particles produced by the disintegration of the radioactive material carried naturally in the air. Subsequent collisions with the molecules of air would produce yet more secondary electrons and they, in turn, would be accelerated still further, leading to the generation of what was known then as 'the penetrating radiation' and thought to be γ-rays, but of very high energy. This radiation had been studied since the turn of the century and was known to possess energies a hundred times greater than those of the fastest particles emitted by radioactive bodies — hence its great penetrating power. All available evidence suggested that its origin lay elsewhere within the solar system but Wilson saw in the thunderstorm a much more local source. Schonland intended to find out and he now had the ideal device with which to try and detect these runaway electrons. Mr Linton, in his workshop in the Physics Department at UCT, was put to work to construct an ionization vessel that would contain the new electroscope that Schonland had made so painstakingly in Cambridge.

While that was happening the scientific community in South Africa was preparing to host the biggest event yet seen in the country's short scientific history. In July, members of the British Association for the Advancement of Science would arrive, *en masse*, for their meeting in

Cape Town. Section A: Mathematical and Physical Sciences, was represented by a prestigious group of eminent men led by no less eminent a figure than the President of the Royal Society, Sir Ernest Rutherford himself [15]. The occasion was doubly significant, for not only was the University of Cape Town celebrating its centenary as an institution of higher learning, but a new building for the Department of Physics was opened officially by Rutherford to much acclaim. The meeting of Section A on 23 July contained a paper delivered by Rutherford on 'The Origin of Actinium'. It was erudite and was delivered in customary Rutherford style. Some thought the subject might be too erudite in a country 'where physicists were limited in number' but Rutherford dismissed that suggestion with disdain [16]. Immediately following him in the printed proceedings of the day, with a contribution entitled 'Some New Electrometers', was one Dr B T G Schonland, the error in his initials not disguising the fact that Basil Schonland, either by coincidence or design, was literally to follow in the footsteps of the master. In his paper Schonland described CTR's 'wee gadget' but did not reveal how he soon intended to use it in his search for 'the penetrating radiation'.

Following Schonland to the podium were some very famous figures whose reputations just twenty years before had so stimulated the young prodigy from Grahamstown in his quest to go to Cambridge. Lord Rayleigh was there, as was the austere A S Eddington, a theoretical physicist who found himself frequently on the receiving end of barbed comments from Rutherford, who was not renowned as an enthusiast for too much theory. There too was R H Fowler, another theoretician but one who was redeemed in Rutherford's eyes, not only because he was the great man's son-in-law, but also because he was blessed with much common sense. Others from the Cavendish, who now sat in the shadow of Table Mountain, were F W Aston, a Nobel Laureate of 1922 for his work on mass spectroscopy, and G I Taylor, the renowned fluid dynamicist of the day whose scientific pedigree as the grandson of George Boole, the founder of mathematical logic, was unimpeachable.

The Schonland name was well-represented, too, for there in Section K was a paper from Dr Selmar Schönland on 'Some South African Plant Hybrids', while one on the 'Springbok Skeleton' was delivered by their great family friend, Dr Robert Broom, whom Basil had visited when Broom, in his guise as an army medical officer, was based in England in 1915. In a matter of just a few years Broom's lasting reputation would be made in the field of palaeontology when he discovered 'Mrs Ples', the first adult *Australopithecus*, at Sterkfontein west of Johannesburg. This followed the equally stunning discovery in November 1924 made by R A Dart, Professor of Anatomy at the University of the Witwatersrand, of the 'Taungs child', the original *Australopithecus africanus* now recognized as the link between man and his non-human ancestors [17].

But the paper that so many had come to hear, and the one that was described by Rutherford as the great event of the morning, was the discussion on 'Life' by another illustrious acquaintance of the Schonlands, General J C Smuts FRS, whose philosophy of Holism was spoken of in almost the same breath as Darwin's Evolution. Smuts, of course, was the remarkable embodiment of a man whose intellectual capacity spanned vast oceans of thought and human endeavour, and if any South African had an international reputation, he did. Amongst his own people, though, views were divided, and while he was held in the highest regard by many there were others who felt very differently about him. One, at least, who belonged to that opposing camp was himself attending the meeting. Hendrik Frensch Verwoerd, then a Professor of Psychology at the University of Stellenbosch, near Cape Town, read a paper on the 'Contribution to the Experimental Investigation of Testimony'. Within just three decades, Verwoerd would become Prime Minister of South Africa and his philosophy of apartheid would rend the very heart of the nation. That meeting of the British Association in Cape Town in 1929 was truly an occasion of some moment in South African history, even if no-one realized it then or if anyone is aware of it now.

*   *   *

While Cape Town offered its inhabitants so much in the way of the pleasures of life in one of the world's most idyllic settings, it lacked the dramatic atmospheric effects that ripped through the late afternoon and evening skies in summer around Johannesburg, almost 1600 km away. To have any chance of success, Schonland's new line of research required an almost predictable season of lightning activity and, whereas Somerset East had satisfied that requirement to some extent, the family farm was not well-provided for as a research laboratory. Johannesburg, though, had both. There, the University of the Witwatersrand (or 'Wits' to its staff and students alike) had become a fully-fledged institution able to confer its own degrees just seven years before, and, as we have seen, its Department of Physics had even been presided over by Alexander Ogg in 1919 during a brief detour after leaving Rhodes and on his way to UCT. Ogg's successor at Wits was a Welshman, Professor H H Paine, whose academic career had taken him from Aberystwyth to Cambridge with the inevitable interlude of the war and the winning of the Military Cross. With the summer lightning season approaching, Paine was asked by Ogg if Schonland might set up his apparatus in one of his laboratories in order to try to measure the effects predicted by C T R Wilson. Paine readily agreed and so began an association with Wits which was to have profound consequences for scientific research in South Africa.

As soon as the university closed for the long summer holidays, Basil and Ismay travelled to South Africa's biggest city amongst the mine dumps; a city with streets that turned all shades of mauve and purple when the blossoms of the jacarandas showered down during the thunderstorms, that were as regular as clockwork almost every summer afternoon. With them on the train was the ionization chamber that John Linton had constructed to Schonland's design, as well as two glass tubes that contained the precious electroscopes, with their minute mica mirrors of the type that he had so meticulously fashioned at the Cavendish the year before.

At Wits, Schonland was given sole use of a room on the top floor of the Physics Department. There, with the assistance of Eric Halliday, his former student from UCT, who was now undertaking his own research on lightning using instruments deployed on the roof, he installed his electroscope. It was mounted upon an insulating rod of quartz and inside a tightly sealed, cylindrical, cast-iron chamber. A glass window allowed an external light source to illuminate the mirror, such that its focused reflection fell on a calibrated scale some 1.5 m away, where it could be read with some accuracy. The cast-iron container was intended

Figure 9. Schonland's ionization chamber containing the electroscope used in his quest to detect the 'penetrating radiation' assumed to be produced within thunder clouds. The cylindrical zinc-coated wire grid shown alongside provided additional shielding against spurious radiation. The whole apparatus was then shielded on all sides except for a narrow conical 'observation window' by being placed on a large iron base and surrounded by a rampart of iron blocks. ([18, figure 2] reproduced by kind permission of the Royal Society.)

to block the passage of all rays and particles except for the highly penetrative radiation that CTR had postulated would come from the thunderclouds. This was just one of many special steps taken by Schonland to eliminate interference from any other extraneous sources of radiation, both within the vessel and in the room. Any negatively-charged β-particles which might be emitted spontaneously from the walls of the cylinder were blocked by a cylindrical cage of zinc-coated wire biased to a 400 V negative potential—Schonland's suppresser grid in action once more—while it also suppressed any photoelectric emissions from the illuminated metal electrode of the electroscope. Finally, all nearby sources of γ radiation were carefully screened by standing the instrument on a large iron slab some 9 cm thick and by surrounding it with a wall of 15 cm iron blocks. The only point of entry for radiation would then be through a cone-shaped aperture directly above the equipment. With the apparatus in place, all that remained was for Johannesburg's afternoon symphony of thunderstorms to arrive on cue.

Arrive they duly did, and some with a vengeance too. Basil had Ismay to assist him and they spent most afternoons and many evenings in their laboratory, sheltered from the pyrotechnics above them by a corrugated iron roof and a thin asbestos ceiling. At the height of the storms, particularly any accompanied by hail, the din within the room not only made all conversation impossible but even obliterated the sound of the accompanying thunder [18]. In all, they observed 22 thunderstorms, seven of which passed directly overhead, with occasional direct strikes of lightning nearby. Every ten minutes they recorded the readings from an array of instruments in the laboratory. The steadily deflected light beam from the ionization chamber indicated the accumulating effects of the background radiation, while any sudden jumps in its reading flagged up the arrival of α-particles, the effects of which were subsequently removed from the calculations so that only the elusive 'penetrating radiation' was measured. In addition, Halliday's capillary electrometer and his electric field sensor above them on the roof provided information about the prevailing electric field conditions as the storms approached from the south-west with almost predictable regularity. To ensure accurate measurements and to allow meaningful comparison to be made with the results of other workers, especially the Germans Bothe and Kolhörster who were active researchers in this area, it was necessary to calibrate the apparatus, and this Schonland did by means of a standard source of γ-rays carefully positioned at a known distance from the chamber, while an estimation of the intensity of the interfering α-particles came from the comparative deflections produced by known amounts of radium, thorium and actinium using the methods that had been so thoroughly developed over the years by Rutherford.

By the time Johannesburg's lightning season had all but exhausted itself, the Schonlands were packing up their apparatus and personal apparel for their return to Cape Town. With them were masses of accumulated data that Basil would analyse and speculate over for many months. However, no analysis was needed to notice the one self-evident result from all their exhaustive efforts: the expected increase in β-particles, those runaway electrons which Wilson had predicted, just did not occur. There was no sudden increase in ionization, no rapid or dramatic deflection of the spot of light, in fact, no measurable effect at all at the moment when this was thought most likely — just at the instant that the lightning flash took place. What had occurred was precisely opposite to what had been expected. The amount of ionization actually decreased when the storm was at its most violent overhead.

This apparent cussedness of nature required careful investigation, but before he could apply himself to it, there was the matter of moving the Schonland home, an occurrence to be repeated at regular intervals over the following years. From Tamboerskloof the family went to St James, a picturesque village overlooking the Atlantic Ocean with the slopes of the mountain behind them. It was there at the house in St Ronan's Road that Basil became the proud owner of his first motor car, a Ford, that set him back the sum of £200. The Schonland's geographical horizons immediately broadened and they were spurred on by the fact that their two children, David and Mary, were now of an age where family outings surrounded by the scenic splendour of the Cape Peninsula became common, and these weekend activities forced Basil to remove himself, physically at least, from his laboratory for a few hours in the week [19]. His thoughts, though, were on the mass of data he had accumulated at Wits over the previous summer, and such was his enthusiasm for his new obsession that his students soon dubbed him 'Lightning Schonland' [20].

Thus began what was to become a dedicated scientific campaign with the specific intention of investigating the relationships presumed to exist between lightning and radioactive phenomena in all their forms. Nothing like this had ever been undertaken with quite such intensity of purpose before, and certainly not in South Africa, where local legend had it that all could be explained in terms of *Umpundulo*, the magical thunder-bird of the Zulus, that dived from the clouds to the earth beneath, its vivid plumage and beating wings producing a blinding flash and a thunderous rumble from the heavens. While Schonland would recount that story many years later, his obsession now was to explain the physics behind the observations he and Ismay had made in Johannesburg.

The process of analysing the data lasted throughout the wet Cape winter of 1930 and soon a pattern began to emerge. Whereas an approaching storm always produced an increase in ionization, a marked decrease occurred when the storm was actually overhead. This was certainly at odds with CTR's prediction that the 'penetrating radiation' would then reach its peak, but it did provide a clue as to the nature of the 'penetrating radiation' itself. There was surely only one explanation. The electrical charges within the thunderclouds, and the immense fields which they produced, were acting in a way just as Schonland's suppresser grid behaved inside his ionization vessel, by repelling the incoming radiation which entered through the tops of the clouds from whatever source produced it elsewhere within the cosmos. The absence of any downward-moving, runaway electrons, despite the very favourable storm conditions, led Schonland to conclude that they were stopped by collisions with other atomic particles before they reached ground—a suggestion that Wilson himself had tentatively put forward in the correspondence that flowed between them.

However, there was one other idea that struck Schonland and it was the most profound of all. Until now the prevailing wisdom on the subject maintained that the 'penetrating radiation' was made up of γ-rays of extremely high energy, so-called ultra-γ radiation. His very careful measurements made in Johannesburg of the change of ionization, and also of the direction of the electric fields produced by the advancing thunderstorms, provided the clues to its actual make-up. This was further reinforced by the earlier results obtained in the Eastern Cape. During the intense activity within a thundercloud at the time of a lightning stroke Schonland observed that the 'passage of a flash ... did not completely discharge the cloud' but produced instead an excess of charge of one sign over the other. Whereas both appeared likely, it was apparent that an excess of positive charge was far more effective in decreasing the radiation below the cloud than a negative one. From this he felt able to suggest that 'the superior stopping power of storms (directly overhead) may be explained if the radiation consists of positively charged particles and the stoppage takes place above the thundercloud.' The excess positive charge within the cloud provided the repulsive force and this then clearly established the polarity of the down-coming particles. In September 1930 Schonland's paper reporting these results was communicated to the Royal Society by Wilson and it added considerable weight to the suggestion, made by the German experimenters Bothe and Kolhörster just a year before, that the rays were indeed 'corpuscular' and not wave-like. But it also went much further, for it identified their positive sign: they were protons.* If true, this finding was of fundamental importance, for not

---

* The 'penetrating radiation' soon became known by the more evocative name of 'cosmic rays' which was more in keeping with their origin in outer space. It has recently been shown that they consist almost exclusively of protons that can attain energies of at least 10^{20} eV. Their source is thought to be supernovae and pulsars, and in their passage through the earth's atmosphere they suffer frequent collisions that give rise to several types of secondary particles such as muons [21].

only had it indicated the true form of the hitherto mysterious 'penetrating radiation', but it had identified the role of thunderclouds at least in its modification if not in its generation.

*   *   *

With his experience of the Cavendish and his own research achievements now well-known around the University, Schonland was held in some awe by those contemplating research careers of their own and his views on the subject were therefore much appreciated. One such student was A L Hales, who encountered Schonland as he walked across the campus one day late in 1930. Anton Hales had just completed an MSc in applied mathematics and was about to take up a scholarship to Cambridge, so he actively sought Schonland's views of his decision to specialize in the exciting field of quantum mechanics then capturing the imagination of some of the world's most eminent mathematicians and physicists. Schonland immediately asked him whether he intended returning to South Africa on completion of the scholarship and Hales said he did. Then, said Schonland, it would be a mistake to go for an area like quantum mechanics where advances were talked about in the laboratories of Europe and England six months before they appeared in the scientific literature. By that time they had virtually been superseded by the next one and so it was almost impossible to keep up to date. No doubt starkly reminded of his own realization of this when he decided to switch his research from atomic particles to lightning, Schonland advised Hales to choose a field like geophysics, where 'you made your own observations and could do research not anticipated elsewhere'. Hales took the advice, and after completing the Tripos at Cambridge with geophysics at its core, returned home and then soon followed his mentor northwards to Wits in South Africa's geophysical heartland [22].

While Ismay had been Basil's research assistant during the lightning storms of the previous summer she now had other more pressing calls on her time, such as a family to bring up, and so he recruited a research student by the name of Viljoen to work with him. Eric Halliday, whom Schonland had supervised in person during the previous summer, and by regular correspondence for the rest of the year, had obtained his PhD and was now lecturing at Wits. J P T Viljoen was a recent MSc graduate with obvious research potential and so Schonland set him the task of investigating further the link between the 'penetrating radiation' and thunderstorms. Instead of the simple electroscope, they made use of a device first produced in 1907 by Hans Geiger when he was a disciple of Rutherford's at Manchester. The Geiger Counter, as it soon became known, provided the considerable increase in sensitivity that further work of this nature demanded. In addition, thermionic valve technology had progressed phenomenally since Schonland last had any contact with these devices in the dying days of the war. Now, their characteristics were well-understood and their use extended well beyond the ubiquitous wireless set. Together, he and Viljoen designed both the Geiger–Müller tube that formed the heart of the counter, as well as the amplifier, consisting of two triode valves, that followed it. The circuit contained some interesting features. Once triggered by even a single pulse of ionizing radiation the Geiger–Müller tube becomes permanently conducting and so is insensitive to any subsequent stimuli. To restore it to its previously non-conducting state required a correcting signal to be fed back to the counter after a suitable time-delay, and a method of doing this had appeared in the pages of *Nature* in 1931. Schonland readily adapted the technique to his requirements, and then had the ever-dependable Mr Linton build the apparatus into zinc-lined boxes to shield it from any sources of electrical noise that might exist in the immediate surroundings. Another, similar amplifier was connected to a 30 m length of wire that functioned as an antenna. Its purpose was to produce an electrical impulse whenever there was a lightning strike within a radius of about 70 km from the equipment. That amplified output, as well as half-second timing pulses from a chronometer, were fed to a paper-chart recorder and the complete system, known as a coincidence counter, was taken to Johannesburg to be ready for the next lightning season.

Schonland and Viljoen set up the new apparatus in the laboratory in which he and Ismay had worked the previous year. Once again careful steps were taken to shield the detector against radiation from all directions except those within a conical zone pointing towards the zenith. Schonland also went to considerable lengths to ensure that the normal operation of either detector did not produce a spurious response in the other, a phenomenon that often bedevils electronic devices when operating in close proximity to one another. When finally convinced that each detector performed its own task satisfactorily, and that no mutual interference occurred between them, he pronounced the apparatus ready for the next measurement campaign just as Johannesburg's lightning season of 1931–1932 was upon them.

In all, Schonland and Viljoen observed the passage of 29 electrical storms, with the accompanying torrential rain and frequent fusillades of hail that turned Johannesburg's streets into cobbled ice-rinks spattered with the battered mauve blossoms from the jacarandas. During every storm there was considerable activity from both pens of the recorder. It was soon obvious that while every lightning flash produced its characteristic electrical impulse, only those more than 25 km away also produced a simultaneous pulse from the output of the Geiger counter. Storms closer than this produced a far more random response from the recording system. Though these nearby lightning strikes caused considerable activity, careful examination showed that the pens no longer always struck the chart in synchronism; frequently the electric field impulse was not accompanied by a pulse from the Geiger–Müller tube. Of course, the possibility existed that the coincident strikes observed from the more distant storms were due simply to chance, and so the only way to resolve this was by rigorous statistical analysis. Between them Schonland and Viljoen dissected the data by projecting an enlarged image of the charts upon a screen and then, by carefully examining the relative pulse inter- vals to find any that lay within 1/10th second of a lightning flash, they were able to plot histograms for every storm showing the counter impulses in time related to the various lightning discharges. The results were illuminating! Of 225 lightning flashes observed within about 8 km of Wits none produced systematic coincidence on the chart, whereas 300 such discharges at a distance of about 30 km yielded 41 coincident pulses, more than ten times the number expected by chance alone. This was surely confirmation of Schonland's earlier observation that no run- away electrons reach the earth within a radius of about 30 km of an active thundercloud. The electrically charged cloud was acting as a shield or barrier by reducing substantially the number of fair-weather ionizing particles that struck the earth beneath. But by what mechanism, then, were such particles reaching the recording instrument from the more distant storms?

The answer it seems lay, once again, in the theory proposed by Wilson. Within a thundercloud the potential difference between the agglomeration of positive charges at the top and the similar massing of negative charges at the cloud-base can be as much as 5×10^9 volts. The intense electric field that then exists between these two charged regions will generate an upward stream of so-called 'runaway' electrons, each having energy as high as 5×10^9 eV. This electron-spray, as Wilson termed it, then encounters the earth's magnetic field and, as required by the laws of electromagnetism, would be deflected from that path in a manner dependent upon its velocity and direction relative to that of the earth's magnetic field. The model was therefore simple and could be applied to various geographical locations around the globe.

In equatorial regions, where the earth's magnetic field is almost horizontal, a vertically-travelling electron would descend to earth to the east of the cloud from which it was sprayed upwards. The situation is more complex at higher latitudes but, in general, such an eastwards deflection would still occur. However, this appealingly simple explana- tion ignored the fact that the intense electric field between the upper layer of the thundercloud and the conducting regions of the upper air would, because of its direction, tend naturally to impede the progress of such upward-moving electrons. Schonland was concerned about this and at some point during this research programme drew Wilson's atten- tion to this apparent impediment. In keeping with the genius of the man, an explanation was soon forthcoming. CTR suggested that if the first lightning discharge to occur was between the upper pole of the cloud and those charged regions above it, the effect would be to cause the retarding field to collapse momentarily, thus allowing the runaway electrons to escape upwards for their encounter with the earth's magnetic field and subsequent bending back to earth [23]. It was a brilliant idea. Such upward strikes had been reported by many observers, including Schonland, and so once again an avenue for research was opening up, and no-one was better placed to exploit it than Schonland himself.

CHAPTER 8

THE WINGS BEGIN TO SPREAD

The University of Cape Town was quick to recognize Schonland's talents and achievements. At its meeting in March 1930 the University Council agreed to award him a special personal allowance of £100 per annum in recognition of the honour he had brought the University through his recent research publications that had appeared in the leading scientific journals of the day. 'Their importance', the scribe recorded, 'was shown by the use made of his results in recent scientific text-books and the literature' [1]. And he had caught the eye of those farther afield as well.

In Johannesburg, the South African Institute of Electrical Engineers had been jolted into action by a paper delivered in April 1930 by one of its members, Mr T P Pask of the Victoria Falls and Transvaal Power Company (the VFP), on the subject of lightning and its effects on the rapidly developing mining industry of the Witwatersrand. As a mine electrician himself before the turn of the century, Pask had experienced the effects of many a lightning strike upon the plant and machinery of a mine and he knew how dependent the production of gold was on a reliable supply of electrical power.

Thirty years later, and now an engineer of some local prominence, Mr Pask called for the formation of a 'South African Association for the study of lightning and allied phenomena' and suggested how the SAIEE could give expression to the idea. 'We South African electrical engineers cannot be content with making use of the research work of other countries, without offering something in return. The study of lightning and atmospheric electricity generally, gives us the opportunity for reciprocal service.' He continued: '... it offers excellent scope for doing a humane work in reducing the death toll from lightning, and further gives a splendid opening for achieving something distinctive in our profession which could enhance our reputation as science workers' [2]. The point was made, and in December the Institute set up a Lightning Investigation Committee with Pask as its chairman. While the Committee's first task was the compilation of statistics of fatalities and damage due to lightning, it was appreciated from the outset that research into the fundamental processes of the phenomenon must be undertaken.

110

Just two months earlier, at the banquet to celebrate the 21st anniversary of the founding of the SAIEE, the Governor-General of the Union, the Earl of Athlone, made a most perceptive and, indeed, percipient statement when he proposed the toast to the Institute: 'To its loss South Africa can claim a painful intimacy with lightning ... we who live in this country have quite peculiar opportunities for observing its behaviour ... and who knows but that a lightning observatory may at some time be established from which the most valuable knowledge will result.... I feel the Institute will have a large say in the matter' [3].

In attendance on both occasions was a former President of the SAIEE who occupied the positions of General Manager and Chief Engineer of the VFP, Mr Bernard Price. Price was a man whose mark had already been well and truly made in South Africa since his arrival from England in 1908 as Chief Engineer of the newly-established power company that was to revolutionize the generation and distribution of electric power in South Africa. A man of great vision, drive and personality, Price was also a philanthropist with a particular interest in promoting scientific research. By his efforts the VFP had established itself as the most innovative and certainly the most successful power generating company in the country, so that by 1922 when the national body, the Electricity Supply Commission (ESCOM), came into being the VFP was producing the cheapest electricity in the world from coal-fired stations [4]. Its transmission lines criss-crossed the Transvaal from power stations near the coal fields to cities, towns and gold mines a hundred kilometres and more away, but almost everywhere those power lines were vulnerable to the lightning that was synonymous with summer in that part of the world. Acting with his usual decisiveness, Price threw his weight behind the Institute's programme to study and understand one of the world's most impressive natural phenomena. Only by gaining such knowledge could his industry hope to contain and mitigate its effects. He saw as his first task the provision of a solid financial base from which meaningful research could be launched and so set about canvassing support from those bodies most likely to benefit from its fruits. Within a relatively short period the Chamber of Mines, ESCOM, the VFP and the South African Association for the Advancement of Science had contributed £500 towards the initial costs of a lightning research programme [5]. All that remained was to appoint the right man to lead its activities and he, as a peripatetic visitor to Johannesburg, was occasionally to hand.

Basil Schonland, alternately in Cape Town and Johannesburg as determined by the seasonal lightning activity, was heavily involved in the investigation of a possible link between lightning and the penetrating radiation that was so readily detected in the laboratory. Both universities realized that he was a man of many talents. If UCT saw him in his all-round role as teacher, researcher and even as an administrator, Wits became aware of a man occasionally in their midst whose contributions to the advancement of science far exceeded those of anyone else in the University at that time. Schonland was therefore a prime candidate for a chair. His colleagues in Cape Town had already elected him as a representative of the Lecturers' Association on the University Senate [6] while Bernard Price, in Johannesburg, had the ear of the Principal of the University there. In England, others were also aware of Schonland's recent contributions to the science of atmospheric electricity and he had been invited to write a monograph on the subject, to follow one on 'Thermionic Vacuum Tubes' then being prepared by Edward Appleton [7].

At home, wherever that happened to be, Basil and Ismay took stock. As she had shown, Ismay was a willing collaborator and assistant in Basil's work and, though not scientifically trained herself, she relished the intellectual challenge it provided and gave him all the support she could [8]. His future would determine hers, but where that would be was less certain. Events elsewhere would play an important part in determining that, and none was more important than a meeting soon to take place in London.

*   *   *

In September 1931 the British Association for the Advancement of Science held its annual meeting in London, and Basil Schonland was a member of the South African delegation. The President of the Association in that, its centenary year, was none other than General J C Smuts, then Leader of the Opposition in the South African Parliament. Smuts was much revered in England both as a soldier and elder statesman, and his philosophy of Holism that linked the physical and metaphysical worlds had taken him into other courts as well. For the duration of the conference it was Smuts, the unaccustomed scientist, who dominated its proceedings, either when making numerous speeches or when presiding over its business. It was, though, his main address, delivered before an audience of five thousand of the world's leading scientific figures, that was so memorable. This was Smuts at his most majestic: an intellect that towered above mere mammon, even in the company in which he found himself that day, and in a capacity which he regarded as the 'crowning honour of his life' [9].

It was certainly an occasion for giants and by happy coincidence this major event in the British scientific calendar straddled the special commemorations that celebrated the lives of two of the world's greatest scientists: Michael Faraday, for his monumental discovery of electromagnetic induction a hundred years before, and James Clerk Maxwell, for its mathematical foundation, on the centenary of his birth. Lord Rutherford, who had himself been ennobled in the New Year's Honours that year, wrote eloquently to *The Times* about both men. Faraday was unrivalled as the greatest experimental scientist ever, while Maxwell, the mathematician and physicist, had exhibited a genius of almost breath-taking proportions when he used the discoveries of Faraday and others to show, by mathematics alone, that light was an electromagnetic phenomenon. 'Few if any men', wrote Rutherford, 'have accomplished so much for the advancement of knowledge or for the ultimate progress of the human race' [10]. So true and written by one whose own name could sit quite easily alongside theirs. Assembling around him now, in what were to be the twilight years of his own illustrious career, were so many of his former students. Amongst them was Basil Schonland, who had chosen not to remain at the Cavendish, as others had, but had returned to the Dominion that had sent him there.

Such decisions are immutable when looked at through the window of history and one cannot even speculate on what might have been had Schonland remained at the Cavendish and had been there when it reached its pinnacle early in that decade. In February 1932, the very next year, James Chadwick announced the discovery of the neutron. The existence of this new fundamental particle possessing no electric charge, that together with protons made up the atomic nucleus, had first been suggested by Rutherford in 1923. Its discovery earned Chadwick the Nobel Prize for Physics in 1935—a remarkably short period between discovery and award in the annals of the prize. That same year John Cockcroft and Ernest Walton, an Irishman working with him at the Cavendish, succeeded in splitting a lithium nucleus by means of a stream of protons that they had generated with a machine soon to bear their names as the Cockcroft–Walton accelerator. This was the first example ever of the artificial disintegration of the atom, and it was achieved with equipment on a much grander scale than Rutherford had ever dreamt of. For this work Cockcroft and Walton, too, became Nobel laureates but had to wait nineteen years for the honour. And there was to be so much more besides as Rutherford's 'boys' took physics by the scruff of the neck and, with it, the Cavendish to its apotheosis.

Many of these Cavendish men were now gathering for the meeting of the British Association as Schonland arrived to present a paper entitled, simply, 'Lightning'. He was nervous because he meant to use this rather special occasion not only to challenge what was very much accepted wisdom in some quarters but to overturn it. It will be remembered that G C Simpson, Director of the Meteorological Office, maintained with much steadfastness that the upper regions of a thundercloud were negatively charged while its lower region was positive. C T R Wilson had challenged that view as early as 1924, only to have Simpson rebut him most forcefully then, and he stood his ground now. Until someone could prove him wrong Simpson would not budge [11]. Now Schonland had arrived with that evidence.

On 25 September 1931 Basil Schonland addressed the assembled scientific community in London. The summary of his paper had previously informed them that he would give an account of 'recent investigations upon thunderstorms in South Africa, and their bearing upon the question of lightning discharges between the cloud and ground. Evidence will be presented which makes it difficult to accept the view that the branches in a lightning flash fork away from the positive pole of the discharge' [12]. We have Schonland's own account of the meeting and the reaction that his paper produced, especially from Simpson, for he wrote a long letter to Ismay just two days afterwards as he relaxed in some splendour at Cranmore Hall near Shepton Mallet, where he had journeyed to spend the weekend [13].

> The talk to the B. Ass. on lightning was a brilliant success. I did it
> very well indeed — the place was packed. And Simpson got the
> knock of his life. He was very decent about it. Said the whole thing
> would be looked into & if, as it seemed, I had succeeded in
> upsetting all work on thunderstorms during the last 30 years he
> would be the first to congratulate me.

Listening to Schonland read his paper, more out of solidarity with a friend than for any genuine interest in the subject, was Ted Allibone, with whom Basil and Ismay had spent many happy hours on the River Cam in 1928. But Allibone soon found himself enthralled by what Schonland had to say because the phenomenon of the lightning discharge suddenly appeared to him in a completely new guise. The similarity between lightning and the long sparks that he had been generating for some time in his Manchester laboratory suddenly occurred to him. What surprised him was to hear Schonland outline the Simpson–Wilson controversy about the polarity of thunderclouds and Simpson's views of the lightning discharge process. Simpson had been unequivocal over the years in maintaining that '... a lightning discharge cannot start at a negatively charged cloud but must start on the ground and branch upwards'. He followed this with the equally forthright statement that 'the branches (of lightning) are always directed away from the region of positive electricity and towards the region of negative electricity' [14]. During his talk Schonland had shown slides to illustrate Simpson's idea and Allibone had immediately become agitated, for he knew from hours of experiments with million-volt sparks in the laboratory that the discharge was just as likely to occur from the negative electrode as it was from the positive one, since it depended not on the sign of the charge but rather on the shape of the electrodes themselves.

During the discussion that followed Allibone was soon on his feet. He spoke up in favour of Schonland's assertion that Simpson's argument was wrong and, to stress the point, invited Schonland to visit him in Manchester where they could make an experiment which would convince all who doubted it [15]. As Schonland left the podium the chairman, Sir Frank Dyson, congratulated him on an excellent paper and said that, in his opinion, Simpson would have great difficulty countering the argument. It had indeed been a significant moment for the South African. His scientific credibility had been tested but was not found wanting. Basil Schonland had just overturned the prevailing model of thunder cloud electrification, as if in a stroke.

As always, a visit to England provided the opportunity to meet old colleagues and many new ones too. As he told Ismay, each day could have filled 72 hours. 'Everybody worth meeting is here and I have had long talks with people who know just the things my book or my cosmic rays or my thunderstorms or my sparks are concerned with'. He had planned a visit to Berlin to see Kolhorster, Bothe and Meitner but it fell through when he discovered that they were all elsewhere on scientific business of their own. Then he considered going across to Ireland to see the laboratory of P J Nolan, an old Cavendish colleague from 1920 who had also moved into the field of atmospheric electricity. Together, he and Nolan had spent a few hours at the BA meeting discussing the subject, and particularly Paddy Nolan's experiences with the Steinke ionization chamber that was soon to become the instrument used in a world-wide campaign of cosmic ray measurement. But that visit was not to be because the Simpson issue rumbled on and required Schonland's more immediate attention.

Dr F J W Whipple, superintendent of the Kew Observatory, called a meeting of all those with an interest in the mechanism of thunderstorms and invited both Schonland and Simpson to attend. Still smarting after he had seen his model of thunder-cloud electrification rather unceremoniously inverted by the South African, Simpson mounted a forthright attack, by querying Schonland's use of the capillary electrometer as an indicator of the polarity of electric fields. He offered an ingenious explanation of his own, but now Schonland stood firm and contended that Simpson was certainly wrong. Visibly on the back foot, but not defeated, Simpson countered that Schonland had told the BA that forks and branches had occurred from a negative pole in *acetylene* when everyone knew, he thundered, 'We are talking about branching in *air*'. Unruffled, Schonland coolly reminding him that 'Whether in acetylene or air, your theory says definitely *no* such branching ever!' To which Simpson gallantly conceded that his was merely a debating point made on the spur of the moment! There was not much more to be said, and Whipple felt the issue had been well aired and that Schonland's proposition seemed to have won the day. Though feeling decidedly 'top-dog', as he told Ismay, Schonland and Simpson parted on good terms and agreed to remain in touch [16].

Some years later, after much dedicated experimental work conducted at Kew using instrumented balloons, Simpson obtained deserved recognition for his major contribution to this field when he published his new model of the thundercloud [17]. It showed that Schonland had been correct — the negative charge certainly lay at the base of the cloud — but, almost in recompense for his unwavering belief that had been so shattered a few years before, he found that there was indeed a vestige of positive charge that lurked in a localized region at the cloud-base. Schonland was fulsome in his praise of this work and, in 1943, when heavily involved in other matters, acknowledged his old adversary in what was probably the definitive paper yet published on thunderstorms and their electrical effects [18].

*   *   *

As the papers at the conference had shown, there was a great deal of interesting work under way in England and, as promised, Schonland received an irresistible invitation from Robert Watson Watt to visit the Wireless Research Station at Ditton Park in Slough. The two men had met for the first time at the BA meeting, and this was now Schonland's chance to see a cathode ray tube in use in an apparatus designed to pin-point the direction from which 'atmospherics' originated. Evidence was mounting that this aptly-named electrical noise that interfered with wireless transmissions could be traced to lightning and so Schonland had more than a casual interest in the work at Slough. The visit proved fascinating, as was Watson Watt's company and, when he left, Schonland took with him the details of the Von Ardenne oscillograph* that Watson Watt was using with a view to obtaining one for himself. Soon he was to receive a letter from Watt recommending a superior model, though not yet in production. 'Let me know' wrote W-W, 'and we will supply you with a Cossor — likely to be better and cheaper' [19]. Also collaborating with Watson Watt in these pioneering experiments in wireless direction finding was Edward Appleton, whose own career had soared since the days in 1915 that he and Schonland, in their uniforms as Royal Engineer signals officers, had worked with wireless in its most basic form. Appleton was now at King's College, London, where he occupied the Wheatstone Chair of Physics. It was there that he had conducted the series of experiments with a young New Zealander, Miles Barnett, that led to the discovery of the ionospheric layer soon to bear Appleton's name and for which he would subsequently be awarded the Nobel Prize. Such, though, was the congestion in Schonland's diary of people to see and laboratories to visit before catching the boat home that

---

* 'Oscillograph' was the term used in England at the time to describe the device now known almost universally under its American name of 'oscilloscope'.

Appleton was just listed as one of the 'sundry people' whom he hoped to see [20]. Sundry indeed!

All that now remained was to travel north to Manchester to Allibone's laboratory in the Metro-Vick works at Trafford Park to make the experiment that would confirm Schonland's assertions about the polarity of the thundercloud. There they pressed Allibone's million-volt generator into service and carefully examined the array of long sparks it produced. It was well-known that the spark discharge between spherical-shaped electrodes produces numerous side-branches that fork away from the positively charged electrode and it was this phenomenon, coupled with his own experimental work on the charging of rain drops, that led Simpson to hold his view of the charge distribution within the thundercloud. Equally, it was known that when a pointed electrode was substituted for the positive pole, and an earthed plane for the negative, the branching always occurred from the point. No such branching ever took place from the negative electrode. However, these were ideal, laboratory configurations that had been used for many years in the study of electric spark and arc mechanisms, but they did not in any way represent the situation beneath a thundercloud where such symmetries were most unlikely. Rather, the projection of bushes, trees and buildings above the ground brought about significant distortion of the intense electric field below a charged cloud and, in addition, they all contributed their own quota of space charge to the region between cloud and ground. As the experiment would so clearly show, this was the key element that caused branching to take place from the negative electrode and hence, in the world at large, from the base of a cloud. To simulate these topographical and natural irregularities on the ground Allibone had driven a number of 5-cm nails into a board that, when connected to the positive terminal of the generator, served as the earthed anode of the system. The negative terminal, the cathode, was a pointed electrode some distance above. When energized with a million volts, he and Schonland photographed short positive streamers rising from the anode to be met by a negative discharge that branched away from the cathode—precisely the situation seen during a lightning storm. These results were no surprise to Allibone, for he had observed the phenomenon many times before and readily understood the process involved in its generation. Schonland, on the other hand, had only ever witnessed nature's massive equivalent of the laboratory generator in action and it intrigued him. Both men were convinced that the processes involved were the same. Together they wrote a letter to *Nature* describing the experiment and backed it up with evidence from Schonland's measurements made at Gardiol and Johannesburg, as well as photographs taken in the laboratory and in the field [21]. The question had surely been settled.

*   *   *

Feeling close to saturation but exhilarated after his hectic schedule, Schonland boarded the boat for Cape Town and home. Buoyed up by his recent scientific success, his thoughts turned to Ismay and his young family, now increased to three with the arrival of Ann, born earlier that year. Another house-move was planned as more room became necessary and, since university affairs were demanding more and more of his time, somewhere closer than the St James seashore made sense. Rondebosch, so reminiscent of the leafy villages of England, was both appealing and convenient and so yet another rented house became the Schonland family home for the foreseeable future.

Back at UCT he welcomed a new colleague, Stefan Meiring Naudé, recently returned to South Africa after a period at the University of Chicago where he had discovered the existence of the nitrogen 15 isotope. Naudé was a product of the University of Stellenbosch and, like Schonland, had displayed intellectual talents far beyond his years. Unlike Schonland though, he was awarded a Rhodes scholarship but, being unimpressed by physics at Oxford, declined it [22] and went instead to Germany where his scientific heroes were Planck, Einstein, Nernst and Von Laue, all Nobel laureates. After earning his doctorate *cum laude* in 1928 he crossed the Atlantic to work in Chicago with two other Nobel laureates, Millikan and Compton. With Naudé now on its staff UCT could surely boast that its was, by far, the strongest Department of Physics in the country. The question was, could Cape Town, for all its unrivalled beauty and serenity, keep two men whose scientific and personal ambitions had already seen wider horizons and whose personalities would demand the space?

The scientific challenge of lightning and all its ramifications had been brought into sharp focus by that 1931 visit of Schonland's to England. His encounter with Simpson had exposed him to the raw edge of scientific debate but he had come through it in some style and with his reputation considerably enhanced. His mental toughness had been tested and had certainly not been found wanting. But most satisfying of all was the challenge offered by the new experimental ideas and techniques he had encountered, amongst which were the wireless detection methods of Watson Watt and Appleton, and the idea of capturing on film the split-second phenomenon of the lightning flash. The initial stimulus for this had come from the scientific literature from which he discovered that C V Boys had developed a special rotating lens camera for this very purpose. Schonland had first met Boys during his Scholarship year at the Cavendish in 1928 and they renewed their acquaintanceship three years later at the British Association conference. It was there that the possibility was first broached of using the Boys camera in South Africa to try to photograph lightning.

The Boys camera is a remarkable device and, since it is central to Schonland's own scientific career, and to the very unravelling of the physics of the lightning stroke, it deserves more than a cursory mention. We will detour briefly from the Schonland discourse in order to examine the Boys masterpiece. Its conception was a fine example of the inquiring mind at work, whether in harness or, as in Boys's case, on holiday. As Lord Rayleigh related in the obituary notice that he wrote for the Royal Society following Boys's death in 1944, it was while on holiday at Bognor Regis in 1900 that it occurred to Charles Boys how the rapidly changing features of the lightning flash might be recorded on film [23]. After some sketches and calculations, done literally on the back of an envelope, he cut short his holiday and returned to London to build his camera. It consisted of two similar lenses mounted opposite one another on the circumference of a circular disc that was rotated by a hand-operated gear and pulley mechanism such that a changing event, like a lightning stroke, would produce two different and displaced images on a photographic film. Since the lenses rotated in opposite directions, successive events would be displaced on the film in opposite directions, but their common features would be obvious on both. By mounting the developed film with the two images side by side, and by measuring the relative distance between successive features, it is possible by simple arithmetic to determine the speed and distance travelled by the various components of a lightning flash, by knowing the speed of rotation and the focal length of the lenses.

For 26 years Boys carried his camera about with him without obtaining a single photograph of a lightning flash. He lent the instrument to Dr Simpson, but he too was singularly unsuccessful in his attempts to photograph lightning and so the camera was returned it to its inventor. Boys was clearly a man of placid temperament and was also a most generous individual, whose sole interest was scientific progress with no thought at all of turning his brilliant idea into financial gain. His persistence and philanthropy were finally rewarded in 1928 when he travelled to the United States as a guest of one Alfred L Loomis, a lawyer and businessman of considerable means, as well as an amateur scientist with connections at the very highest levels of academia and government. Loomis had built a laboratory atop a granite hill in a place called Tuxedo Park in New York State [24]. There, he played host to scientists of international renown whose work had captured his imagination, and Boys was one of them. At midnight on 5 August his patience of many years was finally rewarded when Boys captured the very first photograph of a lightning stroke. From his analysis of the two blurred images he concluded that the stroke consisted of two components, one that progressed downwards from the cloud and another that started at the ground. The two strokes met a mere 1/7000th of a second later. This remarkable

Figure 10. The Boys camera in operation at Gardiol with James Craib, Schonland's father-in-law and the family motor car. (Reproduced by kind permission of Dr Mary Davidson.)

result that showed, for the very first time, that lightning consisted of components that travelled from both cloud and ground might have been trumpeted from many rooftops, but Boys was not that type of man. Instead, when he published his finding in *Nature* a month later under the descriptive title of 'Progressive lightning', a term he had introduced two years before, he simply stated that he attached no importance to the numerical results and concluded: 'Information as to fact can do no harm'! [25]. No harm indeed. In fact, by this single photograph Boys had opened a window on an entirely new area of scientific research and Basil Schonland was the man who was poised to take it forward.

In July 1929 Boys once again wrote a letter to *Nature* informing the scientific community of his intention to have manufactured four sets of aluminium castings, as well as the necessary optical components for his latest design of camera. Recognizing the delays involved in manufacturing such precise instruments, he displayed his characteristic magnanimity towards his fellow scientists by publishing a full description of the camera so as 'to attract the interest of other experimenters, for the provision of extra castings and other gear is more advantageously made all at once than piecemeal'. Instead of using two rotating lenses, it employed two prisms to split the image and to deflect each through a right angle where, after focusing, they fell onto a rotating film — an ingenious idea. These components he would soon distribute to Simpson, Loomis and to Professor R W Wood, an optics specialist, in the United States and would keep one set for his own use. In order to test the new camera Boys decided that he could not sensibly wait for a chance lightning strike near his home in England, so he turned to none other than the Metro-Vick Company in Manchester, in the person of Brian Goodlet, to generate a suitable spark in the laboratory. The result, using a spark some 1.5 m long, showed that the optical performance of the camera was satisfactory, and so the first model was sent immediately to Loomis in the hope that he might obtain the long sought-after photographs of the lightning flash in all its detail [26].

By the time of the British Association conference in the summer of 1931 little if anything of value had come from the United States but Boys was still hopeful that one of the users of his camera might capture a definitive image. In December 1931, he wrote to Schonland and offered him the use of his camera but expressed some concern as to how to pack it for safe posting to Cape Town. Soon after, Schonland was able to arrange for a friend to collect it from Boys and bring it personally to Cape Town and it duly arrived in one piece early in 1932. Schonland had been insistent that he wished only to borrow the camera in order to make a copy of it, after which he would return the original to its most generous owner. Boys happily agreed and from then on progress was quite remarkable, as we will soon note. In July 1933 Boys wrote again to Schonland and expressed his delight at the quality of the photographs of lightning he had just received with Schonland's first detailed paper on the subject entitled 'Progressive Lightning' and he agreed 'with the greatest of pleasure' to communicate it to the Royal Society. Schonland's recent demolition of Simpson's ideas about cloud polarity had apparently not been entirely forgotten because Boys expressed some concern that the paper might be sent to Simpson for refereeing and '(Simpson) has shown a powerful anti-interest'! But the paper duly appeared in the *Proceedings of the Royal Society* (vol 143, 1934) so, 'anti-interest' or not, its significance had been noted and for the first time the intricacies of the lighting stroke were revealed in all their detail. Whatever Simpson's views of the matter may have been, he bore no grudges and many years later, in 1951, was extremely magnanimous in a letter to Schonland in which he told him that 'you have made much more significant findings about atmospheric electricity than I have' [27].

To return to developments in South Africa, we find that by December 1932 construction of Schonland's own version of the Boys camera was well under way in Johannesburg. He had borrowed Boys's camera just long enough to have detailed drawings made of it in the Department of Physics at Wits, where the recently capped Dr Eric Halliday took direct charge of the process of constructing one for use during the next lightning season [28]. But while this was happening in Johannesburg, life at UCT, and certainly in the Schonland home, revolved around Basil's book, his first. It duly appeared in 1932 under the apt title of *Atmospheric Electricity* [29]. When read from the perspective of the present day it is an historical gem containing the considered scientific opinions of the time about the mechanisms and processes of all known electrical phenomena within the lower atmosphere. While displaying the rigour of the physicist, the book also bears the mark of a scientist for whom the ways of the natural world were much more than some ordered process described simply as a catalogue of numerical events. By his own almost infectious enthusiasm for the wonders of world around him, Schonland succeeded admirably in capturing not just data, but also the imagination of his readers. The search by many scientists for the source of the mysterious 'penetrating radiation' naturally received much attention, though his own work to trace a possible link with thunderstorms is just hinted at tantalizingly. He then took his readers into the inner workings of the thundercloud as seen by Simpson, Wilson and others but he was careful to ensure that it was the science and not the personalities that mattered. 'It is not impossible' he wrote, possibly hedging his bets, 'that both the Simpson and the Wilson mechanism may be at work in a thundercloud at one and the same time'. One can only conclude that he phrased this so diplomatically before he demolished the Simpson theory so convincingly at the meeting in London just a few months before!

The monograph concluded with just a few pages on lightning, but very significant they were to be for it was obvious that Schonland saw that the way ahead would be determined by the ability to characterize the lightning stroke in all its complexity. A photograph taken in Germany with a moving camera had already provided startling evidence of some of it, while Boys had published a photograph of the only flash of lightning he had recorded. Others, of Schonland's acquaintance, such as Appleton and Watson Watt, had received much lower frequency electromagnetic emissions on their radio direction-finding apparatus and had concluded that these 'atmospherics' were intimately associated with the lightning process. They suggested that the lightning channel behaved as a 'gigantic wireless aerial 3 km high' with a natural wavelength of 12 km, and their measurements had confirmed this. Other recent work had shown that lightning, while seemingly so ephemeral, was much more than a single flash but was, in fact, a series of current surges that reached immense values as they flowed through a tortuous channel in the sky. So much awaited discovery and here, in the southern reaches of Africa, on Schonland's doorstep, was nature's own laboratory.

*   *   *

In December 1930 the South African Institute of Electrical Engineers (the SAIEE), at the urging of Bernard Price, established its Lightning Investigation Committee. At the Institute's annual general meeting two years later Price drew attention to Schonland's work in Cape Town and at Wits, and on his recommendation Basil Schonland was invited to join the Committee and was immediately made Chairman of its Research Sub-Committee [30]. Soon after, Schonland himself addressed a joint meeting of the Institute and the University in Johannesburg, where he gave a masterly description of all that was known at the time of the development of the lightning discharge and told his audience briefly of the research then under way in the Physics department. Bernard Price in his closing remarks describing the address as 'one we shall not readily forget for it marks a notable advance in the field of lightning research [and] will, I feel sure, attract wide attention in scientific circles, not only in this country, but in Europe and America where so many efforts are being made to unravel the mysteries of the lightning flash' [31].

The Boys camera recently constructed under Halliday's supervision at Wits was driven by an electric motor instead of being cranked by hand, as was the original. In all other respects, though, it was a replica of the Boys device. It was first used by Halliday to photograph lightning from the roof of the Wits Physics block during the lightning season of 1931–1932. The results were published early in 1933, and though to the untrained eye they might have appeared as a conglomeration of whirling blurs, they were, in fact, the first group of photographs that showed the details of the lightning flash from instant to instant as it streaked across the sky within a blink of an eye. By making use of an accurate travelling microscope, Halliday was able to measure the step-by-step progress of the flash. He confirmed an observation first made by Boys himself that it is possible for a lightning discharge to start at the ground and to join up with one that started at the cloud. He also found that successive flashes could traverse the same path, and that forking or branching of the flash could take place both near a cloud or near the ground. Just as Schonland had done some years before, when he brought Ismay to assist him in the reading of instruments and the logging of results, so Halliday acknowledged the vital part played by his wife, for it was she who developed and printed the photographs that, for the very first time, made possible such detailed scrutiny of lightning [32].

Professor Boys was delighted by the publication of the South African results and said so in his letter to *Nature* that followed soon after Halliday's paper had appeared [33]. After so many frustrating years when lightning always eluded him he now saw the pictures that he knew would begin to tell the story that he for so long had attempted to do. Schonland, too, was convinced that this new tool was the means by which lightning could be dissected to reveal in intimate detail the processes that went into making up what to the eye was merely a transient flash of light. To do so, however, required not just a suitable camera but also trained and enthusiastic people to use it. He was in no position

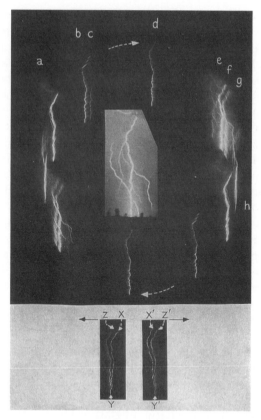

Figure 11. Photographs taken with the Boys camera showing the progressive development of the lightning discharge and a fixed-camera photograph of the same flash for comparison. It should be noted that the strokes are not shown in their correct sequence because of the speed of rotation of the lenses in relation to the duration of the flash. Stroke f was actually the first of the sequence. A subsidiary camera with a single, slow-moving lens determined the actual order. (Reproduced from Schonland's *The Flight of Thunderbolts*, plate IV, 1950 by kind permission of Oxford University Press.)

to play an active part himself, being in Cape Town, and while Eric Halliday had done an excellent job of all the groundwork his own immediate ambition was to follow in Schonland's footsteps and go to the Cavendish as an 1851 Exhibitioner. At hand, though, was an engineer by the name of Collens who worked for Bernard Price at the VFP. Collens was a man with some unusual skills. He had an established reputation both as a shark-fisherman and as a hunter of crocodiles, which he insisted on capturing alive! While neither was necessarily an attribute for capturing lightning on film, it was soon apparent that he was a man of imagination and flair, who thrived on a challenge and for whom inclement weather was no disincentive. Price, knowing of Collens's talents, readily allowed him to devote his time to chasing thunderstorms across the Reef—that spine of rocky hillside that bisected Johannesburg from east to west and which was so synonymous with the gold fields dotted all around the Witwatersrand. And so began the campaign that was to produce the most detailed scientific information yet gleaned about lightning and which was to establish Basil Schonland as the undisputed leader in this field of investigation.

From Halliday's work it was clear that any concerted effort at recording details of lightning would necessitate equipping a special vehicle with a Boys camera, so that storms could be followed and photographed well away from the disturbing effects of city lights. It would also help get the camera as close as possible to the lightning activity to reduce the obscuring effects of the torrential rain and hail that are both common features of a Transvaal thunderstorm. With the financial assistance of the SAIEE's Lightning Committee, three new Boys cameras were constructed, but without the electric motor drive because they were now to be used essentially as portable instruments. A hand-operated crank, just as Boys himself had used, allowed the lenses to be rotated at a roughly constant speed of 1500 rpm by keeping just below the natural, resonant vibration of the camera body—crude but undoubtedly simple and effective. An important addition was the inclusion of a small fixed camera rigidly attached to the same stand as the Boys camera. It served both as a viewfinder and as an important adjunct in identifying the various components of the flash. Over the next four years this combination, in the hands of Collens and those who subsequently joined him, produced about 150 good pictures of lightning flashes to ground which showed around 600 separate lightning strokes [34]. Schonland's analysis of these was quite brilliant, both for its meticulousness and for the accuracy of his measurements of the filamentary traces of light on the photographs. Of even more importance was his ability to recognize and interpret the multitude of features contained in the pictures, from which he was then able to describe the various stages of the flash.

During the early months of 1933 Collens managed to photograph eleven lightning discharges. It had long been believed that the characteristic flickering of a lightning stroke indicated that there was not just one but a number of flashes occurring in rapid succession. The number of strokes, as they were called, varied from just one in some rare cases to as many as twelve in others, but each stroke was made up of two fundamentally important processes: the first was the very faint and comparatively slow downward-moving *leader stroke* which, on striking the ground, led to the second, a much more intense and rapid upward *main stroke*, both bearing the names that Schonland gave them when the results were submitted for publication later that year. He described the leader as being dart-like, while the main stroke resembled an intense flame as it surged back up the channel carved for it by the tentative stab-bing of the leader. He also discovered that the leader occasionally mean-dered from its main track and set up branches, which suddenly died, as if hitting dead-ends in the path. The highly energetic main stroke charged along these, and then it too petered out as if extinguished in some mighty celestial sea, leaving the rest to surge up to the cloud. The amount of infor-mation contained in each photograph was immense. Schonland was soon able to deduce the mean velocities of both the leader and the main strokes, and though subsequent measurements would modify the actual values he obtained, their relative speeds were highly accurate. Finally, and no doubt with quite some satisfaction, he was able to announce that his measure-ments had shown that the 'polarity of the discharges was such as to make the cloud-base negative and the earth positive' [35].

Bernard Price had been quite correct: the world was watching, or at least N Ernest Dorsey of the United States Bureau of Standards in Washington was. It was Dorsey who had first proposed in 1926 that such a dart-like leader was, in fact, an electron avalanche that was a necessary preliminary to the breakdown of the air within the lightning channel. Where he and Schonland differed, though, was that Dorsey had believed that the leader constituted the complete lightning stroke until Schonland showed that it was merely the first part of the process. Dorsey wrote to Schonland in September 1933 to congratulate him on his most recent publication that had appeared in *Nature* [36] earlier in the month. Schonland and Collens had used a pair of photographs to illustrate this two-step process with the utmost clarity. Dorsey enthused: 'Your contribution is undoubtedly one of the most important in many a day' [37]. It was just two years since Schonland had returned from England with the knowledge of the Boys camera that had been so willingly pressed on him by its inventor. In that short space of time the happy coincidence of an ideal geographical location from which to view lightning, and the dedication of two men, Halliday and Collens, had produced more useful scientific information on the physics of the lightning stroke than had ever been gathered before.

*   *   *

Schonland's enthusiasm for his research and the international attention that it was attracting were infectious. He was soon joined by another of his former students and academic colleagues, D J Malan, who had recently returned from France. Dawie Malan was to become a key member of Schonland's team and would go on to make his own mark in the annals of lightning research. He was appointed a lecturer in Physics at Wits and so provided the first permanent link between the University and the lightning programme. But Schonland was the real prize. His frequent presence at Wits and his burgeoning reputation had not gone unnoticed, especially in the office of the University's Principal, Humphrey Raikes.

In 1934 Schonland was awarded a Travelling Fellowship from the Carnegie Institute in Washington and UCT granted him a year's sabbatical to spread his wings by visiting and working in research laboratories in three countries: the United States, England and Sweden [38]. The Rutherford dictum that scientists from the Dominions must continually be exposed to the frontiers of their subjects was partly behind it, but so was the opportunity to seek financial support for his research from further afield, and these latest results suggested that there was no better time to do so. During the Great Depression from 1931 to 1933 the meagre resources of the South African Research Grant Board had been bolstered by a small grant from the Carnegie foundation [39] and Schonland himself had benefitted from it for his cosmic ray research. Now there was a need for support on a much grander scale than the £75 he required then for the building of a hut with a thatched roof and wooden walls to house his cosmic ray recorder in Cape Town [40].

The Governor-General's prediction, made just a few years before, that a lightning observatory would soon be established in South Africa, was fast approaching reality. Bernard Price and Schonland were certainly working towards it as Basil and Ismay set sail for the States in 1934. Their three children went to Gardiol where they were under the expansive wing of their grandmother and the eye of their grandfather, James Craib. On arriving in the United States, the Schonlands established themselves in New York, where the Carnegie Corporation had its headquarters. Basil also visited the Broad Branch Road Laboratories of the Carnegie Institute in Washington, the Smithsonian Institution and the General Electric Company where he examined the photographs of lightning recently taken with a Boys camera by one of its engineers, K B McEachron. However, the main purpose of his mission was to see the Director of the Carnegie Corporation and so, in June, he made an appointment with Dr Frederick Keppel himself. Keppel was greatly interested in Schonland's work and invited him to outline his plans for a geophysical research institute in Johannesburg. Prime amongst these was the fundamental nature of the research to be conducted there. Schonland stressed that though industrial applications could well follow from his research, with benefits for South African industry and its population at large, scientific work must be allowed to proceed 'unhampered by the demands of commercial life'. This was a view that Schonland had long held and in stating it he knew that he had the full support of Bernard Price in Johannesburg. For a man of such practical inclination and commercial acumen, Price believed passionately in the complementary roles of science and engineering and would willingly support investigations which had no conceivable practical purpose if convinced by their scientific merit.

Keppel was a man from the same mould and was impressed by Schonland's forthright proposal. In the terms of Andrew Carnegie's will, grants could be made to benefit the people of Britain, her Dominions and Colonies, and had already been made to Australia and Canada for the support of organizations with similar objectives to those Schonland had just described. Keppel was also impressed by the young man seated in front of him. As their meeting was drawing to a close he looked Schonland in the eye and asked him what he would do if his request for support was turned down. 'I'll do it some other way' was Schonland's immediate reply. That was exactly what Keppel was hoping to hear for, as Schonland related many years later, Keppel was not given to dispensing the Carnegie bequest on a whim and certainly not to someone whose objectives were ill-defined or who lacked the necessary will to fight for a good cause [41]. Schonland wrote immediately to Humphrey Raikes at Wits with the news that Dr Keppel had agreed to view favourably his request for £1000 per year for ten years. It was Schonland's own suggestion that it be limited to ten years because, he believed, if the Institute was not self-funding after that time it should be abandoned. However, the Carnegie grant would not, on its own, be sufficient and so he asked Raikes if Wits itself would agree to contribute an amount of £500 per annum for the same period. Without Schonland's complete involvement such a venture would have been pointless and certainly risky and so he added, by way of a small inducement, 'I should naturally consider moving to a more permanent position when such offered itself' [42].

From the United States, the Schonlands travelled to England and specifically to Cambridge where they renewed many old ties. It was Basil's intention to spend as much time as he could at the Cavendish working on some of the more pressing problems that he had brought with him from the other side of the world, before he attended, as the South African delegate, the General Assembly of the International Scientific Radio Union (URSI) due to take place in London late in September. Before that he planned to visit the Institute of High Voltage Research in Sweden where Harald Norinder was doing pioneering research on measuring atmospherics with the cathode ray oscillograph.

The Cavendish Laboratory was at its peak in 1934. It numbered no fewer than seven Nobel laureates, or those soon be so honoured, amongst its staff and research students. Lord Rutherford was still in charge, of that there was no doubt, with Sir J J Thomson at his elbow, while C T R Wilson was just about to retire from the Jacksonian Chair and return to his native Scotland. George Searle, the scourge of the slipshod in the laboratory, was still an imposing presence, while James Chadwick, essentially Rutherford's right-hand man, was as reserved and punctilious as ever. Amongst the younger men were John Cockcroft, behind whose quiet, enigmatic demeanour lay a flair for organization and a fascination for the 'big machine'. Also there were J A Ratcliffe, a protégé of Appleton and now acknowledged as the Cavendish's expert on radiowave propagation, and Mark Oliphant, ebullient and ambitious, whose own research star was rising and whose path, like that of so many others of that Cavendish generation, would cross Schonland's again.

Schonland basked in the scientific sun during those few months at Cambridge. In a letter to Appleton at King's College, London, he wrote: 'I enjoyed the States very much and found the people very pleasant and kind. At the moment I am doing several things at once—lightning photographs, cosmic-ray tapes, thunderstorm fields—and at the same time have some writing and reading to get through. When I feel bored with it I go and learn something in the Cavendish. It is a very pleasant way of living' [43]. Schonland and Appleton continued to share many common research interests, and previous correspondence between them had touched on the mechanisms at work within the lightning stroke, particularly the atomic processes responsible for the emission of light. It appeared to Schonland that there were three possibilities: excited atoms, metastable atoms or some process of atomic recombination, all of which might produce the intense optical effects so typical of a lightning strike. None had as yet been confirmed and much work remained to be done. Appleton's name was now commemorated in the layer in the ionosphere[*] that bore his name and ionospheric research now occupied most of his time. However, he still retained an interest in the atmospherics that he and Watson Watt had been tracking by radio from Slough, even though the relationship between the two men became decidedly ragged from time to time, as two powerful personalities were forging careers and reputations of their own. What interested Schonland was their use of two mutually perpendicular loop antennae, suitably connected to the horizontal and vertical deflection plates of a cathode ray oscilloscope to display the instantaneous direction of a flash. This was clearly another avenue of research well worth pursuing.

In recommending Eric Halliday to the Commissioners of the Exhibition of 1851, Schonland had a keen eye on those plans he himself had for Wits. He convinced Halliday that much was to be gained by dividing his scholarship period between the Cavendish and King's, and this Halliday duly did by spending 1933 at Cambridge and the next year with Appleton in London. At the Cavendish, under the watchful eye of C T R Wilson, he developed a cloud chamber that was triggered by a lightning flash: a

[*] It was Watson Watt in 1926 who first used the term 'ionosphere' to describe the region above the earth in which 'the main characteristic is large-scale ionisation with considerable mean free paths . . .' [Source: *Biog. Mem. of Fellows of the Royal Society* 1975, **21**, 549–568, written by J A Ratcliffe.]

technique originally devised by Patrick Blackett for another purpose. He then used this apparatus in an attempt to detect the intriguing 'penetrating radiation' first predicted by Wilson and so actively sought by Schonland. Halliday's published conclusions, after some months of measurements during an English lightning season somewhat less violent than he was used to on the Reef, were optimistic. 'The results obtained', he wrote, 'though not conclusive, encourage a continuation of the experiments' and these he planned to do on his return home [44]. The following year he spent working with Appleton on the measurement of the interference between a directly propagating radio wave and one that was reflected from the ionosphere: valuable experience indeed for the work soon to follow in Johannesburg.

The science of radio had, by the 1930s, made enormous strides since those days just twenty years before, when Schonland and his fellow army signallers had used spark transmitters and coherer receivers to communicate from the trenches of Flanders. The thermionic valve was very much in its infancy when the war ended but its potential was obvious, as Schonland and Humby had themselves shown by their small contribution in 1919 to the rapidly increasing store of knowledge of the subject. In England at that time research into wireless was rapidly becoming a subject of great importance, and a major contribution to the understanding of the characteristics and behaviour of the valve had been made by Appleton himself working in close collaboration with a Dutchman by the name of Balth van der Pol. Appleton soon left this field of study to others as his name became inextricably linked with the ionosphere. However, it was his occasionally tempestuous collaboration with Watson Watt on the subject of direction-finding by radio that was profoundly to influence the lives of all of them, Schonland's included, in less than a decade.

In September 1934, Appleton was the chairman of Commission III (Atmospherics) of the URSI General Assembly in London. Much of its deliberations were procedural, involving the setting up of sub-commissions charged with the detailed work on the specialized topics to be discussed over the following week. He appointed Watson Watt to chair the one on the propagation of atmospherics, as well as that on determining their direction of arrival. Schonland was asked to join the group under the chairmanship of Professor Norinder that was looking into the origin of atmospherics. As the conference progressed it became apparent that the world-wide distribution of these effects required the setting up of recording stations on a much wider basis than had been the case before. The northern hemisphere was well-served but the southern hemisphere certainly was not, and so to redress this imbalance two specific resolutions were tabled and immediately accepted. In the light of what was soon to come they bear repeating [45]:

(i)   The Commission, having reviewed the distribution throughout the world of the stations at which fundamental investigations on atmospherics are now in progress, recommends the extension of such work and its initiation in Africa and South America as of special importance.

(ii)   The Commission is of the opinion that full knowledge of the structure of atmospherics originating from thunderstorms and other sources would be of extreme value for their physical interpretation and would, therefore, greatly advance science as well as serve practical purposes.

It was now becoming clear to Schonland that the research institute slowly taking shape, at least in his mind, would have an even wider role to play than he had at first envisaged, for the massive continent of Africa, except for the Union at its southern reaches, remained unexplored by science and scientists. Now he was to take home with him an invitation, stamped with the authority of one of the founding arms of the International Council of Scientific Unions, to do something about it.

Before leaving he paid an almost obligatory visit to Manchester to see Ted Allibone again and to participate in another joint experiment in the high-voltage laboratory of Metro-Vick. Allibone had been following closely Schonland's recent work with the Boys camera and had designed one himself. In view of the much shorter spark produced in the laboratory, Allibone's camera rotated at the rapid rate of 40 m/s in order to capture as much detail of the arc discharge as possible. In addition, he had installed a special wide-aperture lens just before Schonland arrived and together, with much anticipation, they set about photographing a one-million volt, 1.2 m long, flash. As the prints emerged from the developer it was clear that they had evidence of the same leader stroke that Schonland and his colleagues had photographed many times with lightning in South Africa. A tenuous, faint path was seen to snake away from the upper electrode as it made its way towards the earthed ground plane, to be followed by a much more intense, and more rapid, return stroke that itself seemed to follow the track already blazed by the leader, even to the extent of diverting for just an instant down the occasional branch on its way. This was exactly the process of lightning and was an enormously satisfying result which they wrote up and submitted immediately to *Nature* with a note of thanks to the SAIEE for the loan of one of its cameras, no doubt brought over by Schonland, that had guided Allibone in his work [46].

CHAPTER 9

TWO CHAIRS AND THE BPI

Alexander Ogg retired as Professor of Physics at UCT at the end of July 1936 to become Director of the new Magnetic Observatory at Hermanus, about 100 km from Cape Town and well away from all sources of the electrical interference that had made sensitive magnetic measurements there well-nigh impossible [1]. The selection of his successor had exercised the mind of Sir Carruthers Beattie, the Principal of the University, for some while before that. As early as February 1933 he had drawn the attention of the University Council to the particular talents and skills which Basil Schonland had displayed both as a lecturer and researcher, and as an administrator. Beattie informed the Council that 'by character, natural endowments and experience' Schonland had all the attributes of a Head of Department and that he unhesitatingly supported him 'in the belief that he will do justice to his opportunities' [2]. The first of these opportunities was no doubt Ogg's impending retirement two years hence and Schonland had read the situation well in advance. Whether prompted by Beattie or unsolicited we will never know, but he evidently had made his intentions clear to apply for the Chair, when the time came for Ogg to vacate it and the Principal was only too willing to support him.

Since it was in the University's interests that its Professor-elect of Physics should be as up-to-date as possible with developments in his field, no obstacles were placed in Schonland's way when he applied for a year of sabbatical leave in 1934. The appointment of the new professor certainly exercised the Council's collective mind during that period, for on Schonland's return, feeling scientifically much invigorated, he found that events at the University had indeed moved forward quickly, if slightly chaotically. At its meeting on 25 June 1935 the University Council agreed to offer him the Chair of Physics from the beginning of February 1936 and Schonland duly accepted three days later. Then, when it met again at the end of July, having been informed that Professor Ogg was only due to retire at the end of July the following year, instead of six months earlier as had been erroneously presumed, Council took the only decision it could by offering Schonland an Acting Professorship until Ogg had well and truly left. By way of compensation for the embarrassment caused it also promised him a temporary, non-pensionable allowance of £150 until he assumed full responsibility for the Department at £900 per annum, fully pensionable. Once again Schonland indicated his acceptance and turned his attention to more pressing issues closer to hand [3].

As much as his eye was obviously on the opportunities for personal advancement, so was the Schonland presence much in evidence at the forefront of South African scientific activity during this period. In 1935 he served as President of Section A of the South African Association for the Advancement of Science and there was much satisfaction at his election as a Fellow of the Physical Society of London that same year [4]. For his Presidential Address to the S_2A_3, delivered on 2 July in the wine-land town of Paarl near Cape Town, he chose, unsurprisingly, the topic of 'Atmospherics and Lightning'. His intention, though, was not just to recall the advances and achievements of the past but also to indulge in some prophesy, taking his cue from previous Section Presidents, whose abilities as soothsayers were seemingly enhanced on such important occasions. That South Africa now had an accurate map of the geomagnetic field variation across the country had been foreseen by Sir David Gill, Her Majesty's Astronomer at the Cape, at just such a meeting in the closing years of the previous century while, in 1926, the need for dedicated study of the electrical phenomena in the atmosphere, and lightning in particular, had been foreseen by P G Gundry, Professor of Physics at the University of Pretoria. Now, a decade later, Schonland's presence suggested that Gundry, too, had had the magic touch. The study of atmospheric electricity had indeed advanced in veritable leaps and bounds.

In many ways this meeting of scientists in the little town that was the pearl of the western Cape, was a significant moment in the scientific evolution of South Africa. It was the first occasion when Basil Schonland spoke not just of lightning but of his vision for the advancement of science in the country and how that might be achieved. Though viewing the scene through the eyes of a physicist, his conclusions covered the full reach of the scientific spectrum and he laid down some important ground rules. There were, he said, 'three points which must lie at the roots of any successful programme of scientific research in South Africa'. The first was the need to divorce the research from any hint of application: pure science for its own sake was vital. If pure research had any worth at all its applicability to the problems of mankind would follow naturally, but a basic understanding of the forces and mysteries of nature never came from a utilitarian route. Second was the fundamentally important place that geophysical research had in South African science. The country was blessed, indeed it was second to none, as a natural laboratory both above and below its primeval surface and he saw in such riches some compensation for the country's lack of resources 'in men with time for research work and in equipment for carrying it out'. His third point knew no boundaries imposed by scientific discipline, for it maintained simply that the key to all progress lay in the ability to collaborate with others working on related problems. Cooperative investigation, in which limited resources can be directed to yield important results was, he told his audience, vital to the success of any scientific venture. An attack on a broad front offered the best chance of success because the whole was certainly greater than the sum of its parts. The resolutions just passed by the General Assembly of URSI the previous year, which called for just such a concerted effort in order to understand 'the problem of the radio atmospheric', were a case in point. South Africa's unique position at the very focus of lightning activity should enable her scientists to play a hugely significant part in finding the answers. Who better than themselves to attack the problem with the optical armament of the Boys camera, reinforced by the wireless direction-finding methods of Appleton and Watson Watt?

There was much prescience in all this but little did anyone realize then quite how significant those words of Schonland's had been. In little more than a decade Schonland himself would be at the helm of South African science after a world war in which both he and science itself had played massive parts in the fight for freedom. But in 1935 there was an air of optimism about, for economically, at least, the world seemed to be righting itself after the greatest of all slumps and most people had not yet heard of Hitler.

Shortly after returning from England, Schonland had set one of his post-graduate students, D B Hodges, the task of constructing a wireless direction-finder based on the latest ideas that he had seen at the Radio Research Station at Slough. As ever, the skills of Mr Linton in the Physics workshops at UCT were called upon to great effect, while the most important source of technical information was a book that Schonland had brought back with him from England. *Applications of The Cathode Ray Oscillograph in Radio Research* [6] was written primarily by Watson Watt, in his characteristically verbose style, with the assistance of two of his colleagues. It had appeared in 1933 and was, without doubt, the most comprehensive reference on that subject yet published. Little did any of them, authors or readers, realize then that the engineering information contained within those pages was soon to become of massive importance in an area of endeavour that would consume all their time, energy and intellectual effort when the world descended, yet again, into the abyss of war. Now it was scientific research, initially for its own sake, but with practical applications abounding, that was the all-consuming passion of Watson Watt's British team and other small groups of like-minded physicists around the world. Wireless was now becoming very much part of Schonland's life once more, and his stature in the field was soon to be recognized by his appointment, in 1936, as the Chairman of the National Committee on Scientific Radio-Telegraphy in South Africa [7].

Progress had indeed been good, for at that Paarl meeting of the S_2A_3 David Hodges read a paper describing the radio direction-finder and its method of operation, and he also demonstrated the equipment in use [8]. The sensing elements were two 1-m square loop antennas, each containing 100 turns of wire that were brought to resonance at a frequency of 10.86 kc/sec* by suitable condensers (or capacitors, as we know them today). In operation, the loops were very carefully positioned at 90 degrees to one another and aligned, equally carefully, with one parallel to the magnetic meridian. An electromagnetic wave, such as that emitted by a lightning flash, would induce currents in the two loops in such a way that after amplification by two identical four-valve amplifiers, the resulting trace displayed on the oscillograph screen was a line, whose direction coincided with that of the lightning stroke relative to the location and orientation of the direction-finder. Since only one direction-finding or DF station was taking the bearing it was not possible by this method to determine, unambiguously, whether the lightning was, say, NE or SW of the station concerned: this would require either an additional sensing circuit using an auxiliary antenna, which Hodges described, or the use of at least one other station to simultaneously take a bearing on the same source. By plotting the two headings on a map their point of intersection would indicate the position of the lightning flash.

The choice of the 10 kHz frequency on which to monitor the lightning activity stemmed directly from Schonland's observations of the characteristic wave-form of lightning. All users of wireless receivers at that time were well aware of the so-called 'static' crashes that, at certain times of the year, almost obliterated signals in the long- and medium-wave broadcast bands. Precise measurements had shown just how the intensity of the noise was related to the frequency to which the receiver was tuned and it was evident that the noise level was greatest at the lower frequencies. In fact, Schonland was able to be even more specific because his photographs of the leader stroke of lightning had shown it to be made up of a series of very rapid steps as it advanced across the sky. Had it been just a continuous process lasting a few milliseconds, as it seemed to the unaided eye, then any radio emission would have occurred at an extremely low frequency — too low and too weak to have been detected over the sort of distances they had already measured. It was, though, this rapid but staccato advance that determined the dominant frequencies occupied by the energy. 'Our measurements' he stated

---

* kc/sec (kilocycles per second) was the unit of frequency in use at that time. It has now been replaced by kHz (kilohertz).

'indicate that this wavelength should lie between 10 and 30 kilometres' thus the noise lay in a broad spectrum that peaked in the range between 30 and 10 kHz. However, not only did the leader stroke exhibit this stepped characteristic, but the considerably more powerful return stroke had recently been shown in a remarkable series of photographs, taken with the Boys camera near Johannesburg by Dawie Malan, to do so as well [9]. Such an irregular progression of thousands of amperes of current flowing along what could readily be conceived to be an ethereal antenna, 10 km or more in height, would certainly be capable of producing all the 'splashes' heard on his receiver by Watson Watt in England or the 'grinders' and 'roars' as they were described in Australia [10]. Some, though, were not convinced that lightning could be held responsible for all the noise that so plagued low-frequency radio reception and this, as Schonland had told his audience in Paarl, was what made it such a fruitful area for fundamental research. Might there be other causes as well? The recent scientific literature contained a report of work in the United States by a man named Karl Jansky which suggested that there may well also be sources of noise that were of 'interstellar origin' [11]. Not only was this observation of Jansky's well ahead of its time but Schonland's recollection of having read it was to have equally remarkable consequences within the decade.

Moving house had almost become an accepted part of the Schonland family routine and so it was that they moved yet again. This time, though, they would not be renting but would move into University accommodation, because Ismay had been appointed Head of the Women's Residence at UCT from 18 July 1935 and the post brought with it a suitable family abode in the university grounds. There were obvious advantages associated with such proximity to Basil's Departmental office, even though his research laboratory now lay nowhere near the University at all but rather in the skies above, with its epicentre a thousand miles away in Johannesburg. Schonland's absences from the family home, wherever it was, were essentially confined to the summer months, while the rest of year saw him heavily committed to his lecturing and other academic duties at UCT, and the almost continual writing of the scientific papers generated by his research. The quality of their family life was crucially important to both Basil and Ismay and, to ensure that her new role in no way conflicted with it, she asked the University to provide her with an office away from her home, so that she could clearly separate her duties one from the other. She also staked out the dates of the annual family holiday by stipulating that she wished to be granted her leave, of a month, from 20 December each year — lightning in Johannesburg notwithstanding — so that she, at least, could spend it with their children [12].

As the onset of that lightning season drew near, so Schonland and Hodges prepared their DF equipment to measure its effects in Cape

Town. Since theirs was the only monitoring station no cross-bearings could be taken and so it would not be possible to fix the position of the source of the atmospherics. However, the location of UCT, almost at the tip of the continent with only the southern ocean between them and the lightning-free Antarctic continent, meant that there would be no ambiguity in the bearings they obtained: all sources must lie northwards of Cape Town. In addition, the amplitude of the received signals would give some indication of their relative distances since, it was assumed, all lightning sources would be of roughly equal intensity. This work formed the basis of Hodges's PhD research carried out under Schonland's enthusiastic supervision, for it was the perfect example of a dedicated scientific programme aimed specifically at understanding a natural phenomenon with which South Africa was so abundantly blessed. It was soon evident, though, that it also could provide some decidedly practical information that would be of particular interest to meteorologists.

Observations commenced on 3 September 1935, with the DF apparatus set up at UCT and were made on 26 days until mid-October. With the prevalence of thunder and lightning in the afternoons, data were recorded generally between 2 p.m. and 5 p.m., but measurements were also taken on the occasional morning, and at night, in order to gather comparative information of atmospheric activity at those times. To compare the occurrence of atmospherics with known meteorological conditions they sought the collaboration of the Chief Meteorologist at the Irrigation Department in Pretoria and that of his counterpart in Southern Rhodesia,* even farther afield, who agreed to supply them with records of prevailing weather conditions as these came to hand from the few out-lying weather stations on the continent of Africa. It was, though, the very paucity of such information in such a vast country with a sparse population that suggested an immediate practical application for this rather exotic method of weather prediction. Might it be possible, Schonland speculated, to use wireless direction-finding not only to alert the pilots of aeroplanes to the existence of dangerous storms in their path but also to assist in improving the rather less-than-reliable methods of weather-forecasting then available to meteorologists?

This experiment helped answer both questions, and an ideal opportunity at which to present their findings was the meeting of the Royal Society of South Africa due to take place in Cape Town on 16 October; so Schonland and Hodges rapidly put together a paper which Schonland read at the meeting, using his Professorial title for the first time. They found, he reported, that nearly 90 per cent of the thunderstorms identified

---

* Southern Rhodesia was a self-governing British colony until 1953 when it joined in Federation with Northern Rhodesia (later Zambia) and Nyasaland (later Malawi). In 1965 Southern Rhodesia declared unilateral independence as Rhodesia. In 1979 it became Zimbabwe.

by the meteorological network were correctly located, in bearing, by their radio direction-finder and of these, more than half were definitely associated with thunderstorms in regions served by weather monitoring stations. However, about a third of the sources of intermittent atmospherics lay out to sea over the Atlantic and Indian Oceans, while a good 13 per cent, between 5° and 15° east of north, seemed to rumble almost continuously. These, they concluded, were the storms above the watershed of the Congo that had first been reported by the two Australian scientists, Munro and Huxley, on their homeward voyage* from England some years before [13]. The lightning responsible for the more intense atmospherics received in Cape Town was found to range in distance from about 550 to 1600 km, with excellent agreement existing between their radio bearings and the reported positions of storms in the hinterland. These results, Schonland said, supported the conclusion of Watson Watt, Appleton and others, that the lightning flash is undoubtedly the chief source of these atmospherics. In Schonland's view it was the only source [14].

To fully exploit the potential of this exciting new radio location technique, it would be necessary to establish a more widely distributed network of direction-finding stations, so that the bearings of the various lightning flashes could be accurately determined by some means of coincident reporting. Not only would this require the construction of more equipment, but just as important was the need for appropriately trained personnel to operate it. Such tasks required coordinated effort and it was clear that they could only be handled properly within an organization set up specifically for this purpose. But Cape Town, with its temperate climate, was hardly the place to pursue such work, while an annual pilgrimage to the lightning regions of the Highveld around Johannesburg was both impracticable and not in the best interests of his family. So Schonland, once again at Wits to conduct a programme of measurements and to report to the SAIEE on the progress of its Lightning Research Committee [15], was giving much thought to his own base of operations. So too, evidently, was Humphrey Raikes, the Principal of Wits, when he communicated some very important information to Schonland on the last day of 1935 [16].

Just a month before, Raikes had received a letter from the Carnegie Corporation of New York informing him of the recent decision of the Corporation to provide an amount of $54 750 (about £11 000 in South African currency at that time), payable over a three-year period, with the express wish that it be used 'for [the] development of a program in

---

* It was on this trip home to Australia in December 1929, after being trained at Slough in the use of the DF system, that Munro and Huxley used the equipment to observe atmospherics from the SS *Baradine*. They found that the predominant source of these was in the region that constitutes the watershed of the Zambezi and the Congo rivers in Central Africa.

geophysics at the University of the Witwatersrand'. He now requested Schonland to assist him by preparing a scheme for the application of the grant. A similar request was put to Bernard Price, whose own interest in both the University and in Schonland's lightning research were clearly crucial to the success of any such scheme. This magnanimous American gesture had come about as a result of Schonland's visit to New York the year before when he had convinced Dr Keppel that he intended to find suitable resources for geophysics research in South Africa, come what may. Not long after Schonland's visit Keppel himself visited South Africa, and Wits particularly, where he had a lengthy meeting with Raikes on the subject of this geophysics research institute. On his return to New York Keppel took with him a memorandum from Wits which outlined in quite some detail how such an institute might function and what its terms of reference would be, especially as they related to the balance to be struck between fundamental research on the one hand and its practical applications on the other. The hand of Schonland in drafting this key document was much in evidence.

Geophysical research, it explained, covered in its widest sense, the earth's crust and its atmosphere and included 'the radiations of solar, and other origins, which traverse it'. Four distinct functions were identified for the institute, with pure research into all of these phenomena, and more, being its chief function. Consultation and advice would be provided 'for the applied worker' as required and as appropriate, while research on 'problems too difficult for the industrial worker' would also be considered on the same basis. Finally, the training of 'workers operating in the applied field' would similarly receive attention as and when this was deemed necessary. The document stressed that such an organization must be able to pursue these objectives 'unhampered by the demands of commercial life' and should not be dependent upon 'institutions, commercial firms, companies or government departments concerned with the practical sides of the subject'.

Since unity of purpose was clearly important, the document stressed that Schonland, the pragmatic but far-sighted Bernard Price and Mr Raikes were of one mind regarding the role of this entirely new venture for the University of the Witwatersrand. Until then Wits, though a fully-fledged university in its own right since 1922, was still very much a teaching institution with little or no coordinated research [17]. Here, on the horizon, was the flagship that could surely change it into the sort of university that Raikes, an ex-Oxford man and a chemist by training, so desired for Johannesburg. For the benefit of the Carnegie Corporation's governors, the document also contained details of those local bodies and organizations that might offer support for the institute's work and amongst these Schonland and Price had identified the SAIEE and the electrical power companies, the Mining Houses, geological and prospecting organizations, as well as the Meteorological Office and the Postmaster-General, with his responsibility for radio communications.

The memorandum clearly achieved its purpose, for within six months of Keppel's visit, and on the firm scientific recommendation of the Carnegie Institution in Washington, Raikes was able to proceed in the knowledge that he now had a solid financial foundation laid by the munificence of Andrew Carnegie's legacy and untrammelled by any commercial considerations.

Bernard Price, all the while an active participant in the preliminary discussions, was also considering carefully his own involvement in this most exciting venture. His interest was not solely philanthropic, for though he had the fundamental science much at heart, he recognized as well that his Victoria Falls and Transvaal Power Company would certainly benefit from whatever lightning-protection measures might spring from the institute's work. Immediately he saw the commitment from the Carnegie Corporation he acted with characteristic resolve and informed the University that he, personally, would put up the sum of £10 000 towards the establishment of the research institute, but there was one categorical condition upon which his offer depended: that Basil Schonland must accept the appointment as its Director! In so doing Price reinforced the hopes of Humphrey Raikes, who had told members of Council that it was his wish that Schonland should occupy the new Chair of Geophysics that went together with the post of Director of this newly proposed institute [18]. In fact, Schonland had already agreed to this, because Raikes went on in his letter to inform the Council that he would indeed commence his duties at the beginning of 1937. There was, though, the matter of his somewhat delayed accession to the Chair of Physics at UCT that was scheduled to be his from 1 August 1936, but both Schonland's original intentions and UCT's somewhat faltering arrangements had been overtaken by events in Johannesburg and New York. It was obvious to all that Wits and Johannesburg offered by far the greater opportunities for opening up a completely new field of research in an area that was absolutely ripe for the picking and so Schonland's move northwards seemed pre-ordained.

Within a week of the Special Meeting of the Wits Council, convened on 17 April to consider the whole matter of geophysics at the University, Raikes wrote formally to Schonland in Cape Town and offered him the Carnegie–Price Chair of Geophysics and the Directorship of what would now be known as the Bernard Price Institute of Geophysical Research, and forever after as the BPI. The letter contained the usual formalities associated with such appointments but, in addition, it spelt out carefully the particular conditions which applied to this rather special post. The Director would ensure that his Institute conducted pure research in Geophysics under the direction of its governing Board; that he himself would be required to give no more than two lectures a week, and these only to post-graduates working towards their masters degrees in areas of direct interest to the Institute; that the conditions of leave as they applied to the Director would be in line with those elsewhere within the University, with the added proviso that the Board had the power to allow him to travel abroad to attend conferences, and so on, should these be considered to be in the interests of the Institute. As a Professor of the University, the Director would also have seats on the Senate, as well as on the Faculties of Science and Engineering.

One must presume, on the basis of what transpired during the earlier negotiations, that none of these provisions came as any surprise to Schonland because it is almost self-evident that he was instrumental in having most if not all of them inserted in the contract in the first place, for he was in a very powerful bargaining position. The BPI's Board of Control would have as its Chairman none other than Dr Bernard Price, recently honoured by the University with an honorary doctorate, as well as representatives of the Carnegie Corporation, the Principal, Council and Senate and, of course, the Director himself [19]. Some eye-brows were indeed raised at the University, where it was felt that these rather preferential conditions were not entirely conducive to good collegiality, but then this Institute was the first of its kind and clearly was not quite like the rest of the University [20].

Schonland submitted his resignation to the Principal and Council of the University of Cape Town on 4 May 1936, two months before he was due to be confirmed in his appointment as its Professor of Physics, and indicated that he wished to resign his position as from the end of the year. That he was leaving Cape Town for pastures new was, by now, common knowledge but formality required an explanation. 'The step is due', he wrote, 'to my appointment as Director of the Bernard Price Institute of Geophysical Research, Johannesburg' [21]. The Schonland family would once more be on the move, but this time rather farther afield. A worrying complication existed in that Mary, their second child, now 11 years old, was particularly ill. She had been experiencing severe back pain and, while on holiday at Somerset East, was X-rayed and a tubercular disease of the spine was suspected [22]. Basil and Ismay were greatly concerned but Johannesburg was as well-served by good hospitals as Cape Town and so their plans went ahead to move the family home from the grounds of the Groote Schuur estate in Cape Town to the Highveld of Johannesburg.

In the midst of making all the necessary arrangements to uproot the family, while still carrying out her duties as head of the Women's Residence, Ismay Schonland received a most unexpected visitor at the family home. The door was answered by Mary who, embarrassed in her bare feet, ushered in a small, neatly dressed gentleman with a pointed beard known in South Africa as a 'bok baard'.* Her mother immediately took him through to the sitting-room where they spoke for some time. On leaving he gave Mary a farewell kiss and bade Ismay good-bye. He was General J C Smuts, then Deputy Prime Minister and Minister of Justice! It transpired that Schonland's name had been put forward for election as a fellow of The Royal Society and Smuts, a Fellow himself as a result of his especial distinction, had been asked whether he would support Schonland's case. He agreed with enthusiasm and though the formal announcement was still a year away, since the matter was still in suspension, this did not deter him from paying a personal visit to the Schonland home to inform Ismay of this most exciting news [22]. Being elected an FRS was as great an honour as the British scientific community could confer on one of its own; to a South African it was a momentous event and Smuts clearly considered it worthy of a special visit to the family home of the son of his old botanical friend.

*   *   *

When making his donation to the University, the ever astute Bernard Price indicated that it was to be used specifically for the building and equipping of the geophysics laboratory, as well as to provide a sum of £100 per annum to facilitate visits by scientists from overseas—so necessary to maintain contact with the major centres of research elsewhere. In addition, he also requested the University itself to make an amount of £1500 per annum available to the Institute specifically to cover the salaries of the Director and his Mathematical Assistant. Humphrey Raikes moved swiftly so that he was soon able to assure Council that this would be provided 'by sources external to the University'. He also allayed the fears of some on Council that they might be saddled with the Institute long after its useful life had waned, by advising them that these financial provisions would only apply for a period of 20 years. If, after that time, it had not become self-supporting, all financial support would lapse. Finally, it was agreed that the Carnegie bequest would be invested and should yield an annual amount of about £750 over the 20 year period, sufficient to cover other salaries as well as the running and maintenance costs of the laboratory [23].

On 31 December 1936 the members of the Board of Control of the Bernard Price Institute of Geophysical Research met for the first time. Dr Bernard Price was in the Chair, Professor Maingard represented Raikes, Mr Hill and Sir Robert Kotze were Council's representatives, while Professor Paine was there to represent the Senate, with Professor Young, who had been nominated by Price and, of course, Basil Schonland himself. Schonland informed them that it was his intention immediately to commence an active programme of research by continuing the work that had been in train in both Cape Town and Johannesburg on lightning,

---

* 'Bok-baard' or goat's beard, which is an apt description of this style of facial hirsuteness.

thunderstorms and, more recently, on their attendant atmospherics. While stressing that fundamental research was to be the Institute's primary objective, he indicated that he would cooperate readily with the local industry in areas of interest and would also provide training for young research workers in the field of geophysics. He mapped out the fields of research that were of major importance, with his own photographic and radio studies of lightning to commence immediately. Soon to follow would be geophysics on a much wider front, such as the thermal characteristics of rock and especially the reported anomalous temperature gradient noted within the very deep gold mines around Johannesburg. Seismic activity, another feature of deep mining, would also receive attention, as would the radioactivity of rocks in general. Finally, he foresaw that the Institute would begin investigations into abnormal ionospheric conditions as well as meteorological effects closer to the earth [24]. This was a tall order indeed, particularly as Schonland was, at that moment, the BPI's only member of staff!

Plans were already afoot to make two appointments. J A Keiller, a recently qualified instrument-maker in the Department of Physics, whose potential had caught Schonland's eye on his many visits, would become Mechanical Assistant to the Director on a temporary basis for one year from 1 January 1937, while a part-time Mathematical Assistant would be appointed later. Schonland had in mind Dr Anton Hales for that particular post. It will be remembered that Hales was the young man who, in 1930, had followed Schonland's advice and read geophysics rather than quantum mechanics as part of the Tripos at Cambridge. On returning to UCT he embarked on a PhD on the theory of convection currents in the earth, under Schonland's joint supervision, and now was a lecturer in the Department of Applied Mathematics at Wits. Schonland had earmarked him for an honorary appointment to the Institute just as soon as it could be arranged with the University [25]. Where they were all to be accommodated was resolved by a generous offer of space from Professors Paine and Stevens, whose Departments of Physics and Chemistry shared the building on the eastern side of the Wits campus. Schonland, of course, knew its top floor well from his many hours of painstaking work there every summer. Plans for a new laboratory, specifically for the BPI, were being prepared and it was hoped that these would be ready by the end of the year. The Board completed its first meeting by granting the Director leave of absence between May and July 1937 in order to deliver the Halley Lecture at the University of Oxford, and to make 'certain arrangements for the Institute' while overseas.

*   *   *

The Halley Lecture, first delivered in 1910 by Henry Wilde, who endowed it to the University of Oxford in honour of Edmund Halley, one-time professor at the university and Astronomer Royal, was one of the prestigious lectures in the university calendar. Its purpose was to honour the memory of Halley and his important contributions to cometary astronomy and terrestrial magnetism. Within the latter field were included 'the physics of the external and internal parts of the terrestrial globe' and Basil Schonland was invited in May 1936 to deliver the lecture by the Board of Electors, whose members consisted of an array of professors of astronomy, the Astronomer Royal, the Radcliffe Observer and two professors of natural and experimental philosophy [26]. Such invitations were not extended without careful consideration of the scientific merit of the recipient. That Schonland was seen to pass muster was surely the clearest indication yet that his mark had been well and truly made. The date of 28 May 1937 was duly selected for its presentation before an invited audience within the much hallowed halls of the University of Oxford.

At home, Mary Schonland had been confined to bed after a spell in the Children's Hospital, where she had been almost encased in plaster as the only form of treatment then available for her condition; and she was to remain like that for nine months. The problem of keeping a bright and intelligent 12 year old amused and entertained for that length of time taxed both Basil and Ismay, as well as anyone else with suggestions to offer. One of the projects they conceived was a marionette theatre that Basil designed and built with much enthusiasm and great inventiveness. The marionettes themselves were made by the children with the assistance of a neighbour, the Dutch Consul in Johannesburg. He showed them how to model the head with plasticine and *papier mâché* and this activity, followed by careful painting and decoration, occupied many hours of intense and fulfilling activity. Scene-painting and the writing of scripts for such epics as *Jack and the Beanstalk* then followed and all this helped greatly to pass the time as Mary was slowly restored to full health. Basil and Ismay were the audience at many rousing performances [27].

In between times Schonland worked on his Halley Lecture and, if any stimulus were needed to make his contribution to the Oxford calendar no mere footnote, then none was more significant than the name of his immediate predecessor at that podium, one Patrick Maynard Stuart Blackett [28]. Blackett was more of an acquaintance than a colleague of Schonland from his Cavendish days of 1928, but then Blackett was that to most people. His adaptation of the Wilson cloud chamber to incorporate automatic triggering by the very cosmic events he wished to record was a brilliant feat achieved in 1932 with a pair of Geiger–Müller tubes configured as a coincidence detector. Within a year, he had used the technique to demonstrate the existence of showers of positive and negative electrons emanating from space and so confirmed an experimental result published just a while before in America, but based on less tenuous evidence. That such positive particles did indeed exist had been predicted by another of Cambridge's formidable physicists, P A M Dirac, in 1930. For this and subsequent work Blackett would be awarded the 1948 Nobel Prize for Physics. So it was that in 1936 Blackett's subject for his Halley Lecture was 'Cosmic Rays' — the 'penetrating radiation' that Schonland himself had set out to detect at UCT in 1932 using his own form of coincidence detector, and also from Wits before that when their source was thought to be much closer to the earth, in the very lightning discharge which was the title that Schonland chose for his lecture a year hence [29].

But much was still to happen in the interim. Schonland wasted no time in recruiting, as a guest researcher at the BPI, his former student and colleague, David Hodges, who was now a senior lecturer at the Natal University College. Hodges, in his usual enthusiastic way, had already assembled the nucleus of a research team in Durban to use radio direction-finding methods to determine the sources and locations of atmospherics. Under his guidance a DF receiver had been constructed based on the design that he and Schonland had used in Cape Town. However, it now included a special sensing circuit that made it possible to determine the quadrant of the compass from which an atmospheric originated. This receiver was operated by W E Phillips, of the Department of Electrical Engineering, while another recent engineering graduate by the name of 'Sonny' Katz was sent to Johannesburg to work at the BPI under Schonland's supervision. There, Katz spent most of 1937 constructing and setting up an even more advanced receiver. Now that two DF stations existed nearly 500 km apart it would be possible, by means of simple triangulation, accurately to fix the position of the source of the atmospheric as long as it was certain that both stations were measuring the bearing of the same source at the same time, and that it lay roughly broadside to the line joining JB and D, as Johannesburg and Durban were soon abbreviated for purposes of rapid communications.

It was important, of course, to ensure that both stations were observing the same storm and this required a reliable communication link between them. Schonland immediately contacted the Postmaster-General in Pretoria who instructed his Under-Secretary for Telegraphs, Frederick Collins, to assist. The fact that Lt Col Freddie Collins was also Assistant Director of Signals in the Union Defence Force, and so had more than a little in common with Schonland as signalling veterans of the war to end all wars, lent some rapport to the venture. Soon Schonland had the use of a dedicated telephone circuit between Johannesburg and Durban for fifteen minutes every day between 1 and 2 p.m. Collins, in his military capacity, also agreed to look into the feasibility of arranging a special daily broadcast at a different time from the Government Radio

Station at Roberts Heights.* This radio transmission would not only provide Schonland with a second communication link with Durban, but it suggested to him a project of truly international dimensions since, as he reported to the University, '... observations can be extended to similar organisations in Great Britain and Australia' and so he sought to obtain the cooperation of the Radio Research Boards of Great Britain and Australia [30]. However, events not far over the international political horizon were soon to overtake them all and such a grand scheme never came about, but Freddie Collins was now well aware of both Basil Schonland's existence and his scientific ability and, no doubt, saw the drive and determination behind all he did. The two men were soon to meet again.

Meetings with architects and builders, and with the relevant University authorities with interests in such matters, consumed a great amount of time in the early months of 1937. As soon as draft plans were ready, Schonland sent copies to the Directors of the Mond Laboratory and the Geophysical Laboratory in Cambridge, and the Kew Observatory in London, asking for their comments and suggestions. The Royal Society Mond Laboratory was built at the Cavendish in 1933 for work on low-temperature physics, essentially at the instigation of Peter Kapitza, whose presence greatly enlivened the Cavendish during the early years of the decade. When Kapitza was prevented by the Soviets from returning to England after a trip home in 1934, Lord Rutherford appointed John Cockcroft to take over the directorship of the laboratory, since Cockcroft had been closely involved in its design and operation and had the reputation at the Cavendish as an organizer of research on a rather grander scale than Rutherford ever contemplated [31]. Schonland by now knew Cockcroft well and believed that his experience could prove invaluable to him. And so it turned out. Cockcroft responded with useful suggestions that were included in the specifications to Messrs Cowin and Williamson, the architects charged with the design of the building, while Schonland's name was probably included for future reference in the little black book that was Cockcroft's eternal *aide-mémoire* [32].

*   *   *

Before Schonland left for England in May to deliver the Halley Lecture there were other pressing matters that required his attention, and none more so than an experiment that would enable him to photograph the lightning stroke and all its attendant details during daytime. Until then

---

* Roberts Heights, so named in honour of Field Marshal Roberts, British Commander-in-Chief during the Boer War, was renamed Voortrekkerhoogte ('Voortrekker Heights') soon after the Nationalists came to power in 1948. It was essentially a military town on the outskirts of Pretoria and possessed a large radio station on a hilltop known as Uitkyk ('Look Out'). In 1998 it was renamed Thaba Tshwane.

all photographs of lightning had to be taken at night because of the need for the camera shutter to be open before the flash actually occurred, and this caused considerable fogging of the film. But most thunderstorms on the Reef actually took place during the late afternoon and so, to overcome this problem, Schonland had tried using various optical filters to reduce the intensity of natural daylight but, though fairly successful, the lightning discharge itself was only partially luminous at the wavelengths of the filters he used. So he proposed another, indirect, method of triggering the camera that relied on the train of radio frequency, electromagnetic pulses emitted by the stepped leader that preceded the intense return stroke he wished to photograph. A receiver not unlike that in use for the DF work was designed to include a trigger circuit and then built with the assistance of Jock Keiller. It soon proved itself to be so effective that an improved model was planned for use during the next lightning season and a temporary assistant, G A Cruickshank, was appointed in March to work on this project and to assist Katz with the direction-finding apparatus [33].

Others too were taking photographs of lightning but in most cases the sole trigger was chance. In February, Schonland had received one that had been taken quite fortuitously by a member of Bernard Price's staff at the Victoria Falls and Transvaal Power Company in Johannesburg. It showed, in wonderfully sharp detail, a lightning discharge at very close range that had been wafted along its path by a strong gust of wind. The wind had made it possible to photograph, with but a single stationary camera, the various components of the return stroke, just as the Boys rotating lens camera had done, but it also produced some additional details, and all entirely by chance. Schonland was fascinated by this serendipitous picture and immediately set about using it to deduce the actual width of the lightning channel. Eleven clearly defined component strokes had been smeared out by the wind to a width of 3.7 m and there were five branches snaking away from the first of them. From the estimated half-second duration of the lightning discharge, he determined that the average interval between these strokes was in very close accord with the result that he had published with Malan and Collens two years before in the second of their trail-blazing papers on 'Progressive Lightning'. Now, for the first time, he was able to deduce that the average diameter of the lightning channel was about 16 cm. This was a significant new result and he immediately wrote a short paper and submitted it to the *Philosophical Magazine* in London. It duly appeared in March 1937 and was the first publication to appear under the banner of the BPI [34].

The Halley Lecture took place in Oxford on 28 May 1937. Schonland entitled it *The Lightning Discharge* and he used the occasion to paint a graphic picture of the visual and electrical features of lightning, with abundant detail provided by all the experiments he had conducted during the previous decade in South Africa [35]. What to most observers appeared as but a single brilliant flash was in fact made up of a combination of elements: the leader, both stepped and continuous; the return stroke; the dart leader and the still speculative pilot streamer. Each of these optical features that had been revealed by the Boys camera was duplicated by its characteristic electrical signal that had been detected by one or other device, so many of which were inspired many years before by C T R Wilson, with his brilliant grasp of the fundamental processes involved. More recently, it had been men of Schonland's own generation who had begun to use cathode rays to paint pictures of the other features of lightning that could only be detected by the radio waves they emitted. The oscillograph had complemented the camera and together they probed into the very heart of Benjamin Franklin's *thundergust.*

Schonland's lecture was masterly and earned him the plaudits of all who heard it. One who heard it was a young Oxford physicist by the name of Frank Nabarro[*] for whom both lightning and South Africa meant little at the time [36]. Nabarro would encounter Schonland again quite soon, when the very forces of science were being marshalled in the defence of freedom.

The Halley Lecture, if not in name, was given another airing on Schonland's return to Johannesburg when he was called upon to report to a joint meeting of the University and the SAIEE on the progress achieved by his Lightning Research Committee [37]. His audience of engineers and university dignitaries were treated to a *tour de force* through the subject, illustrated with many photographs taken with versions of the Boys camera further developed by Malan, who was now also serving during the summer months as a guest researcher at the BPI. It was, Schonland told them, just as he had done in Oxford, the combination of two instruments (the rotating lens camera and the cathode ray oscillograph) that had enabled such huge strides to be made in understanding the complexity of the lightning discharge. But, even more than the complexity of the instruments, it was the dedication of the men who used them that had really made it all possible. He singled out one man for particular praise, the crocodile hunter turned lightning chaser known only as H Collens, whose Herculean efforts in pursuing 'every storm within ten miles of Johannesburg at any hour of the night' had resulted in photographs of no fewer than 140 lightning flashes containing over 700 separate strokes. That was the rich harvest that enabled Schonland, more than anyone elsewhere, to begin to unravel the mysteries of *Umpundulo,* the lightning of the Zulus.

---

[*] As Nabarro recalled it sixty years later, he remembered Schonland addressing the Oxford University Junior Scientific Club, a more formal gathering than its name might suggest, so it appears that the lecture was actually given twice [36].

Bernard Price's Institute was now about to leave its temporary home for the new, purpose-built laboratory on the south west corner of the University site, its position having been carefully chosen by Schonland so that it offered uninterrupted viewing of the thunder storms as they advanced across the mine dumps that separated Johannesburg from the veld beyond. By July, he was able to report to the Board that 'splendid progress had been made in hastening completion of the building' and he hoped that this would allow 'atmospherics research to begin on September 1st'. Of more immediate importance was the fact that its eventual cost was expected to lie between £8500 and £9000; well within budget. This information, as well as details of the staff appointed to date, was to be provided to the Trustees of the Carnegie Corporation, who had asked Humphrey Raikes for a concise statement that could be tabled at their next meeting [38]. The scientific programme had by now expanded to include matters of direct interest to the mining industry, whose long-term support would be vital to the future of the BPI. A seismic recorder had been set up for the purpose of studying the earth tremors that were as much part of the geophysics below the earth of the Witwatersrand as was the lightning activity above it. The first venture into the world of applied research, necessary for establishing the Institute's 'usefulness' in the eyes of its industrial sponsors, was also embarked upon. It was an instrument to alert shaft-sinkers about to charge their explosives to the imminent arrival of lightning and it did this by sensing the increasing electric field that always precedes a lightning strike. The mining industry was most interested. Discussions were also taking place with the Postmaster General's department regarding the setting up of a section at the BPI devoted to radio research.

Schonland then informed his Board of further topics of a more fundamental nature that he wished to consider. These included the tectonics of the earth's crust, its thermal gradients and terrestrial magnetism. By December the minutes noted that the work of the Institute had been completely transferred to the new building, but that the final fitting of equipment was not yet complete, because this was the height of the lightning season and all available personnel were heavily involved in chasing and photographing it [39]. Humphrey Raikes noted these developments with considerable interest and not a little satisfaction.

But not everyone at the University was quite as sanguine about this new edifice on the Wits campus, particularly as it seemed better placed than most to attract money. One of those rather disgruntled detractors was none other than Schonland's brother-in-law, Don Craib, who was now Professor of Medicine at Wits and a somewhat outspoken member of the University's Senate. Craib complained bitterly that medical research was suffering through a lack of funds, while other ventures had been better provided for [40]. Ironically, Craib happened, at the

149

time to be paying rent to Basil Schonland as the occupant of the Schonland's first home of their own. After years of renting property Schonland, with some assistance from his father, had eventually bought a house at 85 Kilkenny Road, Parktown, but the family were unable to move in immediately because no tenants had yet been found to take over their present house. Not prepared to commit himself to the financial strain of running two properties, as well as the payment of the medical bills following Mary's illness, Basil had allowed Don Craib to take up residence and his monthly contributions eased the financial burden considerably. Though Craib was well satisfied with his personal accommodation, what he thought about the BPI was another matter entirely.

As 1937 drew to a close so the scientific world was deeply shocked and saddened by the sudden death of Lord Rutherford. This great bear of a man, whose energy and vision had built the Cavendish into the finest crucible of physics in the world, died in October at the age of just 66. His influence on a generation of physicists, many of whose own careers were to be as illustrious as his, was probably without parallel in the history of science. In 1928 alone, for example, the annual laboratory photograph shows no fewer than nine existing or future Nobel laureates and many others, though never to reach quite that pinnacle of scientific excellence, who would distinguish themselves in their future careers throughout the world. Basil Schonland was one of them and he recorded his shock on receiving the news in a letter to his father written late in October [41]: 'He was a good friend and a wise counsellor as well as a great man.' Who would be Rutherford's successor at the Cavendish was a question on many lips and Schonland gave his views on this to his father.

> It will not be quite possible to replace him for the centre of gravity of modern physics has moved to America, considerably assisted by many German-Jewish emigrés. Cockcroft won't get the Cavendish post. It will most probably be a joint job now or perhaps a triumvirate: Chadwick, Appleton and Bragg. A very difficult problem for them to solve.*

Intriguingly he did not disclose why, in his view, Cockcroft was not in the running. Elsewhere, many were also speculating on Rutherford's likely successor and one of them was John Cockcroft himself. He suggested Appleton, Blackett and Chadwick as the most likely contenders. He personally plumped for Blackett [43].

---

* W L Bragg was eventually selected to succeed Rutherford after an interregnum when Appleton held the reins and wondered whether the post would ultimately become his. Chadwick had been Rutherford's right-hand man for many years and Rutherford's own choice as his successor but had left for a Chair at Liverpool in 1935 [42].

CHAPTER 10

BERNARD PRICE'S INSTITUTE

The energy and enthusiasm that Basil Schonland brought to his research, as well as to the running of his new research institute, were never more in evidence than during the years immediately preceding the outbreak of the Second World War. As 1938 dawned, all his efforts were directed towards ensuring that the BPI should be seen as the most important centre in the world for lightning research. To achieve this meant that the research that he, his team of associates and guest researchers were doing had to be brought to the attention of the wider scientific world as rapidly as possible by publishing their findings in the leading scientific journals of the day. Undoubtedly, at that time, the *Proceedings of the Royal Society* of London was the pre-eminent record of scientific progress and Schonland's work had graced its pages for many years, with each paper having been communicated by a Fellow of the Royal Society, usually known to the author and certainly well-versed in the subject. This practice of 'communication' had long been the mechanism by which a seal of approval if not an *imprimatur* of both scientific quality and, indeed, of literary style was granted to the work. Schonland's contributions had, since his first offering in 1922, been backed by such weighty advocates as Rutherford, C T R Wilson and C V Boys, each clearly representative of a particular phase of his research career at the time. Now, those men of great scientific distinction had joined together with twelve other Fellows[*] to support Schonland's election to the Fellowship that was announced on 17 March 1938. This was a signal honour for any scientist; for one from South Africa it was a momentous event.

Word of his election reached Schonland that very day by telegram. Almost fortuitously, the Board of Control of the BPI met on the 18th. Immediately on calling the meeting to order, Bernard Price rose and announced that the Director of their Institute had been granted the highest honour by his scientific colleagues in England by his election as

---

[*] H Spencer Jones, C V Boys, F W Dyson, Lord Rutherford, C T R Wilson, E V Appleton, J Chadwick, C D Ellis, G C Simpson, G P Thomson, R Stoneley, S Chapman, J C Smuts, R Whiddington and E N C da Andrade.

a Fellow of the Royal Society. The Board reacted as one man and 'resolved with acclamation to congratulate Professor Schonland . . . and to record its gratification of this distinction upon the Director of the Institute' [1]. Schonland's reaction was one of considerable pride and satisfaction but not complete surprise: General Smuts's unannounced visit to Ismay a while before had let the cat out of the bag!

The acclaim of the BPI Board soon spread beyond their council chamber, as word reached others within the University of the honour done to one of their colleagues. Wits was immensely proud, for Schonland was the first member of its staff to have been accorded such elevated scientific status [2], while UCT, though no longer able to claim him as one of their own, took much satisfaction from the news, for it was there that the Schonland flame had burst into life after Rutherford and the Cavendish had kindled it. But it was in Grahamstown, at the Schonland home and not far away at Rhodes, that the feeling of pride was greatest of all as letters arrived from far and wide. Sir Thomas Graham, Attorney General of the Cape Colony, Judge President of the Eastern District Court and grandson of the founder of the town, wrote offering his warmest congratulations. He remembered Basil, so he informed Selmar Schonland, 'as a boy at St Andrew's [who] used to experiment in wireless from the tower of the windmill which stood in the grounds . . .', while Dr Robert Broom, himself much in the scientific limelight, congratulated the Schonlands on their son's well-deserved achievement. 'How one remembers the time I stayed with you in 1902 when Basil was a little boy'. And there was also a letter addressed to Felix Schonland from Sir Arthur Fleming, Director of Metro-Vick, offering his and his Board's congratulations. He well remembered Basil Schonland's visits to his company's Manchester works and his collaboration with Allibone in the high-voltage laboratory to confirm the crucial features of the branching of lightning [3].

Schonland, too, received many letters and messages of congratulation and one, particularly, brought him much pleasure. It came from Mr Evelyn Shaw, Secretary of the Royal Commission for the Exhibition of 1851. In his reply Schonland expressed his sincere thanks to the Commissioners for the gamble they took in 1928 in awarding him a scholarship at a crucial stage in his scientific career. 'I shall always feel a debt of gratitude to your foundation for what it did for me at a time when I was badly in need of the assistance and encouragement it offered. I should like to take this opportunity of saying how glad I should be to help the Commission in any way' [4]. He remained true to his word and over the years brought the 1851 Scholarships to the attention of South Africa's universities and pressed them to nominate candidates of quality.

Either on his own, or in conjunction with his colleagues Malan, Collens and Hodges, Schonland had by now published five papers in a series he titled 'Progressive Lightning', the term first used by Boys in

1926. All had appeared in the *Proceedings of the Royal Society*, from the first in 1934, and together they constituted the most complete work yet to appear on the mechanism of the lightning discharge process as recorded both optically, with versions of the Boys camera built in Cape Town and Johannesburg, and by means of the oscillograph images of the radio emissions that accompanied it. It was these papers that had shot Schonland to international prominence. Not only was he blessed with a natural lightning laboratory in the skies, that erupted around Johannesburg with almost clockwork regularity every summer afternoon, but he had applied the most modern techniques of the day to extract a mass of information from the microsecond phenomena that, until recently, had been explained more by myth and mythology than by science.

It was in those papers too that he first introduced the terminology, now commonplace, to describe the various stages of the lightning flash: the *leader*, the *stepped leader* and the *dart leader*, the *main stroke* that he subsequently called the *return stroke*, and the *pilot leader* which, he now postulated, preceded them all. He also described the *root branching* that occurred when the leader was within just 20 to 150 m of the ground when the lightning paused, ever so briefly, like a snake, before striking some elevated object on the ground. What followed was a massive injection of charge that spat skywards in the intense optical flash that the eye first saw. From the thousands of photographs that Collens and Malan had taken with the revolving lenses of Boys's camera Schonland had measured the distances that each stage of the stepped leader travelled and then calculated their velocities. The same technique produced values for the velocities of all the other stages of the process and these enabled him to show that they varied from less than a thousandth of the speed of light during the staccato transitions of the stepped leader, to the half-luminal speed of the dramatic return stroke. Then, from the measurements of the intensity of the electric field and the rate of advance of the various luminous tentacles or streamers as they tracked through space, he was able to calculate the magnitudes of the currents associated with each phase of the process. Values as low as 180 A were typical of the speculative pilot leader, while as much as 130 000 A was what the many coulombs of charge removed from a cloud in just tens of microseconds during the return stroke [5].

Schonland's linking of the photographic and oscillographic processes in studying the details of the lightning flash was a master-stroke, for it enabled him to correlate the outputs from the two and so identify those features on the oscillograph screen that were related to others on the photograph. It was his submission, in 1934, of a letter to *Nature* [6], that provided the first evidence of a characteristic difference between the leader of the first stroke and all those that followed it. While these others were continuous and dart-like, the first leader was sporadic and

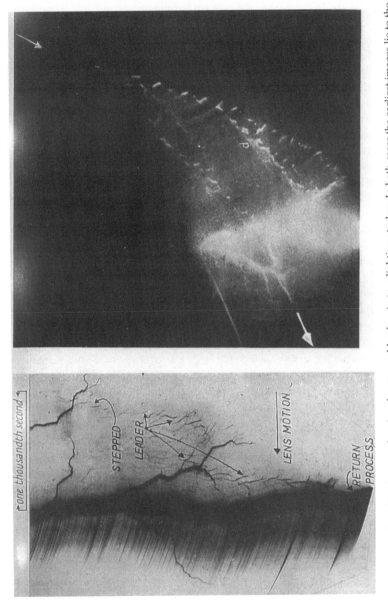

Figure 12. Negative and positive images showing the stepped leaders in two lighting strokes. In both cases the earliest images lie to the right. The very intense return stroke actually caused the film to be over-exposed due to the proximity to the strike which, in each case, occurred only about a kilometre from the camera. (Reproduced from Schonland's *The Flight of Thunderbolts*, plate V, 1950 by kind permission of Oxford University Press.)

proceeded on its way earthwards in a step-by-step manner — hence the name he gave to it. Radio measurements made a decade earlier in England by Watson Watt and Appleton [7], and then a few years later by Cairns [8] in Australia, had shown distinct ripples, with periods of about 100 μs, on some of the waveforms and they immediately concluded that these were the sources of the atmospherics heard on radio receivers: the 'splashes', 'growlers' and 'roars' referred to previously. Then, in 1935, came another letter to Nature [9] from South Africa in which Schonland, Collens and Malan showed that the return stroke too, exhibited step-like behaviour with the time intervals between the components in the range from tens to hundreds of microseconds. The very much larger currents flowing in the return stroke would obviously produce much more intense, broad band electromagnetic radiation, so the search for the source of atmospherics was narrowing.

By 1937 Schonland had equipped a lorry, loaned to him by the VFP, with the most recent version of the BPI camera that used only a single, fixed lens to expose a rapidly moving film mounted within a revolving drum. The first model had been manufactured in London in 1935 to Schonland's specifications and was presented to the SAIEE's Lightning Investigation Committee by Dr Bernard Price. It was quite radical for its time, in that it contained a combination of prisms with just a single lens to produce an image that was projected on to the rapidly moving film. As much as it offered significant improvements in the photographing of lightning, it suffered from a distinct, practical disadvantage because the process of changing films quickly (so necessary to capture a fleeting and unpredictable heavenly subject) was rather slow and so the camera was modified by mounting the film within removable drums that could be interchanged very quickly. In operation, the film streaked past the lens at nearly 2000 m per minute, thus greatly improving the resolution of the instrument, and allowing considerable detail within the stroke to be photographed. The leader alone, with its attendant pattern of ripples, now occupied all of 150 mm of film [10]. To obtain yet more information, the lorry's photographic armoury included two further cameras: one of a slower speed and another that was fixed. Also squeezed into the vehicle was a three-valve amplifier and a Cossor gas-focused cathode ray tube with its battery of dry cells needed to power it, as well as another camera, specifically dedicated to the task of photographing the oscillograph screen. This was driven by an electric motor at about 80 rpm, which meant that the lightning discharge, of about half a second in duration, could be viewed to a resolution of about 70 μs, providing hitherto unseen details of the African storms. This lorry was certainly as complex a mobile laboratory as had ever ventured on to the roads of that continent.

*   *   *

The use of radio techniques to provide still more information about lightning was soon to be eclipsed by an even more dramatic application of radio direction finding. Quite unbeknown to Schonland, as indeed it was to almost everyone else as the war clouds began to gather over Europe, was the highly secret role that his occasional colleague, Robert Watson Watt, was now playing in the development of the only effective means of warning Britain's seriously ill-equipped Royal Air Force of the possible approach of German bombers. What had itself begun in Slough as a method of tracking thunderstorms by radio became the famous Daventry experiment of 1935, which showed that an aircraft would reflect a radio signal with sufficient intensity not only for its presence to be detected but for its range and ultimately its altitude to be measured. This discovery soon led to the most secret activity on a remote, windswept site at Orfordness on the south-east coast of England. There, under the watchful eye of Watson Watt, members of his team from Slough, plus one or two others, began the development of what was to become the Chain Home RDF system that was to lay a radar screen almost around the coast of Britain. The story of radar has been told in great detail and in varying degrees of accuracy in the decades that followed and so will not be elaborated upon here until the South Africans themselves become involved. One fact though is clear. Regardless of who may be credited with actually having invented radar (if indeed any one individual can ever be accorded that honour) no-one seriously doubts that Robert Watson Watt, a mercurial Scot, was undeniably its father [11].

In Johannesburg, work continued on the development of wireless direction-finding equipment. Since the characteristics of the lightning stroke were such that the predominant radio frequency emissions were at frequencies below 100 kHz, the requirements to be met by the electronic circuits were not demanding and components suitable for constructing the three-stage amplifier were readily available. Hodges's previous experience in building his direction-finding receivers held him in good stead when it came to testing this latest equipment and, with Keiller's competence in the workshop, a working system was soon ready for use at the onset of the lightning season. The all-important antenna that collected the electromagnetic fields generated by the lightning was a 2.4 m length of wire, carefully insulated to protected it against inadvertent discharge by the rain, and mounted horizontally a short distance above the roof of the lorry. Its response was made aperiodic by means of heavy resistive damping, while the frequency response of the amplifier was carefully controlled so as not to distort the waveform of the recorded lightning stroke. The procedure followed in every measurement campaign was to drive the lorry to a point in the path of an approaching thunderstorm, having set up the equipment at the University beforehand. Then, a well-drilled series of actions followed, with one member of the team outside the vehicle (and frequently getting wet) observing the lightning flash and communicating with his colleagues inside by means of a system of bell signals that ensured that each electric field change recorded by the instruments was marked and so correlated with the lightning flash that caused it.

Data were gathered in this tried and tested way from more than 30 thunderstorms, and these produced records of the fields from 71 discharges to ground and considerably more within the clouds themselves. The electric field records showed that three major field changes had occurred during the discharge process: the first was characterized by a slow increase in field strength with a distinct ripple superimposed upon it, the second was a large positive spike, while the third was a slowly decaying peak. As the lightning discharges continued after the first in any sequence, so the electrical record showed no ripple during the first period, an increasing number of spikes accompanying the second and subsequent strokes, and a third peak in field strength that was basically unchanged from stroke to stroke. These same three regions had been identified the year before, in England, by Appleton and Chapman [12] and they had classified them as fields of type *a*, *b* and *c*. Now, for the first time, Schonland and his colleagues had produced photographic evidence that showed that these features in the electric field structure could readily be identified with specific aspects of the same discharge when viewed photographically. The stepped leader

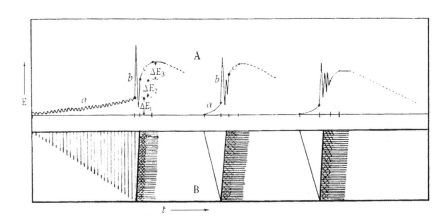

Figure 13. The correlation between the electrical and optical phenomena associated with the various stages of the lightning discharge. The upper diagram shows the recorded changes in electrical field strength, while the lower shows the images that would be recorded on film moving at the same linear velocity as the chart in the field strength recording apparatus. The stepped leader is clearly seen to be associated with *a*, the return stroke with *b*, while the continuing luminosity is related to part *c* of the changing electric field process. ([13, figure 2] reproduced by kind permission of the Royal Society.)

corresponded with part *a*, the return stroke with *b*, while *c* represented the continuance of the luminosity in the lightning channel after the return stroke had reached the cloud.

The data contained yet more intriguing information. Closer examination of the electrical record of the stepped leader showed that it could be divided into two readily identifiable classes. Schonland called these types α and β. Type α, where the field strength increased slowly but steadily before initiating the intense return stroke, was the more common and occurred in about 65 per cent of the discharges to ground. The type β leader was much more rapid in its earlier stages and advanced in long bright steps, but as time progressed so it slowed down markedly, and its visual intensity decreased too, before the sudden, intensely bright return stroke illuminated the sky and generated its potent positive spike on the electrical record. An abundance of photographic strips illustrated these effects with remarkable clarity, particularly the increasing complexity of the type *b* process as the number of strokes increased. To account for this phenomenon, he suggested that they must be ascribed to discharge processes that were actually taking place within the cloud itself, after the return stroke had entered the base of the cloud and this then rendered them invisible to the camera, but not to the radio receiver. Schonland, Hodges and Collens published the results of this extensive study in 1938 [13].

*   *   *

Matters of a more mundane nature, but of real importance to the financial viability of the Institute, consumed much of the Director's time as the final touches were being put to the BPI building. There was the relationship to be struck with the mining industry, for it was by far the most obvious beneficiary of some of the geophysical research either already under way or at least being planned. The Chamber of Mines, whose members made up the bulk of the gold mining industry on the Reef, were first approached by Bernard Price and then, more recently, by Schonland himself. Their discussions about a possible advisory role for the BPI on matters of direct interest to mining led to an agreement being reached in March 1938. The BPI would appoint 'a panel of four experts consisting of Professor Schonland and his Assistant and two University lecturers, who will keep themselves abreast of scientific development in connection with geophysical matters', it said. This panel* would then supply information and advice on these matters to the mining industry and, in return, the BPI would be paid £700 per annum for this service, of which the Chamber's share would be £500, with the balance coming from the Government's Geological Survey, which would also benefit as the need arose. It was also agreed that any practical work undertaken by the

---

* Its members were Schonland, P G Gane, C G Wiles and M A Cooper.

Institute for the mining industry would be handled piecemeal and an agreed fee fixed beforehand in each case [14].

The South African mining industry was very much a multi-headed hydra and each of those had a mind of its own. It was the Mining Houses, essentially finance corporations, that made up the Chamber of Mines and in 1938 there were seven[*] of them. The discovery of diamonds in 1867, but particularly of gold nearly twenty years later, had turned what was almost a forgotten land beyond the mountains of the Cape into a cauldron of frenzy as prospectors and others, many seeking their fortunes, descended upon the goldfields in the closing years of the nineteenth century. What became the South Africa of the next century with its immense mineral wealth mixed with grinding human poverty, racial division both within and across the lines of colour and a strong under-current of national fervour could, to a greater or lesser extent, all be traced to the mining of gold. The Mining Houses that sprang up almost as quickly as the headgear and the gold-tinted sand dumps that dotted the veld were run by men of strong wills and independent minds, while mining itself was an intensely practical and back-breaking activity. Its achievements were immense. By 1937, South Africa was supplying 35 per cent of the world's newly-mined gold. The deepest mines had reached more than 2.6 km below the surface—the greatest depth to which man had yet delved—from which 100 000 tonnes of rock were removed every working day. Gold alone that year earned the country more than £1 500 000 and to do all this the industry employed more than a quarter of a million men who supported many more back home [15]. But for all this most mining men were not much interested in science.

When negotiating with the Chamber of Mines Schonland knew that he was not dealing with another Keppel, but he knew too that turning elsewhere for long-term financial support for his Institute was just not an option available to him. He had to deal with the mining industry as he found it and scientific research with an undefined end was not an activity that industry understood. Many years later, when writing to Anton Hales, he would remark that 'the mining industry is ungenerous in its support of the Institute [because it is] so practical as to do itself some damage in the process' [16]. However, Schonland was himself a realistic man and he sought to serve South Africa's largest industry as best he could, while always hoping that it might, one day, come to appreciate more the value of science, even if initially it was pursued solely for its own sake.

The first request for practical assistance that he received from the Chamber came from Dr A J Orenstein, medical superintendent of Rand

[*] Rand Mines, Gold Fields, Union Corporation, General Mining and Finance, Johannesburg Consolidated Investments, Anglo American, and Anglo-Vaal.

Mines who, early in 1938, asked the BPI to investigate a claim made by a British company regarding a radio communications system reputed to work underground in mines. Orenstein pointed out that such a device, if feasible, would be particularly useful for rescue work [17]. Schonland was sceptical but agreed to investigate. Radio in its many roles was indeed a powerful tool at the BPI, especially for observing natural phenomena, and it had become a central plank of the Institute's work, with Schonland being regarded, at least by those in the country at large, as the authority on the subject. He, though, never considered himself to be an expert in an area of applied physics that was rapidly becoming the domain of the engineer. However, the nature of the research at the BPI made electronics of great importance and an appointment made in February 1938 ensured that it was well served in that important field. Dr P G Gane, a geophysicist with a natural electronics bent, was appointed in a probationary capacity for two years as Chief Assistant to the Director with special responsibility for 'the applied side of the Institute' [18]. Philip Gane soon became Schonland's deputy and was to serve the BPI with distinction for many years. Though afflicted with a stutter and less than perfect hearing that, at times, rendered him immune from events around him, Gane was soon to take charge of all the BPI's interests below ground, while the Director concerned himself with those that took place in the skies above.

*   *   *

On 21 October 1938 the Bernard Price Institute of Geophysical Research was opened officially by General the Right Honourable J C Smuts PC CH FRS in front of an audience of 260 dignitaries and invited guests. Externally, the building was rather stark, resembling in some ways the block-houses that had dotted the South African veld during the latter days of the Boer war. Its function, though, was scientific not decorative and in this it was certainly the most advanced laboratory in the University; indeed, its Director even went so far as to claim that it was 'one of the most modern research laboratories in existence'. Schonland's consultations with his colleagues overseas on matters of detail had paid off handsomely and all were now embodied in this modern, two-storey structure that perched on the south-west corner of the University property, with an almost panoramic view across the city and its southern suburbs to the mine dumps, veld and, most importantly, the sky beyond.

An impressive University brochure was printed for the occasion. It listed the names of the Institute's Board of Control, who actually outnumbered its staff, since Schonland and Gane were the only scientists in the full-time employ of the BPI at that stage, with Jock Keiller, as Mechanical Assistant to the Director, making up the complement. Two Associates, in A L Hales from the Department of Mathematics and G C

Figure 14. The Bernard Price Institute of Geophysical Research (ca. 1938) showing the 'bridge' at the rear which housed the cameras used for lightning photography. (Reproduced by kind permission of Dr Mary Davidson.)

Wiles from Physics, swelled the numbers, while Hodges and Malan were mentioned in their capacities as Honorary Associates; but they were all only itinerant members of the BPI team. There were also three research students whose presence was very much in line with Schonland's view that the training of young researchers was to be one of his tasks, but just as long as that function only took up a limited amount of his time. The rest of the brochure was devoted to describing the Institute itself, with Schonland's precise text illustrated by a number of photographs, plus two of the architects' drawings of the floor plans. The careful thought that had gone into each and every feature of the building was immediately obvious, from the special 'bridge', as he called it, along the southern wall of the building, just like the bridge of an ocean liner, for photographing lightning, to the Oregon pine floorboards mounted on concrete, to allow for the rigid fixing of any furniture to the floor. Pipes and ducts between rooms made for their easy linking by cables and every room could be made 'light-tight' for photographic purposes, while all had electrical timing signals at minute and second intervals fed to them from an astronomical clock. For the seismic work, about to commence in earnest, a special room had been constructed in which part of the concrete floor was rigidly embedded in the rock below. The provision of good support services were of paramount importance in such a laboratory. Schonland ensured by his disbursement of Bernard Price's grant that Keiller's workshop on the ground floor was equipped as comprehensively as was possible, for he knew from the Cavendish and George Crowe, from John Linton at UCT and now from Wits and Jock Keiller that on such men would depend almost everything that he and his colleagues might try and do. With an eye ever on the wider world, and especially on the BPI's contribution to scientific progress within it, he had included within the BPI accommodation facilities for visiting scientists and he would lose no time, once the formalities of official ceremonies were behind him, in appointing its first occupant.

Finally, as if nailing his colours to one of the radio masts that now projected above the Institute's roof, the brochure concluded by listing those areas of research that Basil Schonland considered to be the key aspects of geophysics to which he intended to devote all his time and efforts in the foreseeable future. Atmospheric electricity naturally took pride of place for it, and the name Schonland, were by now synonymous. Seismology, and specifically the study of the Witwatersrand earth tremor and its link to mining, would begin soon, while crustal magnetism would follow. One aspect of research that was actively encouraged by the mining industry because of the all-to-frequent occurrence of rock-bursts encountered in deep-level mining, was an investigation of the elastic properties of rock. And then, of course, there was wireless in all its roles and guises [20].

The first scientist to arrive from overseas was a Cambridge geophysicist, Dr E C Bullard, who was regarded as the foremost authority in the world on the subject of the thermal characteristics of rock. Some work had already been done in South Africa on this problem but the results were controversial in that they suggested that the rock in the South African gold mines exhibited a particularly low thermal gradient, which implied that their thermal conductivity was a lot higher than had been expected, based on evidence from other parts of the world. During the summer of 1938/39 Bullard worked on this problem at the BPI and showed conclusively that the conductivity of the Rand quartzite was some 60 per cent higher than the value generally accepted abroad and this completely vindicated the earlier South African finding [21]. He published a paper on this work that was to form the basis of much important research years later, when the temperature at great depths in the gold mines was a matter of considerable importance to the industry. As was the case with so many of Schonland's scientific colleagues and collaborators, he and Teddy Bullard were to meet again in rather different circumstances as the years ahead unfolded.

Matters above ground saw new developments as well. Schonland had initiated contact with the South African Broadcasting Corporation, the SABC, with a view to the BPI setting up a consultative panel to assist with matters related to radio broadcasting in the country. The work of Appleton, particularly, with which he had kept fully abreast, and the brief but useful period that Dr Eric Halliday of the Physics Department had spent with Appleton at King's College, some years before, suggested that the BPI could provide an ionospheric forecasting service for South Africa. Such forecasting or prediction is vital if a reliable radio service, dependent on ionospheric reflection, is to be provided continuously, over large distances and for extended periods of time. This is so because the ionosphere is a far from static medium: its characteristics vary significantly over a 24 hour period, with the seasons and over a period of nominally 11 years (the sunspot cycle). The decision as to which radio frequency to use for a particular application is based almost entirely on the state of the ionosphere at the time, and relevant ionospheric features have to be measured to make these predictions. Another honorary associate, in the person of G D Walker from the Witwatersrand Technical College, was appointed with the specific task of constructing suitable apparatus for taking soundings of the ionosphere. Walker had recorded such ionospheric echoes with equipment that he had constructed in 1936 and was therefore well-equipped to handle this task [22]. It was Schonland's hope that the new equipment would be ready for initial testing by Halliday and Walker before mid-1939, so that it could then be used in an expedition he was planning to mount in October 1940, when there would be a solar eclipse and much useful information could be gleaned as the ionosphere responded to the sudden solar change.

The nature of the earth tremors that were such a feature of life on the Witwatersrand was to be explored by means of a network of seismographs set up around Johannesburg. By the end of 1938 Phillip Gane had designed an inverted pendulum seismometer that produced a record of seismic activity on smoked paper. A number of these were then built by Keiller and installed at the Union Observatory, at two schools whose science masters had agreed to monitor them, and at Schonland's home, where the solid concrete floor of the wash-room provided the rigid connection to the earth beneath [23]. David Schonland, Basil's son, was an enthusiastic recorder of the results as they appeared on the chart driven by a synchronous electric motor. A timing accuracy of 20 ms was achieved by synchronizing all the recorders each day to the time signals broadcast by the SABC. The data generated by the network were analysed by Anton Hales and H O Oliver of the Department of Applied Mathematics, who then located the sources of the tremors by using a procedure devised by Philip Gane [24]. It was important to establish whether there was any link between these tremors and mining activity taking place almost beneath the streets of Johannesburg, and soon Schonland was able to communicate some of their findings to the Chamber of Mines. The sources of about 150 earth tremors had been located during the three-month period in which the network had been in use and, he told them, this had stimulated the Government's Department of Mines to agree to the installation of recorders over a much wider area. He also informed the Chamber of other research which he believed was directly relevant to its activities, such as Bullard's work on the thermal characteristics of rock and the study of the earth's magnetic field at considerable depths that he was just about to initiate, once he had the necessary staff to assign to it.

Turning to more practical matters, he informed the Chamber that, following Dr Orenstein's request, an investigation had shown on theoretical grounds that a radio system underground could not work, but to convince all concerned he had recommended that the English firm advancing the claim should supply the necessary apparatus for trial. Such a trial, he understood, had subsequently taken place and 'as expected, it was a failure'. But this was by no means the end of that matter, as subsequent events would show. In addition, his panel had given considerable attention to the rock-burst problem and a proposed method of predicting these catastrophic events that had been made earlier that year at a meeting of the Chemical, Metallurgical and Mining Society of South Africa. The BPI's panel were reluctantly forced to conclude, after careful consideration on solid scientific grounds, that the method suggested would not work. On a more positive note, though, he was able to provide the Chamber with a report on the performance of the lightning warning device that had operated very effectively throughout the previous lightning season at the Blyvooruitzicht gold mine, to the west of Johannesburg, where evasive action, when necessary, had prevented any untoward incidents. All was apparently noted with interest. At least the warning device had shown its worth but, for the rest, the magnates of the mining industry, guided no doubt by their hard-headed mine managers, did not feel inclined to follow the example set by the Carnegie Corporation some years earlier by sponsoring any research of a speculative nature. Schonland's suggestion that a technical committee be set up, on which both scientists and mining men would serve, was presumably tabled at some appropriate meeting of the Chamber but nothing came of it [25].

Schonland's other consultative venture with the Broadcasting Corporation had borne some fruit with the agreement that the BPI would set up a panel of experts, who would be available for consultation as the need arose. Once again the Director approached the University for permission to enlist the services of members of its staff who would act as Associates of the BPI when their services were needed. This time Schonland turned to the Department of Electrical Engineering, which had recently expanded its own horizons beyond the world of power engineering and had begun to teach courses in 'wireless'. The lecturer concerned was Guerino Renzo Bozzoli who was born in Pretoria in

1911, the son of Italian immigrant parents, and while still a schoolboy had dabbled with the new-fangled invention of wireless. In 1930 he registered to read Electrical Engineering at Wits and graduated in 1934 with a degree that included no wireless subjects at all, since no university in the land offered any courses remotely resembling wireless or radio, and the term 'electronics' had not even been coined. After a spell as a broadcast engineer with the African Broadcasting Company, the forerunner of the SABC, Bozzoli joined the Wits staff in 1936 as a junior lecturer in what was then termed 'Light Current Electrical Engineering', the subject of wireless having eventually earned its academic spurs. Bozzoli first met Schonland through the Physics Club, set up in that Department as a way of introducing its staff to the array of new ideas then exploding within the world of physics. In 1938 he was invited to lecture to the Club's members on the subject of cybernetics—the use of feedback for the control of electrical and other systems [26]. In June 1939 he became an Associate of the BPI and so began an association with Schonland that would have some far-reaching consequences* [27].

By the year's end, when Schonland tabled his annual report to the University Council, he was able to announce that the staff complement at the BPI had increased substantially. It now numbered five permanent staff members, with the recent appointment of two new research assistants, J S Elder and J W van Wyk, plus fifteen associates from various departments at Wits and other universities and institutions around the country, and abroad [28]. Not only had the BPI's activities attracted the interest of Teddy Bullard from Cambridge, but the wireless direction-finding work that Schonland had initiated for tracking the course of thunderstorms had brought it to the attention of the Rhodesian Meteorological Service, who saw in the method some way of augmenting the very sparse weather monitoring stations that had to serve the vast area of central Africa. One of its meteorologists, R A Jubb, had visited the BPI some while before, in order to see the system in operation. Shortly thereafter a set was constructed and sent to Salisbury where it was to be used, both there and in Bulawayo, as part of a three-station network with the equipment in Johannesburg and Durban, in a concerted effort to determine whether wireless direction-finding could be used in meteorology. The results of this work were duly published in December 1939 in the *Quarterly Journal of the Royal Meteorological Society* and were communicated to the Society by Sir George Simpson, with whom Schonland had first crossed scientific swords at the British Association meeting in 1931 [29].

---

* Bozzoli spent three decades at Wits serving as Head of the Department of Electrical Engineering, Dean of the Faculty of Engineering, Deputy Vice-Chancellor and from 1969 until 1977 as Vice-Chancellor of the University.

The three direction-finding stations, operating on their 10 kHz frequency, were in direct communication with each other through the good offices of the respective Postmasters General. They located thunderstorms at distances up to 900 km away with a position error of about 50 km. This degree of accuracy, or indeed inaccuracy, was due almost entirely to the very long wavelength used, but Australian experience had shown that simply decreasing the wavelength caused yet other problems and so it was concluded that the method was probably as good as could be achieved. To Schonland and his team of geophysicists, the fact that the direction finders produced results in excellent agreement with the those using conventional methods appeared most promising, but he sought an independent opinion from the meteorologists themselves and this came from the Rhodesians. Their view was much more cautious. From the results, as they interpreted them, they did not believe that there was sufficient justification for incorporating wireless direction-finding into the weather monitoring and forecasting processes, not through any inherent limitations but simply because of the operating costs involved. And so an immediate application was not forthcoming, but the expertise gained in the design, construction and operation of this direction-finding system would certainly not go amiss, even though it might not be used to track the movement of clouds and the processes of the weather.

Basil Schonland's reputation as the authority on atmospheric electrical phenomena had spread across the Atlantic. In 1938 he was invited to write a chapter for a book entitled *Terrestrial Magnetism and Atmospheric Electricity* soon to be published by the National Research Council in the United States and so he duly found the time to prepare what was to become a most important review of the state of knowledge in the field of thunder clouds and their electrical effects [30]. This was soon followed by an invitation to accept the position as an honorary Associate Editor of the Council's journal and, once again, this recognition of their Director was noted with approval by the BPI's Board of Control [31].

Work being done on the other side of the world in Australia had caught Schonland's eye too. In 1937, and again the following year, there appeared two letters in the columns of *Nature* that were of particular interest. They came from the Natural Philosophy Laboratory at the University of Melbourne and concerned what appeared to be the oscillatory nature of the atmospherics generated by a lightning discharge, as recorded on equipment designed according to all the principles laid down in Watson Watt's bible from Slough. But investigation by this Australian team under T H Laby suggested that the multiple peaks observed on their photographic records were not all oscillations within the 'atmospheric' itself. Rather, they were the result of reflections from the ionosphere of the signals radiated by the lightning discharge, followed by further reflections between the earth and the ionosphere and so on. Schonland was astounded by this and immediately set up his own programme to investigate the phenomenon for himself.

With his research assistants Elder, van Wyk and Cruickshank he re-examined the thousands of oscillograms accumulated during the recent collaborative work with his colleagues in Durban and Rhodesia and they found ample evidence to confirm the Australian findings. Reflections were indeed occurring at altitudes of about 85 km above the earth and the only possible reflector up there must be the ionosphere. Their interpretation of the photographic records yielded much new information as well. As the distance to the storm increased so it was evident that the amplitude of the first received impulse reaching the receiver became smaller than the first reflected component to be recorded. The Australians had accounted for this effect by suggesting that it represented the inherent complexity of the original atmospheric waveform, but Schonland rejected this view. All his experience of the lightning discharge told him that the duration of the first atmospheric impulse matched very closely that of the return stroke, which was the major component of lightning and remained so regardless of the distance to the strike that caused it. The explanation obviously lay elsewhere and the BPI team believed they could provide it. For lightning nearby, the first recorded impulse travelled in a straight line to the radio receiver, but as the distance between the two increased so the path followed was that of the well-known ground wave typical of all low-frequency radio broadcasting. This ground wave is rapidly attenuated with distance and so it was entirely likely that the first reflected ray (the skywave of broadcasting parlance) would have the larger amplitude. At even greater distances, this first skywave component would itself become tangential to the earth as it propagated towards the receiver and so it too would be attenuated, thus causing the second skywave component to have the largest amplitude, and so on. The masses of recorded data at the BPI supported this explanation, so Schonland immediately wrote a letter to *Nature* setting out their findings and conclusions [32].

This was a valuable contribution to the store of accumulated scientific knowledge, but it went further because it contained an additional and even more remarkable conclusion that Schonland had reached after poring over the data. It was apparent, so he deduced, that in the immediate vicinity of lightning activity the height of the ionosphere itself actually decreased significantly to just 50 km above the earth. Such a feature had never been noted before. The brief BPI communication to *Nature* provided no explanation for this but Schonland presumably remembered the intermittent upward glow from the top of clouds that he and James Craib had seen so often as storms swept across the Karroo all those years before. It is just possible that he now connected that strange phenomenon with C T R Wilson's suggestion that lightning might well strike upwards, even reaching the ionosphere. Might this actually alter the degree of ionization in the immediate vicinity of the storm? He reported as follows [33]:

> The evidence indicates that there is a real lowering of the effective height of reflection for long waves in the neighbourhood of a thunderstorm. Evidence for abnormal ionization of the layer under these conditions has already been found by many workers on short-wave pulse-sounding.

Between May and August 1939 all the efforts of the BPI and Hodge's team at the Natal University College in Durban were focused on this oscillatory phenomenon of the atmospheric waveform. Measurements were conducted mainly at night because the waveforms recorded then were much easier to interpret. Not only were the reflecting regions at a greater altitude than during the day, thus increasing the time interval between received pulses, but the disturbing effects caused by absorption were absent because the lowest D region of the ionosphere responsible for this disappeared after night-fall. This was indeed research of the very highest quality. Schonland, Elder and van Wyk in Johannesburg communicated telephonically through the good offices once again of Lt Colonel Collins with Hodges, in Durban, who had at his elbow Eric Phillips, his electrical engineering colleague. Many hundreds of waveforms were photographed, marked and numbered automatically at the two stations. They did this using the technique developed at the BPI for photographing lightning during daylight, where the brilliance of the cathode ray tube was increased significantly at the appropriate moment by detecting the radio emissions from the stepped leader. Since only the stepped leader, of all the components of the stroke, produces these particular emissions, all subsequent strokes were ignored and so the photographic records were of the first strokes exclusively. Not only was this joint experiment something of a local triumph from the point of view of the sophistication of the equipment used, but it was also an exercise in collaborative working between two teams of scientists some 500 km apart who were tracking and recording, to very high degrees of accuracy, the transient effects of natural events taking place some hundreds to thousands of kilometres away. The operators of the oscillographic equipment in Johannesburg and Durban were in constant telephonic communication with each other, while two more observers, listening in on the same circuits, watched the compass dials of the direction finders and recorded the directions of arrival of the signals from storms up to 3000 km away. Soon such expertise would find another, most urgent, application.

The paper describing this measurement campaign, and the interpretation of the results it produced, took some time to write because of the complexity of the problem, the amount of data to be analysed and the need to develop the theory to explain the phenomena observed. It was eventually published in the *Proceedings of the Royal Society* in 1940, by which time Basil Schonland found himself embroiled in matters of a very different kind [34]. The paper confirmed the finding of the Australians that lightning impulses were indeed reflected from that region of the ionosphere lying between the point of the lightning ground-strike and the two receiving stations. It also confirmed that the point of reflection was at a mean height of 88 km above the earth, thus placing it within the E region of the ionosphere. Of particular importance was their finding that the more complex ionospheric environment that prevailed during daylight altered the waveforms of the recorded impulses. These were less sharply defined because of both the changed nature of the reflecting layer and the increase in absorption due to the presence of the D region below it. Since proper interpretation of the data required a very well-defined pulse-train, Schonland's previous suggestion, made on the basis of daytime data alone, that a thunderstorm could alter the degree of ionization in the region above it was now shown to be in error. With careful analysis the more complicated daytime records yielded the expected result that the point of reflection was about 60 km above the earth, a natural consequence of the lowering of the ionosphere once it had become illuminated by the sun. This explained quite satisfactorily what had seemed previously like an entirely anomalous result and Schonland was quick to point out that his previous hypothesis was, as he termed it, 'unnecessary'.

Intriguingly, this BPI study had confirmed another finding that dated from a much earlier period in the history of radio science, but also one in which Schonland had played no small part. In 1930 H Barkhausen recorded in the pages of the American journal, the *Proceedings of the Institute of Radio Engineers*, a phenomenon that he had observed as a signaller in the German army during the First World War. It related to a sound heard in the headphones that was, as he put it 'like "peou"', a whistling tone that descended rapidly in frequency and it was well known amongst those who attempted to listen-in to the enemy's communications by means of sensitive audio amplifiers connected directly to ground by two, widely-spaced, conductors. Barkhausen had reported this phenomenon in 1919 when such sounds would still have been familiar to Basil Schonland who, in common with all signallers at the time, had heard these strange electrical whistles that some thought were emitted by the shells and grenades that rained down upon them. In his 1930 paper Barkhausen[*] attributed them to lightning impulses being reflected by the Heaviside layer. Schonland's paper of a decade later provided confirmation that indeed they were.

[*] Barkhausen H 1930 *Proc Inst Radio Engrs* **18** 1155.

CHAPTER 11

THE SSS AT WAR

When it met on 6 October 1939, the BPI's Board of Control had its usual full agenda of matters to discuss, but what was termed the 'Present Emergency' set the tone of the meeting when the Director announced that 'the facilities of the Institute had been offered to the Department of Defence for research work of a special nature ...' — but he elaborated no further. After discussing the conditions under which arrangements for this 'work of a special nature' might be made, the Board duly resolved to authorize its Chairman and the Director to conclude negotiations with the Defence authorities [1]. In fact, the only member present who knew any more was Basil Schonland himself; even Bernard Price, the Chairman, was in the dark as to the details [2].

Just a month before, on the morning of 9 September, there was a flurry of activity in the Schonland home as soon as Basil Schonland put down the telephone [3]. The call had come from Pretoria, from Brigadier General[*] Hoare in fact, the Director of Technical Services of the Union Defence Force, the UDF. He asked Schonland to leave immediately for Cape Town on a matter of very great importance to the war effort that was just beginning to gather some sort of momentum in a rather bemused and certainly very divided South Africa. In Parliament on 4 September, after a very tense and heated debate, South Africa's coalition government split on the issue of the war, and particularly whether the country should have any part in it. The Prime Minister, General J B M Hertzog, and his Deputy, General J C Smuts, their military ranks bearing testimony to their roles in the Boer War of some forty years before when they had ridden the veld and had so harried the British, were now irreconcilably split. Hertzog and his Nationalists favoured neutrality while Smuts, speaking with passion, proposed that the Union of South Africa should 'carry out the obligations to which it has agreed, and continue its co-operation with its friends and associates in the British Commonwealth

---

[*] The rank of Brigadier General was abolished in the UDF in July 1940, as it was in the British Army, and was replaced simply by Brigadier. It was reinstated in the South African National Defence Force in 1998.

of nations ...'. The House voted by a majority of just 13 in favour of Smuts's amendment; the Hertzog government resigned and Smuts was, once again, Prime Minister as well as Minister of Defence and Minister of External Affairs [4]. On 6 September South Africa declared war on Germany, just three days after Britain herself had done so.

Schonland's sudden departure for Cape Town three days after this momentous announcement had its origins in a decision taken in London well over a year before. It was at a meeting of the Joint Overseas and Home Defence Committee of Imperial Defence (JDC), held in Whitehall on 20 May 1938, when considering the matter of anti-aircraft (AA) defence, that the degree of the secrecy surrounding RDF, 'this distinguishing apparatus', as Churchill called it [5], was given particular attention. Since those first tentative experiments at Weedon, near Daventry in 1935, the development of what ultimately became radar had proceeded apace under the code-name of RDF.* Its unsurpassed capability for the early detection of enemy aircraft was so important that the JDC agreed that its general introduction into service must not be delayed by the imposition of 'unnecessary security restrictions' [6]. There were those who saw in this remarkable system a secret so important that its very existence should never be discussed but this was, of course, patently absurd because the number of people soon to be involved in its development and operation would increase dramatically from the few who were privy to the details now. Secrecy under such conditions could only be maintained if it was underpinned by common sense. Quite how advanced in this technology other countries were was uncertain. Sir Henry Tizard, Chairman of the Air Ministry's Committee on the Scientific Survey of Air Defence (CSSAD) that had been instrumental in turning the Watson Watt memorandum of 1935 into the Chain Home (CH) radar system, was quoted at the meeting as believing that 'our own apparatus was immensely superior to anything else that had been produced so far'; but its existence was no secret. A recent American newspaper article had claimed that they, the Americans, already knew so much about 'the RDF idea', and had known about it for a long time. To some, therefore, it seemed well worth Britain's while to offer some sort of exchange of information with them, but, as it turned out, that was an idea still somewhat ahead of its time. However, there were some who should be informed and Colonel Sir Maurice Hankey, chairman of the JDC, suggested that Britain's Dominions, particularly, should be made aware of the present stage of RDF development, at least unofficially, through

---

*RDF is frequently assumed to mean Radio Direction Finding or Range and Direction Finding but neither is correct. RDF is not an acronym at all but was merely 'a code name intended to have no identification' according to its originator R A Watson Watt in a note on the *Secrecy of RDF* 28.6.38 (S.40952) AIR 2/4487 [Public Record Office]

what he called 'technical channels': the Chief of the Air Staff's quarterly *Dominion's Liaison Letter* due in September. In August, a minute from Robert Watson Watt, in his new capacity as Director of Communications Development at the Ministry of Aircraft Production, recommended the disclosure to the Dominions of the technique of ship location because it was the considered opinion at the time that the only major risk to the Dominions was an aircraft carrier-based attack. But he too subscribed to the prevailing view about more general disclosure and recommended against passing such information to the USA [7]. On 12 December 1938 the Chief of the Air Staff, Sir Cyril Newall, formally notified his deputy that 'I have dealt with the release of information to the Dominions ...' but he stated that it was 'quite inacceptable [sic]' to release it to the USA or the British public [8]. His brief paragraph read:

<u>Draft Paragraph for C.A.S. Dominion</u>
<u>Liaison Letter</u>

For some years research and development on a radio method for the detection and location of aircraft have been in progress. The experiments, which have been kept highly secret, have been so successful that a chain of stations is now being erected on the east and south coasts of England as an integral part of the fighter organisation. The first group of stations is now in operation. It is not considered that secrecy about the technical and operational details (including the performance data) should be waived unless this is required for the embodiment of locating stations in a fully organised fighter defence system. If, however, you have such an organisation in view for early completion, we are prepared to give detailed information to Air Staff and technical officers whom you may detail for the specific purpose of receiving the information verbally and by demonstration.

However, the matter of secrecy continued to concern many at the most senior levels within Whitehall and before this liaison letter was dispatched Newall had second thoughts and instructed that it be deleted with the intention now to pass on the information to the Dominion governments 'in a separate communication, the circulation of which could be controlled to the absolute minimum'. Opposition to doing even this now reared its head in the form of the War Office and the Admiralty, both of which opposed any form of disclosure and so, to break the impasse, the matter was referred upward to Cabinet level. Eventually the Secretary of State for Air, Sir Kingsley Wood, was given the go-ahead to convene a meeting of the High Commissioners of Canada, Australia, New Zealand and South Africa 'for the purpose of expounding the position as regards RDF'. The meeting duly took place on 24 February 1939 when a statement, prepared in conjunction with the Admiralty and the War Office, was made to them [9]:

Statement on R.D.F. for the Dominions [sic] Representatives

It has been found that wireless waves are reflected by aircraft in flight, and a technique of causing and measuring such echoes has been developed, by means of which it is possible to determine the range, bearing and height of distant aircraft. The system, which is called R.D.F., enables a single ground station to determine the position and height of single aircraft and formations in a wide and deep forward sector, and also to give some information about the size of each formation. The system is particularly suitable for dealing with high-flying raids, and it enables continuous watch to be maintained over the sector under observation.

The information provided by an R.D.F. screen enables the Fighter organisation to effect interception further forward than is possible with ground observation by observers or by means of acoustical apparatus; and in fact it has been shown as a result of tactical exercises that it is possible by means of R.D.F. to intercept the enemy on the coast or, in favourable circumstances, out to sea.

The range at which aircraft can be detected by this means depends on the height at which they are flying. The higher they fly the greater the range of detection. Aircraft flying at 10 000 feet can be detected at a distance of approximately 100 miles to within an order of accuracy of 1 mile. At 1000 feet on the other hand the effective range of the apparatus falls to the order of 20–30 miles. The height of the aircraft can be measured to within 500 feet, and as has already been stated, an estimate can be made of the number of aircraft in formation.

A chain of R.D.F. stations is in the process of construction along the whole of the east coast from Portsmouth to the Tay. Of the 18 stations required to cover this front, 12 are now working and the remainder will be working by the early summer of this year.

There is a number of other applications of the R.D.F. principle which are now under development. A type of R.D.F. apparatus has now been developed which can be carried in reconnaissance aircraft for the detection of surface ships. Its effective range exceeds that of normal visibility in British waters and is independent of visibility conditions either by day or night. This equipment will greatly increase the effectiveness of reconnaissance over the sea, by increasing the area searched by each aircraft in a sweep, by increasing the closeness with which the area can be searched, and by extending to conditions of darkness and of restricted visibility the times at which sweeps can be carried out.

Another form of R.D.F. equipment is being developed for use in ships. Its function is to give long range warning of the approach of aircraft to enable a naval unit either at sea or in harbour to be ready for air attack.

There are other types of R.D.F. apparatus under trials for application to the ranging of anti-aircraft guns and the direction of searchlights on an enemy aircraft. Finally a type of apparatus is being developed for coast defence purposes, by means of which the presence of ships can be detected, and their position determined with sufficient accuracy for barrage fire.

The Dominions Governments will no doubt desire to study the R.D.F. technique, in order that advantage may be taken of our knowledge and experience in connection with the development of their future plans for defence. It is suggested that, as and when the plans of the Dominions Governments reach the point when a study of the technical and operational aspects of R.D.F. appears desirable, facilities to examine the working of the system in this country will be given to an Air Staff and a technical representative from the Dominions Governments.

We do not think that other countries have got as far as we have in this technique. The need for the greatest secrecy in regard to the information that I have just given you is therefore obvious, and I would accordingly ask you to treat it as most secret.

It was then agreed that the High Commissioners would immediately inform their respective governments of this 'security device connected with air defence' and, in view of its special importance and undoubted technical complexity, should request each of them to send to England a physicist at the earliest possible date, in order to study it. Three days later South Africa's High Commissioner dispatched a secret telegram to his Department of Foreign Affairs in Pretoria asking them to inform the Minister of Defence that Sir Kingsley Wood and Air Marshal Sir Cyril Newall had disclosed to him 'under promise of absolute secrecy certain technical developments which are of vital importance against air attack' [10]. As agreed, he also asked that a person with the 'highest possible qualifications in physics' should be sent to England for a period of two to three months in order to become completely familiar with the technology.

The man sent from South Africa was not a scientist but a soldier. Whether it was the need for absolute secrecy, or the parlous state of the Defence Force's coffers or merely a lack of appreciation of quite what was involved is not known but it was decided in Pretoria that the UDF's Director of Technical Services, Brigadier General F R G Hoare himself, would go to Bawdsey on the Suffolk coast to be briefed. To assist him, a South African officer already serving in England, Major H G Willmott,* was told to present himself at Bawdsey Manor, the stately home that had

---

* Some confusion exists between various sources as to the spelling of Willmott's name. Reference [9] gives it as 'Wilmott', without initials, while [10], similarly uninitialled uses 'Willmot' but he was presumably H G Willmott, a South African Air Force officer who had attended Staff College at Camberley in 1938 and had returned to England the following year where he was based at RAF Andover [11].

been the centre of intense RDF activity since March 1936, after moving just down the coast from the bleak site at Orfordness, where this secret work had all started the year before. There too was the group of scientists duly dispatched thither from the other Dominions, all to be enlightened by, amongst others, Watson Watt and John Cockcroft. Cockcroft had himself arrived at Bawdsey just a short while before when, with the threat of war looming, he was recruited from the Cavendish by Tizard to set up an organization that would act as a 'scientific nursemaid' for this precocious stripling, known only as RDF, that was growing rapidly under Watson Watt's enthusiastic guidance [12]. The two South African military men found themselves in the company of the colonial scientists Dr J T Henderson from Canada, Dr D F Martyn from Australia and Dr Ernest Marsden from New Zealand, all sent post-haste to England in response to secret telegrams from London. Marsden was the same young Englishman who, years before, when working under Rutherford in Manchester, had performed the vital experiment that established the existence of the atomic nucleus. Since 1926 he had been secretary of the DSIR in Wellington and now he was his adopted country's representative at the unveiling of one Britain's great scientific secrets.

Watson Watt's opening address, what Cockcroft referred to as his 'Child's Guide to R.D.F.', was an introduction to the subject evidently used with great effect on Cabinet Ministers [13] and Hoare and Willmott would have grasped its significance, if not the technicalities. But soon the subject moved rapidly into realms of physics and engineering beyond their comprehension and it was obvious that a mind trained in such matters was needed to make any sense of it. Their mission to Bawdsey, though, was not wasted, for it allowed the War Office the opportunity to establish South Africa's likely requirements for RDF equipment. On 23 March 1939 Hoare met Britain's Deputy Chief of the Air Staff (DCAS) to discuss the matter with him. They agreed that the South African target most likely to attract the enemy's attention was Cape Town, and Hoare described the rather puny defensive measures then in place to protect South Africa's Mother City. The artillery were manning just two 233 mm (9.2 inch) guns for seaward defence while the Royal Navy's South African base at Simonstown was protected by similar, inadequate, armament, as it waited expectantly for the arrival of 381 mm (15 inch) guns promised for sometime in the near future. From the air, the South African Air Force (SAAF) could provide only six modern aircraft: a single Fairey Battle and a Blenheim bomber plus four Hurricane fighters. For the rest it boasted, for reconnaissance, a squadron of Junkers JU86s recently commandeered from airline service, and another of obsolete Hawker Hartebeest light bombers for more aggressive action in times well past. The South African Naval Service, formed in 1922, to take over the naval needs of the country previously handled by the Royal Naval Volunteer Reserve (South Africa), had been abandoned in 1933 as a result of the financial stringencies of the Depression and so, once again, it was the RNVR (SA) that stepped into the breach and manned two mine-sweeping trawlers and a hydrographic survey ship [14].

None of this was too impressive and so it was imperative that there should be some means of early warning of an attack and this now obviously implied the use of RDF. It was therefore decided to advise the Union Government to make immediate provision for the installation of certain items of British equipment considered sufficient for the protection of Cape Town. These would be three Coast Defence (CD) sets for ship detection from the shore, other sets for ship detection from aircraft, the number required depending on the number of aircraft the SAAF could make available for this role; and a single mobile base (MB) set, mainly intended for cooperation with fighter aircraft, but also usable in an air-raid warning role. It was also readily acknowledged that, in the event of Britain losing its control of the Mediterranean, the sea-route around the Cape would become of paramount importance within 'the chain of Imperial communications' and such a fighter squadron would be a most valuable addition to its defences. The sea-route around the 6000 km of South African coastline would then be of vital importance to the grand plans unfolding in London but it would take some defending. So the complete RDF requirement to include such an eventuality would be have to be increased to the following [15]:

CO* (Chain overseas) stations	2
MB (Mobile base) stations	5
CD (Coast defence) sets	5
ASV (Air to surface vessel) sets	15

General Hoare returned home safe in the knowledge that he had taken steps to draw attention to the defence requirements of South Africa and her consequent need for RDF protection. However, he brought with him none of the technical information that, according to official records, had been given to the Dominion representatives on their departure from Bawdsey [16]. One must presume that only the scientists in attendance were accorded that privilege; however, there was now an urgent need for South Africa to be able to establish, operate and maintain the promised chain of RDF stations around its coast, but without the scientific competence to do so this would be impossible.

---

* CO implied 'Chain Overseas' and referred to the fact that this version of the CH station was semi-tropicalized and intended for service outside of the UK. They were also more modest systems in that they used just two 73 m timber masts as opposed to the considerably more numerous (and larger) CH towers.

If the name Schonland had meant nothing to Hoare when he arrived in England, his association with Cockcroft and Watson Watt during his stay at Bawdsey would have made him fully aware that South Africa possessed just the scientist in whose place he had found himself. This would have been reinforced in no uncertain measure as soon as Defence Headquarters in Pretoria came directly under the control of General Smuts when he took firm hold of the reins of office in September 1939. There can be no doubt that, when Hoare contacted Schonland on the morning of 9 September to set in train Schonland's hasty departure for Cape Town, he did so at Smuts's behest.

*   *   *

Late in August 1939, just before he was due to set sail for New Zealand aboard the SS *Athenic*, which had sealed into its spare mail room an incomplete 1.5 m ASV radar set, two Pye television receivers, some oscilloscopes and reels of coaxial cable, Ernest Marsden changed his booking [17]. Instead, he sailed on 2 September on the *Winchester Castle* which was due to call at Cape Town and he did so in order to meet Basil Schonland there, so that he could pass on to him what he had learnt at Bawdsey about RDF and its capabilities. The diplomatic communication channels between Pretoria and London, and presumably Wellington too, had undoubtedly been busy when it was realized that Marsden could be the courier who carried the scientific message to South Africa. With him, as his most precious cargo, he had what subsequently was always referred to in South Africa as *The RDF Manual*, presumably those diagrams, descriptive data and blueprints supplied to him at Bawdsey.

On 14 September Marsden and Schonland met at Cape Town docks, when Schonland boarded the ship for the next leg of its scheduled journey to Durban. During the three days of the journey along South Africa's east coast these two men, whose early scientific careers had each owed so much to the influence of Rutherford, and whose lives had seemed destined to follow almost uncannily parallel paths, locked themselves in Marsden's cabin, while they worked through the contents of that secret document. Immediately on arrival at Durban they disembarked and made for Natal University College, where Schonland's two associates at the BPI, David Hodges and Eric Phillips, were waiting for them. After Schonland had administered his own version of the Official Secrets Act they worked through the night laboriously making glass photographic slides of selected pages and diagrams from Marsden's *Manual* [18]. These would serve as the blueprints from which Schonland would launch South Africa's own venture into the world of radar.

He rushed back to Johannesburg and then reported immediately to General Hoare in Pretoria. What happened next lies somewhat in the realms of conjecture, but it can reasonably be assumed that South Africa's new Prime Minister was not only fully aware of the highly secret information on its way to South Africa, but had been directly instrumental in having Marsden's travel plans changed at the very last minute so that he could deliver it himself. Schonland now had that information, albeit in much diluted form and supported only by a handful of hastily copied pages on hardly-legible blocks of glass. Arrangements were immediately made with the concurrence of Bernard Price and the University for Schonland to set up a special establishment within the confines of the BPI, but shrouded in secrecy, for this 'work of a special nature'. The secrecy demanded of everybody even remotely connected with RDF took immediate effect when it was decided that this rather irregular grouping of scientific academics should fall under some form of military authority and so on 18 September the BPI found itself under the wing of the South African Corps of Signals [19].

And so it was that none other than Lieutenant-Colonel Freddie Collins who, in his pre-war guise as Under Secretary of Telegraphs, had been of such help to Schonland by providing special telephone facilities between Johannesburg and Durban when he and Hodges were tracking the radio emissions from lightning, now assumed control of the BPI in his capacity as Deputy Director of Signals of the UDF. That occasional pre-war association between them was soon to become very close indeed, as the targets changed significantly and the techniques to detect and track them did too. The role and function of the BPI, however, changed dramatically at this point as it took on a very special responsibility. Seismic research was to continue, but those engaged in it had to be accommodated elsewhere within the University. Anton Hales was asked by Schonland to take over and complete the work on locating the sources of the earth tremors and to write the report for the Chamber of Mines in the absence of Phillip Gane, who was now to join Schonland on other business. All the equipment, plus the mass of records accumulated over many months, were immediately moved to Hales's office in the Department of Applied Mathematics. Hales himself still required occasional access to the BPI's dark-room and so this necessitated his signing the Official Secrets Act lest his eye should fall on the special racks that had been made to support the highly prized glass slides containing the secret of radar [20]. J S Elder, who had been closely involved in the tracking of thunderstorms, declared himself a pacifist and so was transferred immediately to the Johannesburg Technical College to work on the ionosonde being built there by G D Walker.

While Colonel Collins's ultimate objective was to have the necessary organization and personnel in place to install and operate the British RDF equipment when it arrived, Schonland's immediate intention was to design and build an RDF set 'purely to teach ourselves technique' [21]. To do so, however, would require skills that stretched beyond those he had assembled over the past year or two at the BPI. Even though he and his colleagues were fully conversant with the methods of Watson Watt in locating and tracking thunderstorms, Schonland's brief glimpse at the technology that lay behind Britain's RDF chain convinced him that he would need specialist assistance and for this he turned immediately to his electrical engineering colleagues at South Africa's universities. From Natal he had already enlisted the services of Eric Phillips, if only initially as a darkroom assistant, while immediately to hand at Wits was 'Boz' Bozzoli, who had become an Associate of the BPI just a few months before. From UCT came Noel Roberts, another young lecturer willing to offer his skills in radio engineering in the service of the country. They were all recruited without delay. Phillip Gane, too, joined the team. Of course, nothing would have been possible without adequate workshop facilities and a competent technician to use them, so Jock Keiller also found himself a member of this special band whose very existence was so 'hush-hush' that all information related to it remained extremely vague, even to the military who were supposed to be in charge [22]. The need for absolute secrecy required the imposition of military control over the activities now about to begin in earnest at the BPI, but the last thing that anyone wanted, least of all the military authorities themselves, was any hint of the nature of the work that was going on there and so Schonland's special team all remained in mufti [23].

Work commenced immediately on designing an RDF apparatus that embodied as closely as possible the principles used in England, but circumscribed by the very limited availability in South Africa of special components, radio valves particularly, that were suitable for operation at the high frequencies and power levels required. It was soon obvious that the technique to be adopted at the BPI must be based on the so-called searchlight principle of the British coast defence (CD) radars, and the Chain Home Low (CHL) radars that followed directly from them, rather than the floodlighting method of the original Chain Home (CH) stations. Essentially, the difference amounted to either generating sufficient power at some suitable radio frequency and then radiating it over a wide area in front of a transmitting antenna, as would occur with a floodlight or, by using a suitable antenna to produce a much narrower beam, as in a searchlight, that could be steered to illuminate the target.

To achieve its purpose the CH transmitter, which was the immediate outcome of Watson Watt's famous Daventry experiment of 1935, delivered pulses of energy at a peak power of about 350 kW in the frequency range between 20 and 30 MHz to a horizontal dipole antenna array mounted on massive, steel towers more than 100 m in height. The radiated energy essentially occupied a beam that had a width of about 100 degrees [24]. The CH radar receiver, with its cathode ray tube display, was connected to a considerably smaller antenna, or two antennas to be exact, mounted at right angles to one another at about 65 m above the ground. An aircraft within the transmitted beam would reflect a small fraction of the incident energy back towards the receiving antenna, from where it would be fed to a sensitive receiver located in some suitable enclosure on the ground beneath. The direction from which the reflection or echo occurred was determined by the principle of the goniometer first developed many years before by the Italians Bellini and Tosi — a technique with which Schonland was very familiar, both as a signals officer during the First World War and, much more recently, from his own direction-finding work using the crossed loop system developed by Watson Watt at Slough for his research into the mechanism of atmospherics.

The immediate problem faced by Schonland's team when they first sat down to discuss the design issues involved, was how to generate such huge amounts of power, even at the relatively low frequencies used by the CH system, when no suitable valves existed in South Africa and none would be forthcoming in the immediate future from England. In short, it would be impossible and so an alternative solution was sought. By contrast, the British CD set (and the very similar CHL that first appeared in November 1939 [25]), operated at the much higher frequency of about 200 MHz with a peak power of about 150 kW and used a considerably more directional, and rotatable, antenna system to concentrate the radiated energy into a narrow beam that gave the bearing of a target merely from the heading of the antenna, so eliminating the need for the complexity of crossed elements and goniometers. Such antenna directivity also increased the amount of power effectively radiated in the direction of the beam as long as the antenna was large compared with the wavelength. This reliance on the antenna, rather than on high transmitter power, of necessity required the use of a much shorter wavelength if manageably-sized antennas were to be used. So a frequency more akin to that used by the CD and CHL sets, rather than their large CH counterpart, was agreed upon. However, the problem of the transmitting valves remained, for there was no source of valves in South Africa capable of producing anything like the power of the British CD set and no likelihood of acquiring them from overseas under wartime conditions. The fact remained, though, that the chances of achieving some success were markedly better if the CD/CHL approach were followed. And so it was agreed that every effort would be put into producing as much radiated power as possible at the highest frequency that locally-available power amplifier valves could operate [26].

Schonland planned the programme of work with his customary attention to detail and with the enthusiasm that had so characterized every new venture that he had undertaken throughout his research career. He fully appreciated that the electronic aspects embodied in this radar technology were beyond his capabilities, but what he may have lacked in such expertise he more than made up for in his organizational ability and inspirational flair. He delegated to each of his younger colleagues an area of responsibility that best suited their interests and particular experience. Gane designed the transmitter, Bozzoli and Phillips the receiver, while Roberts tackled the numerous timing, synchronizing and calibrating circuits as well as the cathode ray display system [27]. Schonland himself assumed responsibility for the design of the antenna, a crucial element of the RDF system.

It was a demanding task and all they had to assist them were the bald details of the British equipment contained on the precious glass slides, now in semi-legible paper form, that Marsden had supplied. For more fundamental information, as well as the host of practical ideas they could provide, Schonland's team turned frequently to two publications that had assumed almost biblical proportions in the fields of amateur and professional radio engineering. Both were of American origin: *The Radio Amateur's Handbook*, a publication of the American Radio Relay League, and *Radio Engineering* by F E Terman, known to all who used it simply as 'Terman'. For the specialized electronic components, especially the valves, they scoured the local amateur radio suppliers in Johannesburg and relied on Jock Keiller in his workshop for the fabrication of the metal chassis in which to house them. Their progress was remarkable. By mid-November, just two months after Schonland and Marsden met in Cape Town, the individual elements of the system were all operating and by the end of the month they had a working radar-like device but no idea whether it would be capable of detecting any targets [28].

The transmitter and receiver were each equipped with separate antennas and were separated physically from one another to reduce as much possible the paralysing effect of the transmitted pulse on the sensitive receiver. So, for testing, the transmitter was situated in a top-floor room of the main University building, Central Block as it was called, with its antenna on the roof above, while the receiver was in the BPI building and its antenna was located on the capacious roof that had actually been intended, when Schonland designed it, to support such structures but never with such an application in mind. Communication between the two sets of operators was provided via the University's telephone exchange: a technique of coordination in which Schonland's team were already well-versed. Since the intention was that the radar system would be used to detect aircraft, the first experimental target consisted of a fine mesh of copper wires slung beneath a hydrogen-filled balloon that was launched some 10 km from the BPI. Even though at least five pairs of eyes squinted at the cathode ray tube, there was no sign of the tell-tale 'blip' that would have indicated a reflected signal. Not deterred by this, Schonland arranged for a flight by an aircraft of the South African Air Force, the pilot of which was given specific instructions as to the course he should fly but no reason as to why such an exercise was necessary. No echo was received on that occasion either. But there was no scientific flaw to account for this, for it later transpired that the pilot saw little sense in an apparently pointless manoeuvre and so detoured over the house of his girlfriend instead!

Then, on 16 December, an important public holiday on the South African calendar, Schonland and Bozzoli went to the BPI to see whether they could possibly extract just an ounce more performance from their RDF equipment by careful tuning and adjustment—and this they certainly did.

On this occasion no pre-arranged targets were used, instead the antennas were rotated slowly in rough synchronism from north to west. Suddenly a definite signal was observed on the cathode ray tube and movement of an antenna by just a few degrees either side of that bearing caused it to disappear, only to re-appear when the original heading was re-established. This was undoubtedly an echo from a target and with great excitement they scrambled on to the roof to see what it was. At a distance of about 10 km to the north west of the University is a range of hills known as Northcliff and perched atop it is a water-tower, which is a very well-known landmark in Johannesburg. This, they immediately agreed, was the object their radar was detecting and it was the first radar target ever seen in South Africa [29].

Whether it was the water tower, as they assumed, or possibly Northcliff itself, as others tended to believe subsequently, is of little importance for of much greater significance was the achievement itself. Within a period of just three months since Schonland's return from his shipboard meeting with Marsden his team of three engineers, a physicist and a laboratory technician had designed and constructed a working radar[*] set. What makes this feat all the more remarkable was the complete absence at the BPI at that time of even the most elementary laboratory test equipment suitable for use at the frequency at which the radar was operating. They had no appropriate signal generator nor even an effective oscilloscope, and certainly no means of measuring the power that the radar produced, and yet by dint of their collective skill, good engineering practice and, undoubtedly, an element of good luck achieve it they did.

---

[*] 'Radar', as a term, was only coined by an American naval officer in 1940. In 1939 the South Africans would have referred to it as RDF, as did everyone in England. The Australians, for some reason known only to themselves, called their device a 'doover' [30].

In this they were in good company, for the British team under Watson Watt's guidance at Orfordness had themselves developed their first radar in facilities described by Hanbury Brown, one of those pioneers, as being more suitable for bird-watching than for advanced electronics! They too had no signal generator, while their only textbook happened also to be *The Radio Amateur's Handbook* [31].

This prototype radar built at the BPI was christened JB0 by Schonland. In doing this he was simply following the tradition first established there when he and Hodges, in Durban, were collaborating in their direction-finding work on atmospherics. Then, the two stations were always shown on the maps and in the logs that they kept, as JB and D [32]. In his war diary Schonland recorded the details, as he knew them, of this first radar system. It operated at a wavelength between 3 and 3.5 m using identical antennas mounted on steel poles 6 m above the roofs of the two buildings. The antennas themselves consisted of three vertically stacked, full-wave dipoles separated by half a wavelength, with all elements fed in phase. This array was then backed by similar reflectors about a quarter wavelength behind them to produce a single, reasonably directional beam. The special thermionic valves that were so crucial to the performance of the radar were procured from the amateur radio trade, which had just begun to stock components suitable for use at frequencies considerably higher than was used in day-to-day broadcasting, for the amateurs were beginning to explore, what they called, this ultra-high frequency frontier. Gane's transmitter used two type 250TH triodes in a circuit known as a linear power oscillator. Its most distinguishing features were the heavy copper tubes that formed the resonant circuit elements of the transmitter with its two, large, glass-encapsulated valves mounted within a wood-framed enclosure surrounded by fine copper mesh. In appearance it was more like a bird cage than a key part of a highly-secret and technically advanced piece of electronic apparatus, but such curiosities were fully justified by its designer as he strove for as much power as possible at a frequency around 90 MHz. The performance of the radar receiver depended critically on the use of the very new 955 and 956 acorn valves intended for use at frequencies as high as 450 MHz. For the receiver, Bozzoli used the type 956 pentodes as both the RF amplifier and mixer, with the 955 triode as the local oscillator in a conventional Superhet design, with an intermediate frequency of about 9 MHz. Phillips, in designing the IF amplifier, followed very closely the conventional techniques of the time by relying heavily on the circuits used in television receivers that had come into service in Britain, but not in South Africa, just a few years before. Roberts developed the various timing and pulsing circuits with an eye on those elements of the CH system then operating in England, even to the extent of including the so-called 'spongy-lock', a form of electronic shock absorber intended to mitigate against variations in the mains supply.

Such sophistication, though, was hardly needed at the BPI and, ironically, its use would soon present the operators in the field with a major problem. In England it was another matter entirely because of the crucial need to ensure that all the CH stations around the British coast were synchronized; failure to do so caused the traces on the displays to wander about the screen and so make it well nigh impossible for the operators to obtains accurate range readings. To overcome this, all the stations used the 50 Hz mains supply as the standard to which they locked their various oscillators, but the mains supply itself was given to occasional transients and sudden, though brief, jumps in frequency, especially in wartime Britain. To prevent these from affecting the radar chain the 'spongy lock' circuit introduced the electronic equivalent of a mechanical shock absorber into each CH station and the technique worked admirably as long as the radars were all connected to a basically sound source of power, in essence the electrical power grid. Its use with diesel generators in East Africa would, however, turn out to be very different.

While his young colleagues worked with slide rules, soldering irons and sheet metal to produce the various sub-units that ultimately came together to form the JB0, Schonland encouraged and supported them. To some he appeared initially as a slightly forbidding figure, but there was always a hint of a friendly smile on his face. To those who knew him well he was a man of great warmth and friendliness, and the mask was simply a natural shyness towards those who were not yet his intimates. When problems arose he knew when to advise and, equally, when to withdraw, should his presence impose any unnecessary pressure, or even an element of frustration, when neither was intended. For Schonland, more than anyone, realized the magnitude of the task he had set them. Technical problems were bound to arise which, at times, seemed insurmountable but in Bozzoli, particularly, he had an engineer whose earlier experience in the technology of broadcasting was directly relevant to much of what they were doing, and it was those techniques that formed the basis of much of the JB's design and construction. Schonland's own role was very much like that of Watson Watt some time before at Bawdsey. He was the sage and the mentor but, most important of all, he provided the trusted link with the military without which nothing would have been possible. With Ismay ever-present in the background the Schonland home became, over the occasional weekend, the focal point for activities as far removed from radar as he could make them. There, his radar team would gather, some with their wives, to play tennis, to listen to music and to participate in Basil Schonland's passion for charades. Everyone had their part to play and none more enthusiastically than Schonland himself, and his enthusiasm was infectious [33].

*   *   *

The news, when it reached Pretoria, of Schonland's achievement at the BPI caused more than a mild reaction. Until then Colonel Collins had allowed Schonland a completely free hand while relieving him of all financial concerns associated with RDF matters. He now realized that those somewhat loose arrangements made in September required formalizing and so he wrote, on 21 December 1939, to the Secretary for Defence proposing the establishment of a Special Wireless Section (Home Defence) and suggested, for consideration, a suitable Establishment Table of officers 'but', he added, there was 'no need to fill all the posts at the moment'. This Special Wireless Section would fall under the immediate command of Major Schonland with Bozzoli, Gane, Phillips and Roberts all accorded the ranks of Captain, while Keiller and a former Post Office technician by the name of Anderson, recently recruited to assist him, would become Staff Sergeants [34]. Basil Schonland was to achieve the majority that was his, had he wanted it, when in 1918 he was offered the post as Chief Instructor of Wireless in the British Army. Now he found himself once more playing the part of unaccustomed soldier but this time it was shrouded in anonymity.

The Christmas holidays soon intervened, but only briefly. Then, as 1940 dawned, a new man arrived at the BPI. His name was Frank Hewitt and he had just completed an MSc in Physics at Schonland's *alma mater* in Grahamstown where, some time before, Schonland had offered him a position as a member of the lightning research team at the BPI [35]. Though formally unschooled in matters even remotely related to modern electronics, since university physics courses merely touched on such things, Hewitt was well-versed in the complexities of radio receivers, having built his own as an inveterate experimenter at home. With Phillips and Roberts soon to return to their universities where their academic services were once again required, Hewitt's arrival at the BPI could not have been more opportune. Not only would his presence offset the loss of two members of Schonland's team, but it now transpired that the promised arrival of the British radars upon which South Africa was depending for its coastal defence was beginning to look more and more remote and something would have to be done. However, throughout his first week at the BPI, Hewitt was not made aware of the secret work then underway in the laboratory; but he was under some scrutiny. Then Bozzoli, by now chief engineer in all but name, admitted him to the holy of holies and told him of the activities that now consumed their every hour. Suitably sworn-in and inducted into this most unmilitary of military units, Hewitt was assigned his first task,

185

the design of a monitor to be used with Gane's most temperamental of transmitters [36].

By February 1940 the JB0 was detecting aircraft at ranges of about 15 km and, within a month, further work had increased this to 80 km. Then, word reached the BPI that the expected British equipment would be delayed indefinitely. This was a major blow, but Schonland, confident that he had the capability to do more than just learn the techniques of radar, and also very much aware of an impending need, approached the Director of Signals for permission to construct an improved 'field' set, and this was readily granted [37]. The immediate threat that South Africa had to counter came not from the Germans but from the Italians, whose intentions in East Africa had been made abundantly clear five years before when Abyssinia fell to their forces under Marshal Badoglio. Now they were about to advance south from Abyssinia to invade British interests in Kenya, and elsewhere around the equator [38]. Britain herself was stretched to near breaking point; the threat of invasion was looming and the British army was soon to be evacuated from the beaches of Dunkirk. Only South Africa could provide the troops to push back the Italians. Advance warning against attacking aircraft would surely be a necessity and this could be a role for the JB radar, but what was just a laboratory prototype would very soon have to become Schonland's field set.

Based on the experience gained so far, Bozzoli completely redesigned and improved the IF amplifier, greatly simplifying it in the process, by using just three type 1852 pentodes with stagger-tuning to achieve a bandwidth of about 1 MHz. Unaccountably though, he left the 'spongy lock' firmly in place within the timing and pulsing circuits that Roberts had produced. It would soon tax Frank Hewitt in circumstances less ideal than the laboratory. His monitor completed, Hewitt's next task was a device using a so-called 'magic-eye' tuning indicator to show when the spongy lock was actually locked, and by delving into its inner workings he gained invaluable experience, while his general handiness with the soldering iron in wiring-up whole chassis of components served to embed in his memory the value of every component and their precise positions in the circuits. Schonland had noted the prowess of the young man from Rhodes, and it would soon stand both of them in good stead in the months to come. The JB0 was crude. What followed was the JB1, mounted in two racks of equipment no more than a metre wide and about 1.5 m high and sufficiently durable to be able to withstand the rigours of transportation and service in the field.

Schonland soon assured Collins, and through him the Chief of the General Staff, General Sir Pierre van Ryneveld, that he had an equipment ready for service. With Italian air raids on Nairobi, as a prelude to

Figure 15. The BPI building (ca. 1940) in its wartime role as the headquarters of the Special Wireless Section (subsequently Special Signals Services), South African Corps of Signals, with one of the wooden-framed RDF antennas mounted on the roof. (Reproduced by kind permission of Dr Mary Davidson.)

full-scale invasion now imminent, the need existed for some means of early-warning against air attack. On 28 May 1940, the Deputy CGS instructed the Adjutant General formally to appoint Schonland and Gane to the commissioned ranks suggested just before Christmas, with Hewitt to be commissioned as a Second Lieutenant at the same time. All appointments were to be effective from 1 June and the matter was, he minuted, of extreme urgency [40]. On 7 June, Schonland wrote to Humphrey Raikes, Principal of Wits, informing him officially that as from the beginning of the month he, Gane and Keiller were on full-time military service. Presumably he meant that Hewitt was too, though he omitted to mention him. In the interests of protecting the secret of RDF he stressed that: 'It is not desirable that the fact should be made public. The persons named will not necessarily be absent from Johannesburg and when at the Institute they will wear mufti.' He further informed Raikes that in his absence full responsibility for the building and its equipment would rest with 'Mr G Bozzoli', while the running of the Institute had been placed under Dr A L Hales [41]. The process of militarizing the BPI and its staff had begun but the need to shroud it in secrecy clearly kept the disclosure of unnecessary detail to the absolute minimum and Raikes knew no more than that.

Figure 16. A group photograph of the Special Signals Services taken just prior to Schonland's departure for England with some of them in March 1941. Standing (L to R): Staff Sgts Adams, Wadley, Joubert, Schefermann, Clark, Forte, Sgt Maj Keiller, Staff Sgts Hulley, Methley, Sgts Flack, Harrison, Boden. Seated (L to R): Lt Browne, Capt Bell, Maj Roberts, Lt Col Schonland, Maj Bozzoli, Capt Phillips, 2 Lt Guttridge. (Reproduced by kind permission of the SSS Radar Group and the University of the Witwatersrand.)

Rapid testing of the JB1 was now required and, since it was expected that the Italian air force would fly in along the coast, an assessment of the radar's performance over the sea was called for. Schonland's field set was therefore deployed with all haste to Durban, the port of embarkation of the 1st South African Brigade heading for Kenya. The first test over the sea took place at Avoca, just north of Durban, in June 1940. Its operators were its designers, who now found themselves rapidly transformed into real soldiers in uniform. The transmitter, at its frequency of about 90 MHz, was capable of generating 5 kW pulses, a thousand times more, Schonland wrote in his war diary, than had been achieved that day in December when Northcliff was the target. Each pulse was about 20 μs long and their repetition frequency was 50 Hz. The two antennas, mounted on their 6 m steel poles, were linked by a bicycle chain that required some effort from the operators as they cranked them around to point in the appropriate direction—a crucial aspect of the procedure for that was the direction of the target. A motor-driven system, developed in great haste at the BPI, had been found to be unsuccessful and was abandoned.

Schonland, Gane, Hewitt and Anderson spent the better part of three weeks at Avoca assessing the performance of the JB1 and making adjustments. There were very few aircraft on which to test the radar but the shipping lanes off South Africa's east coast were busy and those ships provided many targets, even though this was not the original intention. After successfully tracking craft that lay within visible range, an intriguing phenomenon was noticed sufficiently frequently to suggest that it was worthy of more attention. The cathode ray tube indicated targets, even though none were visible to the eye. All their characteristics on the screen suggested that they were ships but Schonland remained sceptical, even to the point of promising his colleagues a meal at Durban's most luxurious hotel should he be shown to be wrong. And indeed he was, for within a few hours of sighting the echoes the first ships of a convoy duly sailed into the roadstead off Durban. So his men all got their promised meal and Schonland sought an explanation. It was, of course, a case of anomalous propagation, where the radar signal was actually being bent, by some means, such that it was reflected by objects well beyond the horizon. The effect was well-known and is caused by super-refraction in the atmosphere but, being much more common in the tropics than in temperate climates, it could never be relied upon to reveal, with any consistency, targets much beyond the optical horizon [42].

Schonland was a much-relieved man. The JB1 had shown that it was capable of a useful range of about 50 km and could give essential bearing information when it was operated by persons possessing the necessary skill. At this stage, though, such skills only existed within that very select group at the BPI and so they themselves would have to take their radar to war.

*   *   *

Gane, Hewitt and Anderson sailed on 16 June 1940 aboard the *Rajula*, one of three transport ships in the convoy taking part of the 1st South African Division to East Africa. With them was their precious JB1 radar set and three long-established Active Citizen Force regiments, the 1st Transvaal Scottish, the Duke of Edinburgh's Own Rifles and the Natal Carbineers [43]. Three days later Schonland flew to Nairobi by courtesy of the SAAF and was immediately attached to the 1st Anti-Aircraft Brigade, Mobile Field Force [44]. He then set about choosing a site for the deployment of the first South African-made radar to go into action. On 27 July the *Rajula* docked in Mombasa, but the port's Movement Control officer had never heard of this rather unmilitary-looking group and, of course, knew nothing about the JB1, so he dispatched them to Nairobi despite their protestations that Kenya's capital was not their destination. Eventually, after retracing their steps, they arrived at Mambrui, a dusty village on the coast, some 120 km to the north, and were met there by Schonland.

The only source of electrical power with which to drive the JB1 was a diesel generator that they purchased in Mombasa on their way through. Until then all testing of the JB1 had been done with it connected to South Africa's 50 Hz mains network, with which the spongy lock circuit was well able to synchronize. With the single cylinder, low speed diesel generator things were entirely different. Both its output voltage and frequency were most unstable and the various timing circuits of the JB1, on which virtually everything depended, just could not lock-in. This was a matter of great concern to everyone, but most of all to Schonland, for it was at his recommendation that the CGS had agreed to the deployment of South Africa's only operating radar set in an active theatre of war, and its failure at this stage would have been ignominious. It was to Hewitt that he turned for a solution and his confidence was not misplaced. During two tense days Frank Hewitt modified the timing and range calibrator circuits, relying on the oscillators within the JB1 alone to maintain sufficient stability, so that the visual display on the cathode ray tube did not drift or jump alarmingly across the screen as it had done before. To both his and Schonland's considerable relief he succeeded, and for his efforts received the fulsome praise of his CO as well [45]. Hewitt had accomplished this fairly substantial re-design of the equipment in the field without the availability of any of the circuit diagrams. The wall of secrecy that surrounded RDF, and the Unit's obsession with guarding Britain's most precious secret, meant that he had committed to memory all the

190

technical details of the equipment rather than run the risk of having the blueprints fall into enemy hands, should he or his colleagues be captured.

The JB1 was erected and became operational by 1 August 1940, on what became known as South Africa Hill just outside Mambrui. Its operators, now known as No 1 section, had been augmented by five South African signallers and they maintained a 24 hour watch of the skies off the Kenyan coast. Their awaited targets were the Italian bombers that were expected to launch seaward attacks on the military airfield at Malindi just a short distance away. Protecting the airfield was a detachment of the 1st South African Anti-Aircraft Brigade with a battery of rather obsolete guns under the command of Lt Col S H Jeffrey. No test flights using the SAAF's Hurricanes as targets were possible and so the radar was adjusted by tracking the only aeroplane that appeared with any regularity at all: an aged Junkers of the SAAF that flew along the coast at precisely the same time every day. Schonland knew that his colleagues at the BPI were waiting anxiously for any news of the performance of their radar and so he devised a code before he left for communicating such intelligence as the range that had been achieved. Cryptography was clearly a Schonland trait that he had inherited from his father and was one which they had used twenty years before when the UCT appointment was being negotiated. On this occasion the ruse was to add the range achieved by the radar, in miles, to Schonland's age and send that by telegram to Bozzoli. It worked and both amused and enlightened his colleagues.

In the six months that Schonland's first radar was in operation at Mambrui it tracked enemy targets just once, and then there were two of them, the only Italian aircraft to appear. However, they caused more than a little consternation because they came from a direction quite opposite to that from which they were expected, dropped their bombs on and about the airfield and then flew out to sea! But at least the radar did its job, if not that originally intended, as it tracked them for about 55 km as they departed; the two blips on the screen eventually disappearing into the noise. Throughout this period the JB1 had maintained a seaward watch, just as intended. Had the need for 360 degree coverage been required it would have presented the operators with a problem because the two antennas could not be completely rotated without winding their uninsulated transmission lines around the poles that supported them. This fleeting encounter with the enemy now made such a requirement a necessity and this caused the dispatch to Johannesburg of an urgent signal requesting that suitable modifications be made post-haste [46].

*   *   *

While Schonland and his colleagues were engaged in their radar war, efforts were under way within the Headquarters of the Royal Air Force in the Middle East (HQ RAF ME) in Cairo to establish suitable radar cover in the vicinity of Mombasa, and an officer was sent to investigate and report. The man in question was J F Atherton, a Scientific Officer from Watson Watt's Directorate of Communications Development, who had just been commissioned as an honorary Flight Lieutenant in order to be able to perform this task in the various overseas commands [47]. On 29 July 1940 Atherton sent a report to the Air Ministry in London in which he noted with some surprise that the South Africans were already providing 'elementary home-made RD/F [sic] in this area'. He continued that he had . . . [48]

> consulted with Lt Col Jeffries [sic] 1st AA Bde, S.A. Army, and Maj. Schonland, S.A. Corps of Signals, who were making independent arrangements for air warning in Mombasa. Maj. Schonland had constructed an elementary RD/F set on the basis of information supplied by the Home Government to Gen. Hoare, Dir. of Technical Services, S.A. Army; he expects to have this working at Mambrui, about 90 miles North of Mombasa, in approximately 2 weeks' time, and to have another set of the same type available in a further 2 months.

Atherton then said that he had recommended to Schonland that, pending the arrival of British radar equipment in East Africa, the South Africans should co-operate as closely as possible with HQ RAF Nairobi 'in order to make the fullest possible use of available facilities'. Quite what those facilities were and how they would assist Schonland was not made clear, but their discussion had left Atherton in no doubt that the South Africans faced two particular problems and these he spelt out to London.

> The sets built by Maj. Schonland give a maximum warning range of about 35 miles with very rough indication of direction; he is not hopeful of any large-scale production of RD/F in S. Africa, owing to the difficulty of obtaining supplies, particularly transmitter valves. It is also necessary to limit information on this subject to a very few people, as the political situation in the Union causes serious danger of leakages.

Atherton's rather alarming report seemed not to generate any immediate reaction from London other than information, in early September, to the effect that 'No 218 MRU has been dispatched by SS *City of Eastbourne* to Mombasa where it is expected to arrive in 6 to 7 weeks' time' [49]. Atherton himself had completed his survey in East Africa and was about to carry out a similar task at Freetown and Takoradi on the west coast. Before leaving though he pressed HQ RAF ME for further action and, on 3 October 1940, an urgent cipher message was sent from there to the Air Ministry in London. It read, rather confusingly, [50]:

> South African Army now making own R.D.F. sets and training crews for use at Mombasa. 2 sets exist 1 of which now at Mombasa and they propose to continue production. Sets have very limited performance with inaccurate Azimuth, and are of little practical value. R.A.F. provisions will afford all cover that cannot equally well be given by observer screens. Owing to South African doubts of security in their headquarters. Secrecy of R.D.F probably seriously endangered by their active interest. Request policy to be decided, this H.Q. informed, and necessary action taken with Union Government.

There was no immediate response from either Cairo or London. Those within that East African theatre of operations, who were not privy to that conversation between Atherton and Schonland, heard only the trumpeting of Mussolini as he appointed Marshal Badoglio his Viceroy of Ethiopia for having driven into exile the Lion of Judah, Haile Selassie. British East Africa was now seemingly at Il Duce's mercy. Attacks farther south were surely imminent and the British Military Command in Nairobi was steeling itself for action. For all its reported short-comings, Schonland's 'home-made apparatus' would provide the only means of early warning against air-attack until the MRU arrived. There was every likelihood that more radars would be required and so Schonland, who was also unaware of Atherton's alarmist report, returned to South Africa on 22 September to press for the manufacture of additional JBs [51].

The Air Staff in London when confronted by the HQ RAF ME's request for a policy decision based on Atherton's assessment of a perceived security threat to radar from within South Africa gave the matter their immediate attention. The Minute Sheet placed before them by the Air Ministry on 9 October summarized the concerns and offered some proposals. These were based crucially on their interpretation of one critical passage in that signal that even those well-versed in reading tersely-abbreviated military prose described as 'not very lucid'. It was the obvious one: 'Owing to South African doubts of security in their head-quarters. Secrecy of R.D.F. probably seriously endangered by their active interest.'

From this the Air Ministry managed only to conclude, quite erroneously:

> Presumably detailed information about R.D.F. practice may proceed either from some South African R.D.F. Research establishment or from South African units employing R.D.F. equipment in the field.

To Schonland this would have been a horrifying misinterpretation of what he had meant when he confided in Atherton. 'Headquarters' to him never implied the BPI, where the selection of personnel to whom this most precious of British secrets had been entrusted was his responsibility and his alone; it would have meant Defence Headquarters (DHQ) in Pretoria where there were undoubtedly still some whose deep-felt allegiances lay with Germany, which had supported the Boer cause in their war with Britain forty years before.

To a South African this division within the white nation was a scar that had never properly healed, nor would it ever in the hearts of some, even though those of Boer and British stock had joined together in Union, and as a British Dominion, in 1910. It was always dangerous, though, to assume that others born beyond the country's borders understood it, and either Schonland assumed that Atherton did, or Atherton, had he indeed done so, might have assumed that those in Cairo and Whitehall did too! Whether due to misinterpretation, or simply to poor draughting, the Atherton message could so easily have tainted the thinking within the Air Ministry and all South African involvement with radar might have ceased forthwith. However, wise heads within the upper echelons of the Air Ministry had already considered such matters of risk as they might exist within the Dominions and had, though not without argument as we have already seen, decided to release to them the information about RDF. Quite how much was actually disclosed was what really mattered and it was now acknowledged, no doubt influenced by Atherton's rather jaundiced view of South African prowess in this field, that 'the risk of disclosure of any really important information from present South African R.D.F. Units operating in the field is, in view of their immaturity, relatively slight'. It was further concluded that a decision to force an end to all South African radar activities 'would surely dampen their ardour considerably and might have more serious repercussions'. To allay the concerns expressed by HQ RAF ME, however incoherently, three suggestions were made to the Air Staff by the Ministry. They bear repeating verbatim:

> '(i) In view of the availability in increasing quantities and much greater operational value of United Kingdom R.D.F. equipment, this should be used exclusively in all theatres of war;
>
> (ii) This equipment should be manned for the time being, at least, by U.K. units since they are the only trained personnel; and
>
> (iii) Facilities should be arranged for the training of South African personnel (including those now operating South African "lash-up" sets) so that they may be employed with our overseas R.D.F. system in appropriate areas.'

The minute, though, did make two concessions in view of Cairo's apprehensions about South African involvement with radar. The first would require that all British RDF units operating in the same areas as the

194

South African forces should adopt rigorous security precautions 'to prevent the communication to these Forces of information concerning our technical methods and apparatus'. Secondly, only sufficient detail about 'our filtering technique and the information thereby made available' should be provided to the South African forces to allow for their efficient operation [52].

Bolstered by these suggestions it was now the turn of the Assistant Chief of the Air Staff (ACAS) to reply to HQ RAF ME. He duly did by return cipher the next day. His text, though terse in the accepted style, was precise and unambiguous. Based upon the deliberations, he told Cairo, that had led to the High Commissioners of Australia, Canada, New Zealand and South Africa being made party to the existence of RDF, the decision had been that there was to be full and frank interchange of information and research on RDF to all the Dominions. As a result there was no possibility of discouraging any of them from conducting research or operating RDF systems. However, there was one reservation that was based on Atherton's disclosure and Cairo's reaction to it. The ACAS recommended that '... in the circumstances suggest you retain sole control of R.A.F. R.D.F. Stations in areas concerned, thus ensuring secrecy regarding information about our technical methods and apparatus' [53].

Those thousands of loyal South Africans who had so willingly thrown in their lot with Britain in fighting the Nazis would have been horrified to know of the contents and the innuendoes within those ciphered messages between Cairo and London. One who would have reacted with fury was the Prime Minister, General Smuts, and he may well have done just that when he met Anthony Eden, the British Secretary of State for War, in Khartoum late in October 1940 for nothing more was ever heard on the matter. Also present at that meeting was Sir Archibald Wavell, General Officer Commanding-in-Chief, Middle East and East Africa [54]. If anyone in Britain had doubted South Africa's sincerity, or worried about her trustworthiness before, none would have done so after that. What was of the very greatest importance was the need for an urgent strategy for driving the Italians out of Africa. Smuts informed Eden that he intended to send to Abyssinia a fully equipped South African Division to ensure that the Italian designs on central and even on southern Africa would be blocked immediately. Though this was contrary to Churchill's view that all available resources should be brought to bear against the enemy in the Middle East, Wavell supported Smuts and, within six months, there would be 43 000 South African motorized infantry, plus all their supporting arms, deployed in that rather less than glamorous theatre of the war that stretched almost from the Zambezi to the Nile [55]. Amongst them were a mere handful of Schonland's men operating their home-made radars, but they were party to one of the great military secrets of the time and none guarded it better. Essentially part of the Mobile Field Force they were, even then, being loosely referred to as the Special Signals Services or SSS, but their existence was known officially to very few.

What was it that Schonland had said that had triggered Atherton's warning missives to Cairo and hence to London? Schonland's obsession with protecting the secret of radar may well have caused him to state his concern about certain personnel within Defence Headquarters in Pretoria. It was no secret that white South Africa was riven by twin loyalties: there were those who fully supported Smuts and the country's immediate entry into the war to oppose the Fascist onslaught wherever it might occur; while there were others, supporters of the Hertzog–Malan alliance, who did not, and amongst them were many who actively opposed it. There was even undisguised support within South Africa for Fascism itself, even to the extent of resorting to violence, sabotage and active collaboration with the enemy. This came from an organization known as the *Ossewabrandwag* (OB),* a paramilitary movement under the leadership of a dedicated pro-Nazi by the name of J F J van Rensburg who, in the early 1930s, had actually served under Smuts as secretary of justice but had since changed his allegiance radically. The OB had sympathizers in many places, one of which was in the very heart of the military. In March 1940, Smuts was warned that perhaps 80 per cent of the officers at DHQ, as well as many in the police force, were opposed to entry into the war on the side of Britain [56]. Painfully aware of these divisions within his own country, and particularly amongst his own Afrikaner people, he had purposely not resorted to conscription when South Africa declared war on Germany, even though the country's military preparedness had sunk to a pitiful state during the previous, depression-filled decade. Instead, the South African forces that went 'up north' were all volunteers who took what became known as the 'Africa Oath' binding them to serve anywhere in Africa but not beyond its shores. To distinguish them from those of their Permanent Force colleagues who chose only to defend the country's own borders, they wore distinguishing red (or more correctly, orange) tabs on the shoulder straps of their uniforms. These men were as loyal to their country, and to the alliance with Britain, as any soldiers from the other Dominions now going into action across the globe.

---

* The Ossewabrandwag (Ox Wagon Guard) came into being in 1938 at the time of a great outpouring of Afrikaner sentiment on the centenary of the Great Trek. During the war its 'Stormjaers' (storm troopers) were responsible for numerous brutal attacks on servicemen at home on leave and it was heavily implicated in a plan to assassinate Smuts himself.

It was almost certainly Schonland's honesty in telling Atherton of his concerns, but without also giving him a South African history lesson as well, that set the cat loose amongst the pigeons in Cairo, but thankfully did not quite do so in Whitehall.

*   *   *

When Schonland returned to East Africa in November 1940 he will have done so in the knowledge that South African forces were massing for their first major action of the war. His task now was to spend the following month there in order fully to assess the radar situation. The immediate outcome was the decision to install three radars for the protection of Nairobi and one anti-aircraft set, of a different design, for use with the 1st AA Brigade in Mombasa. In addition, arrangements were also made for a filter room to be set up at the Air Defence Centre in Nairobi where incoming reports from the radar stations would be processed. Schonland flew back to South Africa and briefed Bozzoli, after which activities were immediately stepped up at the BPI. Between July 1940 and February 1941 no fewer than six JB radars were built and dispatched from Johannesburg, with five of them intended for service in East Africa, while the other was on its way for evaluation in Egypt. The SSS War Diary shows the situation as it developed over the next few months [57].

Equipment No.	Section No.	Date of arrival in E.A.	Date Working
1	1	July 27 1940	Aug 1 1940
2	2	Nov 1940	Jan 1 1941
3	3	Dec 1940	
4		Dispatched to Egypt from Durban	
5 (JB2)	6 (A.A.)	Jan 1 1941	
6	4	Dispatched 16/2/41	

The British radars in East Africa consisted solely of the MRU or Mobile Radio Unit, No 218, that came into operation in Mombasa on 21 November 1940. The intention was also to provide additional cover by means of a CO station, and 'eventually to replace the two small sets of equipment provided by [the] South African Force' by COL* stations [58]. However, the situation in the western desert was deteriorating and this required that every available MRU be pressed into service there and so No 218 would soon be moved. But the fortunes of war surged back and forth and priorities changed yet again. On 22 January 1941 British and Australian forces captured Tobruk, and while this was a

---

* The COL (Chain Overseas Low) radar was a semi-tropicalized version of the Chain Home Low equipment that came into service in England early in 1940. It operated at around 200 MHz and provided very useful cover against low-flying aircraft, which the lower frequency CH (and CO) radars could not do. It was also useful against shipping.

major victory, the extended commitments had left the Suez Canal dangerously exposed to German air attack, and without any RDF cover, as HQ RAF ME informed the Air Ministry that same day [59]. To provide some it was decided on 10 March to move the 218 from Mombasa to Greece, as the eastern Mediterranean now also became a very active theatre of operations requiring every available radar system to be pressed into service.

Meanwhile, some items of British radar equipment were beginning to arrive in South Africa. As early as May 1940 a single ASV (Air to Surface Vessel) set was unloaded in Cape Town, followed in September by an MRU receiver (but no transmitter) and two months later by a damaged CHL radar without its antenna and masts [60]. The British policy of cooperation with the Dominions was indeed functioning as intended, even if the equipment was hardly in its most operational state on reaching its destination. Then, to further complicate matters, the Director of Signals in Pretoria would not allow the ASV equipment to be installed in an aircraft even though it was designed specifically for use in just such an anti-shipping role. Instead, Captain Eric Phillips was given the task of designing an antenna system so that the ASV could be used in some form of searchlight control (SLC or 'Elsie') as was then being introduced into service in Britain.

Work at the BPI was proceeding apace under Bozzoli. He had designed a 50 MHz transmitter to at least render the MRU useful for training purposes and also addressed the need for a mobile radar set based on the JB. Soon this too would be set in train. The versatility of the designers at the BPI was tested to the full when a new radar, eventually to become the JB2, was proposed. The anti-aircraft batteries in Mombasa under Lt Col Jeffrey had identified a need for a set that would give them a range accuracy to about 200 m with azimuth accuracy of about a degree — considerably better than had been possible with the JB1. Schonland had passed the requirement to the BPI and Captain Roberts assumed responsibility for its design and construction. To achieve the performance required he devised a spiral time-base instead of the linear version in use on the JB1, while the transmitter was fitted with the new VT58 valves that were kept cool by a special air-blast arrangement. The transmitter and receiver were mounted in separate vehicles, each with its own antenna, and the two were to rotate in synchronism by means of a steel cable between them. Needless to say, the JB2, though showing signs of promise in Johannesburg, was very far from operational when delivered in great haste by Roberts on New Year's Day 1941 to the anti-aircraft gunners at the entrance to Mombasa harbour. Roberts soon departed for the Union and Hewitt was called upon to render it useful, but it was a premature idea and was soon abandoned [61].

As the end of 1940 approached, the vulnerability to air attack of the Suez Canal was now a matter of even greater concern in London and particularly in Cairo. Since radar was clearly the vital element of any defensive screen, the RAF appointed, in September, Wing Commander J A Tester as Chief Radio Officer, Headquarters Middle East and RDF adviser to the Air Officer Commanding in that theatre. Tester's evaluation of the radar needs soon showed that not only was there a serious shortage of suitable equipment and trained personnel, but that existing British radars had not been designed to be particularly mobile — even the quaintly named Mobile Radio Unit (MRU), though actually a smaller, more portable version of the Chain Home radar, was decidedly cumbersome. He remembered Atherton's reports from East Africa on the existence of South African radars in the area and felt that this was worth investigating and so, at Tester's request, Schonland flew to Cairo from Johannesburg for a conference [62]. With him was the energetic commanding officer of the 1st AA Brigade, S H Jeffrey. Both were now Lt Colonels, Schonland having been promoted on 1 December. He was also formally appointed as Officer Commanding, Special Wireless Section, South African Corps of Signals, though Special Signals Services* was the name by which that unit was then more usually known within Col Collins's Directorate of Signals in Pretoria. Other appointments and promotions were announced as well. Professor David Hodges whose long pre-war collaboration with Schonland had made him very conversant with radio direction-finding, became Major Hodges, on joining the SSS on 10 December, and he was also appointed Schonland's second-in-command with immediate responsibility for the radar units serving with the Mobile Field Force in East Africa. Bozzoli, Phillips and Roberts were all promoted to the rank of major, with Bozzoli also designated as Chief Technical Officer, while Hewitt gained his second 'pip' [64].

After holding discussions with Tester in Cairo about the immediate radar needs of the Middle East and about the South African experiences when operating in East Africa, Schonland and Jeffrey were taken to Alexandria to see the British GL (Gun Laying) radars in operation there. What had become clear to Tester was that the JB radar was considerably more manageable in rugged terrain than most of the current British sets and he had an immediate requirement for just such equipment. During their discussions Tester also informed Schonland of the severe shortage, within the RAF, of technical officers experienced in

---

*The Special Signals Services of the South African Corps of Signals was only given official approval as a new ACF (Active Citizen Force) Volunteer Unit on 26 December 1941 [63].

the operation and maintenance of radar equipment and Schonland responded by indicating that the South Africans had initiated an active recruiting programme in the universities to ensure that such men would be available when British radars eventually arrived in the country. The response had been heartening, especially at Natal University College where the complete final-year class of thirteen volunteered for service in the SSS [65].

All this appeared as welcome news to Tester and he immediately signalled A P Rowe, who had succeeded Watson Watt as Superintendent of what was now the Air Ministry Research Establishment (AMRE), with the information that [66]:

> We are going to try a more portable type of equipment from South Africa. What we want for Crete, Greece, Aden and such, is something that can built of compact units such as might go on the back of a mule. It can be done. South Africa get 30 miles on theirs with 3 kW though we have yet to see this proved'

By the time Schonland arrived back in Johannesburg, Tester's requirements had reached DHQ in Pretoria. He asked that a JB1 be sent as soon as possible to Cairo for urgent evaluation. And so the fourth set to emerge from what was hardly a production line at the BPI was immediately transported to Durban and loaded aboard the *City of Leicester* on 8 January 1941 to join the convoy heading for the Suez Canal and Cairo. Captain Gane was ordered to fly there directly from East Africa. After reporting to Tester on 2 February, he prepared to coordinate the assembly of the JB1 units as soon as they arrived. On 11 March, Gane signalled Pretoria that the RAF were impressed and had found that the JB1 seemed capable of filling the gap in performance of the MRU and ASV radars then deployed in that area. A week later Tester wrote to Schonland and informed him that there was a chance of all the JBs in East Africa being required in Egypt, and not only the JBs but their operators as well [67]. This was a remarkable turn-around. From being regarded with some scepticism less than six months before these South African 'lash-ups' suddenly found themselves in some demand. Soon they would play an even more useful part in the defence of the Suez Canal than they had already done when operating under General Cunningham's command in East Africa.

*   *   *

While these events was unfolding, Basil Schonland was caught up with others much closer to home. The first was enemy action, not only to shipping around the South African coast but possibly against its sea-ports as well. To counter both an Operational Command of the South African Air Force had been established in December 1940 under Colonel H G Willmott who, early in 1939, had suddenly found himself party to the disclosure of the RDF secret to the Dominions. Now, in much more familiar territory, he was coordinating the patrols being flown by Anson aircraft from bases in Cape Town, Port Elizabeth and Durban [68]. However, without radar, either on land or in the aircraft themselves, such patrols were, more often than not, fruitless exercises and so it was decided that a full assessment of South Africa's radar needs should be made immediately. Schonland would do it and would then travel to England to do whatever he could to speed up the supply of British radars — long since promised but always diverted by more urgent needs elsewhere. In the meantime the BPI, under Bozzoli, would press on with great urgency to develop its own radar equipment for deployment around the South African coast.

Another problem was more sensitive and even more immediate, following as it did so soon after Schonland's return from the conference with Wing Commander Tester. It concerned a British request, just received, for six South African volunteers to be transferred to the RAF, initially for radar training and then for service wherever the need might arise. Schonland had assured Tester that South Africa would do all in its power to assist with the provision of radar specialists but it was now evident that the financial implications of such transfers to the British forces would seriously affect the South Africans since they would be subject to the much higher British rates of income tax while marriage allowances would be paid only to those over the age of thirty. Schonland's concern was for the immediate welfare of his men and their dependants, but the British request had been made at the very highest level and the Chief of the South African General Staff, Sir Pierre van Ryneveld, had not only agreed but had ordered that suitable volunteers be sought without delay. Almost immediately, six young graduate engineers from Durban, all from that class of thirteen, offered their services. Schonland campaigned vigorously that they should remain in the South African forces while rendering their service but the CGS was adamant that they be treated exactly as any pupil pilots had been when transferred to the RAF. To compound his problem Schonland was unable to call upon the Director of Signals, Col Freddie Collins, for support because Collins was visiting the various Signals units, including the SSS, serving in East Africa. However, word did reach him and Collins responded immediately by instructing Schonland to 'carry his view to the CGS that the six Durban men should not go'. If Basil Schonland had any doubts in the matter then that was clearly good enough for Collins! But General van Ryneveld was not to be deflected and preparations were soon under way for these young men to leave for England, initially under Schonland's command whilst at sea, and then to be welcomed into the bosom of the RAF on their arrival in Liverpool.

By the middle of February 1941, Schonland had completed his assessment of South Africa's radar requirements. On the 18th he was ordered to make immediate arrangements to fly to Cape Town by the next available aircraft for discussions with the CGS himself. The six Natal volunteers had all been sent on leave with strict instructions to maintain contact with Captain Phillips who, along with Lt J H Browne, would be Schonland's staff officers during his forthcoming visit to England. True to form, Schonland had arranged that a coded message would be sent to each of them by telegram when the time came for them to report to Cape Town for embarkation. The agreed message read: 'Your leave has been extended by one week' [69]. Having thus dispelled any possibility of their mission being compromised in any way should the telegrams be intercepted, Schonland attended to last-minute arrangements at the BPI by placing Bozzoli in command during his and Hodges' absence, and by appointing his wife Ismay to be the custodian of all official documents. She would also constitute the link through which he would communicate with the SSS while he was away.

Schonland's meeting with the CGS took place on 25 February 1941. It was soon apparent that the SSS was about to make its entrance on to the much wider stage, as he briefed van Ryneveld about the state of radar developments in the country and, as far as he knew them, in England. Since the Italians were now in retreat in Abyssinia, and no German presence had been threatened in East Africa, it was decided that Hodges should close down all except one of the JBs operating there. The rest were to be packed up pending their return to South Africa for coastal defence or deployment to the Middle East. On his arrival in London, Schonland was to make it clear that British CHL radars were needed urgently in South Africa and, if the Air Staff considered it wise, the orders already placed for the older MB radars would be cancelled. Pending the arrival of the British radars, use would be made, if possible, of the JBs from East Africa, but all would depend on the needs of the RAF in Cairo. Finally, the CGS stressed that, 'as a priority requirement' every effort must be made to obtain suitably qualified and competent SSS men for the UK. South Africa had its needs but those of England were, at this time, the greater.

As their conference was drawing to its conclusion, Schonland steeled himself for the moment when he planned to asked van Ryneveld his most leading question. There was nothing precipitate about it, nor was it in any way influenced by matters they had just discussed. It was to do with the further part that Schonland felt he might be able to play in the prosecution of the war, not in South Africa but in England. Basil and Ismay had given it much thought and, though it was painful, for it would lead to a possibly extended separation, she had supported him. Now was the time to put it to the CGS himself.

Schonland asked van Ryneveld if he would allow him to offer his services to the British government if he believed that he could be of real use to them in some capacity. The reply he received was itself almost coded: 'He answered the question by the suggestion that he would like to go there himself! But he did not definitely say no', Schonland noted carefully in his diary [70].

CHAPTER 12

'COULD I BE OF SERVICE?'

Lt Col Schonland, his two staff officers, Captain Phillips and Lieutenant Browne, and the contingent of Natal volunteers* sailed in convoy from Cape Town for England on 21 March 1941 on board a British troopship [2]. They were amongst a large contingent of soldiers, sailors and airmen who were returning home from various theatres of war around the globe and they soon found themselves very much part of this military family that had accepted its lot in life, with a future that was never measured in more than a few hours at a time. These men fascinated Schonland. Most had already seen much action but their demeanours never conveyed any hint of outward concern about what was seemingly a perilous time ahead as Britain prepared for the worst. He saw around him 'Lawrence of Arabia in various styles' and soaked up an atmosphere that he described in his letters to Ismay as 'naturally Conrad with a touch of Kipling'. These were officers whose lives had been moulded by Public School and Sandhurst, and for whom military inconvenience and even imminent death seemed to be just facts of life. He found them magnificent as men, but at times almost tiresome in their bonhomie when surely they should have been concerned by what was about to befall them at home? Occasionally, news would reach the ship of British successes at sea or in the field but their reaction was always much the same. Whether triumphant or vanquished they showed little emotion, just a seemingly urbane superiority. This was surely the English character that had made the Empire — imperturbable and in command [3].

Schonland and Phillips shared a cabin that became almost blisteringly hot as they approached the equator, while the non-commissioned men, most of whom were in the next one along, tolerated it with good humour and great expectations. Life, though, soon became monotonous and even the regular boat drills and emergency exercises assumed a

---

* The six young engineers were K A H Adams, A K F Clark, D R Forte, H Guttridge, R G Hulley and J Methley. Guttridge had recently been commissioned and was about to leave for East Africa when Schonland recalled him to join the group leaving for England. The others were all staff sergeants (a Schonland decision) since the pay was better than that of a second lieutenant! [1].

pattern that made them seem as much part of the day's activities as the meals. Their lifebelts accompanied them everywhere and cabin doors remained wedged open in case of torpedo attack. Schonland's three officers soon found themselves on the ship's roster of guard officers while he busied himself during the days brushing up on the theory of radio or reading whatever he could find in the ship's library. For some reason, possibly just to help pass the time or maybe even to reinforce the impression they hoped to convey that they were just signallers of the South African Corps of Signals, rather than from its most secret Special Service, Schonland had brought along with him a number of Morse buzzers and so regular Morse code practice sessions took place, with the colonel himself being 'top boy' on most occasions. This skill, like that of riding a bicycle, had obviously not deserted him from those days as an RE signals subaltern at Bletchley nearly thirty years before. He had also intended making a start on writing a book in the vein of popular science, but the ship was too crowded and too hot for that and a chance spotting of Lancelot Hogben's *Mathematics for the Million* in the ship's library soon convinced him that 'popular science is horrible anyway' and this rather 'vulgar' book reminded him too that the author was 'a nasty little man' [4]. The fact that Hogben had left UCT some years before in some notoriety, and with many enemies in his wake, may well have had much to do with Basil Schonland's rather caustic views of both the man and his work [5].

Before he left Johannesburg for Cape Town, Basil and Ismay had discussed at great length the matter that was uppermost in his mind: quite how he might offer his services to the British military authorities at some stage after his arrival in England. His immediate task, of course, was to arrange for the delivery of British radar equipment to South Africa and to familiarize himself with all its details. Phillips and Browne would accompany him to the various Army and Royal Air Force establishments, where they would be briefed on technical and operational aspects of the equipment, and as he moved on to the next so he would leave one or the other behind for more in-depth training. This, he envisaged, would occupy a good two months and any thought of remaining in England in some capacity after that would be firmly on ice until this task had been completed. Now, during the long hours in between games of deck tennis and convivial conversations with his most genial but sometimes frustrating British colleagues, he thought deeply about his own future role in the war and most especially about his separation from Ismay and the children, and he wrote to her as often as he could.

What began on that voyage as the letters from husband to wife soon became a remarkable record of Basil Schonland's life as it unfolded throughout the war; and Ismay kept them all. These epistles, as he called them, were no dry catalogue of events, for Schonland never wrote like that. They painted pictures of himself, his companions and of the war around them. Quite naturally, they revealed a side of the man that only Ismay knew, sometimes within the lines but often just between them. She was his sounding-board when he needed one, his adviser often and his closest confidante always. He wrestled incessantly with the future and all that it might hold in store for them all; would he cope and how would his family fare? Did he have the ability, the inner strength and the courage to see it through? But those were his deepest thoughts and only Ismay knew of them. None of his South African contingent, and certainly no one else aboard that ship, ever guessed that this much admired scientist-turned-soldier was, at times, almost consumed with self-doubt and uncertainty about his ability to tackle the tasks that lay ahead. But Ismay knew and bolstered him with her letters that exuded support and considerable insight into the human personality—his and those of their many acquaintances and colleagues.

The crucial decision to stay in England, if his services were indeed required, would hang on one aspect and one aspect alone, and that was the financial security of his family. He spelt out to Ismay that he would offer his services for a limited period of about six months on condition that he remained in the Union Defence Force but attached, as he put it, 'to the English crowd'. If he transferred to the British Army he would be taxed much more heavily than in the South African forces and, if he were killed, his pension would not only be less but that too would be subject to British income tax. The decision was therefore simple: any service that he might render would not be given at the expense of his family's future. Schonland took it without second thoughts and assured his wife that he would be totally unyielding on that point. Should they want him then those were the terms on which he would be prepared to stay. Quite what he thought he might accomplish in six months is not clear, for six months in wartime is either a lifetime consumed by boredom and inactivity or nothing at all in the heat of battle. Circumstances, though, dictate the future and Basil Schonland's right now were those of the little-known venturing into the unknown.

Schonland's inspiration, if any were needed, came from other men and probably none more so than Jan Smuts. Not in any formal way but merely from the manner and authority of the man who had assumed the proportions of a giant, at least in the eyes of some of his countrymen. To Basil Schonland though, Smuts's inspiration was much more tangible. He was not just the Prime Minister, nor a General nor even a man of massive intellect, but he was also a revered family friend and it was his moral courage as much as anything else that made him so. But Schonland also saw in Smuts the sort of courage that he much admired even closer to home in the characters of Don Craib and Pierre de Villiers. Craib was not only Ismay's brother and Basil's former room-mate at Caius soon after they both returned from the Great War, Schonland with an OBE and Craib with the MC and bar, but he was also a hero in the eyes of the Hofmeyr clan. Unflinching in the face of danger as a soldier, and courageous to the point of obduracy in the face of entrenched medical dogma as a ground breaking physiologist, Don Craib was something of a legend [6]. Pierre de Villiers MC was Ismay's cousin and was Commissioner of the South African Police before the outbreak of Second World War [7]. At the commencement of hostilities he had responded to Smuts's call that two police battalions be formed for service within the 6th South African Infantry Brigade and had himself assumed command of the 2nd South African Division in the rank of Major-General. Such heroes cast long shadows over their contemporaries and it was no wonder that Basil Schonland found himself somewhat in awe of both of them, if only in simple military terms. To Ismay he confided [8]:

> I wish I had been born with more moral courage. It is very late to
> find that out at 45. I thought I lacked physical courage and swam
> and dived and played football to gain it. Probably my early piety
> can be blamed for a lot; I confess to having a timid long-haired man
> wearing a crown of thorns as my hero to this very day...
>
> Anyway I would wish to have had the moral courage of Pierre
> or Donal and to have less of the man of sorrows and more of the
> man who chased the money-changers out of the temple.

There was though, he believed, a much deeper-seated explanation for his anxiety and it lay not in any unfavourable comparisons that he might make with others, but within himself and in his vivid recollections of his childhood and the violent quarrels between his parents. On occasions, when these eruptions occurred, his diminutive mother would frequently hurl plates at her gentle and long-suffering husband [9], to the obvious alarm and consternation of the family and their eldest son in particular, for he was old enough to understand and it appalled him. He reminded Ismay that, ever since '... I shudder to hear a quarrel, and will give in to anyone, like Beattie,* who adopts a hectoring manner'. Now all this was at the back of his mind as he contemplated the time when he might volunteer his services in England. Should he, as the 'man of sorrows in me says—forsake all and do any job whatever' or, as Pierre and Donal might say—'if they want you they will take you at a proper valuation & with proper arrangements'.

Fate and one man would soon conspire to ensure that Basil Schonland's services were indeed retained in England. That man was John Cockcroft.

*   *   *

---

* Sir John Carruthers Beattie, Professor of Physics at the South African College (1897–1919) and then Principal and Vice-Chancellor of UCT 1918–1938.

Cockcroft was appointed to the Jacksonian Professorship of Natural Philosophy at Cambridge in the summer of 1939. It was a chair he never actually occupied because, within a few months, his services were required elsewhere as other events unfolded. After familiarizing himself with the underlying science and technology of radar at Bawdsey in 1938, at the invitation of Sir Henry Tizard, Cockcroft very rapidly became the leading player in the introduction of young university physicists to the various radar stations being set up around the south and east coasts of England. They were to be 'the nursemaids', as Tizard called them, whose task would be to ensure that the radars worked as required should war break out. Cockcroft soon displayed a talent for interacting with sceptical military men of the highest rank, who often saw no possible need for this complicated and frequently temperamental apparatus that had suddenly invaded the sanctums of the Anti-Aircraft Command and the Royal Artillery; often, they thought, without any consultation. But Cockcroft had a way with people and he possessed the ability to speak two 'languages': those of the scientist and the soldier, despite his very sparing use of words in the process. Such a gift was invaluable in wartime and it enabled him to encourage both these groups, often poles apart to start with, to work together in the solution of some of their most intractable problems, under the most trying of conditions and in remarkably short periods of time. This mutual trust and cooperation was to lead to what was arguably the most significant difference between the Allies and the Germans during the Second World War: collaboration and trust between soldier and scientist. Cockcroft clearly had this 'bilingual' ability, so too did Watson Watt and it would soon become apparent that Basil Schonland, their former colleague and acquaintance from less violent times, had it too.

No doubt Cockcroft and Schonland's service careers during the previous war helped greatly, since they had both seen action and well-appreciated the conditions under which wars were fought and how men fought them. But there was much more to it as well. Both men listened and heard what was being told to them, and neither was ever dogmatic in their replies, regardless of the listener or the audience. In fact, Cockcroft's replies, when they came, were often so brief as to be almost monosyllabic but, once a problem had been noted in the little black book that accompanied him everywhere, action followed without fail. By the middle of 1940 his astute brain, seemingly boundless energy and a tactfulness that belied his northern roots had been responsible for the installation of four radar stations from Scapa Flow in the north of Scotland to the Forth Estuary in the south. And several more were to follow. In Admiral Sir James Somerville, hastily recalled from retirement at the outbreak of war, he found a formidable ally. Somerville was Inspector of Anti-Aircraft Weapons and Devices who, after visiting

Cockcroft at Bawdsey to familiarize himself with radar, immediately saw in it the means of protecting the all-too-vulnerable seaward approaches and deep-water harbour of Scapa Flow in the Orkney Islands. Things then happened quickly and Cockcroft personally took charge of the erection of three radars to cover the area concerned. The results were impressive. These new radars filled the real gap that existed at low angles of elevation below the area of coverage provided by the lower frequency Chain Home system that had sprung from Watson Watt's Daventry experiment of 1935. For the Royal Navy particularly, low-flying German aircraft dropping magnetic mines into the busy shipping lanes off Britain's coasts were the real threat, but so too were surfaced submarines that could venture close in-shore without fear of detection. The solution was to use a much higher frequency, typically 200 MHz, in what became known as the CD (Coast Defence) and CDU (U for U-boats) sets from which evolved the CHL radars and their semi-tropicalized COL equipment that Basil Schonland was soon to try and secure for South Africa. Their performance was impressive and John Cockcroft's reputation as a man who got things done spread rapidly within the corridors of Whitehall and elsewhere [10].

In August 1940 Cockcroft left England for the United States in the company of six others, known collectively as the Tizard Mission.* This visit by three scientists and three senior officers of the Royal Navy, the Army and the Royal Air Force, plus a secretary, was to be a turning-point in the war, for amongst their baggage was virtually every one of Britain's wartime scientific secrets, including the most precious of all, the cavity magnetron. The purpose of their mission was to disclose the existence of these to the Americans, in exchange for support from the massive American industrial machine and its significant university-based research facilities. Initially, the proposal of Tizard's, made earlier in the year, to disclose these great secrets to a country still steadfastly neutral was met with much opposition from Churchill, then First Lord of the Admiralty, and from members of the scientific community led, not surprisingly, by Watson Watt, whose views on disclosure of the existence of RDF to the Americans were well known. However, with the fall of France and Britain's seemingly inevitable isolation as the only bulwark against the Nazi tyranny, opinions changed rapidly. Churchill, now Prime Minister, spoke directly to President Roosevelt about the need for such a scientific mission and soon even the redoubtable Robert Watson Watt became an enthusiastic supporter.

---

* Its members were Sir Henry Tizard, Professor John Cockcroft, Dr E G Bowen, Captain H W Faulkner RN, Brigadier F C Wallace, Group Captain F L Pearce RAF and Mr A E Woodward Nutt (Secretary).

That the Mission was an unqualified success was due, in no small way, to the part that Cockcroft, Tizard's deputy, played. He it was along with E G Bowen, one of the pioneers of airborne radar, who revealed the miracle of the cavity magnetron to the Americans. This device, born in Mark Oliphant's Physics Department at Birmingham, was the product of the conceptual brilliance and remarkable intuition of two men, J T Randall and H A H Boot. At precisely the time that Schonland's team at the BPI were putting together their first JB radar with just a few watts of power at 90 MHz, so Randall and Boot were attempting to generate very high power at centimetre wavelengths using little more than a heated cathode within a bored-out cylindrical copper block mounted between the poles of a large permanent magnet. In January 1940 the cavity magnetron, as it was called, operated for the first time and produced 400 W at a wavelength of 9.8 cm (or a frequency of just about 3 GHz).[*] Its potential was obvious and so the experience of E C S Megaw at the General Electric Company was called upon to turn this creation into a usable device. By September, Megaw had achieved 100 kW of pulsed power at that frequency and both radar and the war were soon to be transformed, some would say immeasurably [11].

The Chairman of the recently formed Microwave Committee of the US National Defence Research Council was none other than Dr Alfred Loomis, the same man who, more than a decade before, had offered his hospitality and private research facilities at his country home in Tuxedo Park, NY, to C V Boys when he arrived in the States with his innovative revolving lens camera. There, Boys had obtained his first photograph of a lightning flash and, from that single event, would blossom a new field of research that Basil Schonland would make his own. Now, Loomis and his colleagues were to witness an even more remarkable demonstration. Any scepticism that the Americans may have harboured regarding both British scientific prowess or their real willingness to divulge their greatest secrets evaporated within seconds when the capabilities of the cavity magnetron were laid bare in front of them. Its combination of very high pulsed power and extremely short wavelength was immediately recognized as the key that would turn radar into a Titan and, though no one knew it at the time, it would alter the very course of the war [12]. The frankness of the Tizard Mission's disclosures had ensured that the way had been opened for scientific cooperation across the Atlantic on a scale unimaginable just a short while before. Now the British would have access to American research, development and manufacturing capabilities that they could only dream about in England, while the United States would acquire the cavity magnetron and much else besides. As the war progressed, more than a million cavity magnetrons would be

---

[*] GHz (gigahertz): 10^9 Hz.

manufactured in the United States for radar applications on land, sea and in the air. For good reason it has been described as the most valuable cargo ever to cross the Atlantic [13].

Cockcroft returned to England in December 1940 and in April became Chief Superintendent of the Air Defence Research and Development Establishment (ADRDE), based at Christchurch [14]. He immediately set himself the task of bringing the Establishment into closer touch with military operations, particularly those associated with anti-aircraft artillery, and here he found another powerful ally in General Sir Frederick Pile ('Tim' to his closest colleagues), Commander-in-Chief of Anti-Aircraft (AA) Command. Pile was a small, dynamic man with a perceptive brain and a ready appreciation of what science and scientists could do for the army. The complexity of gunnery against high-flying, manoeuvring aircraft that were frequently invisible because of clouds soon made Pile's the most technical and scientific Command in the army [15]. Then, the switch by the Luftwaffe in September 1940 to night-time bombing operations against British cities and other targets, caused him to take the most significant step of all by appointing a scientific adviser to his Command. The man he chose was Patrick Blackett, whose prodigious intellect and naval background equipped him admirably for the role of applying the scientific method to the solution of the far from trivial problems of modern warfare. Of course, it had readily been appreciated that radar would greatly assist the anti-aircraft gunners to lay their guns on the target, but the earliest gun-laying radar, the GL1, was sadly deficient in one key aspect of its performance: it provided reasonable range and bearing accuracy but gave no indication of target height. Without such information anti-aircraft gunnery was well-nigh useless. An adaptation to overcome this problem, known as the elevation finding (EF) attachment, was soon forthcoming but it required careful setting up and adjustment and only worked well when used by L H Bedford, from the Cossor Company, who designed it [16].

It was here that Blackett and those who followed him introduced science and scientists to the arts of warfare. With the proper training and much experience in its use the gunners turned radar into a very powerful weapon. Over the course of the following year the accuracy of these radar-controlled guns improved markedly, as testified by figures quoted in the official history of army radar written after the war [17]. For example, in September and October 1940, 260 000 AA rounds were fired and only 14 aircraft were brought down; a rate of 18 500 'rounds per bird', in the jargon of the time. During 1941, by contrast, the accuracy of the anti-aircraft guns had been improved so much by the use of radar that this figure was reduced to 4100 rounds per bird.[*]

[*] These figures have been quoted by many authors without referring to their primary source which was actually written by Basil Schonland in 1944 in a document that was then classified 'Most Secret' [18].

Pile and Blackett readily appreciated that accurate anti-aircraft gunnery would only be possible if scientists were recruited in sufficient numbers, trained in radar and gun-laying techniques, and then assigned to the gun sites to act as advisers to the gun crews. In November 1940 a radar training school was established at Petersham, on the edge of Richmond Park, where equipment had been set up to test the gun-laying radar sets. The man in charge of this school was yet another of the Cavendish physicists, the very able and energetic protégé of Edward Appleton, J A (Jack) Ratcliffe, who was hastily seconded from the Air Ministry Research Establishment for this purpose. Behind Ratcliffe's appointment was John Cockcroft. Once again Cockcroft had demonstrated his ability to identify a key man for a vital position, for in Ratcliffe he had someone noted not only for 'devastating efficiency' but who was also the most stimulating and lucid of lecturers [19]. Soon the AA Command Radio School had 60 students, mainly biologists from the universities, plus a number of school masters, since all available physicists and engineers had been recruited already for other duties, and so, after some lectures on basic electricity to refresh their memories they were then led through the maze of radar. Not long afterwards they were regarded as competent enough as Radio Maintenance Officers to keep a GL1 set running in the field [20].

Having an operating radar set that had been accurately calibrated on a benign test site such as Richmond Park was one thing; placing it in the hands of troops to track a rapidly moving enemy aircraft, under every conceivable kind of operational situation, was quite another. Blackett recognized this very early on and, with the touch of genius so character-istic of the man, set about establishing a group of highly competent scientists with no specialist training in radar but whose abilities would allow them to approach the problems from the most general scientific points of view. To find these gifted people he called on the assistance of his colleagues in the universities and the government scientific service. Many suggested the names of young scientists of real talent who joined the group, originally based at Savoy Hill House and then at Great Westminster House in London, that was soon to refer to itself, quite unofficially, as the Anti-Aircraft Command Research Group (AACRG), but who would henceforth always be referred to simply as 'Blackett's Circus'. Their diversity of academic backgrounds and skills, which would seem to make them the very epitome of square pegs in round holes, was soon to be their real strength. The first to arrive were two physiologists, D K Hill and A F Huxley, followed by mathematical physicists A Porter and F R N Nabarro, astrophysicist H E Butler, general physicist I Evans and surveyor G W Raybould. Slightly later, another physiologist, L E Bayliss, and two mathematicians, A J Skinner and Miss M Keast, arrived to make up the rest of the team [21].

Amongst their first tasks was to determine the effects of the local terrain around the GL1 radars on both the bearing and elevation readings that were fed to the gun-laying predictors. By using calibration oscillators suspended from hydrogen-filled balloons and the occasional cooperative aircraft, they soon showed that the slope of the ground was by far the most important factor that had to be allowed for, more so than the existence of nearby Nissen huts, trees or ditches or the type of soil. Re-locating the radars and the guns that they served to flat sites was certainly not a proposition that the military would contemplate during the nightly bombing raids on London and other major cities of England and so a more scientific solution was needed. The problem was referred to Ratcliffe, who presented it to the group of scientists and serving officers attending his regular Saturday morning meetings at which such problems of mutual interest were discussed. An idea was soon forthcoming from the fertile mind of L H Bedford, whose earlier suggestion of fitting elevation-finding attachments to the GL1 was providing some degree of success – when in the right hands. He now suggested the laying of wire-mesh mats around each radar equipment so as to provide a highly conducting and level reference plane, against which the radars could easily be calibrated, regardless of the local features of the ground. The task of calculating the dimensions of the mat, and especially the size of its mesh, in order to simulate a solid surface at the frequency of the radar fell to Professor N F Mott, already a theoretical physicist of some renown who had recently joined the group. Assisting Mott in this was Frank Nabarro who, not many years before in Oxford, had listened to Schonland describing the features of lightning [22]. Their calculations soon led to the equipping of every gun-laying radar site in the land with its wire-mesh mat, and though this nearly caused the total countrywide depletion of stocks of such mesh, the accuracy of the guns was improved markedly by its use.

Soon the members of Blackett's Circus found themselves distributed amongst the gun-sites around London and at the AA Command Head-quarters, where they saw at first hand the nightly actions being waged against an invisible enemy high above them. This experience proved to be invaluable and led to the introduction of recording vans, within which the scientists could work in relative peace and with fewer distractions and interruptions from the gun crews outside. They analysed the data fed to them from the gun-laying predictors, logged the number of shells fired in every engagement and became adept at observing the tactics adopted by the enemy and his reactions to the rapidly developing radar-controlled gun-fire. The results of their work were fed back to the AA Command HQ where it soon proved of great importance in determining the anti-aircraft policy that was then evolving so rapidly [23]. Close contact was maintained throughout with Ratcliffe's school in the Old

Vicarage at Petersham where, in the church hall, suitable modifications were made to the mechanical Sperry predictor that measured a target's rate of change in course and speed and then fed the appropriate aiming coordinates to the guns. The training of gun crews required special equipment on which to practise and, once again, it was L H Bedford who produced his 'Trainer' on which the operators could hone their skills and which also allowed the performance of the predictors to be carefully assessed and suitably corrected [24].

In March 1941 Blackett's services were required elsewhere. The U-boat threat to the very lifeblood of the British nation—the sea lanes through which it imported its food and fuel—had reached crisis proportions and the Battle of the Atlantic, as Churchill called it, was about to begin [25]. Air Marshal Sir Phillip Joubert de la Ferté took over the all-important but much under-resourced Coastal Command of the RAF and immediately insisted that he have with him a scientific adviser of the highest calibre. Joubert, as Adviser on Combined Operations the year before, had been in charge of all radar and radio matters of the Air Force and so was an ardent convert to the use of this technology. Without radar, that other battle for the very survival of the nation, the Battle of Britain, would have been lost and he needed no reminding that the ability of the Luftwaffe to bomb industrial targets such as Coventry with almost pin-point accuracy was due to their use of navigational aids that relied on intersecting radio beams known by their code-names of *Knickebein* and *X-Gerät* [26]. Only Britain's use of its own technology, and particularly the brilliant decryption of the German Enigma codes at Bletchley Park, their accurate interpretation and the development of counter-measures by every means possible, had stood between the nation and Hitler's planned invasion with the intention of eliminating 'the English homeland as a base for the prosecution of the war against Germany and, if necessary, to occupy it completely ...' [27].

Blackett, aloof, idealistic and to some almost god-like, moved to Coastal Command and left behind him a huge void at a most critical time. Before departing, he suggested that John Cockcroft should succeed him as scientific adviser to General Pile's AA Command and Cockcroft, ever willing to assume multiple responsibilities, initially considered doing so. But good sense prevailed because he soon realized that it would have been grossly unfair to Pile, were he to try and combine those duties with his existing responsibilities as Chief Superintendent of the ADRDE [28]. Someone else would have to be found to take on this most responsible and onerous of positions. Initially, Nevill Mott, a theoretician of great eminence and an FRS at the age of 31, took over and applied himself with enthusiasm to various problems, particularly the optimum deployment and use of searchlights, but his air of academic detachment just did not sit easily with the military. The advent of radar was rapidly rendering searchlights obsolete and it was apparent that scientists would be better employed studying how best to use radar in the anti-aircraft role, and it was evident that Pile did not see Mott as the man for that job [29]. The immediate solution was to combine the activities of Ratcliffe's Radio School with those of the group of peripatetic scientists whom all knew as Blackett's Circus. Ratcliffe agreed to become Superintendent of this new organisation and a suitable name was required by which this occasionally dishevelled bunch of scientists and school masters might henceforth be known.

It soon came from a term allegedly first coined by Watson Watt at Bawdsey when he saw, better than anyone else, that the use of radar in the tracking of enemy aircraft, the marshalling of defensive measures against them and the control of the myriad of resources thus involved, both human and technical, required very careful *operational research*[*] [30]. Formal adoption of the title Operational Research Group or ORG was delayed for some while, though, because any implied link with military operations was considered by some within Whitehall to be inappropriate on security grounds. Initially, therefore, the newly relo-cated remnants from London, under L E Bayliss, joined Ratcliffe's group at Petersham in April and were recognized officially in July as the ADRDE Petersham Research Group. It took another two months before they could adopt their formal title of ADRDE (ORG) [31].

*   *   *

The ship carrying Basil Schonland and his party of eight young South African engineers arrived in Liverpool 35 days after leaving Cape Town. On Schonland's orders, before disembarking, the men donned berets rather than the pith helmets that were then standard issue to all South African troops, because he did not want the British military or their public to think that they were policemen! [32]. Then, suitably attired, they made their way by train to London. Schonland, with Phillips and Browne, reported immediately to the South African High Commissioner while Guttridge took the rest of the party, destined for training by the RAF as radar officers, to the Air Ministry. There they encountered a problem that none of them, and certainly not their commanding officer, would ever have contemplated — they failed their technical interviews

---

[*] Quite who first conceived the concept of 'operational research' as a distinct scientific disci-pline with military applications is open to much conjecture. Certainly Watson Watt claimed to have originated it and declared it to be one of his 'Three Steps to Victory' in his book by that name; the others were the cathode ray direction finder and radar. However, R V Jones in his *Reflections on Intelligence* (Heinemann, 1989, p 190) gives the credit to A P Rowe; Lord Zuckerman in *From Apes to Warlords* (Collins, 1988, p 370) claims it for his pre-war dining club of the scientific elite — the 'Tots and Quots'; yet others would give the credit to Blackett.

and were informed that they needed two years of instruction in the Signals School at RAF Cranwell before they would be accepted for radar training. It transpired that the officer responsible for the selection of radar personnel had been the editor of a well-known wireless magazine before the war and he was interested in one thing and one thing only: their understanding of the basics of electricity and radio theory.

For two solid months before leaving South Africa, Schonland had insisted that these carefully selected men should virtually be immersed, by Bozzoli, in the technicalities of the JB radar and so they had become well acquainted with the intricacies of its circuits and all its hardware. Now, the man at the Air Ministry probed them for much more fundamental knowledge of the kind they had learnt as students but which had been rather submerged by the more pressing details of an operational radar system. The young South Africans were mortified, while Schonland was incensed, both with them, but even more so with the RAF policy that seemed very short-sighted under the prevailing circumstances. He reacted with characteristic resolve and arrived at the Air Ministry to confront whoever was behind this well-intentioned, though clearly flawed, policy of selection. His reaction was forthright and unequivocal: either the South Africans were accepted for radar training or he would take them straight home again. Given the parlous situation in which Britain now found herself, with every competent scientist and engineer having been pressed into service of one kind and another, the loss of these six volunteers would have led to serious questions being asked at the very highest levels. Capitulation followed immediately and the men were soon issued with RAF uniforms and sent on an officer's training course at Loughborough, from where they then spent six weeks at an operational CH station at Great Bromley in Essex acquainting themselves with every aspect of the equipment and observing the station in operation against enemy aircraft. This practical introduction to British radar was followed by a two-month technical course at Yatesbury in Wiltshire, in the company of 500 others, mainly from the Commonwealth. When the results were announced all Schonland's men found themselves amongst the top fifteen per cent on the course and one of them was placed first. Honour had been satisfied and the colonel was a very relieved man [33].

Early in May, after attending to various formalities at the South African High Commission in Trafalgar Square, including the matter of his more than generous daily allowance that allowed him to survive rather well under the conditions of British wartime rationing, Schonland spent some time at both the Air Ministry and the War Office, where he dealt with the pressing need for British radars to be sent to South Africa. In his first letter to Ismay since arriving in England, he reported that he had been very well received and that the officials in both Ministries had been exceptionally kind and helpful. There was certainly no suggestion that information about British radar and its capabilities was being withheld from him. The decision of the Air Staff to release information to South Africa was being fully implemented, and wiser counsel had clearly prevailed after Flight Lieutenant Atherton had rather stirred things up a few months before. A programme of visits to radar sites and experimental establishments around the country was arranged so that Schonland and his two colleagues could become fully proficient in the operation of the various radar systems then in service with both the army and the RAF.

Schonland, with Phillips and Browne, set off on this tour and once again British morale fascinated him. No amount of desolation in their cities nor defeats of their army in the field seemed to deflect them from their belief that they were going to win the war. They already saw the disappearance of German bombers from the day-time skies as their first victory and were sure that they would have won the night-time battle by Christmas. Food, he wrote to Ismay, though scarce was adequate even though 'orange juice is 6d a small glass and fresh oranges a rarity. No bananas or apples at all. This is a bit hard on the children but doesn't affect me.' He was impressed too by the technical prowess that he encountered and was delighted to discover that there, in the thick of things at one site he visited, was John Cockcroft. This was at Petersham where Cockcroft was busy welding together the radio school under Ratcliffe and the members of Blackett's former circus, under Bayliss, into what was to become operational research in the army. Many of Schonland's other old friends from Cambridge were much in evidence too and one of them, Charles Ellis, invited him to spend the weekend with him when he passed through Bristol on his travels around the country. These were almost like old times and the atmosphere, though bleak and the air dust-laden, contained a tinge of excitement that made him feel almost elated. Should he stay and seek some role with the British Army or would his services be needed much more urgently in South Africa when the new radars arrived and had to be set up? His regular letters to Ismay made her fully aware of his quandary [34].

The first radar site that Schonland and his colleagues visited outside the immediate environs of London, where the GL set was the dominant equipment, was in Wales. The site in question was at Llandudno where Dr M V Wilkes, one of the group of Cavendish physicists who had been introduced to the secrets of RDF at Bawdsey in 1939 and who had then been part of Cockcroft's team that had erected the CD and CDU radars in Scotland, was engaged in radar-siting trials. Wilkes was now a member of Ratcliffe's team at Petersham and was in charge of the radar section. Not far from Llandudno was Great Orme's Head, a promontory of land that jutted out into the Irish Sea with a sweep of coastline and precipitous cliffs around it, all of which made it the ideal test site for a coastal defence radar system. It was also so reminiscent of the scenery around Cape Town, where another Llandudno existed and where the coastal drive around the peninsular and Cape Point was like another Great Orme's Head, but on the scale of Africa. What Schonland saw atop the Orme was a demonstration of careful measurement and protracted calibration of equipment that faintly amused him when he remembered his efforts on the roof of the BPI and in the steamy environs of Mambrui on the Kenyan coast, for here were scientists who, in the midst of a war, were able to plot bearing errors to a few minutes of arc. As Wilkes himself recorded years later, Schonland regarded this as one of the curiosities of radar [35]. However, curious or not, such meticulous attention to detail would soon pay dividends not only in England but also in South Africa, when the JB radars and their British counterparts ultimately dotted its coastline from Baboon Point in the west to the Mozambique border in the east.

The pressure of their schedule allowed no more than two days at this isolated but beautiful spot before Schonland left Phillips behind to absorb the details while he and Browne set off for Somerset and another establishment. Not far behind them was John Cockcroft. Whether just by coincidence or actually by design we will never know, but Basil Schonland and his old Cambridge colleague met up again when he and Browne were familiarising themselves with the CHL radars in operation along England's south coast. Their working week went uninterrupted by the niceties of the weekend and late one Sunday afternoon in mid-May, as Spring eventually began to show itself in all its finery, Schonland and Cockcroft went for a walk after work. South Africa's requirements for radar equipment was just one of the topics they discussed; recent developments at the ADRDE were another. To Ismay, in her role as the go-between with the BPI, Schonland wrote urgently on the 12th and asked her 'to please tell Bozzoli that, owing to delays of various kinds, the authorities here recommend that we do *as much construction* as possible. We are not likely to get much from them for some time though I am going to do my best. I hope to get some of the new micro-pups[*] which will be better than our 250THs by a factor of at least two in power'. And, still smarting from the rather ignominious performance of his young colleagues when initially found wanting in their fundamental knowledge of radio, he put in a special plea that Bozzoli should really stress such details with any further groups that may follow them to England [36].

---

[*] The micropup valve was officially known as the VT90. It was a triode of revolutionary design using a cylindrical copper anode that also formed the valve envelope and was capable of producing kilowatts of pulsed power at metre wavelengths. As such it became the mainstay of the AI (Air Interception) radars.

On this score there were indeed developments. Amongst the many appointments that Schonland had with senior officers and government officials in London, none was more significant than his meeting with Lord Hankey, originally Minister without Portfolio in Chamberlain's War Cabinet and now Paymaster General under Churchill. Hankey was an extremely impressive man, a 'Prince of Secretaries' as *The Times* described him years later. He was a genius for detail, whose quiet demeanour and ready smile belied an intellect that had made him something of a colossus within the corridors of Whitehall. During the First World War he had served as Secretary of the War Cabinet and had the distinction of being the only man who attended every political and inter-allied conference during that conflict. To him, in 1916, must go the credit for first devising the concept and military role of the tank—an idea seized on with enthusiasm by Churchill but rushed into service too soon by Haig. It was Hankey who, in the face of much opposition from the Admiralty, persuaded Lloyd George of the power of the convoy system to counter the submarine menace and it was he, too, who foresaw the use of the flame-thrower as a powerful defensive weapon when Hitler's seemingly imminent invasion of England filled all with massive foreboding. Without actually coining the phrase, all this was operational research. Hankey was now, undoubtedly, the *éminence grise* of Churchill's Cabinet and he had influence far beyond that of any other civil servant. Amongst his array of responsibilities was the provision of adequate numbers of technicians and professional men who could be used to further the war effort, especially in support of the expanding radar systems in all three Services. It was in connection with this pressing matter that he asked Schonland to see him. The fact that Hankey's wife was the daughter of a former Surveyor General of the Cape Colony will have surely ushered in their discussion, but it was soon the man's charm and his grasp of all the issues (operational, technical and political) that rapidly convinced Schonland that South Africa could assist with the provision of, at least some, of that manpower. Schonland made a good impression on Hankey as well, and in the years to come their relationship would bear much fruit. On Hankey's death twenty years later Schonland wrote a wonderful tribute to the man who enabled 'all kinds of scientists in all sorts of places swiftly to be of use to their country in a great emergency' [37]. Basil Schonland himself had certainly been one of them.

By the end of May, Schonland's brood, as he called them, had regrouped in Wiltshire where they were now the guests of the RAF. The cultural change from the Army hit him immediately as cut-glass accents and mannerisms he associated with Eton and Harrow gave way to flying-talk and terminology in all conceivable combinations of dialect and vocabulary. More important, though, was his impression that the RAF was fighting the present war, while some elements, at least, of the 'dear old British Army [were] getting nicely ready for the war after this one'. His own situation, though, intrigued his new colleagues in both services, for here was a lieutenant colonel who actually had detailed technical knowledge of devices which most regular officers of that rank and above had no comprehension of at all. Worse, though, was the feeling that it was not expected that they should. He described it all to Ismay [38].

> Most of them think I must be bogus either technically or militarily.
> In neither Army nor Air Force is anyone with half my technical
> knowledge ranked as high as a Major & the "scientists" are all
> civilians with passes. They think it's the best scheme — perhaps it is
> but I doubt it. It leaves the duds in charge of things & consequently
> things go very wrong indeed. I don't see how I can do a job for
> them over here, for they don't have any "soldier-scientists" at all &
> the idea would be most upsetting.

Upsetting to some, possibly, but not to Schonland himself, who found his dual role most agreeable. He had no problems adapting to the peculiarities of service life and customs as circumstances required. But whether there was any part for him to play as a 'scientist-soldier' here in England was less certain.

An opportunity soon presented itself that allowed an escape from his 'Cook's Tour' around the country. It came in the form of an invitation from the President of the Royal Society to attend a function at Burlington House in London at the beginning of June. There he found himself among a group of recently-elected Fellows who were to be inducted and he was invited to join them. This was an honour indeed, since his own admission to the Fellowship in 1938, though of great moment within the university, had not been accorded quite the same pomp at Wits. Amongst the new Fellows to be honoured were Robert Watson Watt and P I Dee, yet another Cavendish man whose contributions to the development of airborne radar had been so important, as well as E C Bullard, the Cambridge geophysicist who, as Schonland's guest, had done that important work at the BPI on the thermal characteristics of rock just before the outbreak of war. Bullard was now in charge of the vital task of de-gaussing ships to protect them against magnetic mines. Each of them signed the famous book that opened with the signature of Isaac Newton and then, with the minimum of fuss, they ascended the dais and shook the President's hand as he intoned the 350 year-old injunction that they be admitted as Fellows of the Royal Society for Improving Natural Knowledge.* Describing the occasion to Ismay in his weekly letter, Schonland expressed some surprise

---

* On that same occasion the Fellows present were asked to vote on the admission to the Fellowship of one Winston Leonard Spencer Churchill under Statute 12 for 'conspicuous service to the cause of science'. Their decision to admit him was unanimous!

at his lack of elation when it all happened and then continued, rather ambiguously: 'I suppose I ought to have felt a doubt of my worthiness but I didn't at all. A sense of considerable responsibility & gratitude to the country that had encouraged me, that's all'. Which country, we may well ask, had he in mind?

* * *

The weather in England during that summer of 1941 did nothing to improve Basil Schonland's rather concerned state of mind, nor his longing to be reunited with his family. Soon after disembarking from the ship he had contracted a severe head-cold that took him weeks to shake off. Then he developed 'pink eye', which he attributed to the appalling dust that hung over London after the bombing raids that were almost a nightly occurrence. But capping it all was the penetratingly cold weather. Though wearing two sets of underclothes and the thickest socks he could find, he shivered constantly and this was all made worse by the severe shortage of hotel accommodation whenever his duties required him to spend time in the city. Frequent moving from one cold and draughty establishment to another did not help and all was compounded by the difficulty of finding someone to do his washing!

By contrast, life in the various military establishments around the country could not have been more pleasant. Drivers were always waiting at the railway stations to meet him and his party, and to convey them to officers' messes where they were 'looked after like honoured guests'. Batmen then took over and provided almost luxurious service: washing and ironing clothes and catering to every possible need. For all his occasional criticism of aspects of the British military system that he confided to Ismay alone, Schonland himself felt honoured to be amongst men who were adamant that they were going to win the war even though the way they sometimes went about doing it left him feeling utterly exasperated. Quite what his own part would be in this momentous struggle, though, was still unclear, for nothing had, as yet, appeared on the horizon. He was hopeful that the various Ministries involved would soon commit themselves to a date when the radar equipment he had requested would be dispatched to South Africa, so that he could at least arrange his own return to oversee its installation and then set up the organization to use it along the precise lines that he had observed so closely around England. By now his SSS colleagues under Hodges should have arrived in Egypt, as requested by HQ RAF ME, with their JB radars in tow. His likeliest course of action would therefore be to visit them there on his way through, brief them on the latest British technology and organization and then advise them of their future roles in the grand plan of Lord Hankey that was just unfolding. So, if all went according to his plan, he hoped to fly back to South Africa sometime between 3 August and 15 September. 'Five months away from you all' he wrote to his wife [39].

For all the good intentions that were in the air when Schonland had visited Cairo some months before, it was not too clear quite what Hodges and his SSS radars were to do once they had landed in the Middle East. In fact, the RAF there initially denied any knowledge of them, even though all the arrangements had been made by Schonland personally the previous December [40]. Confusion was much in evidence but so too was the lack of radar cover. In January, the Air Ministry in London had received a signal from HQ RAF ME spelling out in no uncertain terms the radar situation that pertained in that theatre of operations [41]:

> All MRUs Egypt fully occupied in view extended commitments
> western desert and German attacks canal zone necessitate early
> cover this area which at present has no RDF cover...

There was a need for urgent action and Watson Watt, now Scientific Adviser on Telecommunications, was consulted by the Deputy Director of Signals at the Air Ministry about the advisability of constructing 50 'pack sets' that could be dispatched by air, the first ten of which would actually be assembled in the Middle East . The man behind all this was none other than Wing-Commander J A Tester, who had accompanied Schonland on his Middle Eastern visit and who had requested that a JB1 set be made available for trials soon after. But Tester's report on the successful trial carried out by Gane in January had not yet reached London and his plans to use the SSS and their equipment had been pigeonholed in Cairo and forgotten about. So, in London at least, there was some considerable uncertainty about the suitability of the South African equipment in this role and the obvious man to ask was Schonland himself who, the Air Ministry discovered, was then on his way to England. It was immediately suggested that Watson Watt should see him [42].

Their first meeting since 1937 took place not in Whitehall but on the very day at the Royal Society when both men added their names as Fellows to the long list headed by Newton's. In conversation afterwards, Watson Watt told Schonland that he had heard how the South Africans had 'done some remarkably good work' and that he would like to hear some details, but the occasion was clearly not appropriate and they agreed to meet again soon. Since no details of any sort were discussed, Schonland had no inkling about the developments, or lack of them, in the Middle East and it suited him not to take the matter further until he had completely familiarized himself with the British radars. So the meeting did not take place immediately.

But a matter that was certainly taxing Schonland's mind was the discussion he had had with Lord Hankey about the transfer of

222

South African technical officers to the RAF. The more he thought about it the less he liked the idea and, as always, it hinged on the issue of their financial security. Could he, in all sincerity, ask the men under his command to accept conditions of service that he himself would baulk at? A better idea, he now felt, would be to train them and then send them to Hodge's SSS Field Force in Egypt from where they could be attached to the British Army as Ordnance Maintenance Engineers charged with looking after all radar equipment in that theatre. In this way they would not suffer financially, nor would they have to serve in someone else's uniform. He wrote immediately to Colonel Collins in Pretoria appraising him of his concerns and of the change in tack that he now wished to adopt. As always, because it was Schonland, Collins agreed without a second thought. Then Schonland asked Lord Hankey if they could meet again to discuss the matter. Hankey listened intently, mulled the matter over and said that he thought the suggestion 'admirable'. On Hankey's nod the War Office was also agreeable and the plan was then communicated to Hodges in Egypt who 'wired back gratefully accepting the idea'. 'Daddy' Hodges, as he was affectionately known by the men some years his junior as they observed his rather bumbling style of command, had already encountered much unease amongst those men. Rumours were circulating, both in Egypt and at the BPI, that they would soon find themselves in the RAF, which was not the course they had intended taking when they 'joined up' back home. Schonland's action saved the day and they would now all remain in the SSS [43].

*   *   *

On 1 July 1941 Lt Col Basil Schonland was appointed officially to the Staff of the South African High Commissioner in London in an advisory capacity with special reference to all matters related to radar [44]. Two weeks later he left his London hotel for the Surrey countryside to spend the weekend with old family friends from South Africa. Sir Harold Spencer-Jones had been H M Astronomer at the Cape a decade before and, now retired, he and his wife offered Schonland a refuge away from the city and the interminable travelling about the country that had become his lot in life over the past many weeks. It was also to be a farewell visit, for he was about to return home in a week's time; his task of securing radar equipment for South Africa was now complete. Together, he and Eric Phillips had booked their passage and the ticket was in his pocket.

Then the telephone rang and the call was from John Cockcroft. The matter was urgent and Cockcroft, as ever, got straight to the point. He wanted Schonland to take over Ratcliffe's job at Petersham! Could they meet in Christchurch as soon as possible to discuss it, but before that would he please see Patrick Blackett in London? Schonland was stunned but not too surprised, for he had begun to read the Cockcroft mind.

On Monday 14 July, Schonland and Blackett duly met and set off immediately to see General Pile at Anti-Aircraft Command. Together, Blackett and Pile explained what they had in mind and then the General, with his usual candour, asked Schonland directly if he would be prepared to take on the job. The initial shock of Cockcroft's equally blunt invitation had by now somewhat subsided and Schonland was better prepared. Here, indeed, was the opportunity of serving the larger war effort that he had long sought for himself, but this was not the moment for precipitate action and he asked for a chance to give the matter some thought. This was readily agreed, for Blackett and Pile knew that a meeting had been arranged for the very next day between Schonland and Cockcroft, and Cockcroft was a remarkably persuasive man, even though he used so few words in the process. On the Tuesday morning, Schonland duly travelled down to Christchurch, where his old Cavendish colleague, as laconic as ever, was waiting for him. The rapid development of radar in all its forms, Cockcroft explained, and its increasing use by the army, navy and air force meant that there was an urgent need for Ratcliffe's services at the Telecommunications Research Establishment (TRE) in Malvern, where he was unquestionably the man to set up a radar school for the RAF and to establish what would become known as the TRE Post-Design Services that needed both a first-class scientist and a first-class organizer to run it [45]. Equally important, though, were the activities that Ratcliffe would leave behind in Richmond; and the Ministry of Supply, the government body responsible, were urgently seeking his replacement. Cockcroft hoped Schonland would accept that position.

Again Schonland requested time to make up his mind and then wrote immediately to Ismay and gave the letter to Eric Phillips to deliver personally. He could now be as expansive as he liked without fear of intervention by the military censor or of betraying any confidences, for Ismay's lips were sealed [46]. He told her that if he accepted Cockcroft's offer he would become 'Director or whatever of ORG' the 'Operations Research Group responsible for testing and reporting on military equipment and for advising the commanders of the units affected by their investigations'. In addition he would be required to 'make suggestions as to new equipment and ideas, improvements to old equipment and to the army's methods of operating it all, as well as to changes in army organisation' should any be deemed necessary in the light of scientific and technical advances. Under him would fall all the army radar systems then in service GL, CD, GCI and SLC, the visual indicating equipment or VIE, as well as the gun-laying predictors and the responsibility for compiling all anti-aircraft battery statistics. He would report to Cockcroft as Chief Superintendent of ADRDE but would have direct access, as well, to General Pile and to Pile's opposite number at Coast Defence.

By the Thursday, Schonland had made up his mind that he would accept the appointment, subject though to their acceptance of five very specific conditions that he laid down, and these he communicated to Cockcroft when they met for dinner the next evening in London. He wrote again to Ismay and gave the letter to Phillips, his courier [47]. That afternoon he visited the South African High Commissioner to inform him of the British invitation and of his conditions of acceptance. By Saturday, all the wheels were in motion and agreement was rapidly forthcoming from all sides. All that remained was a visit was to the Old Vicarage at Petersham to see Ratcliffe and to hear from him directly what the job entailed.

And so, within the space of just a week, an appointment of some considerable significance in the application of the scientific method to military operations was made and Basil Schonland's life as both a scientist and a soldier changed direction forever. It had been a week of some torment as he lay awake at night and wrestled with the biggest decision of his life. The offer was there. Accept it and he would become inextricably bound up in the war in Europe with no possibility of returning home in the immediate future. Reject it and such an opportunity to serve to the limit of his patriotic ability would probably never come again. All, though, had depended on the acceptance of his five conditions and on this he had remained inflexible. And he spelt them out equally clearly to Ismay. They were:

  (i)   that no one else suitable for the job was available;
 (ii)   that the request for his services must be made via Lord Hankey, the South African High Commissioner and General Smuts;
(iii)   that he remain, as at present, a Lt Colonel in the South African Forces, paid at the same rate plus allowances;
(iv)   that he remain as RDF adviser to the Union government; and
 (v)   that the appointment be for six months subject to review at the end of that period, allowing them to dispense with his services if they so wished or he to resign and return home should he so wish.

Schonland's hesitancy had its roots, yet again, in that deep-seated uncertainty as to his own worth, that plagued his innermost soul at times when he was alone and with only his thoughts of Ismay for comfort. Surely there was someone else better suited than himself? But they just laughed when he suggested it. The only possible contender for the post was Nevill Mott and he was never seriously considered. The only issue from his list that caused just the slightest delay was the matter that he remain a serving officer in the South African forces. Until now, such posts had been held by civilians, though there were many brigadiers and colonels within the Ministry of Supply, and so either in mufti or in uniform he would be most acceptable but why, they asked him, should he wish to remain in his South African uniform? The reason, of course, had everything to do with the security of his family. The ORG post carried with it a salary of £1000 a year. Were he to resign his commission and become a British civil servant he would be unable, in terms of British law, to transfer money to Ismay without 'special arrangements of great difficulty'. Even were such arrangements made he stood to lose a quarter or even more of his salary to British income tax while Ismay would then be subjected to taxation on those earnings in South Africa as well. However, by remaining a UDF officer and by retaining his appointment as RDF adviser to the High Commissioner he felt sure that he could insist that he pay South African income tax alone. 'If this is not agreed to I will not take the post because I am not prepared to ask you to make sacrifices ...', he told his wife.

A telegram addressed to Ismay arrived at the BPI on Saturday 26 July 1941. It said simply [48]:

BRITISH ASKING MY SERVICES ON LOAN FOR SIX MONTHS
IN FIRST INSTANCE STOP FEEL I SHOULD STAY BUT HOW
ARE YOU LOVE SCHONLAND

Pencilled on it in Ismay's hand was presumably her reply — 'All very well here. Disappointed but congratulations'.

*   *   *

Having agonized over a matter, before eventually coming to a decision, the Schonland way was not to dwell on it. He immediately set about adjusting rapidly to the responsibilities and challenges of his new position and was soon drawn into an inner circle of those aware of new developments to do with radar. One that had caused even Cockcroft to show just a little elation was the recent success achieved with the new CD radar at Dover in locating an enemy vessel at 80 km range and the resulting action by the RAF that led to its immediate sinking. It was apparent too that all manner of radar developments would cross his desk and life would become extremely busy, but this was just the sort of role that he had sought for himself and he threw himself into it with energy and enthusiasm.

Schonland's first task required him to exercise diplomatic, rather than scientific, skills because already a rift had developed that could cause the British, in all their exasperating ways, to dissipate energy on matters unrelated to the prosecution of the war. In its wisdom the Army Council had seen fit to place the organisation of GL radar, and the maintenance of all the equipment, under the control of the Royal Army Ordnance Corps. This had brought the RAOC into conflict with

AA Command, the prime user of the radars, and matters were reaching the point where a controlling hand had to be laid upon the tiller. Schonland's hand was poised but his only real concern was that he had the necessary tact to negotiate with both sides, while not losing his temper with people who were more concerned with consolidating their own empires than with matters of much greater importance. Ismay, he felt, would handle the diplomacy far better than he would but, as he told her, he had nothing to lose so would confront the problem head-on. This he did and the RAOC backed down.

For all his new-found responsibilities he still kept an eye on the much smaller world of the SSS in the Middle East and back home in South Africa. His telegram to Ismay had triggered a feeling of some concern at the BPI. Bozzoli, who was now acting unofficially as the commanding officer because Hodges was in Egypt, had curtailed construction of any new equipment destined for use outside the Union, in line with an instruction received from Schonland on 23 July. This followed an earlier, somewhat confusing communication that indicated that Browne would soon be returning with just a GL receiver, but that Schonland himself and Phillips would follow soon after 'with a good deal of priceless equipment' [49]. Now, with the bombshell that was the Schonland telegram, it appeared as if the SSS had been abandoned by its commanding officer and had been left with no clear policy as to its future. However, in his letter to Ismay delivered by Phillips, Schonland sent another message to Bozzoli instructing him to 'get Hodges, Gane and, if possible, Hewitt back'. To her, he explained that it had never been his intention to leave them 'up north' and he now felt rather bad about it because the SSS were somewhat rudderless and very much in the dark.

For all this they had been far from idle. Once the Air Ministry and HQ RAF ME had reconciled their paperwork, the SSS found itself very much in action. As a result of the now pressing need for radar cover in the Canal Zone, and following Gane's successful demonstration of the JB1 radar to Wing Commander Tester, it was decided to deploy the SSS radars along the Sinai coast to provide early warning of German and Italian aircraft attempting to lay mines in the approaches to the Suez Canal. Hewitt, by now a captain, had surveyed the sites in the company of the RAF's man, Atherton, whose earlier scepticism at the home-made appearance of the South African equipment had quickly been moderated when he saw how well it performed in practice. By mid-July, three JB1s were operating at El Arish, Rafa and El Ma' Aden and were feeding their plots to the RAF filter room at Ismailia. They regularly tracked targets at distances of 120 km and even had the occasional sighting of the island of Cyprus, all of 400 km to the north—yet another example of the anomalous propagation that they had first seen north of Durban more than a year before. The JB1s, with their favourable siting overlooking the sea, out-performed the longer wavelength MRU radars of the RAF, which were soon withdrawn, leaving the SSS in sole charge of the vital sea-lanes upon which so much of the allied war effort now depended [50]. These three radar stations thus became part of the huge RAF radar network that extended from Malta in the west to Iraq in the east and were given the designations SSS 1, SSS 2 and SSS 3 in the RAF's official list of AME Stations* in the Middle East [51].

On Monday July 28 1941 Basil Schonland moved his meagre possessions into the New Star and Garter Hotel at Richmond and two days later J A Ratcliffe officially handed over to him the running of what was then called the ADRDE Petersham Research Group in the Old Vicarage. Under him Schonland had two sections, the first of which was concerned with radar and was led by Dr Maurice Wilkes, whom he had first met some months before on the wind-swept radar test site near Llandudno, while the second group, under the leadership of Dr L E Bayliss, dealt with gunnery fire-control problems and analysis. Thus, Schonland inherited Blackett's Circus and it was his task now to ensure that they could assist the army in the correct use of radar. It was General Pile though who told Schonland how the army first came to accept the inevitability of such unusual creatures 'in grey flannel suits and without proper haircuts' within its ranks, when what he called 'The Law of the War Office' collapsed in the face of radar. According to this 'Law' no new idea could ever be accepted by the War Office because, if it was thought to be new, it must have been thought of already by the War Office, and if it was not in use it must have been rejected on very good grounds by the War Office. However, radar proved to be the exception to the 'Law' and the War Office capitulated [52].

The War Office though, to its eternal pride, was soon to earn the thanks of the British Nation and indeed, that of the free world, for what was to follow. But without men such as Blackett, Cockcroft and Schonland, who had the ability to talk to soldiers like General Pile and a whole host more in their own language, none of this might have happened at all.

---

* AMES stood for Air Ministry Experimental Stations and as a term was something of a misnomer because it applied to a whole range of RAF equipment that was anything but experimental as it saw operational service in the field.

CHAPTER 13

SOLDIERS AND CIVILIANS

On 1 August 1941 Basil Schonland was appointed Superintendent of the Air Defence Research and Development Establishment (Operational Research Group), or ADRDE(ORG) in its somewhat more abbreviated form. Jack Ratcliffe had now left for his new post at TRE and Schonland was in charge of an impressive group of scientists, mainly biologists and physiologists—all of whom had been carefully selected from the universities around the country to plug the glaring gap that Blackett had foreseen once the physicists were swallowed up by the demands of radar. These men, Blackett's Circus of old, brought, by the very diversity of their scientific backgrounds, a new dimension to the world of applied military research and soon their skills were to be used even more widely, as the concept of active collaboration with the army gained ground, particularly within General Pile's Anti-Aircraft Command. In addition, for a period of a few months, Schonland was also responsible for the technical work of those radar officers at Petersham who were not yet in uniform. The third component of Ratcliffe's 'triple diadem', as Schonland referred to it, the Radar School was pressed on him too, but he steadfastly refused to take it on and so it was placed under the command of Major Pat Johnson, a physicist, rowing blue and sometime soldier before the war. Soon, all these groups would be working closely together as the science of Operational Research found itself on the battlefields of France. Now though, Schonland's most pressing need was to familiarize himself with an organization of many parts, many of which were not yet in full running order. He was pitched straight into an extremely heavy schedule of briefings and conferences with his staff, visits by senior officers and a mountain of paperwork, the details of which he knew very little. To master it as quickly as possible required working to a punishing schedule and so, in the beginning, his working day commenced at 9.00 every morning and went through until 11.00 or even 12.00 at night with just short breaks for meals. Only Sundays were sacrosanct and then he slept and wrote to Ismay.

After the first week he felt close to collapse. By the end of the next he believed that he had come to terms with the more mundane aspects and was rapidly becoming *au fait* with those issues that really mattered. To Ismay he wrote 'I am hopelessly ignorant ... but I shall manage all right and [will] become well-informed. After this I shall be qualified to be Principal of both the Universities of Cape Town & Wits combined. I don't do research but control it & most of my energy goes in organising' [1]. Prophetic words in many ways, though here they merely served to buoy him up as successive waves of people and their demands threatened to swamp him from all sides. One of them was Professor Nevill Mott,* nine years younger than Schonland but already an FRS of some standing. Mott, as we have seen, succeeded Blackett as Scientific Adviser to General Pile but soon found himself somewhat overtaken by events as Cockcroft and Pile saw the need for a dedicated establishment that would offer an array of services and scientific advice to the army. Of all his generation from the Cavendish, Mott was probably the least likely to have much rapport with the military because, as Pile later recounted, '[he] was so absent-minded that he always forgot what he had really come for' [2]. His function at AA Command HQ at Stanmore therefore became redundant and he moved, somewhat reluctantly, to Richmond where he found himself reporting to Schonland. Theirs was not to be a happy relationship even though Schonland, in his correspondence with Ismay, described Mott as 'a delightful person' while referring to him, in some awe, as 'our tame mathematician' [3]. To Mott the application of operational research to almost everything in the army soon rendered it uninspiring. He attempted to generate some enthusiasm in himself by applying it to study the role and use of the tank but, as he later recalled, he 'totally failed to come up with anything' and soon became frustrated and unhappy at Petersham. His mood was not helped by his dislike 'of the man in charge and probably his for me', though this latter sentiment is not borne out by Schonland's own views at the time [4].

To others, such as Maurice Wilkes, then in charge of radar at Richmond, Schonland soon proved to be a brilliant choice in the role that he had to play. It was indeed remarkable that a complete outsider in the form of an unknown South African physicist-cum-soldier, whose experience of radar was limited and whose fame rested on his background, albeit considerable, as an expert in lightning, should now be in charge of one of England's crucial scientific establishments, charged with countering the menace of the German bombers. Yet such was the wisdom of Cockcroft's choice that the transition from Ratcliffe to Schonland was made with almost seamless ease. Though their personalities were so different, Ratcliffe and Schonland possessed the special ability to administer a host of complex activities, seemingly in tandem, and in

---

* Sir Nevill Mott shared the Nobel Prize for Physics in 1977 with the Americans P W Anderson and J H Van Vleck.

a way that inspired the enthusiasm and confidence of most, if not all, who worked for them [5]. Soon Schonland saw the need to formalize, to some extent, the organization he had inherited and he set up the first two Operational Research Sections: ORS1, under Bayliss, with responsibility for investigating fundamental anti-aircraft fire control problems, while Wilkes ran ORS2, which dealt with all technical matters associated with radar in the Army.

While Schonland was well-served by many highly competent people, the ultimate responsibility for the running of his organization rested with him, and he shouldered it with equanimity if not overt enthusiasm for, as he confided to Ismay, 'it is damned hard!' There was the need for constant vigilance when taking decisions on important matters that were referred to him almost every day; and this was compounded by the fact that he felt 'so ignorant' of certain aspects of the work being done at Petersham and yet he had to assume responsibility for it. The pace at which they were all working, coupled with the very long hours, meant that even if he had the time for reading to build up his background knowledge he found himself asleep within five minutes of falling into bed [6].

By late 1941 army radar had come quite some way since the first gun-laying set, the GL1, went into service at the outbreak of war. It was designed to provide early warning of approaching aircraft, rather like Schonland's own JB set in Johannesburg, and was never intended for fire-control, but was pressed into service in September 1940 by AA Command to direct unseen fire in the absence of anything else. Soon, though, Bedford's elevation-finding attachment and Mott's earth mats added significantly to its theoretical performance, but this was only achieved in the field when in the hands of the most skilled of operators. There was, for example, the need to cut the coaxial transmission lines to carefully determined lengths, and the quaint custom of 'stroking the feeders'* required tedious attention to inexplicable detail on the part of the gun crews. Still, such progress and the deployment of scientists at the gun sites had seen the number of 'rounds per bird' decrease fourfold within the year. 1941 also saw the introduction of searchlight control radar, officially SLC but always referred to as 'Elsie'. This development, early in 1940, at ADRDE in Christchurch was a remarkable example of lateral thinking, where the complete radar system, including its array of six Yagi antennas, was mounted on a searchlight. The idea in concept was brilliant; its use, though, under operational conditions posed many serious problems which could only be solved by the application of 'the

---

* The transmission line 'feeders' had to be flexed or 'stroked' at regular intervals in order to break up the concentration of the pest-control compounds that seeped inside these 'tropicalized' cables and upset their electrical performance [7].

scientific method' — a concept that was met with some derision by the occasional battery commander who made it clear that he hadn't joined the army to become an electrician, or words to that effect! It was indeed fortunate that 'Tim' Pile was the General in charge, for no other man of his rank throughout the war had quite the same affinity with scientists and a deeply-held belief in their indispensability to the soldier. With Pile's enthusiastic support Elsie became a most effective weapon in the anti-aircraft armoury.

Since seaward defence was also Pile's responsibility, coast watching radars for plotting the approach of surface vessels were installed in 1940 along the south-east coast of England. These metric wavelength CD/CHL sets were another product of ADRDE and were eventually extended all the way round the east, south and west coasts of the country. In the autumn of 1941, one of the most significant developments in the brief history of radar took place, with the installation at Dover of the first centimetric equipment. It was, in fact, an ADRDE version of the Admiralty Type 271 radar that became the first 10 cm wavelength radar used by the army, the CD No 1 Mk IV, and it immediately proved its worth with considerable improvements in accuracy and range on surface targets. Small vessels could be tracked at distances in excess of 70 km and large ships could even be detected leaving Boulogne. Soon, this equipment and its many variants would play a crucially important part in the defence of Britain, and so much of its husbandry would be done by the teams of operational research scientists then taking shape under Schonland's leadership [8].

Purely procedural matters and the way in which ADRDE(ORG) supplied information to the Army were issues on which Schonland laid great stress. He insisted that reports be written according to accepted military standards, and for good reason, as he informed his staff: 'Our job is to serve the Army and to teach the Army and we must do it in the way that they want to receive it' [9]. One of the first to fall foul of the standards that Schonland insisted upon was a young scientific officer by the name of Stanley Hey, who was Wilkes's right-hand man on the radar side. Soon after Schonland took over, Hey presented him with a rather hurriedly prepared technical report for his comments, before dispatching it to its intended recipient somewhere within the Army. To Hey's amazement the document was returned rapidly to him with a note from Schonland informing him that its shoddy presentation was quite unacceptable and that Hey should re-write it. Somewhat abashed he did so, after making discreet inquiries of his colleagues as to the proper form that it should take. Some while later, and no doubt many reports later, Schonland called Hey into his office and said to him 'You know, Hey, your reports used to be the worst; now they are the best!' [10]. Hey's scientific prowess had never been in doubt and Schonland, now well-satisfied with his performance, was soon to appoint him to a new task that would lead to a discovery of fundamental importance, not just to the war effort but, ultimately, to the science of astronomy itself.

For all his insistence on standards and correctness of procedure, Basil Schonland never lost the human touch that all who penetrated his initially shy and slightly austere exterior soon encountered. If he had a failing in his dealings with people it lay in the other direction, for if his imagination was stimulated by an idea or a suggestion made by one of his confidantes his enthusiasm could sometimes override the level-headed judgement that was his usual watchword. Both Wilkes at Petersham and, later, Hewitt in Johannesburg, encountered situations where Schonland's encouragement of a subordinate could, and some-times did, lead to action that, on reflection, he himself would readily agree had been rather precipitate [11]. But such deviation from a course that was always carefully surveyed ahead of time was rare and it was Schonland the administrator, whether wearing his soldier's or his scientist's hat, who inspired great confidence in those who worked for him and, indeed, in those who watched him from afar.

His immediate superior within the ADRDE organizational structure was John Cockcroft, but theirs was a relationship that extended well beyond the formalities of their positions, for they were also the greatest of friends and, in Schonland's eyes, Cockcroft was almost beyond reproach. They shared a nimbleness of mind that was laced with much common sense, and they pursued their objectives with great dedication and zeal. But where they differed so markedly was in one aspect of their personal make-up. Whereas Schonland was a punctilious man in all he did, Cockcroft exhibited a waywardness when dealing with admin-istrative matters that frequently left a maelstrom of confusion in his wake [12]. Many were the occasions when others had to step in and sort out such matters as the payment of staff* and the formalization of appoint-ments that had seemingly passed Cockcroft by as he busied himself with details in other directions. Even Schonland himself was to suffer from his most admired friend's capriciousness in this regard.

By the end of 1941 Cockcroft's mounting responsibilities were apparent to all and yet still more demands were being made on him, with the result that many loose ends remained untied. It was therefore clear to Dr H J Gough, Director-General of Scientific Research and Development at the Ministry of Supply, that Cockcroft must have a deputy for whom loose ends were anathema and, early in the new year, he called Schonland in to see him [14]. Gough's proposal was simple:

---

* Soon after Schonland's appointment, and after he had unravelled Cockcroft's management methods, Maurice Wilkes jocularly suggested an ADRDE(ORG) coat of arms in the form of 'a pair of dipoles crossed on a field of travelling claims unpaid' [13].

would Schonland consider becoming Deputy Chief Superintendent of ADRDE? Schonland's reaction, as he told Ismay, was once again typical of him in the circumstances: he doubted whether he could do the job but Gough simply 'smiled & said he was satisfied I could'. The size of the establishment, he was told, and the somewhat idiosyncratic behaviour of its scientific community required firm management, and even the need 'to remove dead wood', something that Cockcroft in his benevolent way had never seen fit to do.[*] But what of the ORG? Ismay, once again, shared his thoughts and his concerns:

> Though my present Group will be under my general control as part of ADRDE, I shall have to find a successor and that is extremely difficult. Most scientific men are such peculiarly prickly customers and quite unsuited for the running of an organisation whose whole success depends on keeping one's word and sticking to time.
>
> Anyway, here I am well on the way to controlling the biggest establishment of its kind in England (and perhaps to breaking my heart).

And there the matter rested. Nearly a month passed and no firm appointment was made. Schonland became restive and, at least in his letters to Ismay, suggested wryly that he might have to place the matter in the hands of the South African High Commissioner: after all it was he to whom Schonland was ultimately answerable, being, as he was, just 'on loan' to the British establishment. But, he conceded, there was much reorganization taking place at the Ministry 'and until it is complete things are at sixes and sevens' [15]. Quite where the blame for the official tardiness lay is uncertain but one could guess. Gough presumably had handed the matter over to Cockcroft who, in his usual way, never formalized it and certainly never laid down Schonland's role in relation to the existing staff. Cockcroft was merely acting true to form while Schonland, attempting to be useful but with no clear instructions from anyone, could do little. Some years later, when writing to congratulate Cockcroft on his considerable wartime achievements, and to thank his old friend 'for letting me do something over here', Basil gently chided him about this period when all seemed rather rudderless: 'I except the matter of a chief clerk in 1941!'[†] [16].

The picture, when Schonland looked at those around him at Petersham, was very different. He was much impressed by the calibre of the young men who reported to him and in another letter to Ismay told her of the Cambridge 'biologists' (they were in fact physiologists),

---

[*] Schonland described it to Ismay thus: 'The mess—if mess there is—is not of Cockcroft's making, but he has not done much to clean up an Augean stable'.
[†] Though he probably meant 1942.

D K Hill (the son of A V Hill, 1922 Nobel Laureate in medicine and physiology) and A F Huxley, of even more impressive pedigree and himself awarded that Nobel Prize in 1963. Not only did they possess 'very great mathematical equipment', as he put it, but their life-styles and the fact that they had 'no exaggerated wish for money ... and were fundamentally honest and clear-headed' commended them to him and caused him to tell his wife that he hoped that their son, David, now 17 and about to enter university, would have the chance 'to grow up with such men and Mary (their eldest daughter) with their sisters'. They, and others like them, were the people, Schonland said, who were really winning the war having put aside everything they valued 'to do elaborate mathematical & experimental researches on the infinitely complicated problems of how to shoot down aeroplanes, destroy submarines & protect shipping' [17].

But Schonland, too, was playing a not inconsiderable part and, no doubt, in recognition of that, as well as to reinforce his own position, he was informed by the South African High Commissioner that he had been promoted to Colonel on 1 February 1942. The formalities involved in this elevation of rank were minimal but it was necessary for him to visit a tailor in London to have the red tabs of that rank fitted to the lapels of his uniform, as well a red band around his cap, and to Ismay he confessed that it was a nuisance because he hated being so conspicuous. However, 'having taken flannel', as was the term of the time for such adornments [18], he returned to Petersham and matters of much importance. By now his organizational skills had steered the ADR-DE(ORG) through the Cockcroft eddies and rapids, and its services were increasingly in demand, within a wider sphere than just those of interest to Anti-Aircraft Command. More sections were formed. As well as those dealing with AA, CD and Early Warning, plus Signals, two more concerned with armoured fighting vehicles and airborne operations came into being. By May, the War Office, though always seemingly lagging behind its counterparts in the Air Ministry and the Admiralty, as Schonland had previously so pungently observed, felt the need to have a closer involvement with scientific matters and a Scientific Adviser was appointed to the Army Council. The first incumbent in this post was Sir Charles Darwin MC, pre-war Director of the National Physical Laboratory, and his deputy was Professor C D (later Sir Charles) Ellis. Both were former Cavendish men and, as we have seen, they were Schonland's close collaborators during the early 1920s, as were so many of those now occupying senior posts within the scientific services underpinning the war effort. In July, the Vice Chief of the Imperial General Staff, who was now well aware of the impressive developments on the scientific front, particularly in radar and in the proper husbanding of convoys due to Blackett, proposed the setting up of an Operational Research Group in

235

its own right [19]. Whilst all this was taking place, Schonland had his hands full with matters much closer to operations directly against the enemy, as the radar war intensified on land, sea and in the air.

*   *   *

January 1942 was particularly cold in England. Heavy snow blanketed the countryside and its picture-postcard appeal soon wore thin for Basil Schonland, as it had done a good while earlier for the long-time inhabitants of the country. His daily journey to and from the Star and Garter Hotel and his Petersham headquarters was made doubly perilous by the icy pavements and by his chilblain-affected gait. The latter condition had been caused by the extended periods of time he had spent out in the country, where he was running a training exercise that was classified 'Most Secret'. Its purpose was to lead to a raid on the French coast, in

Figure 17.   Colonel B F J Schonland soon after his promotion in the Union Defence Force on 1 February 1942. (Reproduced by kind permission of Dr Mary Davidson.)

order to capture a German radar set. What became known as the Bruneval Raid was an operation of great daring that had been planned meticulously by Lord Louis Mountbatten's Combined Operations Headquarters. The sophisticated nature of the equipment they had to dismantle and bring back required specialist training and Brigadier A P Sayer,[*] responsible for radar at the War Office, gave Colonel Schonland that task. Not only was Schonland 'a scientific soldier intimately connected with radar' but he had available to him just the sort of unobtrusive facilities for the job of training the Royal Engineers, whose task it was to dismantle and remove the Würzburg radar, perched on a cliff overlooking the English Channel, just outside the French village of Bruneval [20]. Schonland's first view of the target was on a photograph that had been taken with great skill and daring. An RAF photographic intelligence Spitfire, flown by Squadron Leader A E Hill, had produced a remarkable series of images of what the pilot said looked like the bowl of a large electric fire. Closer examination showed it for what it was—the parabolic reflector of a radar that was part of the defensive chain known as the 'Kammhuber line' that controlled the German night fighters, then proving such a menace to the invading RAF bombers. British morale had recently taken something a battering over another radar-related incident involving the German battleships the *Scharnhorst* and the *Gneisenau*, in which Schonland had a peripheral role. There was now a real need for an audacious reaction, not only to provide a much needed fillip, but to press home Churchill's policy of keeping the Germans fully occupied by means of isolated raids into enemy territory [21]. The raid on Bruneval would do much more than that—it would provide priceless intelligence about the state of German radar at that time.

The plan formulated for 'Operation Biting' was for a parachute drop to be made at Bruneval by a detachment of Royal Engineers, whose task was to remove only those elements of the Würzburg that scientists at the Telecommunications Research Establishment (TRE) at Swanage wished to examine in detail. This requirement meant that a radar specialist must accompany the raiding party so that the crucial components could be identified rapidly and then removed without damaging them. Since the radar site was known to be well-defended, the Royal Engineers, under the command of Lieutenant D Vernon, would be accompanied by a company of soldiers from the Parachute Battalion, whose task was to capture the site and then defend it while the Würzburg was dismembered. Once completed, the raiding party were to scramble down a

---

[*] In December 1940 Sayer was appointed to the new post of Deputy Director Anti-Aircraft and Coast Defence specially for RDF policy and development. Early in 1944 he became Director of Radar in the War Office. In 1950, based on his firsthand experience of every aspect of the subject, Sayer compiled the definitive history of radar in the Army.

Figure 18. The site of the Bruneval raid in late February 1942 showing the Würtzburg radar antenna in the grounds of the château just a short distance back from the cliff face. (Reproduced by kind permission of Dr Mary Davidson.)

120 m cliff to the beach below, where a Royal Navy landing craft would meet them and take them to a larger vessel standing just off-shore. Time was of the essence because a parachute drop required a moonlit night, while ease of recovery of the raiding party and their booty needed a high tide to allow the Navy to get in close to the shore. The radar expert selected to accompany them, to ensure that only the key elements of the Würzburg were removed, was D H Preist from TRE [22]. Don Preist had been at Bawdsey in the earliest days of radar development there, and was a man of considerable energy with a flamboyant taste in motor cars. However, on consideration it was then realized that, were Preist to be captured, his extensive knowledge of radar would put not only himself but all British radar secrets at great risk and so, to Preist's great dismay, it was decided that he would remain aboard the Royal Navy vessel just off-shore until the raiding party reached the beach with their prize. If there was no enemy activity he might then go ashore to ensure that the Royal Engineers had all the crucial elements of the Würzburg. In case the raid had to be aborted Preist would carry with him a specially developed radio receiver with which to monitor the transmissions from the Würzburg and so measure the frequency and other

238

characteristics of the signal. The technical man chosen to supervise the actual dismantling of the German radar was Flight Sergeant Cox, a radar mechanic well-versed in the maintenance of British equipment but with no detailed knowledge of use to the enemy should he fall into their hands. It was now Schonland's job to train them. To do so he had constructed a model of the Würzburg, using components from an early GL1 radar, but fitted with a parabolic antenna with dimensions deduced from the RAF photographs. Success would depend entirely on the sappers being able to complete their task in no more than thirty minutes, for the Bruneval site was heavily defended and German reinforcements would undoubtedly be alerted by the first sounds of firing during the paratroop assault. And so it was that Schonland spent three freezing days in late January 1942 on the bleak and windswept Salisbury Plain ensuring that Vernon and Cox could recognize each and every item of importance on the radar, and that their men could remove them with almost surgical care, even when they were working in the snow and under fire [23].

Whereas the sheer effrontery of the raid would more than satisfy Churchill's demand that the Germans be harried on their own occupied ground, its purpose in gathering scientific intelligence was by far the key element in the planning and, to assist in this, Schonland insisted that the raiders capture, if at all possible, a German radar operator and bring him back for interrogation. In this he had the full support of Major General F A M 'Boy' Browning who was in overall command of the operation. After a number of postponements caused by poor weather, the airborne troops eventually left Thruxton Airfield, near Andover, shortly before 10.00 p.m. on 27 February; the Naval force having sailed earlier that afternoon. The twelve Whitley bombers carrying the 120 men arrived over the target as scheduled and the troops jumped, landing within minutes in thick snow. The details of the raid are very well documented and will not be repeated here [24]. As the paratroops, under the command of Major J D Frost, an officer along with Browning himself, soon to become legendary at the battle of Arnhem, attacked the German defences so Cox and his party of sappers set about removing the crucial steel boxes mounted behind the radar antenna, having already sawn off its dipole feed. The Germans defended resolutely once they had recovered from their initial surprise at the assault. The half-hour assigned to the dismantling process soon became no more than ten minutes, but in that time Cox and his men managed to remove almost every piece of equipment that Schonland's training had pinpointed. It was a remarkable display of level-headedness under fire. In a letter that Schonland wrote some years after the war to Professor Leo Brandt, a German wartime radar expert and Schonland's host on a visit to Dusseldorf, he described what happened [25].

> Emphasis was laid by me on the need to bring back the operator
> which was done, though he did not prove a very well-informed
> expert on the German radar. The reflector was too large to bring
> back but it was photographed with a flashlight and some useful
> information was obtained from the facts about the performance of
> the equipment on previous occasions which had been obligingly
> painted on the face of the mirror. The equipment was brought back
> in good order without any broken valves and re-assembled in about
> a week's time in a satisfactory operating condition.

The Bruneval raid was remarkably successful, both in terms of the information it provided about the state of German radar at that stage of the
war, and for the boost it gave to the morale of all those within the
armed services who knew of it. The official intelligence report, written
some while afterwards by R V Jones in his capacity as Assistant Director
of Intelligence (Science), described it as 'an unusual success for which
thanks are due to those officers mentioned individually in the Report
... and to the Units concerned with successful planning and execution'.
Schonland was one of those mentioned individually. As well as returning
to England, having lost just two men killed and six captured, Major
Frost's party brought back three prisoners of their own (one of whom
was the ill-informed radar operator), as well as the bulk of the Würzburg
Type A radar but without its most bulky part, the reflector. Tests at TRE
on the rapidly reconstructed equipment immediately revealed its key
features. The operating frequency lay between 531 and 566 MHz and its
pulse-repetition frequency was 3750 Hz, which, with a peak power of
just 5 kW and a receiver noise level of about 20 dB above the thermal
floor gave it a range of about 40 km. The most outstanding feature of
the equipment was its beautiful construction, with every item individually labelled. However, it contained no electronic features of any technical
novelty and its 1939 vintage suggested that it did not represent the
pinnacle of German radar technology nearly three years into the war
[26]. Of great importance, though, was the fact that this model of the
Würzburg would be very vulnerable to intentional jamming—a crucial
card that would soon be played by the Allies.

Jamming of British radars by the Germans was, by late 1941, proving
to be a matter of great concern and its investigation soon became an
operational research problem that lay squarely in Schonland's court. At
precisely the time that he first became aware of the existence of the
Würzburg radar at Bruneval, his attention was also directed towards
reports that the jamming of British coastal radars was becoming more
severe. What really brought matters to a head, on 12 February 1942,
was the dramatic breakout from the French port of Brest, under cover
of darkness, of the German battlecruisers *Scharnhorst* and *Gneisenau* and
the heavy cruiser *Prinz Eugen*. They had been holed up there as a result

of damage suffered during various engagements the previous year, and their vulnerability to attack by the RAF made it vital that they escape to the safer haven of Hamburg for repair and immediate return to active service. This required a northwards dash up the Channel and then a perilous journey through the Straits of Dover where they would be exposed to the full attention of the British shore batteries, as well as to Royal Naval and Royal Air Force attack. When it came, that breakout was one of the most audacious actions of the war. Just after midnight the three capital ships, accompanied by a large number of escort vessels, left Brest for the Channel. Thirty hours later, after one of the great naval sagas and against all the odds, having passed through the Straits in broad daylight they arrived at the River Elbe and the haven of Hamburg. Hitler was triumphant and Churchill like a thundercloud, for the Royal Navy's planning to deal with just such an eventuality had been decidedly lacklustre [27].

The daring escape of the German capital ships was precipitated by a number of factors, all of which have been subject to extensive analysis and critical comment subsequently. Two related specifically to radar. An RAF Coastal Command aircraft on patrol that very night in the vicinity of Brest, for the specific purpose of watching for a breakout by the ships, suffered a breakdown in its radar and therefore detected nothing. Fortune had certainly smiled on the Germans, but they relied not just on luck. For many weeks prior to the planned departure of the ships, they had been gradually increasing the intensity of their jamming of British coastal defence radars, operating at a wavelength of 150 cm, from their transmitters in the vicinity of Boulogne and Cap Gris-Nez. This use of electronic measures to confound the enemy was an event of the greatest significance because it was the first occasion in warfare that so-called radio countermeasures were used for offensive purposes. Its success was to generate intense activity by the Allies to counter it and soon to use such measures themselves to great effect [28]. And it occurred at a time when a much more powerful source of interference was about to unleash itself upon an unsuspecting world.

The increased level of interference on some of the British radar displays had been observed and reported, but its significance was lost within the higher levels of command. The 10 cm CD radars, however, were not being jammed and it was that equipment near Hastings that first detected the assemblage of ships entering the Channel during the morning of 12 February, but delays in passing the information to the coastal artillery only resulted in the first salvoes being fired at 12:19. The barrage lasted no more than twenty minutes, then the vessels passed out of range. Some damage was done, though not enough to overly affect the progress of the German ships, and they all made it to the relative safety of Hamburg [29].

241

The escape of these three prized ships was viewed as a major débâcle within Whitehall and reaction to it followed swiftly. At a stormy Inter-Service meeting, responsibility for the investigation of all forms of radio (and hence, radar) jamming now passed from the Air Ministry, whose responsibility it had been, to the War Office. Since there was no suitable organization within the Army to undertake this task, it fell to Schonland's Operational Research Group to set up what became the Army J watch equipped with a mobile laboratory, the Army J van, and he appointed J S Hey to take control of it. Though feeling far from enthusiastic about what seemed to him like a 'cumbersome, uninteresting task', Hey had no option but to set about establishing a monitoring and reporting organization that could react immediately on receipt of reports of jamming of Army radars and then issue reports containing his assessment and conclusions within 24 hours. Direct communications were also established with the Coastal Artillery and Gun Operations Rooms, and also with the RAF because the greatest fear was of airborne jamming [30]. What followed soon afterwards not only showed how effective Hey's Army J watch had become at tracking a source of interference but, even more importantly, at interpreting it and determining its origin.

On the afternoon of 26 February, just as the weather was showing signs of clearing, the final preparations for the raid on the German radar site at Bruneval were being stepped up. British AA GL II radars around the coast were operating as usual when, one after the other, they began to experience interference and, now well aware of the possibility of German jamming, the operators passed their reports to Hey's J watch on the nature of the noise they saw on their cathode rays screens. Over the next two days, at the very height of the Bruneval Raid, the interference was almost continuous over the full tuning range of the GL receivers, from about 4 to 6 m wavelength, reaching its peak on the 28th. At nightfall it disappeared. Both bearing and elevation readings of the source of the interference were taken at widely scattered radar sites around the country and, in all cases, the source of the jamming seemed to be moving, yet was always within a few degrees of the position of the sun. Two sites, more than 240 km apart, followed it continuously and even produced optical confirmation that what they were tracking was the sun and not an enemy airborne transmitter. Such a finding was not only unexpected, it was absolutely contrary to all understanding at the time of the sun as a radiator of radio-frequency energy. Hey sought confirmation of this unexpected source by telephoning the Royal Greenwich Observatory and was told that there was a very large sunspot situated almost at the centre of the sun's disc. Since it was already well-known that sunspots were the source of intense magnetic fields and energetic electrons, he concluded that such unusual solar activity was indeed causing the interference observed on the radars and that it was not, as all had concluded a day or so before, due to German jamming. He immediately communicated his finding to Schonland who just smiled. 'Is that so, Hey? How interesting. Did you know that Jansky of Bell Telephone Labs in the USA discovered radio noise coming from the Milky Way?' Hey did not, so Schonland told him of this American work that was first published in 1932 and Hey immediately rushed off to the Science Library in London to read Jansky's papers [31].

While Schonland reacted to this remarkable discovery with much interest, but no real surprise, others were less ready to accept that the sun could possibly produce such intense emissions at such long wavelengths. Maurice Wilkes, who was Hey's immediate superior, was somewhat sceptical, while Edward Appleton, then occupying his wartime post as Secretary of the DSIR, refused to believe that the sun could emit radio waves of such intensity [32]. The ORG report written by Hey that followed this remarkable discovery was only given limited circulation because, it was felt, the information it contained was so sensitive in the context of radio and radar jamming. It was only made known to a much wider audience at the end of the war, first as an AORG publication [33] and then in the scientific literature when Hey published a short note in the journal *Nature* that effectively announced the birth of the new science of solar radio astronomy.

*   *   *

While the stage that he now strode, in those early days of 1942, was so much larger than that he had left behind in Johannesburg, Basil Schonland still maintained a close and even possessive interest in how 'his SSS' was faring. The High Commission in London informed him that one of his former colleagues, Frank Hewitt, would soon be in England to learn all about the new British radars and to place orders for equipment to meet South Africa's needs. Then, with much pride, Colonel Schonland received a copy of an 'unsolicited testimonial' sent from HQ RAF ME to the Air Ministry in London in which the praises of both the SSS and its JB radars were sung. 'And now they want much more', he told Ismay, no doubt remembering well the note of some disdain that had been evident in the dispatches that had reached London from those parts just a few months before, when Flight Lieutenant Atherton had first set eyes on the home-made equipment from South Africa. Now the official view was very different. He wondered too about his former colleagues. 'I hope Hodges is now Lt Col. If so congratulate him for me', he wrote. On the technical side he was better informed. Bozzoli had kept him up to date on all developments taking place at the BPI by sending two monographs that described the way the JBs had evolved to meet the coastal defence needs around South Africa, when it had become clear that British radars would not be arriving in the immediate future. 'I read them with pleasure and pride at his achievements. If we could get a few hundred of those things it would be a very great help.' Praise indeed from the man now so close to the heart of Army radar in England.

Other developments too were now under way in South Africa. Just as had happened in England, when Watson Watt, faced with the dwindling reserves of qualified men, had turned to women to man the ever-increasing number of radar stations springing up all around the country, so the SSS had begun to recruit and train university-educated women for the same purpose. To Schonland's surprise, he discovered that Ismay herself was now in uniform, as an officer in the Women's Auxiliary Army Service (WAAS), with responsibilities for those very female operators, without whom wartime radar stations could not have functioned. And then, as if to add a final touch of lustre to these stories from home, he saw reported in the newspaper that the 2nd South African Division, under the command of Ismay's cousin, Major-General Pierre de Villiers, had captured Bardia in Libya; a key battle in General Auchinleck's great 'Crusader' campaign. This required special mention in his regular letters and he let on that, though he had basked in some of the reflected glory, he felt sure that she 'must be positively refulgent' [34].

Yet mingling with that glow of pride, possibly even accentuated by it, was Basil Schonland's own feeling, at times intense, of loneliness and longing for his wife and for his children. Their photographs sat on the mantel piece above the gas stove in his room where, as he wrote his weekly letter, he felt at least that he was writing it 'in the bosom of his family'. Soon after he had left for England in 1941, Ismay had taken the three children on holiday to Gardiol, where her parents, James and Isabella Craib, provided all the usual home comforts, as well as the solid support that she and her children needed at the time. Mrs Craib was a Hofmeyr, that famous South African family, whose name was synonymous with the liberal tradition in the country and which numbered amongst its members the remarkable Jan Hendrik Hofmeyr — 'Hoffie' to his intimates. Hofmeyr was a child prodigy who was awarded his first degree with distinction at 15, won a Rhodes Scholarship to Oxford and then, at just 22, became Professor of Classics and two years later, in 1919, Principal of what was soon to become the University of the Witwatersrand [35]. In 1939, when the Smuts coalition government declared war on Germany, Hofmeyr was appointed Minister of Finance. Throughout the war, with Smuts frequently out of the country, either visiting his troops in the field or participating in Churchill's War Cabinet, Hofmeyr virtually ran the government and almost *was* the government, as he handled multiple portfolios with ease. But all the while there was the smouldering resentment from the Nationalists, not

Figure 19. Ismay Schonland with her two daughters Mary and Ann (ca. 1941). (Reproduced by kind permission of Dr Mary Davidson.)

only against the war but very much against Hofmeyr and his liberal views of 'the natives' and their place in South African society.

Though such strong family links and connections with the powerful never overtly affected or influenced Basil Schonland's career, they certainly proved useful from time to time. During the First World War, as we have seen, it was Smuts's intervention that prevented Selmar Schönland's internment and later it was to assist in speeding up Basil's own homeward passage. Now, during the second global conflict, Schonland contrived to arrange that Ismay should join him in England and it was his position as RDF Adviser to the High Commissioner that enabled him to do so. As a serving UDF officer she could, the High Commissioner agreed, be posted to London to serve in some RDF-related post, but then Schonland vacillated. Should he ask to her to undertake the risky journey by sea, with the ever-present threat of submarines?; could he expect her to leave the children? His letters reflected his anguish:

'I can only say that I will find it very difficult to go on for another year without you to advise, comfort and share life with me but I can do it if the children's interests demand it. If you think they can be left for a maximum of two years come as soon as you can. Then we'll show them what we can do'.

Quite what he had in mind is unclear, but a matter unrelated to the war, yet very much to do with South Africa afterwards, had come to his attention and it certainly caught his imagination. And Ismay could certainly assist. As early as the previous September, when he was just facing up to his responsibilities at Richmond, Schonland received an invitation from Solly Zuckerman,* the South African-born anatomist at Oxford whose speciality had been baboons and other primates, to attend a meeting of his scientific dining club, the 'Tots and Quots'.† This Schonland did, and Zuckerman recounts how Schonland mentioned that the view then current in South Africa was that Britain had done nothing for the country and particularly had 'allowed the native population to stew in their own juice' [36]. Fair or not, it certainly reflected the perception that nowhere in Africa, not even its southern tip for all its British affinity, was high on anyone's agenda in England and Basil Schonland now saw his opportunity to plead its case, at least in the scientific sense.

From an old friend in the Colonial Office he discovered that, even in its darkest hours, the British Civil Service was looking ahead to the time when peace would be restored and Britain could, once again, turn her thoughts towards her Empire. Not, as one might have thought, to divest herself of the burden, when so much reconstruction was required at home, but rather to invest in their future and, of course, in her own. Of especial interest to Schonland was talk of the establishment of African

---

* Zuckerman, with J D Bernal, was appointed, in 1942, as Scientific Liaison Officer to Lord Mountbatten's Combined Operations Headquarters. In 1944 he became Scientific Adviser on Planning to the Allied Expeditionary Air Forces for the Invasion. After the war he became Professor of Anatomy at the University of Birmingham and then, successively, Scientific Adviser to the Ministry of Defence and finally, in 1964, Chief Scientific Adviser to Harold Wilson's government. He was made Baron Zuckerman of Burnham Thorpe in 1971.

† The 'Tots and Quots' (an inverted abbreviation from Terence's *Phormio*: 'Quot homines, tot sententiae' — 'so many men, so many opinions') was a dining club organised by Zuckerman in 1931. Amongst its members were some who were soon to be numbered amongst the greatest thinkers of the age: Desmond Bernal, Jack Haldane, Hugh Gaitskell, Lancelot Hogben, Julian Huxley, Patrick Blackett and Richard Crossman, to name but a few. The club even contrived collectively to write a book. Following a dinner in February 1940 during which the topic under discussion was the role of science and scientists in war Zuckerman, in the space of two weeks, compiled and edited the anonymous contributions from those present into eight chapters that appeared soon after as *Science at War* (London: Penguin Books). It was remarkable in its prescience and sold more than 20000 copies.

research institutes, and this prompted him to write a memorandum which he submitted to the committee dealing with what was called Imperial Scientific Collaboration. Somewhat to his surprise he was informed that his memorandum would be discussed at a meeting, in early February, at Burlington House. And so, for a few hours that day, he removed himself from the world of warfare and sat through what seemed like interminable discussions on the exchange of overseas professors, scholarships for graduate students and associations for economic improvement. Eventually, the chairman, Sir William Bragg, called on Schonland to speak. He seized the opportunity and, with a directness that, he admitted later, surprised even himself, he laid down the law about what he called 'Lost Chances & Science in Africa'. Expecting at any moment to be 'given a bowler hat' for his rather forthright approach in such august company, he was doubly surprised when the meeting resolved to take some action and instructed its chairman to raise the matter at the level of both the Foreign and Colonial Offices. Schonland now realized that he could exploit the situation to South Africa's decided advantage by dropping an informal word in the ear of government back home and the person to do that was Ismay. So he wrote to her suggesting that she 'talk to General Smuts & Hoffie and Raikes' and then, after her arrival in England, 'as a disinterested outsider' she would be able to tackle the Colonial Office, with no axe to grind, while he attended to matters of more immediate urgency in Richmond [37].

*   *   *

There were many matters of particular and immediate urgency that consumed Schonland's every hour as the war unfolded in its various theatres of early 1942. Quite apart from the *Scharnhorst* incident, and Hey's remarkable discovery that followed from it, and then the sheer effrontery, if nothing else, of the Bruneval raid, Basil Schonland more than had his hands full at Richmond. Operational Research within the Army was now finding applications in areas well removed from radar.

One of the six new sections he had set up was to do with the quantitative assessment of the lethality of various weapons, from bombs to small arms. He gave Frank Nabarro the job of determining which weapons would be most effective in an anti-personnel role [38]. In essence, the problem involved measuring, by some suitable method, the effects on human beings of splinters of steel typical of exploding munitions. Clearly, experimentation with humans targets was not an option but live animals presented the obvious alternative and much work had, in fact, already been done in this area by Zuckerman at Oxford. Soon after the outbreak of war, Zuckerman had crossed swords with the Medical Research Council's experts by rejecting the accepted doctrine of the day, which maintained that the pressure wave caused by an explosion killed or injured anyone within range by forcing its way into the victim's lungs through the mouth and nose. By a series of simple experiments, in which he exposed pigeons, rabbits, monkeys and goats to controlled explosions, he showed that the blast wave produced its effects by direct, external impact with what he called the 'mean projection area' of the body and not by any process involving the forceful injection of air through either nose or mouth. In addition, the prevailing view at the time held that a human being would succumb to a pressure wave of about 35 kPa (5 lb per square inch), whereas Zuckerman showed that the actual value at which death would probably ensue was nearly one hundred times greater. His work on the effects of metal splinters, such as shrapnel, involved a series of carefully controlled set of experiments in which minute steel balls were fired at various velocities into chunks of meat and then, of all things, banks of telephone directories. By the simple device of reading the uppermost name on the page to which the ball penetrated it was possible to calculate the energy lost during the process. From this was obtained the kinetic energy of the projectile and the momentum it imparted to the human tissue; for it was this that ultimately caused the damage and not the shape, spin or even size of the projectile, as was the accepted wisdom at the time within the MRC [39].

Whereas the experiments on animals were conducted as humanely as possible, there was soon considerable public disquiet about them, once word reached the popular press and, inevitably, questions were asked in the House of Commons [40]. Any further work—and much was still necessary as weapon development became more sophisticated—required 'targets' that would not incur even more wrath from the British nation than had been directed towards the Luftwaffe at the height of the Blitz. The Army needed the answers and Schonland was brought in because the lethality of weapons was now within his province. He immediately gave Nabarro the job of devising a suitable, standard target for the various experiments and field trials that would take place in laboratories and on firing ranges across the country. Nabarro's solution was remarkably simple: a wooden model with the effective cross-section of a man, constructed of nothing more than three 25 mm thick planks mounted on a suitable steel support for ease of positioning, either prone on the ground or upright within a trench. Tests showed that his creation was admirably suited to the task, since damage was immediately obvious and easy to measure and so production of these anthropomorphic creatures was soon under way. To Nabarro then fell the duty of writing an appropriate report, as was very much the custom at Richmond under Schonland, and he inquired by what name these mannequins should be known. Now it so happened that Schonland and Zuckerman viewed each other with more than a little rivalry and so, with a glint in his eye, Schonland answered, 'We'll call them "Zuckermen" and let the troops shoot at them!' [41].

These 'Zuckermen' soon appeared in energetic verse. Whether it was Schonland's latent boyhood talent that re-emerged in anonymous guise to produce this piece of poetry on the saga of the 'Zuckermen' we will probably never know, but its appearance some months later in a newsheet for all at Petersham entitled 'The Christmas Memorandum' of 1943 did much to raise a laugh [42].

The Zuckerman's Lament

Fashioned with care, constructed with zeal,
Three planks of wood and a spike of steel.
A coat of paint from a ten-gallon can—
That's what goes into a Zuckerman.

On Salisbury Plain I made my debut,
As an infantryman in 'Fortescue';
They fired all day with two hundred guns,
But they didn't kill me so they won't kill the Huns.

My next campaign was Exercise 'Link',
Though they shelled for an hour I was still in the pink.
A hundred strong when the fun began,
Our only loss was a single man.

On Salisbury Plain I heard my knell,
'Twas the crump from an air-burst five-five shell;
Two troughs in my feet, a deep strike in my head,
Sufficient damage to write me as dead.

Fashioned with care, dismembered with zeal,
Three battered boards, some twisted steel,
How different I looked when I first began!
There's no fun in being a Zuckerman!

*   *   *

By mid-April Ismay was making the necessary arrangements to travel to England. Basil's anguish as to whether he was doing the right thing in asking her to join him had eventually been assuaged by none other than John Cockcroft and, somewhat more loquaciously, by his wife Elizabeth. The Schonland house in Johannesburg would be let, the car put up on blocks and Ann, their youngest daughter, would go to Gardiol to be cared for by her grandparents. David, the eldest, was now at University, while Mary was in her final year at school in Johannesburg and they could therefore both be fairly easily accommodated [43]. All that remained was action from Defence Headquarters in Pretoria to transfer Ismay to South Africa House in London and this was triggered by Basil himself. On 3

June, a cipher telegram on the subject of 'War Appointments—Officers' was sent to DHQ by the Staff Officer in London involved with such things. It addressed itself to the need for an officer to 'undertake RDF and signal liaison duties' at the High Commission. More specifically, it contained a request for the services of 'Mrs Schonland second lieutenant WAAS' who, it was understood, 'was at present performing more or less similar duties under Colonel Collins'. Her knowledge of the Union's requirements would make her especially useful in London, particularly as she 'will be in close contact with and would be able to obtain guidance from her husband'. Any request from Basil Schonland would cause Freddie Collins to move a mountain if needs be. His response was immediate. It indicated that not only would Ismay (rapidly promoted to Lieutenant) be dispatched to England but so would Major Roberts, both as permanent officers attached to the staff of the High Commissioner. In addition, they would be accompanied by Lieutenant Wadley who, together with Captain Hewitt, already there, would 'be looked upon as officers to study RDF technique and operational developments over [a] period to be decided by Schonland'. Their date of departure was set for the first week of July 1942 [44].

And so the Schonland family circle was to relocate itself somewhat, with Basil and Ismay once more together, while the children found themselves rather scattered around South Africa at university, school and at the delightful retreat of Gardiol. David's career was following closely, if not quite as illustriously, along the path blazed by his father at Rhodes many years before. His intention, once he had obtained his degree, was to join the army and then, the state of the world permitting, to go to Cambridge. Mary was Head Girl at Parktown Girl's School in Johannesburg and her immediate target was matriculation with examinations due later that year, while Ann, just eleven and happily oblivious of much that was prevailing in the world at the time, took school and the delights of the farm in her stride. Their parents, reunited once more, found accommodation in a flat near Hammersmith. Inevitably, their thoughts turned to South Africa and their eventual return home even though Basil's six-month loan to the British Army had been extended somewhat beyond that. Ismay hoped that by the Spring of 1943 they might well be able to leave. 'Basil says there seem to be quite a lot of scientists here and perhaps South Africa could do with him after all', she wrote to her sister-in-law [45].

This was no idle boast because Schonland's name had indeed been raised just a matter of a few months before in correspondence between two men whose influence on the world-at-large was nothing less than massive and whose personal relationship had its beginnings many years before in Lloyd George's War Cabinet of 1917. Field Marshal Smuts served then, just as he did during the present war, as the only member of a Dominion government whose advice was so actively sought by the British Prime Minister of the day. And Lord Hankey was its Secretary then just as he was now. In November 1941 Smuts wrote to Hankey with genuine warmth, referring to him as 'a "handy man" who deals with the odd jobs which bring no publicity—but are often most important of all'. He then thanked Hankey for his 'opinion of the value of Schonland's work' and went on to say how Lord Rutherford 'told me that Schonland was one of the most brilliant of his young researchers at the Cavendish Laboratory and when he visited South Africa at once inquired about his whereabouts and visited his Laboratory in the Cape Town university' [46]. That was the memorable occasion in 1929 when Rutherford was in South Africa for the meeting of the British Association at which all three men had delivered papers. Now, amongst Hankey's multitude of responsibilities was the setting up of a scheme to bolster the scientific effort needed to sustain the armies in the field. It was in this role that he had previously approached Schonland regarding the transfer of young South African engineers to Britain for training in radar. He therefore well knew of the activities of the SSS and of its first commanding officer. He knew him too as the man who now ran the Operational Research Group at Richmond and his opinion of Schonland would have carried great weight with Smuts.

It would not be long before Schonland and Smuts met again. In October 1942 the Field Marshal paid his first wartime visit to Britain. As well as attending the twice daily meetings of the War Cabinet, and a multitude of other official functions, he addressed the combined Houses of Lords and Commons on the 21st. It was a memorable speech that ranked with the Churchillian oratory that kept the British nation sustained throughout its darkest days, and it was broadcast by the BBC to an audience of 15 million [47]. The Schonlands attended a special show-ing of a film of the address at South Africa House a day or two later and were then guests at a garden party held in their Prime Minister's honour. Eventually it was their turn to be presented to him and immediately Smuts broke into Afrikaans, 'Maar dis ou vriende die—ons is mos familie, almal Hofmeyrs'.* He then spoke animatedly to Basil Schonland about his work and said he had heard very good reports of it and would like to visit him at his establishment, which he duly did [48].

By late 1942 it was clear to all concerned that the ADRDE (ORG) was rapidly outgrowing its parent organization, as its usefulness was being recognized within a much wider sphere of the Army than had been the case hitherto. Moves had been afoot since July to bring about a significant change within its structure and affiliations, and Basil Schonland's new role was soon to be substantial. Liaison between London and Pretoria ensured that the increased responsibility he would have to carry was

---

* 'But these are old friends—we are even related, all Hofmeyrs'

recognized by his promotion on New Year's Day 1943 to the rank of Acting Brigadier in the South African Staff Corps. Then, on 16 January 1944, the War Office's Weapon Development Committee agreed, with the approval of the Ministry of Supply, to the establishment of the Army Operational Research Group (AORG) as a separate entity with Brigadier Schonland as its Superintendent. Its links with ADRDE were severed and the establishment of 86 scientific staff, of whom about 60 were civilians, moved just across Richmond Park, where it still retained its trials facilities, to their new headquarters at Ibstock Place, Roehampton. Here the AORG formally came into being on 1 February and Schonland found himself in the interesting position of reporting to two masters. The AORG was to be under the direct control of the War Office, which meant he was responsible to the Scientific Adviser to the Army Council (SAAC) who, within a month, would become Professor Charles Ellis after his predecessor, Sir Charles Darwin, had returned to the NPL. But responsibility for the AORG's administration and technical efficiency still rested with the Ministry of Supply and its Controller of Physical Research and Scientific Development (CPRSD), Dr E T Paris* [49]. This certainly called not only for a level head but for a good sense of balance and Brigadier Basil Schonland FRS, the 'soldier and scientist', as the War Office always saw him, seemed blessed with both.

*   *   *

The range of duties of the AORG were carefully spelt out by its founding fathers and they were enumerated as follows:

   (i) to investigate the performance of selected types of service equipment under the conditions obtaining in field operations;

  (ii) to collaborate with design establishments in studying the performance and use of early models of new equipment;

 (iii) to investigate methods of using selected equipment;

 (iv) to analyse statistically the results of selected tactical methods, whether they involve the use of technical equipment or not;

  (v) to advise the War Office and Commands upon the experimental planning of troop trials of equipment or of tactical methods;

 (vi) to be represented by observers at troop trials;

(vii) to carry out any scientific investigations which may be approved by SAAC or CPRSD.

---

* Dr (later Sir) Edward Talbot Paris was the Principal Scientific Officer at the Air Defence Experimental Establishment (ADEE) who, in October 1936, had been attached to Watson Watt's team at the Bawdsey Research Station in order to study the principles of RDF, known then by its unofficial code name of CUCKOO, and to advise on its use by the Army. As such he headed what became known as the Army 'cell' there and, a year later, produced a report 'RDF and Coast Defence' that essentially established the basis for the coastal defence network in the years to come.

In carrying out all these tasks, Schonland was given remarkably free rein in the access he had to the military machine. All departments, establishments, directorates and schools of the Ministry of Supply and the War Office were opened to him, as was every Army Command in the country. His path to the highest echelons of each was to be unimpeded by any red tape and, most importantly, he had direct access to Charles Ellis himself [50].

As soon as the dust of the change-over to Roehampton had settled, Schonland began the expansion of operational research, from an activity that had been almost exclusively in the domain of radar and anti-aircraft gunnery, to embrace the whole gamut of military activity. The six rather ill-defined sections that had made up ADRDE(ORG) now became ten, with their titles and functions unambiguously defined. Thus it was that the Army Operational Research Sections that would soon begin to collaborate with fighting men across the service came into being as:

AORS1   Anti-Aircraft Defence, Radar and Searchlights
AORS2   Coastal Radar and Gunnery
AORS3   Signals in the Field
AORS4   Armoured Fighting Vehicles, Artillery and Anti-Tank Gunnery
AORS5   Airborne Forces
AORS6   Infantry Weapons and Tactics
AORS7   Lethality of Weapons
AORS8   Mine Warfare, Assault Equipment, Tactics and Flame-Throwers
AORS9   Time and Motion Studies
AORS10   Battle Analysis

Each fell under the direct control of a Section Head, some of whom were military men, others civilian, and all were carefully selected by Schonland for their ability to apply the scientific method to the array of problems likely to confront the soldier, and possibly to confound his commander, should these be in any way out of the ordinary or associated with the increasingly complex equipment then appearing on the battlefield. Of crucial importance when selecting those men was the 'personal touch' — that ill-defined ability to gain the confidence of those whose initial reaction was frequently sceptical, and often openly hostile to the intrusion of what were often seen as 'these electrical and wireless mongers' into the domain of the soldier. A broad range of skills rather than sheer brilliance in any narrow field was what he required. Of paramount importance too was the ability to present the outcome of their investigations in the form the military were used to. Not surprisingly, Schonland was emphatic on this point.

Amongst the men who now made up the AORG were many whose names had already featured prominently in the activities of its predecessor. Amongst them were Maurice Wilkes, Stanley Hey and Frank

Nabarro, all men who had worked closely with Schonland since August 1941, and whose expertise he valued highly. Another name, very much from Schonland's past, appeared as well. It was S R Humby, who in 1917 had collaborated with Lt Basil Schonland in a number of experiments that led to their publishing a paper (Schonland's first in the scientific literature) on the subject of the oscillating valve. Now, Spencer Humby had returned, not quite to the colours, but to offer his scientific services as one of the schoolmasters who had been inducted into the mysteries of radar by Jack Ratcliffe in the Vicarage at Petersham. Ultimately, he became the Section Head of AORS2 and then, at the war's end, Deputy Superintendent of the AORG itself [51].

Schonland was very well served by the men, both military and civilian, who worked for him at Richmond and Roehampton. They in turn found him an inspiring leader, blessed with an abundance of good sense and a ready ear who would listen to all points of view. He encouraged them to speak frankly and to voice their opinions, whether they agreed with his or not. To Schonland, men in their position must have initiative and they had to display it. When they did, even if misdirected, as it occasionally was, he would support them. By so doing he gained their confidence and was regarded as an inspirational leader. The AORG was soon melded into an organization that delivered what it promised.

All who knew Schonland as a schoolboy and as a young man whether at Rhodes, the Cavendish or in Kitchener's Army, had seen in him all the potential of the leader. His personality was forged in Grahamstown by the joint themes of service and tradition that were so much part of the English Public School and the British Empire of the time. Lord Methuen's words that had boomed around the Drill Hall at St Andrews, if Schonland remembered them at all some 33 years later, had surely left an indelible mark on his character. But it was the First World War that had forged in him a soldierly mien. Remarkably, but not uniquely, for one destined to hold high rank, his only formal military training was that of a subaltern at the Royal Engineers Signals Depot in 1915, but the active service that followed during the next three years turned a junior officer into a seasoned veteran. Schonland possessed a natural sense of command coupled with a sound mix of humanity, common sense and good humour. As evidence of quite how well he brought all these qualities to bear, when at times he was under great pressure and was carrying all the responsibilities of senior rank, without the benefit of regular army service or any staff college training, we have the recollections of an officer who served under him in the AORG [52]:

> I think the first difficulty for posterity will be to get as good a man
> as the Brigadier to be in charge. He must be a good research
> scientist but not one whose interest is solely in his own line of
> research, he must be young in mind and active, he must suffer fools

> gladly, he must get on well with army officers, and above all he
> must be entirely outside political and departmental intrigues
> (though it is an advantage for him to be conversant with the
> elementary principles of these arts). The Brigadier had all these
> qualities but it will be damned difficult to find them combined in
> one man again for the next war.

This is high praise indeed, but undoubtedly deserved given the achievements that were due to the collective efforts of all those serving under 'the Brigadier'.

One such contribution that was of huge importance to the outcome of the war, and which required collaboration across an even wider front than that now spanned by the AORG, went by the code-name of 'Window'. After the escape through the Channel of the *Scharnhorst* and the *Gneisenau* in February 1942 there was considerable unease within some quarters in Whitehall about the ability of the Germans seemingly to jam British radars at will. The fact that the 10 cm sets were not affected on that occasion, and had tracked the ships plus their escorting armada, provided no comfort, because it was surely only a matter of time before the Germans became aware of their existence and developed suitable counter-measures against them. To counter this, Britain should surely set out to interfere with German radars in some way. There was, though, a strange air of unwillingness in some quarters, particularly evident in the views of Watson Watt, to take any steps to jam German radar. It was almost as if to the father of British radar this would not be quite 'playing the game'. But there was a more serious reason and that was the possibility that any new technique or counter-measure that the British might use against German radars could very easily be used against them, with dire consequences, and this was the particular concern of Watson Watt, of Lord Cherwell, Churchill's scientific adviser, and of some radar specialists within the RAF.

One such technique, which had first been proposed by R V Jones as early as 1937, was the use of metallic reflectors that could be deployed at will to randomly scatter a radar signal and, at least in theory, render it useless. The idea surfaced again some four years later in the form of thin metallized strips of paper about 25 cm long, roughly half a wavelength and therefore resonant at the frequency of the Würzburg radars. They had been shown in actual trials to be extremely effective in masking a target on a radar screen when dropped in profusion from an aircraft. This radar counter-measure was code-named 'Window' and RAF Bomber Command was all set to use it against the German radars. But, in May 1942, the Air Ministry decreed that 'Window' would not be used over enemy territory for fear of the Germans collecting the strips and then suddenly realizing their significance. Such deployment would be sanctioned only after a thorough study had been made of its effects on British radars. In reality, the Germans were fully conversant with the technique, having tested what they called *Düppel* in trials near Berlin and over the Baltic in 1942, and they fully expected the RAF to begin using it soon after the Bruneval Raid had revealed the details of the Würzburg radar. When that did not happen some, such as Goering himself, assumed that only German scientists had understood its significance, and so the Luftwaffe High Command forbade its use in case the British recovered any of their metallic strips, recognized their function and then used them in retaliation against German radars! Paralysis thus set in on both sides [53].

At Brigadier Sayer's instigation, Schonland was asked in the summer of 1942 to investigate in detail how 'Window' would affect the Army's GL II and SLC metric radars then in service, and to comment on its likely effect on the GL III centimetric sets which were about to be commissioned. Joint trials with the ADRDE were therefore conducted well out to sea off the east coast lest any of the metallic strips should fall on land and so betray their existence to all and sundry.* The subsequent report, prepared by Captain D K Hill, confirmed the effectiveness of resonant dipoles as reflectors and worked out the required quantities and their distribution over a period of time necessary to disrupt the enemy's radars. By so doing the AORG were also able to answer the pressing question about the reciprocal effect on British radars. Though it was certainly true that all radars were vulnerable, it was concluded that the Germans were unlikely to use 'Window' against the GL II because this would require strips about 2 m in length and, as such, that would be impractical in the large quantities required. Likewise it was believed that the vertical polarization of the SLC radars would make them immune to the clouds of gently descending, mainly horizontal, strips of foil. Finally, the much shorter wavelength and hence better angular resolution of the GL III should allow it to discriminate target from clutter, so allowing these sets to operate relatively unhindered. Such expectations turned out to be overly optimistic, as shown by a series of tests conducted at the same time by the RAF, but methods to mitigate the effects by means of modifications to the air-interception (AI) radars carried by the night-fighters were implemented by December [55]. As it transpired, by the time the Germans deployed their *Düppel* in January 1944, the RAF had achieved virtual mastery of the air over England and the consequences were far less severe than might have been the case had it been used only a matter of months before.

---

* Schonland used to enjoy telling the story of how he had to convince a rather sceptical police sergeant that the strips of aluminium foil that had inadvertently drifted inland after the trials were of absolutely no importance at all! [54]

Official reluctance to deploy this remarkably simple weapon against radar persisted for many months and was compounded by the decision to close the factory making the metallized strips and to disperse the workers under pain of severe penalty should they disclose what it was they were doing. Whereas concern about German retaliation against British radars was justified, the decision to cease all production of 'Window' in England seemed to Schonland to be sheer lunacy and he made his views very clear to the RAF. He pointed out that it would be calamitous in the national interest were the Germans to start using 'Window' — thus making the ban meaningless — yet the RAF would then be unable to retaliate in kind because the production facilities had been closed down. The logic was inescapable and the factory was re-opened, but it was not until June 1943, and after much protracted argument at the highest levels, that Churchill himself took the decision that 'Window' should be used during the air raids on Hamburg. The results were phenomenal. The German propaganda machine was for once almost speechless, describing the effects of the massed air raids on the city as 'a catastrophe, the extents of which staggers the imagination'. For its part RAF Bomber Command flew 3095 sorties against the city in the space of ten days from 24 July 1943 and lost 86 aircraft or 2.8%. This should be compared with the loss rate of 6.1% suffered during the previous six raids against Hamburg before the introduction of 'Window' [57]. But for Schonland's intervention they might have had no 'Window' at all!

AORG's purview was wide and this required in its Superintendent a flexibility of mind and attitude that would enable him to control the research activities of the many disparate groups of scientists who grappled with problems thrown up by the vicissitudes of war. Schonland was showing himself to be such a man. He was always in command, but never rigidly so, and always tolerant of the individual whose talents sometimes caused him to stray beyond the narrow confines of a project, just as long as his efforts were well-directed and there was some likelihood of a useful outcome, however uncertain it may have appeared at the start [58]. A case in point was the use of VHF radio communications by the Army. It began with the acquisition by Cockcroft of some American type SCR 610 FM radio sets, which he passed to the AORG for evaluation. The equipment operated between 27 and 39 Mc/s[*] — a frequency band deemed by the War Office to be of little practical use since the achievable range was thought to be limited to line-of-sight. The Army used HF (3–30 Mc/s)[†] almost exclusively. However, as the AORG discovered, there

---

[*] Mc/s (megacycles per second) is now MHz (megahertz).
[†] Modern practice is to define the various frequency bands in decades of 3, thus HF is 3 to 30 MHz while VHF is 30 to 300 MHz and so on. However, military HF systems can often extend as low as 1.6 MHz.

was growing evidence, particularly from the police in Birmingham, that similar frequencies to those used by the SCR 610 met all their mobile communication needs, in and around the confines of a large city, where line-of-sight between communicating police vehicles was very much a rarity. Schonland therefore whole-heartedly encouraged AORS3 in its tests with the SCR 610 sets in the hilly countryside in Somerset, through woods in the New Forest and over mountains in north Wales. In every case the radios produced both reliable and effective communications in situations precisely like those encountered by soldiers in the field. Signal strengths, though somewhat lower than those at HF, were entirely satisfactory and the almost complete absence of any interference from either man-made or natural sources actually yielded a greater signal-to-noise ratio: the crucial factor. By contrast, at that time, almost all the Army's communications were squeezed into a narrow portion of the HF spectrum between 7 and 9 MHz during daylight hours, and at even lower frequencies at night, as dictated by the peculiarities of the iono-sphere, which reflected the signals back to earth beyond line-of-sight. Not only did this lead to serious mutual interference between users but the veritable cacophony of sounds from stations far afield, which propa-gated very effectively via the ionosphere, seriously degraded the quality of service. Such long range propagation also increased the likelihood of the enemy both over-hearing the traffic being passed and of jamming the radio nets at will. The complete absence of such long-range interfer-ence, and the greatly reduced risks of eavesdropping at VHF, made the use of those frequencies very attractive.

In his report the officer in charge of AORS3, with the full concur-rence of his Superintendent, stated that 'Army Operational Research Group are of the opinion that, for short-range work under army condi-tions, the theoretical objections to VHF have little substance'. Warming to his theme he continued 'It is this freedom from receiving and produ-cing interference which provides an overwhelming advantage for the use of VHF at short distances. The argument would, in fact, still be cogent even if VHF sets had only a fraction of the efficiency of HF sets, instead of being their equal. Between 30 and 50 Mc/s there are more channels usable than between 2 and 9 Mc/s and, as the ranges will not exceed 10 miles, the whole of the frequencies can be allotted to fairly small areas with no danger of interfering with the same frequencies in neighbouring areas.' [59].

The report was duly forwarded to, amongst others, Sir Edward Appleton, in his capacity as Secretary of the Department of Scientific and Industrial Research and also closely associated, as one would expect, with the Inter-Services Ionospheric Bureau, the body responsible for the allocation of frequencies within the HF bands to the multitude of military users. AORG's report offered not only a superior alternative to

HF for certain applications, but evidence to back it up. But there it rested due, no doubt, to the ruffling of some feathers which it would have caused since the interests of HF were well and truly vested in certain official quarters at that stage of the war, not least of which were Appleton's. To his increasing chagrin, Schonland found his recommendations all but ignored for the better part of a year. Only in November 1944 were the first tentative plans being hatched to run the comparative tests between HF and VHF that AORS3 had recommended more than a year before [60]. John Cockcroft was later to remark, somewhat wryly, that though AORG fought a great battle for the use of VHF in the forward areas they lost because of strong opposition from the War Office, though subsequent experience in Normandy 'showed how right they were' [61].

While it never was, by any stretch of the imagination, the function of the AORG to determine Army policy, Schonland certainly believed that its purpose would not have been served were AORG not to influence it in those areas where they had particular expertise to bring to bear. His own scientific and military experience equipped him rather well, when it came to assessing the part played by radio communications in warfare and he took a personal interest in one of the most demanding requirements of all—communication within the so-called 'skip-zone'. This particular region is well beyond line-of-sight and beyond the ground wave range, thus too far for VHF but too close for the conventional long-range HF signals which 'skip' over this zone as they are reflected from the ionosphere. However, this region, within a radius of some tens to hundreds of kilometres of the transmitter, is the crucial area for an Army commander, because it is there that he fights his battles and so he must be in communication with all the subordinate units under his command.

With all eyes beginning to focus on the next great land battles to come, in the heart of North West Europe and in the jungles of Burma, good communications would be of paramount importance. The recent campaigns in North Africa, culminating in Montgomery's victory at El Alamein, had shown this, yet problems still existed with 'signals in the field' and Schonland saw their solution as one of AORG's major tasks. To a geophysicist it was perhaps ironic, but not surprising, that the ionosphere and the sun that controlled it should affect the fortunes of war but they had already done so more than once, as we have seen, and would certainly do so again.

The eruption of sunspots that produced the electromagnetic noise which so effectively blinded the GL radars across England in February 1942 had occurred when the 11-year sunspot cycle was in rapid decline and, therefore, was somewhat anomalous, but such is the complexity of the process of sunspot formation that no geophysicist was too surprised. Much more readily predictable, however, is the effect of the number of sunspots on HF radio communications and techniques for doing this were already in use, though somewhat uncoordinated, before the war. The need of the various Services for more precise forecasting of the optimum frequencies to use at specific times and over defined paths led, in some haste, to the establishment of more formal prediction procedures and the AORG acted as the body that dispensed these forecasts throughout the Army. The severe problem of communicating within the skip zone required both careful choice of operating frequency, based upon knowledge of the critical frequency of the ionosphere, and the use of appropriate antennas at both ends of the link, where the energy had to be radiated almost vertically skywards. AORS3 made a detailed study of the problems from September 1943 onwards and, over the course of the following twelve months, issued a series of Memoranda that provided precise information about the peculiarities of this communication problem and on the design of simple antennas for easy deployment in the field and on military vehicles. The hand of Schonland was much in evidence in these careful theoretical and experimental studies, and especially so in the style and quality of the reports which resulted from them [62]. On paper, at least, the British Army was well-equipped to cope with the vagaries of the ionosphere and its implications. Events on the ground though, particularly at Arnhem a year hence, would show quite how many slips could occur 'twixt the scientific cup and military lip.

*   *   *

Even in the very thick of war the British scientific establishment insisted, almost as if in disdain of Hitler, on conducting its affairs in as normal a manner as those most abnormal of times would allow. The Physical Society met regularly in London to hear papers read by its members and occasionally to honour those whose contributions to the subject were considered meritorious. One of them was Basil Schonland, who had been elected an Ordinary Member of its Council in 1943 and was honoured in July by the award of the Society's Charles Chree Medal and Prize of £500. This prestigious award had been instituted in 1939 in recognition of distinguished research in the fields of terrestrial magnetism, atmospheric electricity and related subjects. Its first recipient was Sydney Chapman, whose theoretical work with Appleton on the ionosphere was particularly significant. Schonland was next to be honoured. Following them over the course of the next few years would be some notable contributors to the state of knowledge in these areas such as Appleton himself, G C Simpson and E C Bullard all, at one time or another, colleagues or even competitors of Schonland's in their particular world of scientific endeavour [63]. And so, war or no war, Schonland saw it as much a duty as an honour to read a paper on the occasion of his award, which was presented to him by the Society's President, Sir Charles Darwin, yet another familiar face and an old friend.

On 16 July, Brigadier Basil Schonland OBE FRS, resplendent in his uniform and with Ismay at his side, duly appeared very much amongst friends and colleagues at the Royal Institution in Albermarle Street and talked most eloquently about thunderstorms and their electrical effects. His paper was a veritable *tour de force* through the field that he had almost made his own, but he also used it to honour those who had gone before him, such as C T R Wilson and C V Boys, and to make special mention of his South African colleagues Collens, Malan and Hodges, as well as Allibone in England. His address was essentially the definitive review of the physics of lightning, its generation, propagation and electrical effects as known at that time. To have prepared it when his thoughts were very much elsewhere was no mean feat but then that was much the measure of the man. The paper was published that same year in the *Proceedings* of the Physical Society, complete with a photograph of its author in uniform, and it remains a fitting record of the duality of Schonland's career at that time [64].

Yet another honour followed quickly. Basil Schonland's long involvement with the South African Association for the Advancement of Science culminated that year with the award of its 'South Africa Medal'. His duties naturally prevented him from receiving it in person but not from receiving the personal acclaim from his colleagues in Johannesburg who had watched, with much pride, his progress into the seemingly stratospheric heights of the British military establishment. The Board of the BPI met in June at a special meeting under the chairmanship of Bernard Price, but held at the Johannesburg Country Club, for they were not privy to the secret radar work taking place almost under their noses at the university. They resolved to 'convey the Board's congratulations to Professor Schonland on this distinction and also on his acting appointment to the rank of Brigadier', and Bernard Price duly did both on their behalf [65]. Others were following Schonland's progress too.

In September, Field-Marshal Smuts paid yet another visit to his troops 'Up North' and then, after dining in Tunis with General Eisenhower, who was soon to step on to the world stage as supreme commander of all Allied Forces in Europe, Smuts flew on to England to confer with Churchill. Whilst there he visited military establishments and headquarters across the length and breadth of the country and one of those was the AORG at Roehampton [66], where Schonland introduced him in a warm and gracious speech to the assembled scientists and soldiers, and Smuts replied in his peculiarly characteristic Afrikaans accent with its burr, known to South Africans as the 'Malmesbury bry'[*]

---

[*] Pronounced 'bray'.

[67]. Idiosyncrasies of speech aside, the Field Marshal commanded great attention wherever he went and Winston Churchill certainly regarded his old Boer War adversary as one of his closest colleagues and most revered advisers. If Smuts had a blind spot, though, it was in his sometimes flawed choice of key men to fill important positions [68] yet, in one instance at least, his judgement was soon to prove perfect. As his visit to England drew to a close he sent a secret telegram to J H Hofmeyr who, as we have seen, held the reins of government in Pretoria whenever the Prime Minister was away. Hofmeyr also took particular responsibility for scientific research in South Africa. In his telegram Smuts said [69]:

> I think I have found our future director of scientific research ...
> Brigadier Schonland who is in charge of all scientific work at the
> War Office. Both as administrator and as scientist he is held in
> highest esteem here, and risk is that unless we appoint him in time
> he may be offered post-war appointment by War Office. Do you
> approve my approaching him for post-war appointment as
> Director?

Though Smuts rather exaggerated Schonland's authority within the British wartime scientific establishment, he undoubtedly sensed the contribution that the son of his long-time botanical colleague was making towards winning the war. One who would have been fulsome in his praise of Schonland and his work was no less a figure than the General Officer Commanding-in-Chief, Anti-Aircraft Command, General Sir Frederick Pile, who was Smuts's guide during his visits to the gun sites around London. 'Tim' Pile's affinity with scientists and his understanding of the role that science was playing in warfare made him almost unique amongst his senior military colleagues and he missed no opportunity of singing Schonland's praises [70]. The Schonland star was certainly in the ascendant and Smuts read the signs more quickly than most.

Contact between them during the Prime Minister's visits to England extended beyond the formal occasion. Smuts was looking ahead to the future of South Africa when the war was finally over. He, like Churchill, saw only an Allied victory as its outcome but he knew too of the costs that his nation would have to bear to re-build itself after those years of conflict. Science would be a vital element in that programme of reconstruction and advancement and Schonland's part in it would therefore be crucial. These matters often brought them together, usually after the working day when they sat together in Schonland's office at Roehampton, or occasionally over lunch at the Hyde Park Hotel [71]. Smuts knew that Schonland had the ear of men such as Tizard, Cockcroft, Appleton and many others, whose marks had already been well and truly made in formulating British scientific policy and it was this, as well as Schonland's knowledge of the structure of British science, which determined that they should meet. Plans were already afoot for the establishment of some sort of scientific research body in South Africa as soon as the dust of war had settled and their discussions revolved around the structure that the organization should take and where in the country it should be situated. The most appropriate examples that existed elsewhere were obviously in England but Australia, Canada and, to some extent, the United States also offered interesting alternatives. Schonland was surely well-placed to make the necessary inquiries into their operation and management.

He acted immediately. In December 1943, shortly after Smuts had returned home, Schonland wrote to him conveying the views of Tizard as to the viability of the Council of Scientific and Industrial Research in Australia that had been established in 1926. Whereas Smuts had been led to believe that it did not function very effectively because of inter-state politics and rivalry, Tizard's opinion was that it was a great success and provided the best model for South Africa to follow; a view also held by Schonland himself [72]. And so the next phase of the Schonland career was beginning to emerge but there was still the matter of the war, as the greatest invasion force ever mounted was about to land on the coast of France.

CHAPTER 14

MONTGOMERY'S SCIENTIST

The turn of the year from 1943 to 1944 saw the Allied military commanders massing their forces for the invasion of Europe — Operation Overlord — intended to drive the Germans from France and the Low Countries and force them, in retreat, into the heart of their homeland and defeat. The most elaborate seaborne assault ever planned would be an Allied operation involving three Army Groups under the supreme command of General Eisenhower. British and Canadian forces were combined within General Montgomery's 21 Army Group, while the Americans formed the remaining two and all would land on the coast of Normandy just north of Caen. The date of the invasion was finally fixed for early June, after the most detailed assessment of every conceivable option ever undertaken for a military operation. The United States 1st Army would go ashore on beaches code-named Omaha and Utah on the Cherbourg peninsula while the 2nd British Army and the 1st Canadian Army would land to the east of them at Gold, Juno and Sword. The sheer size of Overlord necessitated a degree of planning and preparation that greatly increased the demands made upon every arm of the Service in terms of manpower, equipment and, most crucially of all, the use of the most sophisticated technology to outwit and overwhelm the enemy. Science was well and truly at war.

Whereas the degree of cooperation that existed between soldier and scientist within the Allied forces was one of the truly remarkable features of the Second World War, its absence within the German military system was no less so, given the sophistication of so much of the technology that was produced in the laboratories and factories of Berlin and the bunkers of Peenemünde. Its use, though, by the German forces was frequently uncoordinated and its incorporation as a key element within military planning was sporadic. The 'Sunday Soviet', those informal meetings between senior serving officers and scientists that were so much a feature of life at both TRE and AORG (though not by that name), is now acknowledged as a major factor contributing to the success of the Allied war effort. By contrast, such informality was certainly anathema to the Third Reich and possibly not even in character with German social traditions [1]. In England, as the fourth year of the war dawned, the underlying philosophy of warfare had become highly professional. Gone was any vestige of the First World War general, aloof and so often remote from the real horrors of the battle, and with little or no appreciation of how warfare had changed since the Boer War had ushered in the new century. By 1918 it was gas, the submarine, the tank and the aeroplane that were the products of science and technology, though by no means definitive in the final outcome, that showed how great a part modern technology now had to play when nations went to war. By the end of the Second World War, Tizard, when commenting on his role as Chairman of the Defence Research Policy Committee, summed up the change in thinking that had taken place when he said that 'No major recommendation of defence policy is made by the Chiefs of Staff without his [i.e. Tizard's] knowledge and assistance' [2]. Science and warfare were therefore inextricably linked in England and this was never more evident than during those last months, when the invasion of Europe, the 'Great Crusade' according to Eisenhower, was about to begin.

By late summer 1943 the British Army had begun specialized training under Lieutenant General Sir Bernard Paget, Commander-in-Chief Home Command. The complexity of a massed landing by hundreds of thousands of men on defended beaches, along with all their weapons, equipment and supplies necessitated planning, staff work and training of the very highest order not only to ensure success in capturing their immediate objectives but also to minimize the effectiveness of the German counter-attack. To Professor Charles Ellis SAAC and Brigadier Schonland at AORG it was immediately evident that operational research expertise and personnel must be involved at the earliest possible opportunity, to assist in the training and to advise on problems as they saw them, and so they visited General Paget to discuss the matter with him. Paget readily agreed not only that an Operational Research Section should be formed within his command but also that Schonland should act as his scientific adviser throughout the period of training. What became known as No 2 ORS thus came into being on 14 August 1943 under the command of Lt Col Patrick Johnson, who had recently been in charge of operational research in the 8th Army in Egypt, with his headquarters in Cairo and the task of assessing the effectiveness of anti-tank weapons on the field of battle. However, the refusal of General Montgomery, the then 8th Army Commander, to allow anyone from GHQ, other than his Commander-in-Chief, to visit the battle area virtually emasculated them and the section disintegrated [3]. Johnson's experience in army operational research, though, was well-founded because, even before Schonland's appointment to ADRDE, he had taken over the running of the AA Command School from Ratcliffe and so was well versed in the arts of scientific soldiering. He threw himself into his new role with enthusiasm and Schonland immediately placed all the resources of the AORG at his and Paget's disposal.

Initially, Johnson had with him only one other officer, a signals specialist by the name of Hennessey, but as the need arose, he called upon the AORG to provide others to address specific problems. Most were civilians and this made things awkward for the army in terms of the protocols in military messes and such seemingly mundane matters as the appropriate accommodation to be provided when on manoeuvres. However, the much greater problem was the understandably high degree of secrecy associated with Overlord itself. Neither Schonland nor any of his staff were privy to any of the details and this rendered their role some-what superfluous. So, even though Paget had willingly accepted No 2 ORS* within his command, they soon found themselves excluded from the very exercises they were supposed to observe, and so no information was forthcoming about the performance of infantry, artillery and tank weapons, and none about battle damage from the array of 'Zuckermen' that had been so carefully positioned for just that purpose. Instead, Johnson found himself occupied with the provision of suitable anti-aircraft equipment and the testing of phosphorus-filled bombs, while Hennessey became technical adviser on all manner of signals problems, especially cross-channel communication links and the setting up of a wireless guidance system for the flail-carrying tanks that were to lead the advance to the beachhead, destroying all mines as they did so. As useful as they were, Schonland was not that happy with the highly technical turn that the Operational Research Section had taken, when he perceived that its real role was to observe the bigger picture and to advise on the solution of scientific and technical problems as they encountered them [4].

He himself had certainly taken a much wider view of the war and, after reading the account of Operation Torch, the Allied landing in November 1942 on the coast of French North Africa, was alerted to the very serious danger of possible mutual interference between the myriad of radio and radar systems that would be operating almost cheek by jowl on the beaches of Normandy as the invasion forces established their beachhead, before starting their fighting advance inland. Torch had suffered greatly from this problem, described by Field Marshal Alan-brooke, Chief of the Imperial General Staff, in his characteristically composed way as 'a certain confusion—even chaos' [5]. In October 1943, Schonland therefore alerted Brigadier Sayer at the War Office to

---

* There was a clear distinction between the operational research sections under Schonland at Roehampton, always indicated as AORS1 or simply ORS1 etc., and those formed from May 1942 onwards for service overseas. The first of these was SD6 in the Middle East, but all subsequent sections were known as No 1 ORS in Italy, No 2 ORS with 21 Army Group etc.

the possibility of such interference and suggested that a trial be held without delay involving all the types of equipment likely to be used during the landings in Normandy. The trial should simulate as closely as possible the reality of the situation from the points of view of the various equipment, their relative positions one to another and the frequencies on which they would be operating.

None of this would, however, be possible without precise knowledge of what was planned for D-Day itself so it was inevitable that Schonland and Sayer were both soon to be 'bigoted' [6]. This was the code-word that referred to the special briefing, sanctioned only by Eisenhower himself, that was given to those senior officers and anyone else with a need to have full knowledge of the details of Overlord. Bigoted with them was John Cockcroft, brought into the fold because of the special role that ADRDE would have to play in providing so much of that equipment to be used in this greatest of all invasions. All were then made members of a special committee set up by Eisenhower under the chairmanship of Sir Robert Watson-Watt* and charged with advising the Army, the Royal Navy and the Royal Air Force on how radar could best be used during the landings and how various counter-measures could be employed to mislead the enemy [7].

The trial that Schonland proposed was a major undertaking. Known as Exercise 'Feeler', it took place over ten days from the beginning of December 1943 at a carefully selected and closely guarded test area at White Waltham in Berkshire. Examples of every type of radar and radio communications equipment to be deployed on the British beaches of Normandy, as soon as the invading forces landed, were set up within an area of about 8 square kilometres. Their relative positions and the distances between them were carefully chosen to model as accurately as possible their disposition on D-Day. All the equipment, representing Army, Navy and Air Force units, was then brought into operation such that every conceivable combination of military eventuality could be simulated and the effects of mutual interference between them carefully noted. The objectives of Exercise 'Feeler' were spelt out as follows:

'(a) To ascertain the degree, nature and origin of mutual interference which is likely to be experienced if Naval, Military and Air Force radar and communications equipments are operating on land under congested conditions in the early stages of an opposed landing.

(b) To make recommendations regarding possible modifications to equipment and methods of their employment with a view to minimising such interference.'

---

* Watson-Watt was knighted in 1942 and adopted his hyphenated surname from then on.

The Army radar equipment used in the trial were the SLC, Baby Maggie,* LW and the GL Mk III, all British sets, whilst to simulate the effects of American radars operating on the British flank an SCR268 was also appropriately deployed. The Royal Navy provided its type 277 set and the RAF set up its GCI and Types 11 and 13 radars in their appropriate positions. Every radar installation was equipped with radio communications facilities, as were the heavy and light anti-aircraft gun emplacements that would be set up on the beachhead, and all were linked to the main beach signal station, the anti-aircraft operations room and to searchlight control. Likewise, the RN and the RAF equipments communicated with their respective signal stations and sector operations rooms and, via rear links, to the field force headquarters and ships offshore. Since the exercise took place well inland in the interests of secrecy, suitably equipped military vehicles played the part of 'ghost ships' to lend as much realism to the proceedings as possible. A mixture of HF, VHF and even UHF radio sets, reflecting every type of equipment to be deployed in Overlord, was used; most were fixed but some were mobile and hence capable of being moved in and amongst the various radar and communications installations of the mock up of the beaches soon to be the focus of maritime landings the likes of which had never been seen before.

The trial lasted for ten days. Just as had been the case during Torch, a year before, the exercise revealed numerous cases of electromagnetic interference: radars interfered with each other and with radio equipment nearby, while radio sets caused clutter on radar screens and jammed each other. In all, a total of 30 specific cases of mutual interference were sufficiently severe to warrant detailed investigation. With the assistance of ADRDE and SRDE, as well as active collaboration from the technical wings of both the Navy and the Air Force, the major sources of interference were all identified and a set of preventative measures was devised to ease, if not completely to overcome the problems, since some of them represented major design faults that could not be corrected with so much equipment now already in the field and the invasion just months away. By far the most practical solutions, however, lay in the careful choice of frequencies and sufficient physical separation between any source of interference and its vulnerable counterparts. The AORG report specifying these minimum distances and the various modifications required to equipment was issued on 19 February 1944 [8]. For D-Day's planners it became a vital document.

---

* 'Baby Maggie' was an emergency fire control 1.5 m radar known, more formally, as the AA No 3 Mk III. It incorporated a form of automatic following based on the 'Magslip' principle and followed closely the design of the SLC radar or 'Elsie', hence the name Maggie, of which Baby Maggie was a smaller and more readily transportable version.

February also saw the publication by Schonland himself of a document that was classified Most Secret [9]. It was a retrospective look at the performance of all the radar equipment then in service with the British Army at home and abroad, and was the very first report to cover this highly classified subject in any detail. Subsequently, it became a much-quoted source, especially when the definitive history of Army radar was published in 1950 [10]. In it Schonland disclosed how the accuracy of anti-aircraft gunfire had improved so markedly from the almost prodigal figure of 18 500 'rounds per bird' typical of the months of September and October 1940 to the stage when the GL1, now fitted with its elevation-finding attachment, saw the number of aircraft downed per round fired drop precipitately to 4100 for the year 1941. These figures have since become part of the folklore of radar. The anti-aircraft gun thus became both a weapon feared by the Luftwaffe and a great morale-booster for the local population, a fact enthusiastically reported by General Pile himself. Schonland reported too on the use of 'Elsie' for the control of searchlights, on the performance of the coast watching CD/CHL radars, and on the dramatic results produced when the first centimetric equipment, the CD No 1, came into service in the autumn of 1941. He also discussed, in some detail, the phenomenon of anomalous propagation, by which ranges as much as four times the norm were sometimes reported. While such extreme range was of much interest both to the Royal Navy and to those manning the coastal guns, it was a random event that remained essentially unpredictable and was therefore of limited operational use. However, the scientist in Schonland saw this not as an irrelevance but as a challenge awaiting concentrated investigative effort, leading ultimately to its unravelling and its prediction. He would have remembered well the same occurrence in June 1940 when, north of Durban with his South African colleagues, he had seen on the screen of their primitive JB radar evidence of shipping beyond the horizon. Schonland's scepticism then had cost him a dinner at Durban's Cumberland Hotel, when a convoy sailed into view; now he was a confirmed believer. Those early experiences with radar in South Africa, and his recent receipt of Bozzoli's monographs, prompted him to make special mention of South Africa in this AORG report when he related how the radar chain then in operation around its coastline had 'assisted materially in saving 95 000 tons of shipping from going aground'. Small beer, maybe, in the massive global conflict, but the SSS were helping to keep open a vital sea-route and Schonland flew their flag with not a little pride.

Coastal radar in Britain, and the guns that it served, had scored another great hit with the development of equipment able to observe the fall of shot to an accuracy greater than that possible with visual methods. This, coupled with some most important work done by M V Wilkes and J A Ramsay at the AORG, on the determination of target size from the size of the radar echoes they produced had, most recently, led to the Straits of Dover virtually being closed to enemy vessels over 2000 tons. And so, by the circulation of this Schonland report, the veil of secrecy, that had so shrouded British radar since that February day in 1935, when Watson Watt and Wilkins had confirmed their predictions, had now been lifted, if only by one corner, to reveal a remarkable catalogue of success by the Army. Its performance both in the air and at sea was equally impressive.

While the technical matters of radar and radio were occupying the minds and the time of some within the Army and its associated scientific services, issues of a much greater magnitude were exercising those of Churchill, his CIGS Field-Marshal Sir Alan Brooke* and their American allies. Crucial amongst them was the choice of the man to command all British and Dominion forces in Operation Overlord, and ultimately the land army following the invasion. The appointment was made on 3 January 1944 and, as always in view of the personality involved, it caused reverberations that extended well beyond the corridors of Whitehall. In the face of considerable opposition from the Americans, who were providing not only the majority of the troops, but the Supreme Commander General Eisenhower as well, and from Churchill himself who much favoured General Alexander, it was the will of Field Marshal Brooke that prevailed when General Sir Bernard Montgomery — 'Monty' to every soldier who had ever served under him — was appointed to command 21 Army Group for the invasion of Europe [11]. Under him now would be the 1st US Army, the 1st Canadian Army and 2nd British Army, plus various allied contingents from Europe, all to be known as 21 Army Group. In addition, the Allied Expeditionary Air Force and the Allied Naval Command of the United States and Britain would provide an invasion force, the like of which had never been seen before. Against them would be ranged an army that, in the minds of many, contained elements that were the finest ever to go into battle [12]. Its *corps d'élite*, the SS Panzer Divisions, were certainly the most fanatical and probably the most effective battlefield forces of the Second World War. To defeat them would require much more than sheer force of arms. The key lay in deceiving the Germans into believing that the invasion, when it came, would take place across the narrowest reaches of the channel separating Britain from France, with a landing in the Pas de Calais, when really it would occur in Normandy, almost an ocean away by comparison. This would call for the use of highly innovative techniques, most of which relied on technology and the clever application of scientific principles.

Once landed on D-Day, Montgomery's 21 Army Group would strike out for Paris, the intention being that it should be in Allied hands by

* Later Viscount Alanbrooke KG, GCB, OM, GCVO, DSO.

D-Day + 90 [13]. Between Montgomery and that objective stood a German Army of thirteen Infantry corps and fifteen Panzer divisions, and the *bocage*—the wooded and broken countryside of small, twisting lanes bordered by hedgerows and steep embankments—easy to defend but difficult to negotiate, especially by troops and armour trained on the flatlands of East Anglia and Dartmoor, and led by Generals whose recent memories were of a very different place: the north African desert. A crucial advantage though, and one already held by the Allies, was their almost total command of the air. As a result of massive Allied air attacks on the factories producing German aircraft, and the mounting losses of its pilots suffered in defending them, the Luftwaffe was capable of nothing more than token resistance once the invasion had started [14]. However, nobody, and least of all Montgomery himself, underestimated the ferocity of the battles that lay ahead. Victory was by no means certain and so every means available had to be employed to ensure that the Allied Armies were capable of inflicting maximum damage on the enemy. Fortunately, there were some who appreciated that, to achieve this with minimal loss of life, while using all its resources to their maximum efficiency, would require that 21 Army Group had considerable scientific support. In February 1944, Schonland was visited at Ibstock Place by Charles Ellis, in his capacity as Scientific Adviser to the Army Council. They were friends of long standing from their days together at the Cavendish and Ellis was now Schonland's immediate superior. The purpose of his visit was simple: he wanted Schonland to become Scientific Adviser to Montgomery.

Basil Schonland's initial reaction was, as he later recalled, a mixture of joy and trepidation, but Ellis will have seen neither, for the Schonland façade betrayed little, and their discussion moved rapidly to the implications of the appointment and the role that Schonland might conceivably play in influencing the thinking of Montgomery's staff officers, if not, directly, that of the General himself. The part that science itself could play in affecting the course of a battle was by no means clear, but there was, in the view of Ellis at least, an obvious need for an operational research capability in the headquarters of the Army Commander.

However, the commander in this particular instance was a man of formidable ego and very definite views, especially as they pertained to the art of soldiering, and none of them encompassed having a scientist in his Mess or, indeed, anyone else for that matter who had no direct business there. Ellis well knew this but persisted and, a short while before, had sent a signal to Montgomery asking him whether he would like a 'small team of scientists to observe his battles'. Montgomery replied somewhat witheringly 'I observe my own battles'! Fortunately, good sense and subtle persuasion prevailed and ultimately he acquiesced and the process of appointing the right man to that daunting position went ahead, albeit rather late in the day, and Basil Schonland was seen to be that man. To Schonland himself, though, it was immediately obvious that a scientific adviser was only as effective as the expertise available to him and so, before agreeing, he laid down his own conditions. These he discussed carefully with the man whose sage-like wisdom he valued above all others: John Cockcroft.

Cockcroft, whose advice Ellis himself may have sought, encouraged Schonland to accept but suggested that he might request that the appointment should terminate three months after the landing in Europe [15]. Quite why he did is not clear, but he presumably felt that by that time the advance on Germany would be relentless, with the German army in retreat and the function of a scientific adviser would surely have served its purpose. Suitably fortified by this advice, Schonland replied to Ellis that he was prepared to accept the appointment, but only if Ellis would take the necessary steps to make available to him on the Continent any member of the AORG whom he wanted, in uniform and at short notice. Ellis saw this as entirely reasonable and agreed to see that it would happen.* Whether Schonland was aware at that time of the hornet's nest that he was to stir up is uncertain. The fact that he achieved his objective was due, in no small measure, not to any softening of Montgomery's attitude and steadfast views but to the most accommodating and reasonable attitude of his Chief of Staff, Major General F W de Guingand, the man most responsible for mending the personal bridges that his chief so high-handedly demolished at almost every turn. And so Brigadier Basil Schonland was duly appointed Scientific Adviser to General Sir Bernard Montgomery in March 1944.

This appointment was unquestionably the greatest indication yet of Basil Schonland's worth in the eyes of those around him and, as he later confessed, the fact that he had been invited to go as a Brigadier on Monty's staff was the most exciting thing that had happened or would ever happen to him. But the realization that he was to give advice 'to a bunch of regular soldiers who had fought triumphantly from Alamein to Tunis and from Sicily far into Italy turned my heart cold' [16]. Matters now moved quickly, as indeed they had to, for the invasion juggernaut was rapidly gaining momentum. He was soon to join Montgomery's staff and so he began the process of handing over command of the AORG to his deputy, Lt Col Omond Solandt.† Solandt was a Canadian

---

* According to Schonland's post-war notes, the history of this matter and the negotiations required to make it happen 'would fill a small book'.
† Omond Solandt was thirteen years younger than Schonland. He was a Lecturer in physiology at Cambridge at the outbreak of war, joined the AORG in 1943 from the Medical Research Council's Physiology Laboratory and became Deputy to Schonland that same year. He was Superintendent from mid-1944 until the end of hostilities and then served as Chairman of the Canadian Defence Research Board from 1947 to 1956.

physiologist who, in the multi-disciplinary tradition of the AORG, had established himself as an expert on tanks and armoured vehicles in general. Now, on stepping into Schonland's shoes, he inherited an organization of ten operational research sections, some of which he had helped create, covering fields as diverse as radar and anti-aircraft fire-control, infantry weapons and tactics, time and motion studies, and the analysis of battle records. Schonland himself immediately set about strengthening Johnson's No 2 ORS, which was now formally part of 21 Army Group. From the AORG he selected two of his best officers, Major M M Swann,* from AORS6, and Captain H A Sargeaunt,† soon to be promoted, from AORS4, with the intention that they would assist Johnson and himself to turn the section on to 'real operational research' [17]. Swann's wartime career started in Intelligence but he was soon transferred to the newly formed Royal Electrical and Mechanical Engineers (REME). It was there that Schonland discovered him and, after an interview that impressed Swann as much for the searching nature of the questions as for Schonland's innate friendliness and charm, he joined the AORG and spent much of his time at Barnard Castle near Durham on infantry matters — diversity indeed for a zoologist [18]. Sargeaunt's forte, by contrast, was armour and Schonland's thinking in selecting these two specialists was clearly influenced by the role he saw operational research playing, in support of Montgomery's armies, during the major battles to capture north-west Europe from a formidable foe.

One of Schonland's last duties at Roehampton before he joined the headquarters of 21 Army Group was to arrange yet another conference on the increasingly important subject of 'Window'. In the nine months since Churchill himself had stepped in to break the log-jam of conflicting views at Bomber Command by decreeing that this simple yet highly effective method of confusing radar should be used, it had become obvious that these strips of metallized paper had offensive as well as defensive potential. To fully evaluate the effects on all radars then in service, a series of trials had been run by ADRDE, as we have seen, while both AA Command and the AORG had been actively accumulating operational data, as had many others within the scientific intelligence community then serving the war effort. The conference was held at Roehampton on 4 March 1944 under Schonland's chairmanship. It was attended by more than 60 scientists and serving officers, including a

---

* In 1962 Professor Michael Swann, having returned to the Department of Zoology at Cambridge after the war, was elected FRS. Subsequently he succeeded Sir Edward Appleton as Principal and Vice-Chancellor at the University of Edinburgh in 1965. He was appointed Chairman of the British Broadcasting Corporation in 1973. On his retirement in 1981 he was ennobled as Baron Swann of Coln St Denys.
† Sargeaunt became Superintendent of the AORG in 1947 and SAAC in 1952. In 1962 he was appointed Chief Scientific Adviser to the Home Office.

delegation from the United States led by the US Military Attaché. The first speaker called by the chairman was Professor John Cockcroft, who was asked to brief the meeting on the scope of the ADRDE investigation. Fuller details, presented in presumably more expansive fashion, were also provided by Cockcroft's ADRDE colleague, R G Friend. From these it was evident that the effects of 'Window' on all types of radars to be deployed on D-Day were now well understood and various counter-measures had also been devised should it be deployed by the enemy. Lt Col Northey of AA Command and Stanley Hey of the AORG presented results of their studies of the *Düppel* dropped by the Luftwaffe and the methods devised to mitigate its effects, while Dr F C Frank, from the Air Ministry's Directorate of Intelligence, provided details of the German methods of dispensing *Düppel* and also the latest intelligence on the briefings given to German aircrews on its use [19]. Warfare had clearly assumed a significant electronic dimension and 'Window', used either defensively or offensively, was a key component of it.

From the time that he had been 'bigoted', Schonland had been involved with aspects of planning the invasion from the point of view of devising radio and radar measures intended to confuse the enemy's defences, and to protect Allied ships and aircraft from the enemy's radars. This was all part of the great decoy operation, under the code-names of Taxable and Glimmer, intended to divert German attention and defences away from the invasion beaches of Normandy. These would involve, amongst many other measures, RAF aircraft flying carefully-controlled 'race-track orbits' that advanced towards the occupied coastlines in the vicinity of Boulogne at the speed of a convoy while dropping quantities of 'Window' at precise rates as they did so. The intention was that the few German Freya, Seetakt and Würzburg radars, intentionally left operational following the RAF's attacks against the 100 or so along the French and Belgian coasts, would interpret this offensive use of 'Window' as the invading armada while, in reality, the actual invasion fleet was more than 100 km to the west [20].

*   *   *

Late in March 1944 Brigadier Schonland arrived at St Paul's School in Hammersmith, long-since evacuated of schoolboys and now Montgomery's headquarters. In fact, it was his Rear HQ, for the Main HQ had moved to its first 'battle' position in Portsmouth — the hub around which the invasion would revolve. Of the two, Rear HQ was by far the larger. As Schonland himself later recorded: 'There was probably no member of Montgomery's staff who joined with as much expectation of being snubbed, cold-shouldered and perhaps politely told that he was not wanted as I'. He was wrong. Montgomery's Chief of Staff, Major General Freddie de Guingand, turned out to be extremely friendly and, after the morning's conference, introduced him to the heads of the various branches that made up the top echelon of 21 Army Group. Though none of them was too sure what his precise function was, except that they had heard that he was there 'to solve difficult problems for them', they made him very welcome. Schonland realized that he would have to establish his own credibility and so he paid personal visits to each of the senior staff officers, all Brigadiers, to explain what his function was and to inquire where he might assist them. It was vital that he should rapidly make himself familiar with both the structure and the procedures of an Army Group Headquarters, but fortunately such formalities as there were melded well with his own style of operating and the transition into the ways of Montgomery's empire turned out to be much easier than he had at first imagined.

Rear HQ contained the main administrative sections of 21 Army Group and dealt with such matters as logistics, transport, supply and medical services plus a whole host of others, including pay. It was soon obvious though to Schonland that he should not be there if he was to be any practical use at all, and his request to de Guingand that he be transferred to Main HQ was heard sympathetically and, more importantly, was granted. In itself that may not appear to have been a major achievement but given both Montgomery's very rigid policy on who should have any access to him at all, and de Guingand's own insistence on keeping the Main HQ populated only by those directly involved in fighting the war, it most certainly was. The Chief of Staff had obviously been sufficiently convinced of the importance of operational research to agree that Schonland was in the wrong place. He must be much closer to the hub of operations.

The character of Montgomery imposed its own, almost idiosyncratic demands on his Staff and no one knew this better than 'Freddie' de Guingand. Alternately massaging the ego of his Chief or cajoling him for his frequent and intemperate demolition of all around him, de Guingand was a master of his trade and, as it would turn out, was the linchpin of 21 Army Group. He and Basil Schonland were soon to become close friends. De Guingand informed Schonland how Montgomery would distribute the elements of his HQ once they had landed in Normandy. First of all, Montgomery would isolate himself from the detailed planning of the operation in hand by setting up his Tactical ('Tac') HQ in a caravan that would be served only by his selected group of young liaison officers, his signallers and the few soldiers charged with his personal sustenance and protection. It was his Chief of Staff, always based at Main HQ, whose function it was to turn the plans formulated by the C-in-C, in seclusion where he was free to think, into operational orders for the armies in the field. Assisting de Guingand in this were the specialist advisers in the areas of artillery, armour and signals, plus his General Staff Officers, with responsibilities for Operations, Intelligence, Staff Duties, Plans and Air Support. Clearly it was with these various BGS(Ops), BGS(I) and so on that Schonland would have to collaborate most closely and that would require that he should cultivate the greatest degree of harmony and trust with them as soon as possible. As much as the Montgomery legend may have been punctured by some historians blessed with much hindsight, there is no doubt about his ability to select the best men to work with him. In Belchem (Ops), Williams (I) and Herbert (SD) he had individuals of the very highest calibre. Belchem and Herbert were regular soldiers but 'Bill' Williams* was an Oxford don who served as an infantry officer in the Western Desert and then, after rapid promotion and a DSO, became Montgomery's head of intelligence in the 8th Army. Schonland the physicist and Williams the historian, 16 years his junior, instantly became close friends. While they may have made the somewhat unusual transition from university common-room to Generals' Mess with some foreboding, each had a commanding presence and their abilities were beyond question. Their natural kinship within a world of regular soldiers no doubt owed much to their academic backgrounds, while their friendship was the stronger too because both were men who used the English language with precision and not a little flair and both saw in their respective functions the need to communicate with soldiers 'in readable English without superfluous jargon' [21].

*   *   *

It soon became apparent that Schonland's appointment as Scientific Adviser to the land force commander should have taken place much earlier. The planning of the invasion was far advanced by the time he arrived in Portsmouth and, from discussions he had with the General Staff, it was evident that many problems of a scientific nature had either been ignored or tackled in a somewhat haphazard fashion. For example, no real thought had been given to how the DUKWS, the American amphibious vehicles used to transport troops between the landing ships and the shore, would locate their off-loading points at night. Infra-red devices would have provided an obvious solution but there was no time now to provide such things, nor to arrange the necessary training of the crews. Other problems of a tactical nature to do with the complex interaction between the army, navy and air force all needed some scientific input but none had been considered at the time. Even the part to be played by

---

* Brigadier Sir Edgar 'Bill' Williams CB, CBE, DSO, DL had a distinguished career both during and after the war becoming, successively, a member of the UN Security Council secretariat (1945–46), Fellow of Balliol (1945–80), Warden of Rhodes House, Oxford (1952–80) and Secretary of the Rhodes Trust (1959–80). He was the UK Observer at the Rhodesian elections (1980).

Pat Johnson's Operational Research Section within the Overlord plan was ill-defined, simply because there had been no scientist of sufficient rank within 21 Army Group to define their role and to husband their resources. While it might have seemed obvious that No 2 ORS would now report to Schonland himself, it had already been decided that operational research fell within the ambit of Staff Duties under Brigadier Herbert, BGS(SD), amongst whose manifold responsibilities was the Army Group's equipment policy, its supply and performance in the field and the training of troops in the use of new weapons. Most, if not all of these, were things that an ORS could study and so, quite naturally, Johnson and his section found themselves under Herbert's wing. However, this came as a surprise to Schonland but the pace at which the planning for the invasion was running, and his own rather novel position on Montgomery's staff led him, wisely, to accept what was obviously now very much a *fait accompli*. To challenge the decision now would have led to 'an unholy row' and would have done little to consolidate his own position within the hierarchy at Main HQ.

Soon, though, it became clear to him that this had, in fact, been the right decision, for the Scientific Adviser's role was very different from that of an operational researcher. It was to him that the General Staff, and particularly the Chief of Staff, turned for advice on all manner of problems and he had therefore to distance himself from the detailed investigations that the ORS were conducting, while never losing touch with them. Indeed, he persuaded Brigadier Herbert that it was surely logical that the Scientific Adviser should 'exercise indirect control over the ORS, [by] advising the BGS(SD) as to what it should do, whether it was doing its job correctly and what changes were necessary in its staff'. In addition, he took particular pains to ensure that Johnson had considerable flexibility in how he planned his investigations, while having the full support of the BGS(SD) when it became necessary for the ORS to liaise closely with the fighting forces in the field. As Schonland noted: 'the BGS(SD) ... is a very powerful man who can help considerably, more than any SA, since he has almost daily personal contact with the forward units concerned and they trust him'. There was, though, one facet of Schonland's relationship with No 2 ORS in which, somewhat anomalously, he certainly had direct control and that was everything to do with 'the Air'. From discussions with the General Staff it became clear that, while an ORS should be vitally concerned with air cooperation and bombing in support of the army, these were rather grey areas with which the BGS(SD) was not directly concerned and so it was agreed that Schonland, as Scientific Adviser, would control No 2 ORS directly in all Air-related matters [22]. Such a foray into combined operations, as it undoubtedly was, would later lead to some friction with those in the RAF who were also charged with assessing the effectiveness of bombing and close-support operations, and this required in the Scientific Adviser a healthy dose of diplomatic skill as well as a robustness of spirit.

Once settled into Main HQ the question soon arose about Schonland's own staff. De Guingand's policy regarding the size of his headquarters was rigidly enforced and frequently reviewed. Some sections were made to move back to Rear HQ while others had their staffs drastically cut. Schonland was most conscious of this but, since he had not joined 21 Army Group with the intention of building an empire, his requirements were modest. Other Brigadiers had a GSO1* and perhaps two GSO2s, plus a few clerks, but Schonland indicated that he wanted nothing more than one GSO2, a single clerk and a driver. Such parsimony of personnel was therefore well received and the appointment of his GSO2 immediately went ahead. Major D K Hill, who was amongst the first members of Blackett's Circus appointed in 1940, was on the staff of Sir Charles Ellis at the War Office when Schonland called for him. Previously they had worked closely at Petersham and then Hill had gained considerable experience in operational research in North Africa, and with the 2nd Tactical Air Force (2 TAF), when it was preparing for the invasion. In addition, he was a man much to Schonland's liking in his approach to the business of scientific soldiering. Thus began an association that would see them together from May 1944 until Schonland answered Smuts's call and returned to South Africa in December.

In David Hill, Schonland had not only a physiologist who had proved himself to be a most adaptable scientist and an imaginative soldier, but he also found an agreeable colleague who served him well as his personal Staff Officer throughout those months as Montgomery's armies fought, and occasionally stumbled, through north-west Europe. Hill's own recollections of Schonland are of a good man who shared with J A Ratcliffe the remarkable ability to grasp almost immediately the details of what everyone around him was doing. In addition, he found Schonland's skilful handling of staff meetings impressive and admired his gracious way with visitors, especially dignitaries. Schonland's natural reserve and innate privacy coupled, no doubt, with an ingrained sense of good form, meant that their social contact was relaxed but never familiar. David Hill could only guess at the inner soul that lay behind Schonland's urbane and benign exterior [23].

Their first task was a study of the probable effects of Allied bombing on the beaches of Normandy before the invasion, and to do this they planned an experiment to find out. By using American aircraft dropping bombs of various sizes, fitted with a range of fuses, on an east coast beach they were able to provide information to the commanders on the likely effects of such concentrated bombing on the movement of vehicles. To

---

* GSO1: General Staff Officer grade 1.

Schonland's regret this was the only operational research, in the true meaning of the term as he always understood it, that he was able to do in the short time between his appointment and D-Day itself. However, he solved many 'conundrums', as he called them, the sorts of scientific and technical problems that confronted the Staff as the plans for the invasion were being laid. While these were in no way in the nature of operational research, they formed what he considered to be a subsidiary but important function of the Scientific Adviser's role. Amongst them was a question put him by Brigadier Bill Williams that had its origin in intelligence emanating from France. The Germans had put it about that, to deter any attempt at a landing on the coast, they intended to 'electrify the sea' by running cables from a local power-station directly into it. If true, and such a rumour were to spread amongst the troops, the effect on morale would be serious. How feasible was such a bizarre scheme, Schonland was asked? He did an elementary calculation, consulted a 'distinguished physiologist' (for there were many around him to choose from) and then pronounced it impossible, to much relief in the Mess. Another conundrum concerned a possible threat to Montgomery himself. The signallers who provided the communication between the C-in-C and the outside world had their W/T station only a short distance from his Tac HQ: with what accuracy could the enemy locate it by direction-finding and thus locate the HQ itself? This was indeed an interesting question of no little significance, as well as one that had much resonance with Schonland's pre-war BPI and its research on 'df-ing' atmospherics. Quite how reassuring Schonland's answer was to Monty is not known for his reply is not recorded! And there were many more conundrums like them to exercise his mind [24].

At South Africa House the appointment of Basil Schonland to Montgomery's staff was received with great pleasure by the new High Commissioner, Colonel Deneys Reitz who, as Smuts's most loyal lieutenant during the dying days of the Boer War, well knew of the fluidity of warfare and the need for more than just the sword to win the battle. That a South African should find himself as scientific adviser to a commander-in-chief was certainly a matter of great pride both in London and in Pretoria. There was, though, the need for some adjustment of Schonland's previous position *vis-à-vis* the British Army. Until then he had merely been 'on loan' from the UDF but now it became necessary to second him to the British Forces in the temporary rank of Brigadier to serve as 'Scientific Adviser, Passive* Defence, at Headquarters of

---

* Passive Defence, or more particularly Passive Air Defence, refers amongst other things to measures taken 'to nullify the effectiveness of attack by aircraft' and, in Normandy particularly, to the classification of bomb damage and eventually the assessment of the effectiveness of the V weapons.

21 Army Group' [25]. This duly took place on 19 May. Just four days earlier the final conference before the invasion was held at St Paul's School in London. There, in the presence of King George VI, the Prime Minister Mr Winston Churchill and Field Marshal Smuts, Montgomery presented to the high-ranking officers from all three Services of both Britain and the United States the overall plan for OVERLORD. It was a masterly performance and, as the official history records, at the end 'a sense of sober confidence pervaded the room' [26].

*　　*　　*

D-Day, originally scheduled for 5 June, was postponed due to bad weather. A successful landing of the huge number of men and material depended critically on tide, phase of the moon and, of course, on the weather. The first two were predictable, the third very much less so and General Eisenhower, as the Supreme Commander, had to take the monumental decision. He trusted both his meteorologists and his luck and decided that the invasion of Europe would begin just before sunrise on Tuesday 6 June 1944. Montgomery left Portsmouth that evening aboard HMS *Faulknor* and, to the amusement of some, ran aground briefly on a sand bank just off the landing beaches. By the 8th, he had set up his Tac HQ in the grounds of the château at Creully about 10 km inland from the beach code-named Juno [27]. It was not intended that any operational research activity associated with the invasion should commence until two months after D-Day, but radar was so crucial to the defence of the beach-head, and during the following advance, that Pat Johnson in his capacity as radar adviser to the anti-aircraft brigade was ashore by D-Day + 2. He accompanied what was called the Weapons Technical Staff with Field Formations, known rather more conveniently as 'Wheatsheaf', whose function was to monitor the performance of weapons in the field, while Johnson's was to do the same with the radars then coming ashore in profusion. For the next three weeks he observed the multiplicity of radar and radio apparatus crammed into that narrow area to ensure that all were performing according to the guidelines so carefully laid down after Exercise 'Feeler', the previous December. An advance party from No 2 ORS consisting of Majors Swann and Sargeaunt, their drivers plus vehicles, went ashore about three weeks after the first Allied troops had landed and similarly attached themselves to Wheatsheaf near the headquarters of the British 2nd Army. Brigadier Schonland, meanwhile, was still in England, but the Main HQ of 21 Army Group was preparing to relocate itself to France and he would accompany them [28]. With war having reached this advanced stage, and Schonland now preparing to go into action himself, this was obviously the time for Ismay to return to South Africa to be with the children and so arrangements were made for her to do so as soon as her passage home could be booked.

Montgomery's original intention was to capture the city of Caen on D-Day itself and by D-Day + 90 to be in Paris. He achieved the second — in fact he was 11 days early — but his plans for Caen were very wide of the mark. German resistance in and around the city proved much more resilient than had been expected, and by early July it was decided that the support of the RAF would be enlisted by using heavy bombers, for the first time ever, in tactical support of an Army. On the night of 7 July over 450 Lancasters and Halifaxes of Bomber Command attacked German positions just north of Caen. This was followed by a further wave of light bombers and then by an artillery bombardment that commenced at 11 p.m. At 4.20 a.m. the next day, with accompanying further shelling, two British divisions attacked. Following very heavy fighting throughout the day, British and Canadian armour, supported by infantry, eventually occupied the high ground overlooking the city. Caen was finally in Allied hands by late afternoon on the 9th of July [29].

The members of No 2 ORS, along with all the forces still in the vicinity of the beachhead, had witnessed an awe-inspiring sight as the hundreds of black-painted aircraft had thundered overhead on their way to the target. They soon saw the raid as an opportunity to view at first hand the effects of such an attack, particularly when the conflicting accounts of the Army and the RAF of the damage done began circulating. Without a directive from anyone, for no-one on that side of the Channel had yet conceived of a role for operational research, Johnson, Swann and Sargeaunt spent several days surveying the ruins of Caen while 'attempting to reconstruct and assess the battle'. Their subsequent report generated some interest within 2nd Army circles but considerably more when it was received by Schonland in England. Though flawed in many respects, since only the physical and structural effects of the raid were noted, and no attempt was made to evaluate its effect on the morale of the enemy, it turned out subsequently to be a seminal document that laid the foundation for much of the work that No 2 ORS were to do for the next year [30].

There was also another report on the devastation of Caen. It was written by David Hill who, on Schonland's specific instructions, had gone across to Normandy to assess, quite independently, the effects of the bombing. The sight that met him was one of sheer devastation: a city all but destroyed, with much loss of life amongst its 20 000 remaining inhabitants, but very little evidence of the enemy themselves, or of their alleged headquarters, which had been the focus of Montgomery's attention. But German snipers were still around and David Hill, when surveying this scene, escaped death by inches when a bullet passed through the webbing of his jacket but left him miraculously unscathed. His report to Schonland described the tragic business of what, to his eyes, was a battle that had gone horribly wrong [31].

Also there, clambering amongst the rubble of the city and its bomb-cratered roads, was Solly Zuckerman, who had as direct an interest in the outcome of the raid as any man alive since it was he who had worked out in detail how the bombing offensive should be mounted, using bombs with 6 hour delay fuses designed to destroy the concrete defences constructed, so intelligence reports maintained, to the north of the city. When Zuckerman later visited the scene he found no such defences, nor any signs of enemy gun positions, tanks or even enemy dead [32]. Though carried out with great precision by Bomber Command, this massive attack on Caen was a catalogue of confusion at the very highest levels of command and its planning, coordination and even its purpose have been fruitful areas of heated debate amongst historians ever since. Montgomery regarded the 'capture' of Caen as a great victory in the face of what he saw as almost a conspiracy being waged against him personally within the corridors of Supreme Headquarters; others saw it as one of the most futile attacks of the war [33]. Amongst the many lessons learnt was one of crucial importance in all further massed bomber raids: smaller bombs with instantaneous fuses, rather than the delayed-action 'one-thousand pounders' (455 kg) that caused so much cratering and the destruction of so many buildings, must be used. The subsequent proliferation in Caen of operational researchers in various uniforms but with no obvious sign of coordination between them only added to an air of confusion, as personalities as much as military objectives muddied the waters of Normandy in the middle of July 1944.

The battle of Caen was followed almost immediately by Operation Goodwood, the attack by the British 2nd Army on German positions south-east of the city. This was not only one of the biggest set-piece battles fought by 21 Army Group but it was the cause of considerable dissension between Montgomery and his American allies, and not a little from many of his own countrymen such as Air Chief Marshal Sir Arthur Tedder, Eisenhower's deputy, whose animosity towards Montgomery was well known and undisguised. The pattern of the battle was now familiar: a massed air attack followed by the advance of three armoured and four infantry divisions across the River Orne, with the purpose of destroying the German armour and thereby opening the way for the break-out from the bridgehead and the advance on Paris. But it was not Montgomery's intention to break-out, rather he would 'hold the enemy armour away from the American front' by attacking the point of greatest strength in the German line, while Lt General Omar Bradley's US 1st Army wheeled to the east around the 'hinge' thus provided for them, and thence to Paris. Ultimately, as the official history records it, 'Goodwood ... achieved its main purpose [but] it had not attained all that was intended' [34]. Montgomery's perceived reluctance to ask his British and Canadian forces to take the casualties that would naturally have occurred had he attempted to break through the German line was seen by the Americans as tantamount to asking *them* to do so. This perception was hugely reinforced by the sheer enormity of the air strike that he called for from the force of 1600 RAF and USAAF aircraft that attacked German positions at first light on 18 July. Such a massive bombardment, with thousands of tonnes of bombs, seemingly implied more than just 'holding the enemy', for its effect must surely have been devastating, but was it? Little quantitative evidence existed to support the claims made by the air forces involved, yet reputations and even national honour were at stake. One reputation that had attracted the closest possible scrutiny, particularly from his American allies, was that of Montgomery himself. By refusing to commit the British 2nd Army to all-out assault at that stage he sealed his own fate in the minds of those who either had misinterpreted his intentions or were overwhelmed by his bombast.

The effectiveness of massed-bomber raids was thus an issue of great contention during that penultimate year of the war and became even more so when civilians, in their tens of thousands, became the main casualties within a year. Schonland, though still at 21 Army Group Main HQ in Portsmouth, was particularly interested in the effect of bombing on the outcome of a battle and his action in sending David Hill to comb the ruins of Caen suggests that he wanted first-hand information from a colleague whose judgement and ability he had come to trust. That Johnson, Swann and Sargeaunt of No 2 ORS had acted on their own initiative to do the same thing provided him, ultimately, with even more useful and corroborative evidence. When Operation Goodwood was imminent, Schonland sent an urgent message to Johnson asking him to put his team on the ground immediately the dust had settled and to do so after every similar battle in future. Soon there was a new requirement. The RAF were using the Typhoon, a fighter-bomber that carried either rockets or a conventional bomb load, in its attacks against German troops and armour. As a psychological weapon it was formidable. To Allied soldiers the sight of the Typhoons circling high above in their 'cab rank' patrols, while awaiting target directions from the Forward Air Controllers before peeling off into near-vertical dives to unleash their rockets, was both stirring and morale-boosting; to the Germans their mere presence signalled imminent danger. However, little quantifiable evidence existed to indicate quite how effective these formidable aircraft were and Schonland arranged, through the 2nd Tactical Air Force, to have the matter investigated thoroughly. Sent over from England especially for this task was Major David Pike, who joined No 2 ORS just at the time when the Staff of 21 Army Group were beginning to understand, and then to appreciate, their function. This was due undeniably to the part that Schonland himself was now playing at the highest levels within

Montgomery's Main HQ, that had itself moved to Normandy late in July and was fully established there by 4 August. From its position at le Tronquay, just south-west of Bayeux, he was now able to follow the progress of the battles as the reports came in and so was well placed to advise de Guingand on all matters of a scientific nature that had any bearing at all on operations.

Schonland very seldom saw Montgomery himself; few did because of the almost hermit-like existence the Army Group Commander chose for himself at his Tac HQ. All contact with him was through de Guingand, who reigned supreme at Main HQ, and it was in the Generals' Mess that Schonland began the process of informing the three key brigadiers on the General Staff precisely what role operational research could play in the multitude of situations they now faced in this mammoth land-battle. He was now also in close contact with Pat Johnson and the members of No 2 ORS who had set up their base camp in an orchard in the nearby village of Noron la Poterie, and he used every opportunity to get to know all Johnson's officers and to understand precisely what it was they were doing. To those officers Schonland was soon viewed as a close, almost personal, friend who had considerable influence at the court of the Chief of Staff. Their relationship with him was described by Swann as 'cordial and productive' and though they were never sure whether their virtual freedom to choose where they went, what units they visited and how they worked was the result of deliberate planning on Schonland's part or just 'an oversight of the military machine', they valued it greatly as independent-minded scientists pitched into the maelstrom of war. In fact, it was certainly no oversight, nor even an accident, but was the result of Schonland's own doing. He had convinced Brigadier Herbert of the need for such flexibility and Herbert agreed to sanction it, on the understanding that Schonland saw that it was not abused [35].

*   *   *

The account so far might well convey the impression that the scientific pursuit of operational research within a military context was a fairly well ordered affair, carried out with the full support of the fighting men. That it was, by and large, in 21 Army Group was very much Schonland's doing. It was certainly not always the case elsewhere. Those involved in operational research in both the Royal Navy and the Royal Air Force found numerous obstacles in their paths. Some just fell in the way, as such things tend to do, but others were placed there, with every intention of making life less than pleasant. One who experienced such obstruction was none other than Patrick Blackett. Even his redoubtable presence and reputation counted for little in the heat of battle. Just before D-Day Blackett found himself very much out of favour with the Navy and some more perceptive or informed individuals may well have wondered what the future held for them now that the father of operational research, and the best-known scientist in the Admiralty, suddenly found himself without a job. He had been 'given the sack by a bunch of sailors who didn't want to be "fussed" when they were going into battle', so he told Schonland when he left 'in high dudgeon' and gratefully accepted Schonland's offer of 'board and lodging while he was tidying things up'. Feeling equally maligned at much the same time, and presumably for similar reasons when he confided in Schonland, was Wing Commander Michael Graham, then in charge of operational research for the 2nd Tactical Air Force. He had just been informed by his Air Officer Commanding that the RAF were not prepared to take any operational research people abroad at all. Graham eventually arrived in France, but by that time David Pike and Tony Sargeaunt of No 2 ORS had already done much work in the field of bomb-damage analysis, and it was Schonland's impression that Graham never really won the acceptance of his Staff in France after that [36].

Major David Pike's study of the effects of the attack by the RAF's Typhoons at Mortain was as revealing as it was controversial. On 25 July the break-out from the Cotentin Peninsula by the 1st US Army, Operation Cobra, commenced with heavy casualties inflicted by the USAAF on its own troops: an all too frequent occurrence. German resistance was fierce and the American advance was, at first, creepingly slow. But the German lines were thin, with the great body of their troops still massed against Montgomery's front line to the east of Caen. The Americans wheeled eastwards and what occasionally seemed like a rout was punctuated by periods of courageous and determined resistance as German soldiers, disciplined and well-trained, threw themselves into what was clearly a last-ditch battle. Retreat, though, was inevitable and the roads and tracks in the vicinity of Mortain were packed almost solid by German tanks, half-tracks, trucks and even horse-drawn wagons, plus thousands of soldiers on foot, all heading east. American artillery was firing on the columns as the Typhoons from 2 TAF wheeled in to attack, firing their eight 27 kg rockets into the rapidly obscured mass of machinery, men and horses below as smoke and dust billowed skyward. The effect was devastating. Men jumped from their vehicles to join those already clambering blindly through the hedges of the *bocage* into the woods and fields beyond. Anything to get away from the screaming aircraft and their lethal cargoes. What had started as a German counter attack to cut off the Americans south of Avranches was soon reduced to mass slaughter as vehicles caught fire, some exploded, and both soldiers and horses died in the midst of great carnage. In its subsequent report of the battle, the RAF called this 'the Day of the Typhoon' and the reputation of this fighter-bomber in its ground support role was indubitable, since they were credited with destroying 89 tanks, 56 tracked vehicles and over a hundred motor vehicles of various descriptions. To some commanders, in both the RAF and the USAAF, this was convincing proof of the invincibility of air-power, even in a role which many of them still saw as almost demeaning when compared to the massed bomber raids on German cities — their real objective. To others, though, these claims seemed very wide of the mark. The Allied Expeditionary Air Force's own conclusions on the training of pilots for close support work were that rockets could not be delivered accurately, by average pilots, against targets such as these on the winding roads and tracks that laced through this dense countryside. Sufficient scepticism certainly existed within 21 Army Group to justify conducting an independent survey of the scene of the battle, so Schonland instructed Major Pike to carry out a full investigation.

Pike spent eight days from 12 August examining, in the closest possible detail, every German vehicle of any description in the vicinity of Mortain. To broaden the area of his search as much as possible, an aircraft was laid on to do an aerial survey of the scene of the battle. His findings were illuminating. The total number of armoured vehicles of all types discovered was 78, while only 30 trucks were found. The discrepancy with the figures claimed by 2 TAF were intriguing enough, but even more remarkable was the fact that only seven tanks showed evidence of being hit by rockets, while two had been disabled by bombs and another 19 had clearly been destroyed by US Army weapons, either artillery or infantry-fired Bazookas. Some bore not a mark on them and had simply been abandoned by their crews in their panic to escape the aircraft, while four others had been deliberately destroyed by their occupants as they would have been trained to do. The fate of only three tanks could not be unequivocally ascribed to some deliberate action or another. Pike's report caused a degree of consternation within the RAF, especially as its own somewhat dismembered operational research section, which had also been to Mortain, had been persuaded by the absence of any cast-iron evidence on the ground that the Germans must possess an extremely efficient tank recovery system! This was a conclusion that was directly contradicted by Pike's report. Where there was no disagreement between them was in the effect that the Typhoons had on the morale of the soldiers below them on the ground. As already noted, their appearance was always a fillip to Allied troops and, equally, a very bad omen to the enemy even if their targeting ability was not quite as trumpeted by some.

Vested interests, and the antagonism that existed between Air Marshal Sir Arthur ('Mary') Coningham,[*] the commander of 2 TAF,

---

[*] Montgomery always referred to Coningham as 'Maori', probably more accurately, since he was a New Zealander.

and almost everyone else meant that the battle for credibility between Schonland's operational researchers and their associates from 2 TAF led to many a protracted argument and not a little acrimony. Further evidence, after a similar battle in the area around Falaise, where the Germans narrowly avoided being trapped in the 'Pocket' that Montgomery was trying to close between there and Argentan, confirmed the scientific findings. Here, in a region of such devastation that they referred to it as 'The Shambles', the complete No 2 ORS team laboured amongst hundreds of damaged and destroyed vehicles, with the almost over-powering stench of decaying corpses of horses and men hanging in the hot August air. Their findings provided yet further evidence that by far the greatest percentage of armoured fighting vehicles lying around them had been destroyed by 'land action' and not by the attentions of the Air Force. Only 11 out of 171 vehicles of all types showed clear evidence of having been hit by bombs or rockets.

Schonland reviewed the evidence presented to him and stoutly defended the claims made by his operational researchers against the forthright assertions of the RAF, in whose eyes the Typhoon was well-nigh invincible . Whereas it was not his intention to seek a confrontation with 2 TAF—indeed he very actively sought the closest possible collaboration with them—he was not about to be browbeaten by anyone. When No 2 ORS's findings at Mortain were so aggressively called into question, he countered with vigour that 'unless there were fairies in Normandy who could remove a large formation of tanks' it was now surely time to accept the evidence: fact not fiction must determine future actions. To this 2 TAF gave their somewhat grudging support, since the facts presented to them were becoming incontrovertible, but they stubbornly resisted any suggestion that changes in weapons or tactics might follow. However, it was obvious that only by cooperating would any useful purpose ultimately be served by operational research and Basil Schonland's persistent but firm encouragement led finally to the setting up of a joint investigative team for all further air operations against ground targets [37].

By late August there was an air of some expectancy in many quarters that the war might be soon be over, as the Allied armies relentlessly pushed the Germans out of France. Certainly, within 21 Army Group the mood was very positive. According to the intelligence available to Montgomery, German resistance in western Europe was on the verge of collapse [38]. Talk in the Mess was that within a month the war in Europe might be over. In a letter to Solandt at the AORG in Roehampton, Schonland suggested 'evens for Sept. 15th, three-to-one Oct. 15th' [39].

Such was the optimism amongst the Allies in general that Basil Schonland's own contribution to the war effort was seen to be nearing its conclusion, at least in the mind of the South African Prime Minister. In fact, as early as July, General Smuts had requested that Schonland should relinquish his post at 21 Army Group and prepare to return to South Africa, but de Guingand was unhappy with this and immediately requested a short postponement until the end of October. Schonland himself was very keen to remain where he was for, as he told Ismay now back in Johannesburg once again and, as Captain Mrs Schonland, was working for Lt Col Hodges at the BPI: 'I like the people and feel I can do useful work. I don't think I have been happier in my work at any time during the present war than I am at present'. It was not just the intellectual challenge as scientific adviser that appealed to him but the personal challenge to prove to himself that he had the moral and physical courage to put himself at risk and then to know that he had done so. He continued [40]:

> I like a life which forces me to do fairly uncomfortable & not very
> safe things at short notice. I like also to prove to myself that I am
> not a coward, even though I do not have to do anything outwardly
> brave. I find it has an extraordinarily invigorating effect on one's
> personal relations with others to know (unconsciously) that
> yesterday I walked through a minefield or lay in a crater while we
> were being shelled or bombed. You can't really be afraid of people
> after a few experiences like that, because it is not people you are
> ever afraid of but really of yourself.

These occasional sentences from husband to wife provide just rare glimpses of the depths of the Schonland character, but fascinating they are. To those around him, especially David Hill who saw him most often, Schonland was very much at ease amongst his military colleagues at 21 Army Group. The General's Mess had a most convivial atmosphere and it was clear that Schonland not only enjoyed the company there but was very much accepted as a key member of Montgomery's senior staff. His opinion was sought frequently on all manner of scientific and semi-scientific issues that affected the huge military machine that was grinding towards the German plain and the cities of the Ruhr, and his thoughtful but down-to-earth replies were always well received. The confidence that de Guingand and his colleagues had in Schonland also meant that the work of No 2 ORS was given increasing credence by soldiers who, if not hostile, had certainly been rather sceptical of the value of the work done by scientists masquerading in uniform. As Brigadier Sayer expressed it in his masterly post-war review of army radar: soldiers, with their gain in understanding, soon realized 'that scientists were not necessarily wild and woolly nuisances, but [were] capable of helping them in action' [41].

This was certainly true of the men under 'Johnny' Johnson in No 2 ORS, even though he himself was given to some erratic behaviour in his manner of command that caused a degree of concern back at Roehampton, where Omond Solandt and many of his colleagues rather fretted at their seeming isolation from the action and harboured the impression that Schonland had gone off and forgotten them. Solandt began to bombard Schonland with letters and requests to get various members of the AORG across to France, including himself. This eventually served to try Schonland's patience, particularly when Solandt, writing in his rather long-winded style, informed him how he had to explain to 'these local enthusiasts' why under present circumstances it was impossible to do any operational research in 21 Army Group. The 'local enthusiasts', amongst whom was the energetic Frank Nabarro, had seized on Schonland's stated intention to send for them once he had assessed the situation in France. A considerable time had passed since the Brigadier left them but no invitations were forthcoming. Schonland's reply was rather irascible (it later transpired that he was recovering from a bout of 'flu) but his ire was up at the suggestion that he had seemingly failed in his task to introduce appropriate science to Montgomery's commanders in the field. 'What the Hell is biting you?' he asked Solandt. 'First you wrote me an extraordinary letter about being eaten by cannibals or going native or something. Today I get another ... I wish you would now explain all this, if you are serious about it'. He then gave Solandt chapter and verse about the performance of Johnson's section and a footnote about his own [42].

> Sargeaunt's work on tank penetrations has been of considerable value and he has done a lot besides which will never get on paper, lecturing to Tank formations & so on. They [No 2 ORS] are now with the Canadians & doing, I hope, battle analysis while Tony [Sargeaunt] is up in Holland with the Armour. Their work on bombing in close support and air attack on transport has been very laborious and will be historic.
>
> I have taken up the question of reporting on Overlord with the Chief of Staff & will let you know in a few days what the form is. It is only now possible to consider a small analytical team with me. I've been spending some time on V1 and V2 sites, just for general information. They certainly intend to attack London in a very big way.*

---

* Schonland and Hill visited the site where a malfunctioning V2 had fallen to earth without exploding in a field near Waterloo with the only casualty being a cow. There, Schonland was delighted to discover that he could converse in Afrikaans with the local Flemish-speaking population, who thronged around the downed missile.

It was by now fairly common knowledge that Schonland would soon be leaving 21 Army Group to return to South Africa and the question of his successor had been exercising many minds, not least Solandt's who, quite naturally, saw himself as the most likely contender for the post of Scientific Adviser. This was no idle speculation on his part because, according to yet another of his many letters to Schonland, he said that the scientific triumvirate of Chapman, Ellis and Kennedy who were advising the Army Council had told him as much. Such, though, was not the view within 21 Army Group and de Guingand was not to be imposed upon by any mandarins in London. Rather, he sought Schonland's views about his successor and set great store by them.

Schonland had by now formed a very clear perception of the role and function of the Scientific Adviser to a Commander-in-Chief and he saw the personality of the incumbent and his standing within the General's Mess, as the key elements in any appointment. Without military experience, preferably in a campaign, the rank to go with it and the manner and presence to win the confidence of experienced staff officers the Scientific Adviser would be nothing more than 'a dressed-up civilian'. Based on these criteria there was no obvious successor waiting in the wings. Schonland took soundings from amongst the members of No 2 ORS and heard their unsolicited views as well. Solandt, for all his sincerity and conscientiousness, tended to be pedantic and was certainly given to verbosity; the Generals and Brigadiers would soon lose patience with him. On Schonland's advice de Guingand solved the problem by deciding to appoint Lt Col Johnson as the 'de facto' Scientific Adviser without giving him such a title, while leaving him in command of No 2 ORS, and thus still under the command of the BGS(SD) [43]. As with all compromises it was not the ideal solution but it was by far the most pragmatic in the circumstances.

The seemingly unstoppable Allied tide of armour and infantry, backed by the formidable combined air forces of the RAF and the USAAF, which were poised to sweep all before them just a month before, suddenly encountered resistance and then a counter-attack from the Germans that both called into question some of the intelligence provided to Montgomery, and his own judgement when assessing it. As a direct result of mounting public opinion in the United States that the US Armies should be led by an American general, particularly as the perception had been given that they were doing all the fighting, but also because Montgomery's own penchant for self aggrandisement had so antagonized his allies, Eisenhower assumed direct and overall command of all the Allied armies in Northern France on 1 September. Montgomery was promoted Field Marshal but was effectively demoted as well, for under him now in 21 Army Group were just the British Second Army and the First Canadian Army. His objective of a single, concentrated thrust north of the Ruhr, and then onwards via the north German plain to Berlin, had been rejected by the Supreme Commander in favour of a two-pronged assault, with 21 Army Group, assisted by the First US Army, following the northern route, while the newly-landed Third US Army under Lt General George S Patton would advance on Berlin across the Rhine and through the Saar valley. For all Montgomery's pique at these developments, all evidence pointed to the Germans being in total disarray and any disagreement between Allied commanders seemed of little consequence.

By 4 September Montgomery's forces had reached Antwerp and vast swathes of France and Belgium had been liberated without opposition; the end of the war seemed well within sight. Ahead of them though lay three rivers, the Maas,* the Waal and the Neder Rijn and, though intelligence sources had either missed it or had disregarded its capabilities in the light of reports of a routed German army, there too was II Panzer Korps, refitting after hastily disengaging from Normandy. The operation to secure the river crossings and so open up the advance on the Ruhr was to be centred on the town of Arnhem and its road and rail bridges across the Neder Rijn just to the south. A massed airborne assault by paratroops and glider-borne infantry would capture the bridges and hold them until XXX Corps of Dempsey's Second Army, advancing along the road through Nijmegen, arrived at Arnhem and so established the bridgehead for the march into Germany. That was Montgomery's intention: what followed as Operation Market Garden became a tragic débâcle and ended in surrender and retreat. The story of this epic battle in the annals of the British Army has been told at length elsewhere and so will not be repeated here [44]. However, since the inadequacy of the radio communications is often cited as one of the reasons for the calamity that befell the First Airborne Division that aspect will be examined, particularly as it relates so directly to the sterling efforts just a few months before of Basil Schonland and the AORG to introduce new ideas on radio communications to the War Office.

*   *   *

On 17 September 1944, the largest assault by airborne forces yet seen in warfare began with the departure from various airfields in England of over 1500 transport aircraft and nearly 500 towed gliders. The gliders, carrying infantry and light vehicles of the British First Airborne Division, landed, as intended, in broken and wooded country 10 km west of Arnhem. Farther south, paratroops from the American 82nd and 101st Airborne Divisions landed between Eindhoven and Nijmegen: in all over 35 000 airborne troops arrived on or near their intended drop zones over a period of three days. Finally, after being delayed by

---

* The River Maas in Holland is known as the Meuse in France and Belgium.

bad weather the First Polish Parachute Brigade landed some few days later.

All communications, and good communications were surely crucial, between the various formations of this disparate force deployed in the area were based, as was British policy, on high-frequency HF radio. The actual radio equipment assigned to the operation was very limited, both in the number of sets allocated, because of the load-carrying limitations of the aircraft, and in terms of the power output capabilities of the sets themselves. First Airborne Division were equipped with four types of set: No 68P, which provided both speech (RT) and Morse-code (CW) between 1.75 and 2.9 MHz at a power output of 250 mW; No 22, RT and CW between 2 and 8 MHz at a maximum of 1.5 W; No 19(HP), the 'high-power' version of the most ubiquitous of all British wartime radio transceivers, had the same frequency range as the No 22 but was capable of about 15 W on RT and 50 W on CW; while the No 76 was a CW-only transmitter intended for long-range communication between 2 and 12 MHz at 9 W output. It required a separate type R109 receiver, which made it very much a fixed installation. The No 68P, by contrast, was designed to be carried by one man and, with its 4 m rod antenna, could be operated when on the move but its output power was particularly low. It was, because of its relative portability, the only radio equipment in the British Army intended primarily for use by airborne soldiers but was never meant to be used beyond a range of about 8 km using the ground wave[*] mode of propagation. All the other sets had the capability of also exploiting sky wave propagation, if appropriate antennas were used, but were much larger and required the use of large 12 V accumulator batteries which, together, allowed only static or vehicle-based operation [45].

Ideally, in order to exploit the element of surprise to the full, the first airborne troops should have landed as close as possible to their objective—the Arnhem bridges. However, the topography of the land made this impossible and so the First Parachute Brigade were all of 13 km away, with the prospect of a forced march ahead of them. The task of dashing to the bridges with all speed fell to the First Airborne's Reconnaissance Squadron equipped with heavily armed Jeeps and 28 of these set off just two hours after the first British paratroops had touched Dutch soil. They were supported by a detachment of the 2nd Battalion that moved off in that direction on foot. Because of the impending retaliation by the Germans it was absolutely crucial for the headquarters of the

---

[*] The 'ground wave' follows the contours of the ground itself but is rapidly attenuated. The 'sky wave', if launched at appropriate angles of elevation, will propagate over considerable distances after reflection from the ionosphere but usually with the existence of a 'skip zone' beyond ground wave range and the point of signal return from the ionosphere.

First Airborne Division at Groesbeek, some 8 km south of Nijmegen, to maintain contact with these lightly armed paratroopers charged with capturing and holding the bridges at Arnhem, about 20 km away, until such time as the armour of XXX Corps arrived from its start-line some 100 km to the south, but problems occurred almost from the outset.

The numerous accounts that have been written about the battle of Arnhem place considerable emphasis on the allegedly poor performance of the radio communications equipment used by the British forces. In the main, these descriptions lack technical depth and ascribe the woefully inadequate communications to a combination of low-powered transmitters, the sandy soil and the heavily built-up and wooded areas around Arnhem, thus implying only ground wave propagation. Major Lewis Golden, the adjutant of First Airborne Divisional Signals during the operation, attempts to set the record straight in his thoughtful account by discussing the many issues that contributed to the failure of Market Garden, with poor signals performance being only one of them. But poor radio communications were a major factor underlying the affair and there were two fundamental reasons for this. Golden alludes to one, the antennas used; but he makes no mention of the other, the sunspot count. Since the distances, in the main, between the various elements of the forces converging on Arnhem greatly exceeded the ground wave range it was vital for sky wave propagation to be used, but this had to be at almost vertical angles of incidence because of the relatively short distances involved. To be effective, though, this mode of operation requires careful choice of the frequencies used by day and by night, *as well* as the use of appropriate antennas at each end of the link. None of these factors seemed to have been given the necessary attention during the planning of Market Garden, and this is particularly ironic because they had been considered in detail by the AORG just months before, and the two definitive reports on the communications problem and proposals for its solution had been written by Schonland himself in September and November 1943, and were given wide circulation [46]. The abstract accompanying his second report summed up succinctly the situation soon to prevail at Arnhem:

> The Report points out that wireless communications by means of ground-waves in the H.F. band is a particularly difficult problem for Army mobile sets. *In some theatres of war the ranges are bound to dwindle to insignificance.*[*]

The failure of communications at Arnhem — for failure it surely was, since crucial situation reports and orders did not get through at many critical times during the week-long battle — can be traced to a combination of the sunspot cycle and the use of completely inappropriate antennas. The

---

[*] Author's italics.

nominally 11-year solar cycle reached its minimum in February 1944. The implications of this natural phenomenon on radio propagation at HF were well understood. Most important of all was the need to use significantly lower frequencies to communicate over a given distance using sky waves than would have been the case if operating in the vicinity of the sunspot maximum some five or so years previously or in the future. In July 1944, when the sunspot-count had hardly begun to rise from its February minimum, the AORG issued to a large group of recipients including, amongst others, the Director of Signals, the School of Signals and the Military Signal Officer at Combined Operations HQ a report entitled 'Frequency Prediction for Short and Medium Distance Sky-Wave Working during September 1944' [47]. It listed, at hourly intervals, the upper and lower frequencies best suited for reliable communications over sky wave paths in Western Europe at transmitter power levels of 5 W and 100 W. In view of the equipment used at Arnhem, only the lower power level was relevant. To maintain communications over a 24 hour period would require the use of frequencies as low as 2.2 MHz around 4.00 a.m. rising to 4.7 MHz at about 5.00 p.m. Any frequency higher than that would have penetrated the ionosphere and communications would have been impossible. What is not mentioned in this AORG document, but is undoubtedly implied by the propagation mode involved, is that successful communication is dependent upon the use of an appropriate antenna which, in this case, should be the long horizontal wire of the type recommended previously by Schonland. The 4 m vertical rod that was used specifically on the orders of Major General Roy Urquhart, in command of the British forces, was entirely inappropriate — in fact, it would have been downright useless.

General Urquhart's order was fully justified under the circumstances, since it was intended to ensure that the mobility, and indeed the safety, of his troops was in no way compromised by the need to attach the necessary horizontal wire antennas to suitable supports while exposed to enemy fire [48]. No-one could possibly have done that under the hail of lead that filled the Arnhem sky that day. Later, when the battle became attritional and the First Airborne Division was fighting a desperate rearguard action as it retreated, both rod and wire antennas were tried, but with limited success. As Golden describes it: 'The distance between divisional headquarters and corps headquarters (about 20 km) had proved to be too great for the use of the ground wave, but not enough for the use of the sky wave. It was not an unknown phenomenon. Greater care in planning might have avoided it, although it was seldom possible to anticipate the characteristics of a particular radio frequency in a unique location' [49]. Greater care in planning would indeed have avoided the problem, for *two* solutions had already been proposed to the War Office by the AORG. Clearly, the Royal Signals troops at Arnhem had not been made aware of this work, which indicated precisely how communications could be provided within that zone between the ground wave limit and the arrival of the more usual, obliquely incident sky wave. Strangely enough, there was one military unit there that was able to communicate. It was the GHQ Liaison Regiment known as Phantom. This small group of highly trained specialists operated in an intelligence-gathering role in which good communications were of paramount importance. They evidently knew all about how to launch sky waves at vertical incidence 'using a special kind of antenna', referred to in one account as a 'Wyndham' (better known as the Windom, a long and predominantly horizontal antenna) which was well-described in the Royal Signals Training Pamphlet Part 2 and ideal for just the mode of propagation Schonland had indicated [50].

The other AORG proposal was stillborn, since it had encountered the deaf ear of the War Office. It involved the use of VHF radio communications that had much exercised the mind of Schonland and the expertise of Major E W B Gill at Roehampton the year before, and would have been a godsend at Arnhem. Schonland's AORS3 had proved conclusively that VHF could provide more than adequate communications well beyond the line-of-sight and within obstructed areas such as those in and around Arnhem. But such are the ways of war, and so embedded was the Law of the War Office (which General Pile had warned Schonland about two years before), that it took the tragedy of Arnhem before either of these recommendations was implemented with any degree of urgency.

*   *   *

Brussels was liberated on 3 September and 21 Army Group's Main HQ soon established itself there in rather more comfortable surroundings than they had endured until now. Schonland and David Hill shared a suite of rooms in an hotel and much enjoyed each other's company, despite all Hill's attempts to persuade the Brigadier to reduce the number of cigarettes he smoked per hour! Schonland agreed to try but soon found himself a prisoner of the habit. No 2 ORS were also in the city and for the first time since its members had left England two months before, they found themselves in close touch with their HQ and its senior officers. Whereas in England before D-Day many of them had the feeling that they were treated with tolerant indifference by the General Staff, the situation had changed dramatically since their arrival in Normandy. Their battle-damage reports, and their interrogation of German prisoners and local civilians to glean whatever they could about the effects of bombing on the morale of the enemy and on the dislocation of the inhabitants, were seized on particularly by Brigadier Herbert the BGS(SD), because he said 'it was only the O.R.S. that told him what really went on and what really mattered'. However, the gradual strengthening of their position had not a little to do, as well, with the efforts of the Scientific Adviser, for not only had Schonland persuaded Herbert to allow them maximum flexibility in the way they operated, and so tapped the rich vein of scientific independence on which they all thrived, but he had also taken a great personal interest in them and their work.* To all these young officers Schonland appeared as both a friend, who was always willing to make time for them, and as a powerful interlocutor amongst the generals and brigadiers in Montgomery's head-quarters [52]. But such easy entrée to the nerve centre of the Army Group had not long to last, for Schonland's date of departure from 21 Army Group, and from the Army itself, had been agreed because Smuts wanted him in Pretoria by the new year.

But before leaving there were some urgent matters that needed attention from Solandt at Ibstock Place. All concerned methods of deal-ing with concealed enemy gun and mortar emplacements which had proved to be extremely troublesome during both the advance of XXX Corps towards Arnhem and during the battles fought by the First Canadian Army as it secured the Channel ports of le Havre, Boulogne and Calais before pressing on to the real prize, which was Antwerp. Tanks and infantry had suffered severely as they advanced past care-fully camouflaged guns concealed in the woods, while the increasing use by the Germans of an array of mortars, particularly the multi-barrelled *nebelwerfers*, were taking a heavy toll and severely affected morale because of their high rate of fire and silent approach. Some means of detecting their presence was required urgently and Schonland wrote to Solandt asking him to 'put some heat on these questions', but threw in his own suggestions as well. He ruled out sound ranging and radar for purely practical reasons but much favoured flash and heat detectors or even 'any optical device, or such, which will distinguish dead chlorophyll' In the same letter he sought to clear the way for his successor by informing Solandt that Lt Col Johnson 'is now doing an excellent job and I have confidence in his ability to work on the right lines'. This was not really what Solandt wanted to hear because he had Schonland's job very much in his sights and never missed an opportunity to try and further his cause. But the decision had been taken and Johnson it was who would step, somewhat gingerly, into Schonland's shoes.

*　　*　　*

The time had come for Basil Schonland to bid farewell to his many friends in uniform, as well as those who had served the armed services so

---

* Some years after the war Schonland told an audience of engineers and students at Wits how Montgomery, 'who was so satisfied with what they had to tell him ... that he would be glad if he would bring over as many more of these queer creatures, within reason, as he liked' [51].

admirably as civilian scientists pressed into unaccustomed roles by the demands of modern warfare. There was one man to whom he owed a special word of gratitude for the opportunities that had come his way and that was John Cockcroft. Basil duly wrote what was clearly a letter between old friends. Though undated, its contents suggest late September 1944 [53].

> Main HQ
> 21 Army Group
> B.L.A.

Dear John,

D + 3 months is over & we are on the borders of Germany & even inside it. I doubt if you could have foreseen that when you suggested I should stay till D + 3 months. Or did you? I think I must stay a bit longer; it is fascinating to see armies carry out operations as planned instead of, as in most of the last war, face dismal failure after dismal failure.

All your particular babies have played their part; though AA has not been called on to do much here it certainly paid dividends with Tim [Gen. 'Tim' Pile] who finished up in a blaze of glory, well-deserved. One of the later things you started, down on the plain, has done extremely well too.

The O.R.S. under Johnson began stickily but has now achieved quite a place in the sun. I have battled for it successfully. But I think the thing I am most proud of is that I have imported two efficiency experts to report on signals procedures & traffic control in signal offices! They have been well received.

The Americans have proved magnificent both in organisation & fighting quality & they are very adaptable and quick. I am most impressed with the staff here and wish I could tell you more about it.

I saw Charles E. [Charles Ellis, SAAC] the other day & from what he said I concluded that Gough has succeeded in getting a ruling from above that whatever reorganisation takes place the man who does the big administration is the senior. So Charles is rather depressed & I fancy that he will fade out of the picture after receiving his reward. In my opinion after trying to be as fair as possible to him he has been a pretty complete failure. I have never had any real help or backing from him & have had to live down a certain amount of prejudice which has arisen from his theories.

But the other side of the picture is that I myself am regarded, I think, as a missionary who has gone native. So perhaps it is not fair for me to criticise the Archbishop who sent me out. Anyway I find the natives most pleasant & intelligent & <u>not</u> unwilling to listen to the true gospel provided it does not involve the psychological & mathematical study of morale. They eat all that kind of preacher. They also would like to boil Zuckerman & Bernal & possibly Blackett. Yes, I think I must have gone native.

297

> Give my love to Elizabeth & the children. If you can let me
> have a line, send it to S. Africa House as I must be back there by
> Oct. 15th at latest. 'Old Global', as he is irreverently called by one of
> my colleagues, wants me back.
> I had meant to finish on this page but I must tell you how
> grateful I am for letting me do something useful over here & for all
> the support & constant help & friendliness you have shown.
> (I except the matter of a chief clerk in 1941!).
> I shall leave for S. Africa with a greater admiration for you
> than I had before & that is saying a lot.
>
> > Yours
> > Basil

and then he added a postscript:

> Ismay is still busy in the Army & is inspecting Radar stations
> to report on the discipline & morale of the browned-off girls.
> Ostensibly she goes on demobilisation matters. She says that Smuts
> is too occupied with world problems to run his Govt properly; that
> people say that all their plans are paper sops to the electorate &
> nothing ever eventuates; that he won't delegate authority to his
> Cabinet. So that I may find it impossible to do what I want to do.
> Fortunately I have the BPI in reserve for the first three years.

This letter is particularly revealing both for its whimsical humour,
showing a side of Schonland that those who got behind the formal
exterior had come to know well and for his views of personalities and
events. That he was most proud of the performance and acceptance of
his two 'efficiency experts' in the signal offices indicates quite what he
saw as the primary role and function of operational research: the study
and analysis of military operations and procedures and the provision of
scientific solutions to any problems that may be uncovered. It was not
the function of operational research to delve deeply into technical matters
that were the province of other specialist organizations. Of course there
had been occasions, most vividly illustrated by his own involvement in
the radio communications problems of the Army, when he gave his
technical experts free rein to conduct experiments that led to improvements in techniques and possibly even to tactics. It is also evident how
well Schonland himself and his ideas had been accepted by those staff
officers with whom he was so closely involved in Montgomery's Main
HQ. He had won their confidence when others, of most impressive
scientific pedigree, had not. Basil Schonland felt very much at ease
amongst the soldiers whose task it was to turn Monty's plans, formulated
in the virtual isolation of his 'Tac HQ', into the operational orders around
which the battle was to be fought. And they felt so with him.

Of course, none of this would have happened had it not been for
John Cockcroft's canny sense of judgement in 1941 when he plucked

Schonland from the obscurity of the SSS to run, successively, the ADR-DE(ORG), the AORG and then, at the very pinnacle, to become Scientific Adviser to the Commander-in-Chief. But that was Cockcroft's way and his views, however taciturn, were never treated lightly. Who else could have persuaded the Ministry of Supply that Schonland was the man to step into Ratcliffe's shoes at Petersham when the demands of centimetric radar in aircraft made it so important that Ratcliffe should return to TRE? After that Schonland's worth was never in doubt and John Cockcroft looked on from many vantage points as his own multi-faceted career moved almost inexorably towards the project that would develop the most fearsome of weapons and the most potent source of power. The Schonland–Cockcroft relationship that had started two decades before at the Cavendish had been cemented into a lifelong bond during the last three years of the war.

We gain a glimpse too of Ismay Schonland's perceptive summing up of the mood of the South African nation, or at least those of European origin. The country had become disenchanted by the long privations of war and was looking to a new future. Of course, the Nationalists had long made much play of the fact that Smuts was an internationalist, more interested in the affairs of the world, particularly the British world, than of those of his own people. 'Old Global'* would have been but a mild appellation for him in some parts of South Africa. The demise of a successful wartime leader was, of course, not unknown, as Winston Churchill himself was soon to discover, but Ismay's political antennae were especially sensitive and Basil had long valued her judgement in such matters. She detected troubled times ahead. At least, as he said, there was always his laboratory at the BPI should the plans that he and Smuts had been hatching ever since 1942 not come to fruition.

On 18 October *The Times* carried a small item under the headline 'Research in South Africa' regarding an announcement made the day before in Pretoria that 'Brigadier Basil Schonland, of the South African Corps of Signals, who is now serving on General Eisenhower's staff, has been appointed head of a new industrial research body for the Union of South Africa'. It was, of course, no secret within the circles in which Schonland had been operating that Smuts wanted him back in South Africa, but now it was public knowledge. Amongst that interested public was the University of the Witwatersrand, for Schonland was still Professor of Geophysics there, but this matter had been carefully considered by Smuts and Schonland during their frequent meetings in London and Roehampton on every occasion that the Prime Minister visited England. As ever, Schonland had been particularly careful to ensure

---

* It was probably Brigadier 'Bill' Williams, who referred to Smuts in this jocular though, indeed, perceptive way.

that his own position, and especially his family's positions, were secure before he accepted any appointment. In this instance that security lay in retaining his Chair at the University, at least on a part-time basis, and Smuts had readily agreed to this on terms yet to be worked out with the University. So the die was now cast and Basil Schonland prepared to leave for London, where he would spend a month or so tying up loose ends at South Africa House, before returning home. Two days before leaving Brussels he wrote his last letter from 21 Army Group to Solandt at Roehampton. For all the Canadian's occasionally rather irritating ways, he and Schonland remained good friends and Solandt had generously offered him accommodation at his home for the duration of his stay in London. Schonland declined this but promised to visit the AORG as soon as he was able to do so. Then, with characteristic thoughtfulness, and as magnanimous as ever, he asked Solandt to pardon his truculence for replying with such vehemence to what must have seemed at the time like a litany of lament from the AORG. As if in compensation he announced that Frank Nabarro, the most energetic of Solandt's men, could indeed visit 21 Army Group, and even though No 2 ORS had not shown much enthusiasm, Johnson had agreed to get him out [54].

Basil Schonland left Brussels for London on 30 October. As a parting gift to David Hill who, as his sole staff officer, had served him so well he left his personal copy of the *Oxford Dictionary of Quotations* which he inscribed 'To David Hill, with best wishes to recall $D - 40 \rightarrow D + 150$, B J Schonland, Brussels 1944'. Soon to follow him was a letter from Montgomery's Chief of Staff, Major General Sir Francis de Guingand, the man with whom Schonland had worked most closely throughout those 190 days. Given the trepidation with which he had entered the portals of Montgomery's headquarters that short while before, this letter was indeed an accolade [55]:

10th Nov. 1944

My dear Schonland,

Although I have expressed to you personally our thanks for all you have done for us whilst you were a member of this Headquarters, I feel I would like to say something in writing.

As you know, when the suggestion was first made of having a Scientific Adviser attached to this HQ, we were all a bit frightened of finding a very learned, but theoretical professor in our midst. As luck would have it, we got the learned professor but also one who contributed to a tremendous degree towards the practical solution of many of our problems.

I still think however, that this may be an exception—to find someone of your standing, with his 'feet on the floor'.

We were all extremely sorry to see you go, and all we can do is to wish you all the very best of fortune and happiness in your new task. I only hope that I have the opportunity of meeting you in your own country at some not far distant date.

Yours ever,
Freddie de Guingand

More formal and official recognition of Basil Schonland's service was not long in coming either. On Friday, 1 December 1944 the Supplement to the *London Gazette* carried the following announcement:

The KING has been graciously pleased to give orders for the following promotion in theMost Excellent Order of the British Empire, in recognition of gallant and distinguished services in the field: —

To be an Additional Commander of the Military Division of the said Most Excellent Order: —

Brigadier (temporary) Basil Ferdinand Jamieson Schonland O.B.E., South African Forces.

So, the 'parting gift' for service to King and Country in 1918, that had so delighted Selmar Schönland, was now elevated to the CBE(mil) as Basil Schonland was about to resume his career as a civilian and a scientist.

CHAPTER 15

WHEN SMUTS CALLED

'Whether the Director should be a scientist, historian or sociologist, he must have a statesman's vision of the future of South Africa and such a vision should not be impeded by a soulless mass of technical details'. These were the ringing words of Professor H R Burrows of the Natal University College in a memorandum tabled at a meeting of the National Research Council held in Pretoria in November 1942. They helped to defuse a near explosive situation and certainly focused the attention of those present on their immediate future. In 1938, the old Research Grant Board set up at the end of the First World War had been reconstituted into two bodies—a National Research Council, essentially a policy advisory body, and a National Research Board charged with administering research grants and translating the decisions of the Council into action. The objectives were good, their implementation under the chairmanship of Dr J Smeath Thomas, Master of Rhodes University College, less so. Then came the war, priorities shifted dramatically and many of the Board's members found themselves otherwise engaged. Amongst those left there was a degree of restlessness and much concern at their sense of impotence and, in the mind of one, even an air of mutiny. Illtyd Buller Pole Evans, a botanist and a close friend of both Smuts and Selmar Schönland, had been a pioneering figure in South African botanical research. By nature a man driven by the need to get things done, if at times somewhat intemperate in his manner, Pole Evans not only castigated his colleagues for failing 'to carry through expeditiously the work entrusted to them' but then called on them all to resign 'to leave the Minister to start *de novo*' with the appointment of a new organization entirely [1]. Not surprisingly his colleagues failed to vote for their own demise but the need for change was evident. The war, and the needs of the armies 'up North' for ammunition, weapons, vehicles, boots and a myriad other impedimenta of battle, had placed huge demands on South Africa's manufacturing industries, which responded magnificently, but the absence of any structured research organization to buttress them was a glaring omission that had to be redressed. Just such a body had been the vision of H J van der Bijl,[*] brought back to South Africa in 1920 by Smuts from Western Electric in the United States where he had made discoveries of fundamental importance in the field of thermionics, but his proposals fell on fallow ground when, in 1924, the new government under General Hertzog clearly had other priorities [3].

Nineteen years later, with Smuts once more Prime Minister and with the increased urgency caused by the pressures of war, the search for the right man to take charge of South Africa's scientific destiny was renewed with some vigour. By 1941 Smuts was moving within the inner circle of the British government and so it was natural for him to seek the guidance of colleagues like Lord Hankey in order to find the man best suited to this task. Hankey, chairman of the British Cabinet's Scientific Advisory Committee had, as we have already seen, already identified Basil Schonland as someone worth watching when he saw, from a distance, how operational research had begun to take wing under Schonland's leadership at Petersham. By late 1943 it was certainly firmly fixed in Smuts's mind too that he had found the country's future scientific supremo and this he confidently reported during his absence abroad to J H Hofmeyr, acting Prime Minister and the *wunderkind* of South Africa.

Hofmeyr, van der Bijl and Schonland — a triple diadem indeed! One can only wonder at the path that South Africa might subsequently have followed had Hofmeyr's untimely death at the age of just 55, and van der Bijl's just two weeks before, not coincided by some strange twist of fate with the defeat at the polls of Smuts himself in 1948. But such a triumvirate was not to be.

In 1943, Schonland, now acting very much as Smuts's unofficial scientific adviser in England, extricated himself occasionally from events at the AORG to make inquiries about the way in which scientific research was organized at a national level in England itself, and particularly in the two Dominions, Australia and Canada, that in many ways were comparable with South Africa. Both had started similar ventures: Australia in 1926 and Canada, even earlier, in 1916 and their experiences would prove useful if only to highlight pitfalls to be avoided along the way. For example, the charter of the Canadian body, on coming into being, made it difficult for it to take control of existing research organizations, and since there were already a number of these in South Africa, Schonland favoured the Australian model. This view he communicated to Smuts and backed it up in December with Sir Henry Tizard's firm

---

[*] Van der Bijl directed his considerable intellect and energy towards the establishment of the Electricity Supply Commission (ESCOM) in 1922, the South African Iron and Steel Corporation (ISCOR) in 1928 and the Industrial Development Corporation (IDC) in 1940. In 1939 he became Director General of War Supplies, an operation so successful that South Africa was soon dubbed the 'Repair shop of the Middle East' [2].

conviction that the CSIR* in Australia provided the best scheme for South Africa to follow [4]. But others were not as convinced. Dr S H Haughton, Director of the Geological Survey, wrote to Schonland in October 1944, just as he was preparing to leave 21 Army Group prior to his return to South Africa. He raised two points which he hoped Schonland would consider. The first was that the proposed National Research Council (an interim name pending a final decision) 'should not fall under the control of the Public Service Commission but should become a "body corporate", as is the Fuel Research Board at the present time', and the second was his view that the Australian State governments were in no way deflected from setting up their own research bodies in various fields, if they saw the need, even though the national organization existed, allegedly, for the use of all. By contrast, he felt, the Canadians had good liaison even with those research departments which they did not control. Haughton therefore believed the Canadian approach should be followed but conceded that, whichever was ultimately selected, the success of the South African venture would depend 'largely on the calibre of the man who is at the head of it' [5].

It was clear by now that the calibre of the man Smuts had in mind was beyond doubt. Schonland's fame had spread around South Africa and the members of the National Research Board were well aware of it. At their General Meeting on 14 February 1944 they confirmed a previous minute in which 'Brigadier Schonland indicated that he was prepared to accept the appointment as Chief Executive Officer on a three-year basis', with a review after that. But they noted as well, with some concern, that he also wished to 'continue his association' with the BPI by remaining as its Director. And that was not all. The Brigadier was insistent, amongst other things, that the Chief Executive Officer must have direct access to the Minister responsible for scientific matters, as was the practice in England, Australia and Canada. He was not prepared to operate within some tortuous process dictated more by the civil service than by science. The Board, of course, would be kept fully informed [6]. Smeath Thomas and his colleagues were a little taken aback. However, they considered Schonland's proposals carefully and remarkably dispassionately. While agreeing that the Chief Executive Officer should have unhindered access to the Minister, they balked at the idea that Schonland, or indeed anyone else occupying such an onerous position, could possibly devote the necessary attention to its affairs while also functioning in the dual capacity he wanted. It was vital, they believed, that the Chief Executive Officer be appointed on a full-time basis with no other commitments or

---

* The Australian Council for Scientific and Industrial Research (CSIR) was renamed the Commonwealth Scientific and Industrial Research Organisation (CSIRO) in 1949.

responsibilities. And they surely had a point. But then the matter did not even rest there; they next considered the man himself.

Of particular concern was the very idea that a scientist as eminent as Schonland would be lost to research by becoming an administrator, albeit of the body serving the scientific interests of South Africa. That he was the most able and suitable candidate there was no doubt, but the Board believed that were he to be appointed it would be 'a waste of an outstanding scientist capable of making great contributions to physical science'. Such ability and certainly such international distinction were rare in South Africa and, rather than see him embroiled in administration, they queried whether 'it would be better to create a post where his talents as a researcher could be more directly utilised than as Chief Executive Officer of the National Research Council' [7].

Though well-intentioned, these views were almost as unrealistic as they were impracticable. Scientific progress depends as much on the resources being available to conduct the research as on the people who do it, and this demands that someone of exceptional scientific ability should be at the head of any research organization worthy of the name. If proof were required, they need have looked no further than the remarkable mobilization of scientific effort during the war to find examples, in abundance, of brilliant scientific minds occupying positions within the administrative stratosphere. In their defence, such information was by no means common knowledge in South Africa as the war entered its dying days in Europe; but Smuts knew. And he also knew what part Schonland had played, even before the first shots had been fired, and so he was not in the least deflected by these arguments and simply ignored them. Hofmeyr, his Minister of Education, supported him while the rest of the Cabinet were so much in awe of them both that no dissenting voice was raised.

In the meantime, Schonland himself had not been inactive. After leaving 21 Army Group he spent a month at South Africa House during which time he sought out Appleton, Blackett, Cockcroft and Tizard on matters to do with his imminent new career in South Africa. He also visited the AORG, to bid farewell to his numerous friends and colleagues there, before eventually leaving England by air for home on 4 December 1944. The journey was punctuated by a stop-over in the Middle East and, then, after four-and-a-half days of travelling, he touched down in Pretoria to a joyous reunion with Ismay. Before setting off for Gardiol, the family retreat, and the holiday that they had been promising themselves for nearly four years, there was one urgent matter that he had to attend to: he had to see the Prime Minister. Without the absolute agreement of Smuts to what Schonland called his 'basic principles' he would be unable to accept the invitation to change the whole structure of scientific research in South Africa. These principles were those that he had previously set out for Smeath Thomas's Board: the continuation of his Directorship of the BPI; the establishment of a CSIR outside the Civil Service; the guarantee of adequate financial support to enable him to do so and, finally, his own direct access to the Minister. Most pressing of all was the first, for on it hinged all the others. Now it was vital that he obtain the Prime Minister's formal sanction before pressing ahead with anything else. As Schonland subsequently told Sir Keith Hancock, Smuts's biographer: if the Prime Minister was convinced by the power of argument he would concede and offer his considerable support and backing, but 'it was my funeral if it didn't work' [8]. In this instance Schonland saw the BPI not only as his insurance, should Parliament or anyone else make the CSIR plan untenable, but he also saw his active involvement there as a duty that he owed to its staff, the University and, most particularly, to Bernard Price.

The BPI, opened with some grandeur by Smuts himself just a year before the outbreak of war, hardly had time to spread its wings before it withdrew behind a shroud of great secrecy to become the fountain-head of South African radar—at Smuts's behest. Now, on Schonland's return from the very heart of the radar war, he saw it as his personal responsibility to ensure that the Institute re-established itself as the leading laboratory for the study of the earth and its atmosphere, at least on the African continent if not world-wide, and that would require him still to be at its helm, at least in some part-time capacity. Smuts agreed and the Schonlands duly departed for what soon seemed like another world.

At their next meeting on 25 January the members of the National Research Board were informed by Smeath Thomas that the Government 'had taken steps to create a new National Research Council and that Brigadier Schonland had been offered and had accepted the post of President of this Council, as well as Chief Executive Officer of the organisation' [9]. There was no further discussion.

*   *   *

For a month Basil and Ismay did nothing but lie in the sun, play tennis, swim and fish occasionally. All this, Schonland told Omond Solandt in the first letter he wrote since arriving home, made him feel as if he was in not just another world but on another planet, so different was everything from the one he had just left behind. But being the man he was, with pressing matters ahead, meant that he intended to be fully equipped for his new position and that, close to the heart of government, meant being as bilingual as possible. So he used the opportunity to brush up on his Afrikaans—the few Flemings clustered around the downed V2 some months before being the only people in the last many years with whom he had conversed, however falteringly, in that tongue [10].

On their return to Pretoria, the Schonlands found themselves once more in rented accommodation, while Basil had an office in the Department of Commerce and Industries in Central Street. There, for the better part of the next month, he worked on the details of the Bill that would bring the CSIR into being. His own appointment as Scientific Adviser to the Prime Minister was confirmed[*] and so was his joint appointment with the University of the Witwatersrand when, soon after Christmas, Humphrey Raikes and Dr Bernard Price met the Prime Minister's secretary to discuss the details. These could not have been more simple. Schonland would 'render approximately one-fifth of his time to the work of the Bernard Price Institute'. How this was done would be left to him and the University to decide [12]. In view of his rather unusual dual responsibilities Schonland thought it incongruous for him to occupy a seat on the University Senate and so he wrote to Raikes asking that he be allowed to forego that privilege in the interests of University harmony. He assured the Principal that he was glad still to be a servant of the university, though no longer in the guise of the military officer he had been when he left it in 1941, for he had just been demobilized on 20 January [13]. These matters now settled, Schonland turned his full attention to the details of the draft Bill that would go forward to the Government's Law Advisers and ultimately to Parliament.

His review of the existing structures for scientific research in the country and the general competence of the people actually carrying it out caused him some concern. Vested interests abounded, while the quality of the science was not very good. There were, though, exceptions and most prominent amongst them was the Veterinary Research Institute at Onderstepoort, near Pretoria, which fell under the Department of Agriculture. It had an international reputation under the inspired direction of P J du Toit, of whom much will be heard later. Elsewhere things were less healthy. What was surprising was the sheer proliferation of bodies nominally involved in research that had sprung up in some sort of collaboration with the diverse industrial activities they claimed to support. In a country like South Africa, not overly endowed with scientific resources, it was indeed remarkable to find so much 'science' already under way and its patrons so insistent that they retain their independence and control. At times, Schonland felt quite despondent and saw no way around the impasse that confronted him but Ismay, as always, cheered him up and he ploughed on through a field of institutes and laboratories dealing with such things as diverse as fuel, sugar, leather, timber, fish and even dust.

---

[*] Smuts asked Schonland what title he should have during this interim period and when Schonland said 'Scientific Adviser' the General 'smiled gaily, amused that he should need such a person.' [11]

South Africa's economy, until the demands of war had so changed everything, had been based on mining and agriculture. Locally manufactured goods had been a novelty and those that did exist were of doubtful quality. Suddenly, in 1939, the country's traditional suppliers in England directed all their efforts towards staving off the effects of the Nazi onslaught across Europe and South Africa's needs had to be met through her own industry and resources. During the course of the next six years it would be the production of military equipment that provided the manufacturing impetus now just waiting to be harnessed in the post-war ploughshare industry. But the political machine was at work as well. Both the United Party government and the Nationalist opposition sounded off about the future and industry's role in it, though their longer term aims and ambitions were rather different and beneath it all were some sinister undertones. On the international stage Smuts was being fêted as one of the great wartime leaders and his Preamble to the Charter of the United Nations, written in May 1945, was a work of considerable intellect and great statesmanship. South Africa, along with all its wartime allies, was revelling in the mood of optimism that followed every reported advance of the Allied armies into the heart of Germany and her ultimate capitulation. But Schonland was also looking farther ahead and what he saw worried him. Smuts's government was tired and Smuts, himself, doubly so. The Nationalists, for their part, were exploiting the mounting euphoria by deftly playing to the public mood after quickly disavowing their previous overt sympathy for Nazism [14].

Near the end of February, while temporarily in Cape Town as part of the Prime Minister's Parliamentary entourage, Schonland wrote a long letter in reply to one he had received from John Cockcroft, who was now in Canada, where he was in charge of the Chalk River Laboratory, part of the Manhattan Project. It read in some ways like a progress report but it was also a communication between friends, with snippets of family news interspersed between items of some gravity. He reported on the progress of his Bill that he had so carefully drafted. It was now at the mercy of the politicians: 'Whether it will go through this session I do not know, but it is hoped to put it through'. Then he expressed his concern about the over-abundance of research laboratories in the country. He was struggling to deal with a multiplicity of interests and the 'vanity and obstinacy of government departments' that controlled them even though, in many cases, their quality left much to be desired. It was frustrating and depressing. Turning his attention to matters well away from science he looked ahead to the South Africa of some future time. 'Politically, I feel we are heading for a republic in the British Commonwealth but I don't know how long it will take'. Those words clearly echoed what Ismay had noted during the closing months of the war and how perceptive they turned out to be, at least in one sense if not the other. Here was a war-weary nation under a Prime Minister who, to many of his fellow-countrymen, seemed no longer to be one of them. While Smuts was consumed by the war and all the ramifications that peace held for the new world-order that would follow, his political opponents had manoeuvred themselves carefully into positions of influence and were soon ready to offer the South African nation their vision of that future. The Schonlands and their many liberal-minded friends could see all the signs and portents of a future under a Nationalist government, and they did not like what they saw.

Ismay, whose inclinations veered towards the socialist at times, watched the political barometer in the Schonland household and it was her reading of it that alerted Basil to its trends and certainly sharpened his own perceptions of developments around them [15]. It would, of course, be fifteen years before the Nationalists got their republic and by then the character of both the Commonwealth, and that of South Africa too, would have changed markedly. In closing his letter to his old Cavendish colleague, Schonland parried a suggestion of Cockcroft's that he might write his recollections of having been Scientific Adviser to Montgomery by claiming war-weariness himself. But, as was the case with so many things that John Cockcroft suggested, this was not the end of the matter by any means because a seed had been sown. The letter ended, almost by way of a postscript, with a request: 'There will be a vacant Chair of Physics in Johannesburg at the end of the German war. Local candidates are weak. Who is likely to be available who can run a department with 600–800 students?'

Cockcroft's reply was not long in coming. Again he pressed Schonland to reconsider writing his reminiscences of his time in 21 Army Group. 'I would love to have heard of your experiences with Monty. It's too bad that you aren't writing up your experiences on operational research. I fear that if you don't, the next generation may get quite the wrong ideas! So think again.' He also followed the birth-pains of the CSIR with particular interest. 'I shall like to know about your blueprint for research when you've got it approved. If you take your projected tour and come to London you will have to stay with us and let me hear about it. How do you propose to avoid Civil Service sterility? We have thought anxiously about this in connection with our future places. Amongst our points are administrative freedom; no hierarchy of D-C-Ds in London; promotion strictly of [sic] merit and not by age and experience; throwout [sic] possibilities; interchange of staff with Universities and Dominions.' [16]. These few lines from Cockcroft were the equivalent of paragraphs from the more loquacious and to Schonland they were like the cardinal points of the compass, defining the path which he himself was intent on following in setting up an organization answerable to Government but not governed by its bureaucracy.

Cockcroft had also given some thought to the question of a suitable physicist for Wits. 'Your best chance of getting a good one', he wrote, 'is to go for someone young'. He then offered Schonland a string of names, either junior colleagues of his at the Cavendish or men who had served in the early days of the war installing radars in remote places, or in aircraft, and who were now very anxious to get back to the research they had dropped in order to offer their services in the defence of the realm. As it turned out, Wits made no appointment from the available pool of talent in England, or anywhere else for that matter. Instead, the unsatisfactory situation in the Physics Department continued until 1953 when Schonland was called upon once again to play a part in locating someone to lift it out of the doldrums, and this he did from within the ranks of his own AORG [17]. But we will come to that later.

With the fluidity of the South African political situation very much on his mind, Schonland now gave serious consideration to the attitude that a new government, particularly a Nationalist Government if he was reading the signs correctly, might take towards the CSIR in particular and to research in general. Mere speculation on his part would serve no useful purpose and so, with Smuts's approval, he contacted the Leader of the Opposition, Dr D F Malan, and requested a meeting with him and some of his senior colleagues, particularly those within whose political orbit the CSIR would fall. Malan agreed and the meeting was arranged for 23 April 1945.

Daniel Francois Malan was a former *dominee* in the Dutch Reformed Church, the founding editor of South Africa's first major Afrikaans newspaper, *Die Burger*, and now was the Prime Minister-in-waiting. With him that day to meet the scientific adviser of the Prime Minister were Dr A J Stals and Mr Eric Louw. Stals, a medical man, was the shadow Minister of Education while Louw, an outspoken opponent of South Africa's involvement in what he called 'England's war', was in line for the Ministry of Economic Affairs [18]. Until now, Basil Schonland's exposure to the political animal had been limited almost entirely to his many meetings with Smuts, and their relationship had always been rather special. There were the family ties through Ismay to the Hofmeyrs on the one hand, and the link between Smuts and Selmar Schonland, as kindred botanical spirits, on the other. Now, he was to enter the world of party politics of a particularly intense kind, practised by men who saw themselves following divine destiny. It was also to be the first test of Basil Schonland's recently polished Afrikaans.

After a cordial welcome and formal introductions Schonland explained that he had approached Malan for two reasons: firstly, 'to ask whether, in the event of a change of Government, his Party would be prepared to give any sort of assurance that they would support the CSIR as a national institution' and secondly, he said he was prepared 'to explain any doubtful questions with regard to the Bill'. He assured them too that he was not expressing any personal interest to himself but was thinking solely of the staff that he would engage and the commitments that he would be asking them to make. Malan and his colleagues listened with interest as Schonland provided a general account of the Bill and answered their various questions; some posed in English, some in Afrikaans, as if to establish the parity that they would insist upon between the two. Malan then replied in Afrikaans. He and his Party would require that the CSIR should have *'n Suid-Afrikaanse kleur'* — a South African colour — and should show itself to be dedicated to the interests of South Africa and to no other country. His colleagues, too, seized upon this and insisted that research should not be carried out on behalf of 'overseas countries'. From both the tenor of their remarks and the vehemence of their delivery it was clear that Schonland had touched a fundamental tenet of Nationalist philosophy: not only South Africa first, but South Africa alone, if needs be. Then, in more conciliatory mode, Malan stressed that scientific research, and the CSIR that Schonland envisaged, fell within the framework of the economic council that his Party would create, and he understood the need for an assurance of the kind Schonland was seeking so that staff would have security of tenure and thus the same measure of security as the academic staff at the universities. Schonland could rest assured, Malan said, that he would have his support provided that the 'organisation was properly conducted', and there the meeting ended [19].

Quite what Basil Schonland made of the Nationalists at close range he did not record but he was surely in no doubt that the official view from Pretoria after the election due in three years' time would be very different from the outward-looking vista that so typified the style of government under Smuts: Bill Williams's 'Old Global' indeed.

But there were matters of more immediate importance to worry about. On 16 April Schonland had sent some notes to Hofmeyr, once again acting Prime Minister during Smuts's absence abroad. These set out in some detail the purpose of the Bill that would launch the CSIR and provided background information about the existing scientific bodies in the country and how it was intended to consolidate them under the umbrella of a single corporate body headed by himself. A week later he wrote a detailed letter to Sir Edward Appleton, still Secretary of the Department of Scientific and Industrial Research in England, and asked him for his advice on various matters that required immediate attention before the Bill could be tabled in Parliament. He also asked Appleton if he would pass the letter to Sir Charles Darwin, Director of the National Physical Laboratory, for his comments [20]. Schonland's former colleagues in England were only too happy to oblige, for he was surely one of them — men who had all spent the better part of the last five years working in scientific collaboration against a common foe and who were now about to lead their countries into the post-war world, where the potential for scientific advancement had never been greater. Collaboration, not isolation, would be just as important now as it had been then.

A matter of particular concern to Schonland was the position of the Standards Council, already in existence, and its laboratory facilities, that were about to be ratified by Parliament. In a country like South Africa with limited scientific resources any duplication of laboratories within the government sector would be nothing less than a profligate waste. He acted swiftly by drawing this to the attention of the politicians and so managed to have the Bill setting up the South African Bureau of Standards altered accordingly. Now, the establishment of its laboratories would be a matter for joint agreement by the CSIR and the Standards Bureau, but their inter-relationship at the operational level worried him, so the views and possible guidance of Appleton and Darwin would be very useful. At issue, too, was how he might go about persuading the country's none-too-enthusiastic industrial sector to invest resources in research. Would some form of tax relief, as was the case in Britain, be the way? Schonland felt it might, but quite how the inevitable borderline cases should be adjudicated was a problem which he knew had been tackled successfully by Appleton and his advice would be much appreciated.

And then there was the matter of South Africa's uranium resources, which had already attracted the interest of the Americans and more recently that of John Cockcroft at Chalk River.* In 1941, G W Bain, Professor of Geology at Amherst College, Massachusetts, had visited South Africa to collect samples of gold ore from the Rand for his collection. Soon after the United States entered the war Bain became geological consultant to the Manhattan Project and when cataloguing all possible sources of uranium he examined the South African ore with a Geiger counter and found it to be slightly radioactive. This was a discovery of great importance[†] and it immediately led to a highly classified visit to South Africa by another American geologist by the name of Weston Bourret, who spent four months during 1943 in the country examining samples. Only Smuts and the President of the Transvaal Chamber of Mines, informed personally and in the utmost confidentiality by the Prime Minister himself, knew of the significance of Bourret's visit and

---

* Cockcroft presumably asked Schonland about uranium but only two pages of his letter have survived and require some reading between the lines [16].

† This was not the first time this has been noted. As early as 1915 it was reported that radioactive deposits had been found in South Africa, while the presence of uraninite in Rand concentrates was confirmed in 1923 [21].

his subsequent report [22]. The political, economic and scientific implications of American and British interest in South African uranium were clearly enormous and Smuts intended to handle the matter very carefully, particularly as Britain saw her special relationship with him, and hence with South Africa, as a trump card in her dealings with America [23].

The discovery of nuclear fission in Germany in 1939 revealed the devastating power bound up in the atomic nucleus and this led to intensive research, first in Britain, and then particularly in the United States, under the code-name of the Manhattan Project, from 1941 onwards. The man who was essentially in administrative control of the huge British enterprise code-named 'Tube Alloys' was none other than Appleton, and Schonland felt it most necessary to put him in the picture, if somewhat obliquely [24]:

> I had a word with General Smuts on the matter in which Cockcroft
> was interested. He has instructed me to take no action whatever.
> If you feel that I can be of any assistance I should be glad if you
> would raise it yourself but otherwise I shall keep out of the whole
> thing.

There the matter rested, at least as far as Schonland was concerned, and his efforts to concentrate on South Africa's post-war scientific development were not deflected by the project that within four months would bring the Second World War to a cataclysmic end.

*　　*　　*

The recruitment of senior staff to establish and run the various laboratories of the CSIR was obviously a crucial element in the programme and the President-designate saw this as one of his most important functions. He had identified the immediate need for three national laboratories based very much on similar facilities that had been in operation in England for some time. These would be called the National Physical Laboratory, the National Chemical Research Laboratory and a National Building Research Institute—the subtleties of distinction in their titles being subject to change with time. Already in existence, amongst the plethora of research establishments (or 'island laboratories' as Schonland called them) around the country, were the Fuel Research Institute, the Government Metallurgical Laboratory and the South African Institute for Medical Research, and formal relationships with them would be set up by negotiation. He also had a vision to see established as soon as possible a laboratory dedicated to research into telecommunications and all the modern developments that had sprung from wartime research in areas such as radar. The model for this, in his mind's eye, was the Telecommunications Research Establishment (TRE) in England. An idea also lurking there, and one stimulated very much by the success achieved in South Africa under Dr Simon Biesheuvel in the selection of aircrews during the war, was to set up a facility for research into industrial psychology. When he first put this suggestion to Smuts, the Prime Minister responded by telling Schonland that it sounded like 'pretty good nonsense' but Schonland argued that Biesheuvel had developed scientific techniques of selection that were well in advance of those employed elsewhere and that their application to industry would be of great benefit in the longer term. Though Smuts's own assessment of people was sometimes criticized [25], his philosophy, once he had selected the right man, was to allow him full rein and so, having put his faith in Basil Schonland, he now accepted his judgement and allowed him to proceed. Schonland's faith in Biesheuvel would indeed be vindicated in the years to come. Now he was to trust his own judgement again in making his choice of the man to run the National Physical Laboratory — the first scientific appointment that he would make to the CSIR.

Stefan Meiring Naudé had been Professor of Physics at the University of Stellenbosch, some 50 km east of Cape Town, since 1934 and he and Basil Schonland knew each other well from even earlier days, when they were both on the staff of UCT. In April 1945 Schonland, while in Cape Town, telephoned Naudé and asked if he could visit him in Stellenbosch at the University's Merensky Institute of Physics. They duly met and after a brief tour of the Institute sat down together to lunch. Schonland then informed Naudé that he was about to form the National Physical Laboratory within the CSIR and asked whether he had any suggestions as to who would make a suitable Director. Naudé mentioned a name but it was soon clear to him that Schonland already knew whom he wanted — Naudé himself. If he refused, he was told, then that would leave Schonland no alternative but to 'import a scientist from overseas'. To an Afrikaner whose roots ran deep amongst the traditions of his people this would have been unthinkable, especially as 'overseas' probably meant England and Naudé, no Anglophile, had already made his views of some centres of British physics quite plain, as we have already seen, when he turned down the Rhodes Scholarship awarded to him in 1925, in favour of an opportunity to further his own early career in Germany. Schonland's loaded question, delivered almost certainly with a glint in the eye, was all the stimulus Meiring Naudé needed to help him make up his mind. His answer was soon forthcoming and he accepted the appointment in May [26].

Basil Schonland was a good judge of people. In Naudé he had a man who not only had an international reputation as a spectroscopist, for his discovery of the nitrogen 15 isotope when working at the University of Chicago in 1930, but he was also a man of some character and personality who generally got what he wanted. Scientifically, therefore, he would hold his own in any company, while his drive and ambition would ensure that the NPL would not flounder as long as he was given a nucleus of good young scientists, around whom to develop his own ideas. Schonland seemingly also had another, longer-term, objective in mind when he selected Naudé. It was to do with his own tenure as the head of a national organization, which he felt duty-bound to found, to lead and to serve with all his ability — but not forever. The urge to return to his own research at the BPI was still very strong, especially after the break of five years forced on him by the war now to be extended still further as he fulfilled what he saw as his duty to Smuts. In Naudé he believed he had someone with all the potential ultimately to succeed him as President and such a man, who should be a staunch South African, must be in place at the very birth of the fledgling CSIR. With Naudé working closely alongside him he knew that no charge that the CSIR was serving broader interests than South Africa's alone could ever be levelled against anyone.

The CSIR now needed a home with facilities for laboratories and offices so that work could commence as soon as possible. Schonland saw equal merit in siting it in either Johannesburg, South Africa's largest city and the hub of most of its industrial activity, or in Pretoria, close to the seat of government, but Smuts wished it to be in Pretoria and, though they discussed the matter at some length, the Prime Minister had made up his mind and was not to be swayed by any counter argument [27]. It so happened that Meiring Naudé, when asked where he thought the CSIR should be based, was emphatically opposed to Johannesburg, so much so that he indicated that he would not accept Schonland's offer of appointment were it to be situated there! [28]. So Pretoria it was to be and a temporary site in the now disused munitions section of the Mint in Visagie Street was selected, on the understanding that its suitability would be reviewed after three years. However, it was scientists of calibre rather than the geographical location of the laboratory that was much more important and Schonland soon turned his attention to finding the right man to lead the National Chemical Research Laboratory.

The person he approached was another former colleague at UCT, a New Zealander by birth and an Oxford man, W S 'Bill' Rapson. When Schonland was about to take over the Physics Department in Cape Town from Ogg in 1935, Rapson was in the throes of establishing the Department of Chemistry at the University. There was little interaction between their disciplines in those days so professional contact between them was limited, but Rapson's wife was Assistant Dean of Fuller Hall, the women's residence at the University, of which Ismay Schonland was Dean, and so their wives knew each other well and Ismay also knew Bill Rapson. Since Ismay's views mattered very much to Basil Schonland he would have considered them before writing to Rapson to ask what he thought of a chemical laboratory within the framework of the CSIR, that was now beginning to assume some sort of shape, at least in his mind if not in concrete. Rapson's detailed reply was both convincing and well put and clearly to Schonland's liking, and so this early correspondence was soon followed by a more formal letter offering him the Directorship of the National Chemical Research Laboratory. Rapson accepted and, though still employed by UCT, agreed immediately to spend his holiday during July in an office provided for him in the CSIR's first home in the Mint. There, nearly frozen by the sort of Transvaal mid-winter that surprises many, he started to draw up plans to equip a laboratory for a task as yet undefined within an organization as yet unofficial. But, as he described it, 'there was a missionary spirit about in those early days: science was everything and we were all fairly young, enthusiastic and keen' [29]. Schonland's enthusiasm was certainly that of the mission-ary—a role he was by now well used to playing—and no-one was in any doubt that he meant to change the place of science in South Africa.

Both Rapson and Naudé were given simple objectives: to establish laboratories 'for the better application of scientific research, pure and applied, to the development of the natural resources and industries of the Union'. In fact, these very words formed the basis of the Bill to go before Parliament later in the year. Coupled with these high ideals were two others: the effective coordination of research throughout the country and at every level, from universities and technical colleges to the state-aided institutions and government departments; and the training of those who were to carry out this work. South Africa had certainly lagged some way behind its Dominion partners in all three areas, for though research had been conducted by dedicated individuals and small groups of scientists from the time that the Abbé de la Caille had laid the foundations of Southern Hemisphere astronomy in the Cape in 1751, subsequent work, though highly distinguished in any number of fields, remained individualistic and certainly uncoordinated. So, by these first two appointments, Schonland set in motion the process of science in South Africa as a national activity, soon to be backed by the power of Parliament and, most important of all to him, with the full support and patronage of the Prime Minister. The synergy that existed between these two men, Smuts and Schonland, was the key. It undoubt-edly produced the force that drove forward the ideas that they had kindled, originally during snatched moments in London, when they had been able to extract themselves from other more pressing matters to discuss South Africa's scientific future. Now, Basil Schonland had taken the first steps to turn those ideas into reality.

*   *   *

While these events were unfolding on the larger stage, the BPI at Wits was slowly returning to its previous incarnation as a university research insti-tute. Military uniforms disappeared, the guards who nightly walked its perimeter to protect the secrets of radar were stood down, never to return, and all the remaining vestiges of the SSS were slowly being removed to Defence Headquarters in Pretoria.

On 27 March 1945 the Board of the BPI met, almost five years to the day since Schonland had attended for the last time, before leaving for England and events of a different magnitude. He was warmly welcomed by his old colleagues, who included amongst them the imposing presence of Bernard Price himself. Schonland, once again in the chair, then welcomed many of them back from their own exploits in uniform. He then moved quickly on to the real purpose of their meeting which was to resuscitate, just as soon as they could, the scientific programmes of the Institute. One matter not discussed was the part played by the 'Beep',* and its staff, in the defeat of the Nazis, but that would come later. To Schonland what really mattered now was the research that lay ahead of them, not the diversions, however important, that they had left behind.

It was soon evident, once discussion commenced, that seismology and the characteristics of the earth beneath the Witwatersrand were of particular interest. Whereas lightning and other atmospheric phenomena had dominated the BPI's work in those few years before the war, an ever-increasing interest in what was happening beneath their feet had begun to emerge just before the outbreak of war. The Witwatersrand was renowned for its earth tremors that were almost a daily occurrence from one end of the Reef to the other and the development of methods to measure and characterize them had now assumed particular impor-tance. Work on this had been started in 1938 and was then continued during the war by those few members of the BPI not otherwise engaged, and by associates from other university departments. The magnitude of these tremors, and their frequency of occurrence, were such that they could be monitored at considerable distances and this suggested that the information that could then be obtained about the structure of the earth's crust from the resulting seismograms made their study of very great importance. To do this effectively, however, meant that a number of seismic monitoring stations was required and they would have to be linked by radio in order to synchronize and control the equipment. Such engineering requirements now fell almost naturally within the ambit of men like Philip Gane and Jock Keiller, whose talents had been so successfully diverted from the design and construction of geophysics equipment to much the same thing with radar during the war.

Schonland, who had long seen the need to widen the scientific front over which the BPI was operating, enthusiastically supported this excit-ing work. However, he reminded the meeting that the BPI 'is now

---

* The wartime name given to the BPI by all those who worked or were trained there.

regarded as the most important centre in the world for research on lightning and thunderstorms. It is desirable that we keep this position'. So, even though his presence might be sporadic, his influence was certainly not and lightning research would continue, but now the daily inspiration and leadership would come from his old colleague Dawid Malan, who, coincidentally, was a cousin of the Leader of the Opposition, whom Schonland had called on just a while before.

Malan had spent the war years at UCT but was now about to return to the BPI as a Senior Research Officer to assume Schonland's mantle, if only temporarily, in the on-going search to unravel the lightning flash in all its intricate detail. There was also a new method of lightning investigation that was soon to be added to their armoury and this was radar. Schonland informed the meeting that he was about to consider its use as a 'general tool and also [to use it] to plot automatically the positions and paths of storms on the Witwatersrand' [30]. For this purpose he had in mind employing, either at the BPI or within the as yet non-existent telecommunications research facility of the CSIR, members of the SSS whose skills in both the design and use of radar had been so effectively honed in the most recent past. Foremost amongst them was Major Frank Hewitt, still in London at South Africa House, but already working closely with Schonland by providing him with the liaison he needed from time to time with colleagues and officials in England, as the plans for the CSIR were developing. Radar had, by now, a short but glorious history. In the opinion of many, its use had effectively won the war, but to those scientists who had been so intimately involved with it from its earliest beginnings, whether CH, GL, ASV and even Schonland's own JB, it was a technology that offered so many other possibilities in the world of scientific research.

Others, too, had realized this and it was not at all surprising that, with the smoke and chaos of war still in the air, efforts were under way in various places to turn radar into a scientific tool. One was A C B Lovell, who at TRE had been the driving force behind the development of H_2S, a truly remarkable application of radar in aircraft. In April 1945, Lovell accepted a lecturing post at Manchester so as to resume the work he had been doing there before the war with Patrick Blackett on the detection of cosmic ray showers. Very soon after taking up his appointment in September he acquired, from J S Hey at the AORG, a GL radar which, operating as it did at a wavelength of 4.2 m, was ideal for detecting these showers, as Lovell and Blackett had predicted in a paper published in the very early years of the war. Bernard Lovell soon installed the radar on a piece of land some 15 km south of Manchester, known as Jodrell Bank, and there set in train the beginnings of a new branch of astronomy [31]. Stanley Hey, remembering only too well his momentous discovery of 1942, which so confounded some but intrigued

Schonland, would soon point skywards the antenna of his wartime radar equipment at the AORG, to once again detect radio emissions from the sun, and the cosmos in general, and so start his own career in this most exciting new field of scientific exploration. Radio astronomy, discovered almost by accident by Jansky in the United States in 1931, was about to explode. In Australia, just as in South Africa and New Zealand, radar equipment had been developed following Britain's cryptic announcement to its Dominions in 1939 about 'certain technical developments which are of vital importance against air attack...'. As soon as the war ended, plans were being made at Dover Heights in Sydney to use a 200 MHz army radar to confirm Hey's finding that enhanced solar radiation was linked to the presence of sunspots, and this they duly did early in 1946. Soon the Australians would join the three groups in England under Lovell, Hey and Martin Ryle of Cambridge, a former TRE man, as leaders in radio astronomy and, in every case, it was the stimulus provided by radar that had brought this about. So Schonland was in good company and very soon Hewitt would join him to design a radar to look specifically at lightning.

With his other hat on for four fifths of the week, Schonland's major concern was the Scientific Research Council Bill being piloted through Parliament by Hofmeyr. The process started in May and the Bill was finally passed in June. As he had done so often before, Hofmeyr had taken on the mantle of acting Prime Minister with Smuts away attending the San Francisco Conference that saw the birth of the United Nations. As an exercise in inter-party harmony, the passage through the House of the CSIR Bill was remarkably free of rancour. All parties saw in it the higher ideals that were not always the currency of the place. At the end of July, Schonland received a letter from, of all people, the Secretary for External Affairs, informing him that permission had been granted by the Public Service Commission for the temporary establishment of five posts as from the beginning of August. 'If the Scientific Research Council has not yet been constituted by the beginning of January 1946 the matter should be raised again'. The letter than informed him that 'the following temporary units [sic]' would be paid in the interim while the Civil Service turned Parliament's decision into the Act that would formally constitute the CSIR. Joining him as 'a temporary unit' on the Government's payroll were an Administrative Assistant, a Shorthand Typist, a Woman Clerk and a 'European Junior or Native Messenger' [32].

Schonland appointed D G Kingwill as his Administrative Assistant, but very much in the role as his personal assistant. Major Denys Kingwill, who was also a physics graduate from Rhodes, had just been demobilized as a meteorological officer, with service in almost every theatre of war in which the South African Air Force had operated. From now on he would work very closely with Schonland as they steered a flimsy creation through, at times, very unfamiliar territory. Kingwill's appointment, as well as those of both Naudé and Rapson, though still to be confirmed officially, were to be cornerstones of the CSIR establishment. Both Naudé and Rapson were to rise to the highest levels within the organization, the former as its President with Rapson his deputy, while Kingwill would soon initiate the function that Schonland believed to be so very important, the development of close contact with scientific bodies beyond South Africa's borders in the UK and the USA particularly, but also with many countries in Europe, the Commonwealth and in Africa itself. As his secretary Schonland appointed Miss M A Murray, who was to serve him for many years [33].

*   *   *

On 13 July 1945, the main committee room at Wits was the venue for a meeting of the Board of the BPI for which the agenda was somewhat different from those of late. It was the first meeting of the Board since VE Day and attending by special invitation were Colonel Freddie Collins MC, Director of Signals, South African Army, Lt Col D B Hodges, Major G R Bozzoli and Major P G Gane. For this rather special occasion Dr Bernard Price took the chair with Basil Schonland at his side. The University was represented by Humphrey Raikes and he was accompanied by members of Council and Senate. The first business of the Board was to confirm the minutes of meetings from as long ago as March 1940, since which time much had happened, and it was with all that very much in mind that Dr Price then addressed them. He began in a style to which many had become accustomed when listening to some stirring wartime broadcasts [34]:

> Never had there been a greater victory or one pregnant with greater
> possibilities. Whilst not unmindful of the many difficult problems
> confronting the world a unique opportunity now presented itself
> for a great step forward in the march of civilization. I hope the
> Institute will be able to play a useful part in the new era now
> opening up before them.

Impressed by this burst of Pricean oratory, those of the audience who had played no part in the events of the SSS were now treated to Price's version of the origins of the SSS and of the subsequent career of the BPI's Director at the very heart of affairs in England and in north-west Europe. When recounting the background, Price was on fairly firm ground but, when he came to describe Schonland's activities in more distant territory abroad, his eloquence and understandable lack of real acquaintance with the facts rather run away with him:

> The magnificent achievements of the Director was a fascinating
> story. One of his great achievements was to convince the military
> chiefs as to the value of Science as applied to field conditions in

war. This was very difficult on account of the tradition of set military organisation. His success in this was borne out when Field Marshal Montgomery asked him to join his staff before D-Day and the Director became one of only four persons who knew all that was planned. Dr Schonland has been awarded the CBE, and very much higher honours were due to him were it not for certain political difficulties in connection with such awards.[*]

Who it was that briefed the Chairman before he regaled the meeting in this way is unknown, but we can be sure that Basil Schonland found it rather embarrassing to hear his contribution to the war effort blown up in quite such exaggerated terms. The South African press, no doubt in keeping with the decidedly euphoric mood then sweeping the country, were sometimes given to rather grandiose statements and Bernard Price was caught up in it as well. Their cue may well have come from a statement made in 1944, allegedly by a 'prominent member of the British War Office'. In it this anonymous individual, when referring to the contribution made by the Dominions, was quoted as saying, 'Brigadier Schonland is, in my opinion, South Africa's outstanding individual contribution to our war effort. If the Union had done nothing else except send him over to help us, we should still be very much in her debt'. Much paraphrased, when subsequently picked up by the newspapers, these few words of hyperbole caused Schonland 'to squirm', according to Ismay many years later, for he was not a man to puff himself up in any company and, as Ismay made clear, it was 'so untrue and so unmindful of the lives sacrificed to secure the final victory' [35].

Bernard Price then congratulated Hodges, Gane and Bozzoli on the work they had done 'as executive officers responsible to Col. Collins for the running of Special Signals' and, before calling on Colonel Collins himself to address the Board, he hoped that 'it would be possible to preserve in the Archives of the Institute a permanent record of the activities of the SSS in the form of a collection of photographs, drawings, documents, literature and other interesting records'. Major Bozzoli willingly agreed to put this matter in hand.[†]

Colonel Collins then rose to speak. A man of consistently pleasant disposition Freddie Collins was, of course, not just Schonland's closest contact with the South African military but also an old friend of the

---

[*] Quite what Price meant by this is not clear. He may possibly have had some titular honour in mind in which case the 'political difficulties' would have been the decision taken by the South African Parliament in 1925 in requesting His Majesty the King no longer to confer British titles on South Africans [Union of South Africa—Debates of the House of Assembly 24 February 1925, 222–331].

[†] A magnificent photographic record of the work of the Special Signals Service was indeed produced and now resides amongst the SSS collection in the Archive of the University of the Witwatersrand.

Institute. He was also fulsome in his praise of the recent work done in the greatest secrecy at the BPI but then he too became somewhat disconnected from the facts when describing how Schonland had found himself in England in 1941 and what he had done once there:

> Brigadier Schonland was called to England by high officials
> including Mr Churchill, Field Marshal Smuts and the Air Ministry.
> I am not capable of saying what he had done except that he put
> South Africa on the map in regard to Science in the field of radio
> location. This had added to the reputation of South Africa.

Indeed it had, but Collins's version of quite how Schonland came to be in England was pure fantasy. He had not been called to England by anyone; if anything he had sent himself with the tacit agreement of General Sir Pierre van Ryneveld, the CGS, and presumably with Freddie Collins's full support because South Africa was in urgent need of British radar. While Smuts may have been aware of this, since he was also Minister of Defence, he played no part in it at all and Britain's Prime Minister had certainly never even heard of Schonland at that stage of the war. The Air Ministry, on the other hand, particularly in the person of its Scientific Adviser on Telecommunications, Robert Watson Watt, knew of the South Africans and their 'elementary home-made RDF' in East Africa but had made no call for Schonland's services, though Watson Watt was keen to renew old acquaintances when Schonland eventually arrived. Collins, too, had simply allowed the heady atmosphere to run away with him and Basil Schonland had to endure the flattery.

The record of the SSS itself, as presented to the meeting, was more factual and the Director of Signals used the opportunity to present in quite some detail how it had functioned, where its various elements had been deployed and how this very special unit within the UDF had eventually recruited women to become radar operators in the finest tradition of their counterparts in England. 'I wish to pay a tribute to the women of this organisation who were of the very best', he said, and indeed they were, for without them none of the radar stations around the country's coastline would have been adequately manned. Finally, Lt Colonel Hodges was called upon to address them all. Dave Hodges stuck closely to the facts and paid particular tribute to Schonland for his leadership in launching the SSS and to the University for housing it. He quoted some interesting figures. During its existence the SSS had developed the first radar set in South Africa; it had built the equipment for 25 complete radar stations, six complete mobile radar sets and nearly fifty radar beacons. It had also made spare parts for engines, generators and aerial tuning mechanisms and had constructed 20 wireless transmitters and receivers. Then there were the more than a million microfilm negatives that were enlarged and printed, the drawing and printing of 650 engineering drawings and the management of stores to the value of more than £1 500 000. Finally, the SSS had trained 500 women operators and 300 radar mechanics who could repair 'every conceivable part of a radar set'. It was an impressive record. He then announced that the 'B.P.I. was being returned by the Defence Department with a great deal of regret, but the SSS had endeavoured to preserve the spirit of the Institute'. Schonland could feel justly proud of what they had achieved and he himself added the footnote that the technical achievements of the SSS were certainly appreciated in Great Britain. And so the curtain fell on a remarkable interlude in the short life of Basil Schonland's Institute.

The BPI had now to regenerate itself within a world that was rather different from the one into which it had been born before the war. The Director would appear in person only on Fridays, in terms of the agreement reached between the Prime Minister and the University. Lightning, under the guidance of Malan, would soon take second place to seismology, the province of Anton Hales, as the research priorities altered and every effort was made to encourage the mining industry to participate more fully in the programmes of the BPI. Some relationship with the stripling CSIR would have to be struck: a matter that could be left with the Director, who was clearly in a rather powerful position to bring that about. Of particular concern, though, was the BPI's continuing financial viability.

In his *alter ego* as Scientific Adviser to the Prime Minister, and as soon as he became its President, Schonland had insisted that the financial viability of the CSIR was secured by means of an annual Parliamentary grant. The situation at the BPI, by contrast, was rather different. Its primary sources of funding were the one-off contributions of £11 000 from the Carnegie Corporation of New York and £10 000 from Bernard Price, plus an annual contribution of £1500 from the University. The terms on which these had been agreed almost ten years before required that the BPI be self-supporting within 20 years; if not, then these financial provisions would lapse [36]. Schonland therefore had every reason to seek support from those South African industrial and commercial organizations that were likely to benefit from the BPI's research. Foremost amongst them was the Chamber of Mines, and he undertook to negotiate with them for the continuation of their grant of £2000 to £3000 that had been agreed in 1938 [37]. While there were numerous avenues of research which he believed to be of great interest to both the BPI and the mining industry, that of particular importance was the study of earth tremors and whether their investigation should be made over the full extent of the Witwatersrand. Since the mining of gold was very much South Africa's premier industry, Schonland took the entirely reasonable view that the Mining Houses, which together made up the Chamber, would view both the BPI and the CSIR, when it came to fruition, as great assets in the hugely demanding operation of extracting the gold-bearing ore from the very deep mines so typical of the Rand goldfields.

Their interest, however, was sporadic and often only lukewarm. Mining men at even the most senior levels of management, he soon discovered, saw themselves and their industry as being so intensely practical that research, with its aura of the white-coated scientist in his laboratory, was either ignored or even dismissed as irrelevant. Schonland will surely have noted many parallels between that attitude and one he had encountered often during the past five years in uniform. The significant difference, though, was that there was no Cockcroft or Watson Watt within the highest echelons of the Chamber of Mines, nor was there a 'Tim' Pile amongst the 'generals' who ran the mines. This uneasy relationship that seemed always to exist between the BPI and the South African mining industry would continue to plague Schonland for years to come.

*   *   *

By late August, the formulation of the structure of the CSIR was almost complete. On the 27th a meeting was held at the Union Buildings in Pretoria under Schonland's chairmanship. In attendance were eight of the nine members representing the worlds of government, academia and industry who had been proposed to serve on its Council. Schonland, who now bore the titles and responsibilities of President, Chairman and Chief Executive Officer, was joined by Dr F J de Villiers (Industrial Adviser, Department of Commerce and Industry), Dr J Smeath Thomas (Master, Rhodes University College), Dr Bernard Price (Chairman, Victoria Falls and Transvaal Power Company), Dr R W Wilcocks (Rector, University of Stellenbosch), Dr H J van Eck (Chairman, Industrial Development Corporation), Mr T P Stratton (Consulting Engineer, Union Corporation), Dr S H Haughton (Director, Geological Survey) and Mr J E Worsdale (Chairman, Cape Portland Cement). Not present was Dr P J du Toit (Director, Division of Veterinary Services), who at the time was in Kenya. This group constituted what was called the Exploratory Committee on the CSIR and it would function in that guise until the Council came into being [38].

Schonland opened the meeting by reminding them that the Scientific Research Council Act had received the wholehearted support of all Parties in Parliament and it was now their task to implement it. An undeniable need existed for state-supported scientific research in South Africa along with the active encouragement of fundamental research in the universities while, at the same time, the development of industrial research within organizations and individual firms was to be encouraged. But in drawing up the proposals that he had put before them he told them that possibly he himself 'may have taken too close a view of the problems [and] may have laid too much emphasis on the organization of particular laboratories'. It would therefore be a mistake to consider the proposals as rigid. They were there to advise him and he looked forward to receiving their comments and criticism. He feared that he had erred on the conservative side in drawing up his list of financial requirements, since they were decidedly modest in relation to other countries, both on the basis of income and population. However, he continued, 'South Africa had made a late beginning and it was impossible to hasten development, particularly as there were not sufficient men with the necessary qualifications to fill the posts'. It was therefore most important that they 'provide encouragement to men and women to take up the necessary research training' since, he felt, it would take at least five years to 'train better men than were available at present'. An important principle, which he believed would be fundamental to the success of the CSIR, was that it must 'avoid the dangerous security of the Public Service', where low rates of pay and promotion based solely on seniority would snuff out the very enthusiasm and creativity on which any vibrant scientific organization was totally dependent for its success. Here he was echoing the views of John Cockroft, who had spelt out precisely such a warning to him when they corresponded some while before. Though Cockcroft saw the dangers, his style of management was apt to overlook the very problem even when it stared him in the face; by contrast, Schonland faced it head-on and insisted that the CSIR should operate outside the stultifying atmosphere of government bureaucracy. His putative Council agreed and so, ultimately, did the Government when it affirmed that the CSIR scientific staff would have the same measure of security as their colleagues in the universities: tenure with promotion on merit.

Matters of detail now dominated the meeting. The chairman had tabled a number of papers and these were dissected almost line by line. Why, for example, should the proposed National Industrial Chemical Laboratory have its industrial links emphasized, while the National Physical Laboratory was seemingly untainted by them? Why was there a need for a National Building Research Laboratory? Surely the most pressing demands for both Medical and Defence research should see them included within the ambit of the new Council, while neither Economic nor Social Sciences should have any part in it? Who would be responsible for studying and solving the physiological problems, which had already been encountered when mining at very great depth? What level of commitment was the Government prepared to make towards research grants to universities? Post-doctoral fellowships should be set up as a matter of urgency for persons of great distinction and their holders should be encouraged to spend time overseas. What provision would be made for liaison with similar organizations overseas so as to enhance South Africa's international standing? The list went on. Much discussion followed and consensus was reached quickly on some, while others were held over to another time or for the attention of special sub-committees which sprouted like the very leaves of spring then just evident after a dusty Transvaal winter.

The thorny issue of the relationship between the CSIR and the South African Bureau of Standards (SABS), known then simply as the Standards Bureau, still remained. As he had told Sir Edward Appleton some months before, Schonland had been most concerned by the duplication of staff and equipment that would surely occur once these two bodies began operating and this was something that South Africa could ill afford. Through his efforts, the Bill to set up the SABS had been amended so that there would be negotiations between the two bodies before any decisions were taken on Standards laboratories. To his way of thinking, the Bureau should only be responsible for the routine testing of products and appliances to ensure their compliance with some agreed specification, whereas the NPL might well determine any primary standards involved. Ideally, though, he would have liked to have seen just a single entity doing this work, and that would be the CSIR, but such was not to be, or at least, not immediately.* The Board agreed with him, but since the die was well and truly cast, all that was now possible was vigilance and close collaboration with their sister organization.

The Chairman was on home ground when it came to the setting up of a Telecommunications Research Laboratory. Not only was this an area with which he had much recent familiarity but he already had in mind the very nucleus of the staff to bring it into being almost immediately. 'He proposed', he told the Board, 'forming a military team under the ablest men available to deal with this work'. By military he really meant his former colleagues in the SSS who were now awaiting demobilization. From amongst them he would select those who had displayed exceptional ability during the previous six years when they had done such sterling work in designing, constructing, installing and even operating the various versions of the JB radars around South Africa's coast, in East Africa and in the Middle East. Many of them had also become highly proficient in the operation and maintenance of the British equipment as it became available, and some had even spent some time in England at various secret establishments, such as the Telecommunications Research Establishment (TRE) in Malvern, learning the latest centimetre wavelength techniques. It was, he told them, his intention to use this talent to 'cover the entire field of telecommunications', by which he meant both radio and radar. Since there were already other bodies in the country with an interest in some of these matters, such as the South African Broadcasting Corporation (SABC) and the GPO, it was particularly important, in the light of the

---

* In 1956, long after Schonland had left the CSIR, the two bodies were amalgamated under the control of the CSIR only to be separated once again in 1962 [39].

Figure 20. The South African Prime Minister, Field Marshal J C Smuts, speaking at the first meeting of the CSIR Council held in the Union Buildings in Pretoria on 8 October 1945. Schonland is seated immediately to Smuts's left with P J du Toit alongside him. (Reproduced by kind permission of Dr Mary Davidson.)

clash of interests with the Standards Bureau, to ensure that the same thing did not happen here as well. It had already been mooted that both the SABC and the GPO were intent on setting up their own research units but Schonland made it clear that he would oppose this most strenuously. Again his Board agreed.

The CSIR was legally constituted on 5 October 1945. Three days later its Council met formally for the first time at the Union Buildings in Pretoria when their proceedings were opened by General, the Right Honourable, J C Smuts, Prime Minister of the Union of South Africa. It was not only an undeniably important occasion for South Africa but equally so for both Smuts and Schonland. For Schonland it marked his coming of age: everything he had achieved in his life and career so far, and there was much, was as if preparatory to this occasion on which he would be identified with the birth of the new scientific infrastructure in his own country. His fascination with science had led him from the boyhood experiments with telephones, in the old coach house in Grahamstown, to St Andrew's and then to Rhodes. All were the stepping stones across which he seemed to fly — and then there was Cambridge. His timing was perfect, for there, at the Cavendish, the genius of Rutherford had just launched what is probably the most glorious period of modern British science. Basil Schonland was part of the pantheon of physics that was the Cavendish Laboratory during the two decades before the war. Now, at the dawn of a new age of South African research, this scientist-turned-soldier was, as a scientist again, poised on yet another threshold.

To Smuts the occasion was no less significant, for it marked the birth of an ideal to which he had first given voice many years before, when he had appointed H J van der Bijl as Scientific and Technical Adviser to the Government. In 1921 van der Bijl proposed a government research institute but, as was so often the case in South Africa, national politics and the forthcoming election intervened. Smuts was defeated and the new government turned its attention elsewhere. Then, after a period in the political wilderness when he too turned his mind to other matters, Smuts delivered in 1931 his Presidential Address to the British Association for the Advancement of Science on the theme of 'The Scientific World Picture of Today' and enthralled an audience of luminaries with his remarkable grasp of the state of scientific knowledge at the time. This was the man who, himself at Cambridge in the last decade of the previous century, had swept the boards with double firsts in the Law Tripos and then, soon after, had waged war against England as a youthful Boer general in the saddle. All the while his philosophy of *Holism and Evolution* was hatching in his mind and it was to become his *magnum opus* when, in just eight months in 1924, he wrote the book of that title in which he expounded his views on both the physical and metaphysical worlds. Geology, archaeology and above all botany fascinated him and it was this last, as we have seen, that brought him into close contact with Schonland's father. That link, the BPI and, of course, the scientific imperatives of the war almost pre-determined the existence of the CSIR with Schonland at its head.

In terms of the Scientific Research Council Act No 33, Basil Schonland was appointed President of the CSIR for a period of three years from 5 October 1945. Other members of his Council received appointments or between one and three years [40]. The CSIR, as had been agreed, was a corporate body outside of the Public Service, which would report to Parliament directly through the Prime Minister. This was an important stipulation that had come about as a direct result of Schonland's many discussions with Smuts. It meant, of course, that Schonland himself had direct access to Smuts and this was of enormous benefit to the new President because, once he had obtained Smuts's blessing, he was able to operate without hindrance in the knowledge that he had his Prime Minister's complete support and authority behind him.

However, this direct Prime Ministerial patronage was not to last forever. In less than three years the Nationalists would win the General Election by the most slender of majorities and the Smuts era would be over. The new government under Dr D F Malan, for all their assurances given to Schonland previously, saw the CSIR from rather a different perspective. Of course it was still a national institution but no longer would it have a special place in any Prime Minister's heart. That it should stress its *Suid Afrikaanse kleur* was fundamental to Nationalist thinking and Malan watched over it, but only obliquely, for two years after taking office to ensure that it did. Then its representation in Parliament became the lot of the Minister of Economic Affairs and the status of the CSIR within the eyes of South Africa's new government changed significantly. No longer would it be seen as the coordinating and advisory body to which the Government would turn on all matters of science and technology, since these, as with virtually everything else in South Africa, were now enmeshed within an overriding philosophy of very different proportions. Those early halcyon days under Smuts were to be short-lived.

In selecting the staff who would form the nucleus of the CSIR's various laboratories it was natural that Schonland should look first to those many colleagues with whom he had worked over the previous ten years or so. Eric Halliday, his student at UCT and active collaborator at Wits well before the establishment of the BPI, accepted an appointment in the NPL. He would eventually see out his years there as one of the CSIR's most loyal servants. The SSS had already been earmarked as the source of men of outstanding ability for the Telecommunications Research Laboratory (TRL) that Schonland was intent on setting up without delay. Within four months of the launching of the CSIR, two laboratories were already engaged in useful investigations [41]. The National Building Research Institute had a head-start on all the others, as it has been in existence even before the CSIR under the auspices of the Department of Commerce and Industries. The other was the TRL, which found its first home in the Department of Electrical Engineering at Wits under the guidance, for a brief while, of the recently demobilized Mr G R Bozzoli, Senior Lecturer in the Department. Soon, Frank Hewitt, who had also just returned to civilian life, would take over the running of the Laboratory and would start its first research programme with a staff of four former SSS colleagues.

The TRL certainly had a most auspicious start. Two research programmes were soon under way: one to apply the knowledge of radar that was so fresh in all their minds to the location of distant storm clouds, and the other was the development of an automatic ionospheric sounder. Not only would the information about the features of the ionosphere be of basic scientific interest, but there were immediate practical benefits to be had as well because of the ionosphere's key role in long-distance, high frequency radio communications. The ability to predict its characteristics, based on regular measurements, was vital for the planning of such communications circuits and, while many laboratories elsewhere had been doing so for quite some time, it was the novelty of the method adopted by the TRL that made its contribution so important.

The new technique was due entirely to the remarkable ability of one man, Trevor Wadley, an engineer of somewhat iconoclastic frame of mind, who, with the much more reticent Jules Fejer, was regarded by Hewitt as the most brilliant member of the SSS. Schonland had ensured that both were now offered positions at the TRL. Wadley's fertile mind had been behind many advances in electronic technique made at the BPI during the war when the JB radars, in varying degrees of sophistication, were being produced under Bozzoli's technical direction. At the TRL his first task was to design an improved ionosonde, as it was called, which did not involve the complicated mechanical switching between wavebands typical of such systems in use at that time. With astonishing speed Wadley conceived of a method of doing this and then, even more remarkably, constructed the equipment based on his ideas in a matter of weeks. The ionosonde proved so successful that another four or five were built by the TRL and their performance was as good as, if not better than, any in use worldwide. But more was to follow. During the war, Wadley had considered various problems associated with conventional communications receivers, that relied on the mechanical switching of various oscillators and their associated banks of filters, in order to cover the 30 MHz tuning range typical of such receivers. In what seemed to his colleagues a stroke of genius he devised a solution, based on his work on the ionosonde, that within a decade would be recognized as one of the most important advances made in the art of radio receiver design.[*]

In essence, it involved generating a radio signal, of variable frequency and over a very wide range, with a degree of stability and ease of tuning hitherto unknown. The principle used an ingenious triple mixing scheme in which a selected harmonic from a crystal-controlled oscillator was mixed with a sinusoidal signal from a free-running VHF oscillator that was also mixed with the signal being received. Any frequency drift in the VHF oscillator, the main tuning element of the system, would automatically be cancelled in a third mixing stage, where the two signals were subtracted from one another while always maintaining the stability of the crystal oscillator. Interest in this revolutionary advance in the art of radio engineering was immediately forthcoming

---

[*] The British company RACAL owes its very existence to the RA17 receiver, which was developed in 1954, entirely on the basis of Wadley's so-called 'triple loop' system of frequency generation invented at the TRL in Johannesburg.

from the GPO, the South African Broadcasting Corporation and from the Union Defence Force, and over the next few years a number of receivers were built, both within their own workshops and by a company in Durban, for their evaluation. The reaction was universally favourable and, remarkable as it may have seemed to some in the country, the possibilities of exploring possible markets overseas looked most enticing [42].

Schonland watched these developments as they unfolded with growing interest and not a little pride since it was 'his SSS' who were responsible for them. His own contributions to the advancement of science were recognized yet again when, in 1945, two institutions with which he had had a long association chose to honour him. On 10 November a telegram arrived from London advising him that the Council of The Royal Society had awarded him the Hughes Medal for that year 'in recognition of his distinguished work on atmospherics and of his other physical researches'. The award was recorded within the columns of *Nature* in December with a precise and succinct summary of his scientific and military contributions. Particular mention was made of the work that Schonland had done on studying 'the relation between penetrating radiation and thunderstorms' and it pointed out that this work was of special importance 'since few such systematic observations have been continued over long periods in the southern hemisphere' [43]. Then, just a month later, he and Ismay attended a graduation ceremony at UCT where the Chancellor, Field Marshal* the Rt Hon J C Smuts, conferred on him the honorary degree of Doctor of Science *Honoris Causa*. The accompanying citation was duly laudatory as details of his life and career were read out by the university's Public Orator. Only on matters of much more recent origin was the scribe somewhat reticent when he said [44]:

> The details of his work during the late war have not yet been
> released but it is known that he was in the confidence of the
> military chiefs and was called on, with good success, to resolve
> scientific problems whose solution was necessary for the success of
> the war effort.

Such muted testimony was much more to Schonland's liking than the rather overblown oration that had embarrassed him just a few months before, when both Bernard Price and the Director of Signals had sung his praises with rather more fortissimo than fact.

*   *   *

---

*The question of Smuts's rank depended very much on the company he kept. He was promoted to Field Marshal in 1941 by King George VI and became the first and only South African to hold that highest rank in the British Army. But 'General' was as much a title as a rank to Smuts and he maintained that he was still 'General Smuts' to his friends and indeed to most of South Africa.

A source of real pleasure to Schonland were the letters that reached him during the course of that first year of peace from his many friends and colleagues in England, with whom he had spent the better part of half a decade collaborating in their various campaigns against Hitler. Generals de Guingand and Pile wrote offering their congratulations on his CBE while David Hill saluted his old boss and provided much local gossip. Omond Solandt and Pat Johnson expressed their pleasure at the award and stressed how well-deserved it was. And there too was a letter from the man with whom Schonland had had the longest military association of all: Spencer Humby, his fellow signals officer and scientific collaborator during an earlier conflict. Now, more than a quarter of a century later Humby, one of the AORG's stalwarts and its Deputy Superintendent, expressed his sincere good wishes to his former colleague through two world wars. There was also a note from one Dr R V Jones, the man whose own contributions to the war effort were still shrouded in secrecy. In years to come Jones's part as Assistant Director of Intelligence (Science) at the Air Ministry would be recognized as being absolutely crucial to the eventual Allied victory. Now, along with so many others who were privy to the greatest secrets of the war, he kept his own counsel. Jones paid a very genuine tribute to Schonland. 'It is pleasant to see someone, who we know from experience not to have been out for himself, to receive a well-deserved honour'.

Solandt and Johnson, for all their unstinting praise of their former colleague and commanding officer, made slightly barbed comments about each other—a vestige of those days in 1944 when, in Solandt's eyes, Johnson had the run of Europe while he was stuck at the AORG HQ in England. By contrast, Johnson could only compare Schonland's precise and direct style of operating with the rather pedantic approach of the Canadian colonel. Where they did agree, though, was on the debt of gratitude that both they and all their colleagues owed Schonland for his leadership, friendship and highly competent performance, first at the AORG and then in 21 Army Group. Johnson, already pipe-dreaming about his intended return to Magdalen College, Oxford, speculated on what might have been achieved by Operational Research had Schonland remained instead of returning home to attend to other matters as required by his Prime Minister. They had soldiered on without him but it was not quite the same. 'When the final banquet of No 2 ORS is held', he dreamt, 'there will be an empty chair to which we will all look with sorrow that father is not there'.

A letter had also arrived from the other side of the Atlantic. It was written by Edward L Bowles, then Expert Consultant to the Secretary of War in Washington and one of the most influential men behind the development of radar in the USA. He wrote to offer his best wishes to Schonland on his appointment as Scientific Adviser to Smuts and then proffered some advice. Bowles's terminology was that of the engineer engaged on the important problem of coupling his radar equipment to its antenna by means of a transmission line. 'My own experience here in the War Department has led me to a simple but convincing conclusion, that impedance matching is essential if we are to couple our technical minds and those interested in military and government affairs with anything like an acceptable reflection loss' [45]. This was precisely the problem that so bedevilled discussions at that interface between science, with its logical and systematic approach, and the world of politics that so often involved neither of them. This engineering analogy was a good one and Schonland long remembered it.

Positioned now as he was between the political machinery of the state on the one hand and the scientific visionaries and industrial pragmatists on the other, called for all the impedance matching Schonland could manage to bring into being the scientific organization that would best meet South Africa's multitude of needs. To achieve this meant appointing the right people to the key posts and his criteria in selecting them were simple. They had to be the most able scientists available and they had to be men capable of leadership. The two did not necessarily sit happily together but his task was to find the best people, around whom to build the laboratories that would form the nucleus of the CSIR. Some were already taking shape while others were no more than points for discussion, argument and even some rancour. On the horizon was a laboratory that would concern itself with the problems of deep level mining—the greatest challenge facing the South African gold-mining industry. Already in existence, as a result of the perceived need by the various Mining Houses, was a Research Institute for Deep Level Mining and the CSIR had been granted what was termed 'assessor membership' on its Board but without any say in its detailed operation. To many members of the CSIR's Council it was unacceptable that the national body responsible for research should be denied voting rights by one of its client industries and the CSIR's President was requested to put this to the Prime Minister for clarification [46]. But almost more pressing than mere voting rights, as Schonland saw it, was the need to address the physiological problems that mining at such great depth might cause. Whereas the industry had amassed considerable practical experience about mining techniques at great depth, the implications of the significantly increased rock temperature, dust and other environmental factors on the health of the miners involved were unknown. Such problems must be addressed and he viewed them as the CSIR's responsibility. A physiologist of some renown was obviously needed to lead the research in this area, and he certainly knew a few.

Wartime operational research in England would have been a very different beast had it not been for the large number of physiologists within its ranks. They had made up the core of Blackett's Circus, then the ADRDE(ORG) and, finally, the AORG itself. Bayliss, Hill and Huxley of the original Circus were all physiologists as was Solandt, who ultimately succeeded Schonland in command of the AORG. It was therefore not surprising that it was to Omond Solandt that he turned for recommendations and suggestions for a good physiologist for South Africa and even added a tailpiece in case Solandt himself might have been interested. Almost by return of post Solandt replied and indicated that indeed he was! However, for a variety of reasons, none scientific and all to do with the strongly entrenched position of the mining industry in the country at the time, the idea of a mining research institute within the CSIR did not prosper and all such matters went into limbo [47]. But medical matters in general continued to exercise Schonland's mind as he sought to position the CSIR as South Africa's all-embracing research organization.

By 1945 medical research in the country already had a long and proud history. The South African Institute for Medical Research (SAIMR) was formed in Johannesburg as early as 1912, with considerable support from the mining industry, and its production of prophylactic vaccines and therapeutic sera was of major importance to the general well-being of the country. Profits from the preparation and sale of these were ploughed back into research but, in general, financial support for medical research in South Africa remained poor [48]. At its first meeting the Exploratory Committee for the CSIR believed that medical research should fall within the ambit of the new organization but, after discussion, it was realized that this could lead to vigorous opposition from the medical profession, who were well-served by their own body of researchers in a number of small facilities around the country. Schonland explained that, though it was very much the intention of the Government that the CSIR should undertake the development of medical research, it was generally agreed that the matter required careful consideration and consultation with the South African Medical and Dental Council. At this juncture it was glaringly obvious that the medical profession was not even represented on the CSIR's Council, so the President suggested that Dr P J du Toit, the highly distinguished director of the Onderstepoort Veterinary Research Institute, should, at least in the interim, represent the Medical and Dental Council's interests in this field, however incongruous that may have appeared to some! Eventually, Professor S H Oosthuizen, its President, joined the CSIR's Council and reinforced the view that medicine was well able to look after itself, but he felt that the CSIR could become a very useful conduit for the funding of promising research at those existing institutions already set up for this purpose [49].

*   *   *

With his feet firmly planted in two camps, at the CSIR in Pretoria for four days a week and at the BPI in Johannesburg on the fifth, Schonland saw the development of scientific research in South Africa from at least two perspectives and he felt the associated tensions. However, it was those Friday visits to his old Institute at Wits that gave him the opportunity to think, not about the politics and economics of research, as was almost his perpetual lot in Pretoria, but actually to participate in it by engaging in discussions with his colleagues. And so began one of the BPI's rituals: the attendance by all scientific staff, regardless of rank or station, at morning and afternoon tea every Friday. Those gatherings were the occasions when the Director actively provoked discussion about particular topics of interest that stemmed from the BPI's own research programme or from his wider perceptions of the field as viewed from his more lofty perch elsewhere. Informality and the freedom to challenge any view and even to desecrate sacred cows, if needs be, were the cornerstones of these occasions and Schonland nurtured them very carefully. It had taken no great leap of imagination on his part to conceive of such an unrestrained meeting of like minds, for the prototype had long existed. The 'Sunday Soviets' at TRE in Malvern and his own informal gatherings at the AORG in Roehampton were the source of so many remarkable ideas that had occasionally even changed the very course of the war. Such methods must surely work in peacetime too and indeed they did. He also seized the opportunity to bring the staff of the BPI and the TRL together each Friday, since they were just a stone's throw apart at Wits, and the benefits were surely mutual.

By May 1946 he was able to report to the BPI's Board that Malan was designing a new lightning camera at the BPI, while Hewitt and Bob Meerholz, another ex-SSS man at the TRL, were heavily involved in developing their centimetric radar, in the hope that it might be ready for use at the onset of the lightning season near the end of the year. Also under way, as an extension of the seismic research that had been run throughout the war by Hales and Oliver, was the development, under the direction of Philip Gane, of a 45 MHz radio telemetry system that would enable a central recording station at the BPI to receive the seismic signals from six remote stations to be located in the environs of Johannesburg. From these recorded data, synchronized on a single paper chart, it would be possible to determine the epicentres of the tremors that were undeniably associated with the mining activity on the Reef. Whilst all this was happening, the BPI's laboratory was itself being extended to include more research rooms with a special lightning facility on the second floor. As always the matter of finance to support such ventures taxed the Director, but injections of cash to cover both the extensions and some new equipment were forthcoming from the Government, while Bernard Price donated a further sum of £2500 to assist. Schonland advised the Board that the Chamber of Mines grant was due to expire in October, but that he was hoping to have it extended because of the relevance of the seismic work to the industry's own 'Deep Level Research Organization', the same body which had held the fledgling CSIR at arm's length just a while before. Though remaining doggedly, or even pig-headedly, parsimonious in its support of research, the mining industry could never claim to have been unaware of the BPI's activities, for Schonland, regardless of the hat he was wearing, waged an unceasing campaign to coax it into some form of fruitful cooperation [50].

The birth of the CSIR had not gone unnoticed by the South African press and articles and editorials soon appeared announcing the arrival of a strapping youngster. Though somewhat younger than its Common-wealth cousins it had, they said, inherited only their better characteristics and was expected to show great promise at an early age. In Basil Schon-land it had a father of distinction, whose personal achievements in the worlds of science and soldiering were drawn to the attention of an inter-ested South African public with due acclamation. With careful nurturing and adequate financial support his creation should soon fill the nation with pride in all its achievements. So ran the eulogies.

Schonland, like many fathers of illustrious offspring before him, was asked for his aspirations and expectations of this prodigy. His plans for it were simple and the first audience outside the inner sanctums of govern-ment to hear of them were the good citizens of Germiston, an industrial town to the east of Johannesburg. There, at a luncheon in the Town Hall in November 1945, he told them what the future held for science in the country. The CSIR had five major objectives: to develop South Africa's natural resources; to encourage industrialists to form their own research associations; to act as the 'back room boys' within a number of very specialized laboratories; to set up, on a rigorous scientific footing, a laboratory for the selection of personnel by appropriate aptitude testing, and, finally, the development of fundamental research in the universities. In some of these a promising start had already been made, while others would still take some years to mature. The country was enthusiastic and even felt flattered. Their own scientists and engineers would be lead-ing them into the next decade and away from the two greatest disasters that had befallen the nation in recent times: the war that had so sapped their resources and poisoned their relationships, and the economic depression of the decade before that had come even closer to bringing the nation to its knees.

It was, therefore, an air of confident expectation that greeted every announcement made by those who heralded the birth of the CSIR in South Africa. South Africans who, one hoped, had been welded together by both the privations and the victories of the war, were now ready to take their places in an international arena, where science would play an ever-increasingly important role. But this was true only in some circles in the land. In others a very different set of priorities was rapidly gaining ground and enthusiastic support. The Nationalist opposition saw, not international collaboration, but an introverted isolationism as the only future for the white Afrikaner nation. Nobody had even thought to ask the black majority what they wanted.

CHAPTER 16

NATIONALISTS AND INTERNATIONALISTS

Even while the British nation had its collective back to the wall in 1941 there was talk within the Royal Society of holding an Empire Scientific Conference as soon as the war was over. The assumption of an Allied victory was seemingly taken for granted and was inspired, no doubt, by the regular bursts of Churchillian oratory that stiffened the resolve of the country as often as he felt it necessary. In October, representatives of Canada, Australia, New Zealand, South Africa and India attended a meeting in London and formed the British Commonwealth Science Committee with the intention of framing plans for a post-war Conference. Discussions with the various governments followed and the Committee reported back, indicating general support, in April 1943. A proposal to hold such a Conference then went to the War Cabinet's Scientific Advisory Committee and, in July 1944, His Majesty's Government approved it and invited the Royal Society to be responsible for its organization [1].

When, in November the following year, the South African government received official notification of the Conference it came as no surprise to Basil Schonland, for he had been his Government's representative at all the preliminary meetings and he and Smuts had effectively dealt with the matter between them. As Scientific Adviser to the Prime Minister, Schonland then wrote to all those Departments of State which, he believed, should be made aware of the Conference and its relevance to South Africa and he recommended strongly that there should be 'top-level South African participation' [2]. The delegation to represent South Africa was led by Schonland himself and, to ensure that it would have sufficient influence, its members were drawn from the highest echelons of the country's various scientific disciplines. He himself, as well as being its leader, would take special responsibility for all matters concerned with physics and its applications from atomic energy to radio and radar, while also keeping close tabs on all issues to do with national defence. To accompany him to London he chose Dr E H Cluver (Public Health), Dr L T Nel (Geology), Professor S F Oosthuizen (Medical Research), Dr A R Saunders (Agriculture), Dr J Smeath Thomas (Chemistry) and Dr P J du Toit (Veterinary Science). Between himself, Smeath

338

Thomas and Oosthuizen, both the CSIR and the universities would also be adequately represented while the diverse interests of the rest of the delegation would, he hoped, enable them to make useful contributions and establish international contacts outside of their immediate fields of specialization. In P J du Toit, his right-hand man, he had a scientist of sparkling brilliance with a personality to match. As a linguist, committee chairman and after-dinner speaker 'PJ' was unrivalled, while his reputation as a veterinarian was second to none. He would undoubtedly make his mark in any company; by contrast, the rest of Schonland's delegation contained no men of international stature; the events soon to come would prove to be the testing time and he would watch them closely.

While London itself in June 1946 still showed all the signs of the air raids that had been its lot since 1940, the English countryside was resplendent in the hues of summer and the faces of the British population showed, for the first time in years, that the tension had gone and their eyes, once again, were bright. Such were the perceptions that greatly struck Schonland on his arrival and which he communicated to Ismay in the many letters that he wrote, in between representing his country's interests, first at the Royal Society Conference that took place over three weeks and then at the Official British Commonwealth Scientific Conference that occupied the next two. The prestigious gathering was opened by His Majesty the King, accompanied by Queen Elizabeth, in the Beveridge Hall of the Senate House, University of London, on 17 June. The leaders of all delegations then had the honour of introducing their colleagues to the Royal Party, which Schonland duly did with some aplomb, even though he managed to interchange the first names of two of them, much to his subsequent embarrassment and their enjoyment, at a rare example of his fallibility. To the Queen he said how much his fellow South Africans at home were looking forward to the Royal visit the next year and she indicated her own excitement at the prospect [3]. Similar warm exchanges between the British sovereign, his consort and the scientists of the Commonwealth served to underline the close ties that had bound them all together during those previous years of turmoil and, undoubtedly at times, near despair. The Conference thus opened in an air of grand optimism. Nations that had until now been comrades in arms or competitors at games were about to explore together the challenges of science.

This first ever gathering of Empire scientists in England *en masse* was an occasion to be marked by due ceremony and by the awarding of honours to those most deserving amongst them. If any institution in the land could lay claim to having been the genesis of modern science, especially physics, it was the University of Cambridge and its Cavendish Laboratory. It was not even thirty years since Rutherford had returned there as a Nobel laureate with the intention of solving the mysteries of

Figure 21. Schonland and Dr Ernest Marsden visiting the Telecommunications Research Establishment, Malvern, during the Empire Scientific Conference 1946. (Reproduced by kind permission of Dr Mary Davidson.)

atomic structure. Since then his flair for experimental physics and his forceful, but engaging, personality had been behind some of the most remarkable discoveries in the modern age that placed Cambridge and the Cavendish at the very pinnacle of scientific excellence. The fact that Rutherford was himself a colonial added another dimension to the story, for in the eyes of his fellow-countrymen and their colonial cousins he was one of them and so his achievements took on a very different hue. Here was a son of the New Zealand soil who had become arguably the greatest experimental scientist since Faraday. What an incentive that was to others from the other hemisphere to follow him to England, and follow him they certainly did. In the two decades before the war Cavendish scientists from far flung corners of the globe helped to move physics forward in giant leaps. Then, in the conflict that followed, all their attentions were focused elsewhere. Now, Cambridge took the opportunity to honour some of them who had returned to England as leaders of their national delegations.

On 24 June the University held a special Congregation at which it conferred honorary degrees on these men of distinction. Intoning in Latin his words of introduction the University Orator struck a resonance: 'Since men desire nothing more strongly than the harmony of nations and pray for this today, it is of the greatest importance to free states that British society provides an example of mutual goodwill and common service. Consequently we are happy that so many men, excelling in genius and scholarship, have come to England at this time from all corners of our Empire and have assembled to debate amongst themselves freely and at ease about their studies and learning.' Turning to Cicero for a pithy maxim that would elevate all above the basest of human inclinations he enjoined them to 'Refellere sine pertinacia et refelli sine iracundia'.[*] Thus was the mood set not only for the five honorary graduands to present themselves to the Chancellor but also for the deliberations that lay ahead of them all during the next many weeks. Leading them was the conference chairman, Sir Henry Tizard, whose mission to the United States in August 1940, with the most precious cargo ever to leave British shores, had enabled American technology to be applied 'with common zeal to overthrow the common enemy'. The Orator then introduced Basil Schonland.

> Although he has come from a far distant land, nonetheless we rejoice that this man, whom I next present to you, was once a famous alumnus of ours. In his native South Africa, he presides over a council whose purpose is to apply the science of Physics (in which he himself excels) to the advantage of human life. In the war just recently ended, he served our cause not only in a variety of ways, but also particularly in this, namely that he played the role of mediator between scientific experts and military leaders, being experienced in each branch.
>
> I present you the most distinguished son of Gonville and Caius College, to be counted among the ornaments of the Royal Society.

This was as succinct a résumé of Schonland's recent past as one could imagine. It summed up, in its most gracious and eloquent Latin phrasing, the career of a man whose life, of late, had been one of service to a greater cause than any driven solely by personal ambition. Unlike some recent flattery which he had had to endure, this would not have caused him to cringe.

Following Basil Schonland in this parade of men of science were the Canadians, C J Mackenzie and C H Best and the Australian, F Macfarlane Burnet.[†] All were honoured with words of similar eloquence that rang in their ears, and it was indeed an auspicious occasion conducted in the style that Cambridge reserved for just such gatherings [4].

The two conferences that followed were organized on a grand scale which, given Britain's most recent past, was quite remarkable, but in their own way they typified the mood of the British nation as it picked itself up from nearly six years of war. Schonland's participation in the various

---

[*] 'To refute without obstinacy and to be refuted without anger'.
[†] Sir Frank Macfarlane Burnet was the co-winner with P B Medawar of the 1960 Nobel prize for physiology and medicine for the discovery of acquired immunological tolerance to tissue transplants.

deliberations was especially taxing. Formal meetings on matters of policy, visits to research establishments, where many of the wartime miracles were on public display for the first time, and the reading of three papers to various sections of the conference proved to be a punishing schedule. And then there were the elaborately arranged events in the evenings where selected speakers found themselves almost entertaining invited dignitaries with light-hearted after-dinner speeches. Basil Schonland was one of those asked to enlighten and to amuse and he envied his South African colleagues as they set out at the end of the day for London's theatres while he put the finishing touches to yet another speech. It was, though, all in a good cause and he shouldered the responsibility, ever mindful that it was an honour not accorded to all. His facility with the pen undoubtedly helped and, once completed, he wrote frequent notes to Ismay to keep her abreast of all the events. David, their son, was now at Cambridge and it had been some time since father and son had last seen each other. Basil's absence in England for much of the war coincided with David's three years at Rhodes University and a short period in the SSS just before the war's end. Then, when father returned home so son followed the well-worn path to Cambridge and Caius. Their meeting in London was a joyous occasion as Basil described it to Ismay. 'You have a very, very nice son — good looking, fresh & clean & intelligent & never bored. I was delighted with David and he was, as you said, a most delightful companion'. Together they caught the train one Sunday and travelled to Lewes from where they then walked 12 miles across the South Downs in countryside that Basil and Ismay knew so well. The younger Schonland was a mathematician and his future career was naturally a topic of mutual interest. England offered him all the opportunities and his father had many contacts with eminent men, who had once again resumed their places in academia, after diversions into the science of fighting a war. A career at a British university soon beckoned for Schonland minor.

The conference itself provided numerous opportunities for renewing old acquaintanceships, both civilian and military, and to make new ones. Blackett and Bernal, the duo who, at Mountbatten's invitation, had introduced the scientific method with such success to the planning of combined operations, were amongst the first Schonland saw. Then, there was Ernest Marsden, the man who in 1939 had introduced him to radar and then led New Zealand's own remarkable radar developments on his return home. Marsden, as Secretary of the New Zealand Department of Scientific and Industrial Research since its formation in 1926, was leader of the New Zealand delegation and had much hard won experience to impart.

Soon John Cockcroft arrived from Canada looking 'extremely fit and fat'. Schonland's long-time friend and colleague had just been appointed Director of the Atomic Energy Research Establishment about to be built at

Harwell, near Oxford, and was much involved in the machinations that go with trying to develop a research establishment under government patronage while at the same time keeping at bay all government bureaucracy. At Cockcroft's instigation, Schonland also met, for the first time in person, the man whom he had known well by reputation as Chief of the Air Staff. Lord Portal of Hungerford, after retiring from the Royal Air Force, became Controller of Atomic Energy, in which position he was responsible for the production of uranium and fissile material for use in Britain's embryonic atomic research programmes. Quite naturally Portal had a great interest in South Africa's uranium reserves and he invited Schonland to dine with him so that they might exchange views on matters of mutual interest and concern. Underlying Portal's elevated wartime rank and considerable authority was a man of great charm and simplicity and there was an immediate rapport between them that made open and frank discussion easy. Such untrammelled communication was due not only to their compatible temperaments but almost certainly to the good word that Cockcroft — equally impressed by Portal's informality at their first meeting a few months before — would have put in the former Air Chief-Marshal's ear about South Africa's leading scientist.

Coinciding with the Empire scientific conference in London was another gathering where the Dominions were well represented. Their Prime Ministers had arrived for the Victory Parade that took place on 8 June and Brigadier Schonland received an invitation to be present on the saluting base. The only problem was that the invitation reached him the day after the event! It turned out that it had been received well ahead of time at South Africa House but a South African public holiday, which saw the High Commission closed for a day, had intervened and all onward postal deliveries were delayed. However, Schonland was a special guest on an upper balcony at South Africa House, from where the view of the marching troops was spectacular. Accompanying him was Omond Solandt, awaiting discharge from the Canadian forces and the AORG, who, that evening, then extended his hospitality to Schonland at the Hyde Park Hotel, from where they watched the last ceremonial event of the war as fireworks and searchlights lit up the London sky [5]. This was now the end of it all. Uniforms and marching bands seemed strangely out of place as the world tried to come to terms with peace — however fragile that might prove to be. Inevitably, with Jan Smuts in London, a meeting was soon arranged between him and his scientific adviser. Uranium was also on Smuts's* mind, and particularly

* General Smuts, as a member of the War Cabinet, was fully aware of the Quebec Agreement of August 1943 between the United States and the United Kingdom that established formal cooperation between them on all matters to do with atomic energy. Only Canada, of all the Dominions, knew of its existence because of her crucial role in producing plutonium at the plant in Chalk River near Montreal [6].

how South Africa might stake out her position in a world of uncertainty, as the United States sought to preserve hers as the only nation on earth capable of producing the atomic bomb. Britain, for all her fundamental contributions to the underlying science, was as yet in no position to do the same, while Russia was a cause of much concern.

It was Schonland's intention to return to South Africa on 1 August 'unless something extremely high-up keeps me'. As it turned out the most likely source of such delay, the 'Ou Baas',* had already flown home so all that remained was the social round of farewells and the garden party at Buckingham Palace. On reflection, Schonland concluded that the conferences had been a great success. His South African delegation had made some useful contributions to the various scientific debates, while the opportunities presented to him for discussions with men of stature, both from England and elsewhere, had been especially important. None was more so than those with Sir Edward Appleton and John Cockcroft. Not only were they friends of long standing, but both were now in positions of great influence within the world of post-war British scientific research. As for himself, Schonland had been able to draw the delegates' attention to the peculiar problems of Africa, for he saw himself representing not just South Africa and its wholly white scientific establishment, but also the countries in the vast hinterland to the north. He called for the establishment of an African research committee to look closely at the special requirements of the continent, or at least that part of it south of the Sahara. Intensive scientific research could markedly improve the general health of the inhabitants of some of the world's poorest countries, while their economic futures could be secured by applying modern scientific techniques to the exploitation of the vast mineral resources that lay untapped beneath its surface. National boundaries should not stop the investigation of fundamental scientific problems, for such frontiers were just stumbling blocks and impediments to progress. Then, changing his CSIR hat for the smaller BPI variety, he saw geophysics, physics and geochemistry as obvious subjects in which South Africa should take the lead. Recalling, too, the fruitful collaboration he had established with the meteorologists in Salisbury in the late thirties he suggested that Southern Rhodesia should become the natural home for fundamental work on the weather systems of the sub-continent.

While these were just ideas, a very tangible start had already been made in the field of veterinary medicine by P J du Toit and his colleagues at Ondersdepoort. Their pioneering work on the development of prophylactic control measures against the variety of stock diseases that made Africa the graveyard of the cattle farmer, was acknowledged as far

---

* Lit. 'Old Boss' from the Afrikaans but essentially a term of endearment used particularly by Africans when referring to an elderly (usually white) father-figure.

north as Kenya and was soon to follow in Abyssinia, where the emperor, Haile Selassie, remained in South Africa's eternal debt for driving out the invading Italians in 1941. Schonland's pan-Africanist views accorded well with the objectives of the Conference, even though these may have appeared somewhat myopic in those countries outside the British sphere of influence. They also sat comfortably alongside Smuts's vision of Africa – 'the forgotten continent' with its vast resources, yet inexplicable backwardness – that must soon 'make its contribution towards its own and world progress' following the initiative of President Truman to promote the development of the deprived corners of the world [7]. But to some these ideas must have verged on heresy. Where, murmured the Nationalists back home, was the *Suid-Afrikaanse kleur* if Schonland could suggest that 'British research workers must share in any such centralised project in South Africa, not as guests or visitors, but as partners in the enterprise, as would South African workers attached to projects situated outside the boundaries of the Union'? [8].

On returning home it was clearly apparent that South Africa too was enjoying a post-war revival. European immigrants were arriving in large numbers and the economy seemed about to right itself after years of sluggishness, caused initially by the great depression of the early thirties and then stunted still further by the war. However, to those blessed (some would say cursed) with the longer view, dark clouds lay on the horizon. The United Nations, founded on a set of principles drafted by Smuts, was, ironically, about to debate a motion which saw South Africa in the dock. Charges of discrimination against the Indians in Natal had been levelled against the country by the Indian Government, which reinforced its case by imposing a trade boycott and recalling her High Commissioner. Smuts himself attended the General Assembly in October to present South Africa's case. The mounting tensions on the sub-continent between Hindus and Muslims, and the huge loss of life that followed when they clashed, only served to reinforce his view of his own country 'where so many races, cultures and colours come together ... the Union is doing its best on fair, decent and wise lines to keep the different elements, as much as convenient and possible, apart and away from unnecessary intermixture, and so prevent bloody affrays like those in India or pogroms such as those we read of in other countries'. The matter of the Indian population in South Africa, he insisted, remained an internal, domestic affair, but his Government had consented to its being referred to the International Court. However, he reminded the General Assembly, he would not agree to any investigation by an outside political body. When it came to the vote, Britain and the United States fell on opposite sides of the divide that now appeared amongst member states. Attlee's Labour Government supported the South African argument that a dangerous precedent was being set if internal meddling became the norm, but the

345

Americans did not [9]. And so, with the word yet to become common currency, 'apartheid' had fallen under the international spotlight and immediately became synonymous with South Africa. Henceforth the lives of every one of its citizens would be influenced to a greater or lesser extent by a philosophy that became increasingly untenable as the post-war world, shaken to the core by the recent genocide in the very heart of Europe, began to abandon the old empires and all that they stood for. South African politics now became a focus of international attention.

Within the country, such bonds as still existed with Britain were to be weakened considerably in little more than a year, when Smuts would be defeated at the polls and D F Malan's Nationalist Government would snatch power by a hair's breadth. Before that happened, though, there was one final occasion when, after nearly 150 years of British influence, a reigning monarch paid the first visit to the southern tip of Africa. The British Royal Visit in February 1947 had two purposes: to thank the people of South Africa for their loyal support and far from insignificant contributions to the war effort, and to honour the man whose personal standing in England, over such a prolonged period, was arguably higher than anyone's. The visit was almost a farewell to Smuts and it culminated in a State Banquet in Pretoria to which Basil and Ismay Schonland received invitations. While neither professed to be monarchists, and Ismay especially had strong socialist views, they, along with many of their compatriots, felt a strong kinship with the British nation and its figureheads who had endured so much in the recent past. As Basil had written in a letter to Ismay just a few months before when he met the King and Queen in London: 'The King looked very nice in Admiral's uniform and spoke extremely well—really well. The Queen won all our hearts and was most charming. This institution of monarchy undoubtedly has its points!' [10]. But the royal visit was but a brief and happy interlude. It was apparent that were now emerging tensions in South Africa that none but the most insensitive could ignore.

*   *   *

Schonland's weekly visit to his laboratory at Wits provided him with the opportunity to divorce himself from both the affairs of State and the nation, as well as the frequently tedious administrative deliberations at the CSIR that so often ensnared him. At the BPI he could join his colleagues in discussing their research and also indulge in some of his own. There had already been progress in a number of directions during the first year of peace, with Malan's new lightning camera with its rotating prism being an outstanding success. In addition, the construction of the radio-linked seismometric equipment was proceeding under Philip Gane's direction, while Anton Hales, whose work on earth tremors had been disrupted by the building operations in the BPI, was concentrating on theoretical work to do with the ionospheric reflection of signals emanating from a lightning discharge [11]. Schonland himself collaborated with Malan in the analysis of data obtained during the summers of 1937 and 1938 that had been stored away for the duration of the war. It represented some 37 lightning discharges to ground from 199 separate strokes, most of which were within a distance of 6 km from the measuring apparatus. Of especial importance was the fact that Malan's new drum camera was triggered electronically by the impulsive noise that emanated from the leader stroke of the lightning itself. This allowed the photographs of the lightning and the accompanying electric field emissions to be recorded and correlated—something that had never been done before, especially from lightning activity that was almost overhead.

The results, published in 1947 as the seventh paper in the series from the BPI on 'Progressive Lightning', were intriguing. By being so close to the source of the discharge to ground it was assumed that the electrostatic fields would dominate naturally over the induction and radiation fields from the discharge and so yield new information about the nature of the very processes themselves, especially any that may take place within the cloud or even upwards into the upper air, as had been seen on occasions in the past by both Schonland and Malan, and by C V Boys even earlier. Whereas previous measurements of lightning activity at much greater distances had shown the field changes to be positive, these new records from clouds almost overhead produced predominantly negative-going fields—an effect which had been reported in England just before the outbreak of war. Malan and Schonland now sought an explanation for this interesting phenomenon. The new photographic evidence, when combined with the electric field measurements, should enable them to deduce what was happening, because it was assumed that every measured change in field intensity would correlate with a corresponding variation in luminosity. However, it was the occasional absence of correlation that was interesting, particularly during the period of the first leader from cloud to ground. They concluded that this implied that the pilot streamer was moving continuously, as suggested by the smoothness of the electrostatic record, while the optical flickering was caused by some other mechanism that was 'secondary from an electrical point of view'. They believed they could also explain satisfactorily the change in polarity of the field by using the elegantly simple electrostatic models that C T R Wilson had first employed so many years before and which Schonland himself had used to such good effect in 1931 when he overturned the prevailing view of the polarity of a thundercloud. And, to cap it all, they even had an observation to make about discharges from the upper, positive pole of the cloud. Since almost every recorded

347

stroke carried negative charge to the ground, there had to be some other mechanism by which the positive charge could be removed from the upper regions of the cloud — as it must be in order to maintain the overall state of neutrality. There had surely to be a process of discharge into the conducting layers of the upper air. They postulated two mechanisms by which this could occur and believed that they had evidence amongst their accumulated data to support both, but it was not sufficient for any definitive pronouncement to be made. Wisely, they concluded that 'further work is needed before these observations and arguments can be regarded as fully establishing the existence of a regular upward discharge of the upper positive pole of a thundercloud immediately after the return stroke to ground has occurred'.

The paper describing their observations and analysis appeared in the *Proceedings of the Royal Society*. Schonland decided to use it as the vehicle to publicize to its international readership Malan's visual observation in 1937 of lightning discharges above the clouds [12]. Malan's Huguenot heritage had inclined him as a student towards the French and their institutions rather than the route followed by many of his fellow South Africans to England, Cambridge and elsewhere. He therefore completed his doctoral studies in Paris and so had, quite naturally, directed his earlier papers to French journals. Now, he and Schonland reminded the scientific community at large of the phenomenon that previously Malan had described so elegantly as *'une longue et faible bande d'une lueur rougeâtre. Deux fois le phénomène se présente comme un rideau auroral composé de cinq ou six de ces bandes'**. This may well have been the first accurate description by a skilled observer of a remarkable phenomenon, but it would be fifty years before the technology existed to allow these ethereal effects to be examined in any detail and even then the processes that gave rise to them would be but poorly understood [13].

While fundamental research continued unabated at the BPI, there was also work under way of a very practical nature that had immediate application in the local industries. The attention of those concerned with the manufacture of dynamite at Modderfontein, some 20 km north east of Johannesburg, was drawn to a device described in yet another publication that year from the BPI [14]. Written jointly by Schonland and Gane it described a lightning warning instrument that used the same technology as the device that triggered Malan's lightning camera, and was based on work started at the BPI as early as 1936. This elegant electronic system was a tribute to Gane's design skills and consisted of

---

* 'a long and weak beam of faint red light. Twice the phenomenon appeared like a curtain of sunlight at dawn composed of five or six of these beams.' Was Malan describing in accurate detail, and possibly for the first time, what would become known very much later as a 'red sprite'?

an aperiodic amplifier, connected to a vertical rod antenna that sensed the electric fields generated by lightning activity in the vicinity. A rapid-acting trigger circuit would initiate an alarm signal, once the intensity of the field exceeded some preset threshold. To be useful, and especially to prevent false alarms, with the inevitable loss of confidence in the system that would ensue, the device had to be capable of discriminating between lightning activity at a distance and that within a predetermined range of the warning instrument. This feature was achieved by means of an accurately calibrated attenuator that allowed the operator to select the range at which the warning system should be initiated, while a circuit that measured the actual corona discharge between objects on the ground and the clouds above would provide warning of the risk of imminent lightning discharge. The instrument was installed at the dynamite factory to the acclaim of the local press, who hailed it as the 'world's first lightning alarm' and its inventors as possessing the 'finest scientific brains in Southern Africa'. Such hyperbole was only in keeping with the mood of the times — even Schonland noted in his Director's Report on the BPI for the year 1946 that it was 'one of the finest laboratories in the world', while Smuts's son and biographer was also much caught up by the mood [15]:

> The years 1946, 1947 and the first half of 1948 were the most glorious in South Africa's history. Never had there been such prosperity; never had there been such goodwill; never had our name and affairs stood higher in the eyes of the outside world. It seemed like the golden millennium after the long years of struggle.

And there was even more to come from the BPI. Dubbed the Ceraunometer,* its purpose was to record lightning activity as an aid to weather monitoring and forecasting. In essence, it was yet another variation of the lightning-triggered device, with minor refinements to suit the application and it went into immediate use with the Rhodesian Meteorological Service in Salisbury [16]. While it was never Schonland's intention that the BPI should become an applied physics laboratory to service the needs of local industry and others, it was soon obvious that some of the fundamental research conducted within its walls could well be applied in just that way. Fortunately, the South African branch of the Dutch electronics company Philips of Eindhoven were more than willing to take on the manufacturing of these novel pieces of equipment, sensing as they did the potential markets in the mining industry, where the risks of premature detonation of explosives during shaft-sinking was always a matter of considerable concern. It took little to convince the University of the

---

* From the Greek word for thunderbolt.

financial advantages which might well come its way as a result, and so contracts were soon entered into between them.*

Almost next door to the BPI stood the TRL, and it too was beginning to show its potential. Still in its temporary home in the Wits Department of Electrical Engineering, the CSIR's Telecommunications Research Laboratory fell under the guidance of Frank Hewitt, while it remained very much in the eye of Schonland. This neighbourly proximity of the two laboratories was no coincidence, for it had always been Schonland's intention that the BPI and the TRL would collaborate most closely since, in essence, the one had spawned the other through the intermediary of the SSS. By the end of 1946 Hewitt had installed his 50 cm radar on the BPI's roof and was about to start an investigation of thunderstorms using this novel method. Shortly afterwards, Trevor Wadley resurrected an old problem and completed an investigation, at the request of the Chamber of Mines, into the feasibility of providing radio communications for rescue purposes underground. Lack of effective communication between the 'Proto'† team fighting a fire underground and its 'fresh-air base' farther back severely hampered such operations and the Chamber sought a solution. Though Schonland had doubted its feasibility when the question was first put to him almost ten years before, the matter was once again urgent. After making a number of measurements in mines throughout the Witwatersrand, Wadley believed it could be done and wrote a detailed report presenting his findings and recommendations for a suitable system.‡ However, the Chamber, for reasons known only to itself, chose not to pursue the matter and the proposal languished for more than decade before the idea was once again revived [17].

It was this reluctance by the mining industry in general, and by the Chamber in particular, to embrace new ideas that proved so frustrating to Schonland. Here, in their very midst, was a laboratory whose sole objective was to peel away the layers of mystery that shrouded the geophysics of the earth and its atmosphere, and yet the largest industry in the land was so unwilling to really commit itself to exploit the fruits of this scientific research. Only in one area did the mines show more than glimmers of interest and that was the seismic research programme at the BPI. From time to time the Chamber provided the occasional injection of cash to support it and did so once again in 1946 to the tune of £1500. Schonland

---

* Both the lightning alarm and the Ceraunometer were patented by the University. Subsequently there was some concern expressed within the BPI that their function was to pursue 'pure research' and not to collect royalties accruing from patents. It was agreed that all research results would be published 'for the benefit of mankind in general' (minutes of 17th ordinary meeting of the BPI board: 11 March 1948).

† So named because of the type of breathing apparatus they used.

‡ Wadley's report 'Radio communication through rock on the Witwatersrand mines' appeared as an official CSIR report from the TRL in August 1949.

was duly grateful and used it to set up a chain of outstations across the Witwatersrand, from where seismic data were transmitted by radio to the BPI, where they were then stored on a magnetic tape delay unit for up to six seconds, while the automatic smoked-paper recording equipment was switched on and brought up to speed. The technique worked exactly as intended and the results obtained were most encouraging and helped to enhance the stature of the BPI, in scientific circles at least. The Director, certainly, was quick to grasp the opportunity to make some capital from this. 'It may be said', he announced in his annual report, 'that the two techniques, here applied, namely telerecording by radio and the "memory device" are believed to be in advance of seismological practice anywhere else and have aroused interest in Cambridge and in Canada.'

*   *   *

On 5 September 1947 Basil Schonland found himself back in uniform—at least nominally. On that day, a Government Gazette from Pretoria announced the formation of a new unit within the UDF, to be known as the South African Corps of Scientists and its Director would be Brigadier B F J Schonland, whose temporary wartime rank became substantive two months later [18]. To many in the country this strange creation must have come as a surprise when a brief announcement appeared in the daily press. However, its origins can be traced back to just a few months after the end of the war when the role and function of the CSIR were at their most fluid. In January 1946 there was correspondence from the British Government to the Secretary for External Affairs in Pretoria inquiring whether South Africa wished to participate in joint work on 'guided and propelled missiles'. Evidently a committee had been set up in London to discuss the matter and it would be meeting shortly. Rather peremptorily the Chief of the General Staff replied that 'it was not considered necessary for the Union to be represented'. However, his decision had clearly been taken without much consultation, for just a few days later, he received a sharp rejoinder from the Secretary of Defence advising him that it would indeed be in South Africa's interests for a member of the UDF staff in London to attend meetings of this committee. Much vacillation then followed. It was soon appreciated at both DHQ and the Union Buildings that such representation must be at an appropriate level and the only man of sufficient standing in the country, both scientifically and militarily, was Brigadier Schonland. But by then Schonland was fully committed elsewhere. As South Africa's pre-eminent scientist he was to lead his country's delegation to the Empire Scientific Conference at much the same time as his services were now in demand by DHQ. He therefore informed the Secretary for External Affairs (the lines of communication having widened considerably) that he was

351

unfortunately unable to attend either the meeting of the 'Committee for the Coordination of Research and Development of Guided and Propelled Missiles' or that of the Defence Research Council to follow immediately afterwards. The CGS was left with no alternative but to inform his Minister that, since there were 'no research scientists in the Permanent Force' and that he could only be guided by Dr Schonland who was unable to assist: 'we should pull out altogether'* [19]. Such, then, were the somewhat hamstrung deliberations between various Ministries of State and the Defence Force that led ultimately to Basil Schonland seeing the need to take up yet more reins of responsibility in his country's interest.

He had already foreseen a role for the CSIR in the field of military research when, in September 1945, he took it upon himself to raise the matter with the Chief of the General Staff in Pretoria. Schonland concentrated particularly on the subject of aeronautical research and informed the CGS that he envisaged the CSIR providing two types of assistance to the Defence Department, 'one secret and one general' [21]. Discussions, no doubt all secret, soon followed. By July 1947 a Defence Scientific Committee had been formed under the chairmanship of Major General W H E Poole, the Deputy Chief of the General Staff, with Schonland representing the, as yet, non-existent new Corps of Scientists. Alongside them was the Quartermaster General, as well as senior officers from the air force and the navy, plus Schonland's recently appointed deputy of this fledgling corps, Professor B L Goodlet† from the Department of Electrical Engineering at the University of Cape Town. The function of this new body, as Schonland saw it, was quite simple, at least on paper. It would study the application and implementation of scientific methods to warfare and would advise the General Staff accordingly. The matter was now left in the hands of the military to decide what to do next.

It was not only the mining industry that found the prospect of allowing scientific interlopers into its innermost sanctums rather less than compelling. The South African military, for all their exposure in the war just ended, to all manner of scientific wizardry, were not about to have their doctrines determined for them by scientists—regardless of their credentials. Thus it was that Schonland's patience was to be tested by another of those agencies whose procedures and practices were almost set in stone. But, in this case, the scientist in question had more

---

* In the end South Africa did have a representative at the meeting of the Defence Research Council. Mr Eric Boden, a former officer in the SSS, who ran the CSIR's scientific liaison office in London attended on behalf of his government [20].

† During the war, Goodlet had played a major part in South Africa in the programme of degaussing ships to reduce their susceptibility to magnetic mines. In 1949 John Cockcroft invited him to become Deputy Chief Engineer at Harwell.

than a jot of experience himself of the part that science had played in turning the tide of war and the generals, at least, knew that too. It would, though, be another three years before this new Corps of Scientists received more than just passing interest within DHQ. As well as a corpus of scientists to advise on and provide solutions to military problems, Brigadier Schonland saw the need for the newly formed Committee on Defence Research to be expanded to include representatives of the universities and the various scientific bodies within the country. It was, he reminded the CGS, 'very desirable that this committee be called together at an early date so that I may obtain its advice and guidance in connection with the development and expansion of the Corps of Scientists in an emergency ...'. That emergency was not long in coming. By late 1950 the Korean war, that had broken out in June, was threatening to become a conflict with global consequences and Schonland's persistent call for a scientific presence within the military superstructure began eventually to bear fruit. His request for the expansion of the Committee on Defence Research was approved in principle by the Minister of Defence and in December 1950 the CGS asked him to recommend those bodies and their individual representatives that should constitute the Committee [22].

However, we are getting ahead of the story. In October 1947 South Africa was on the verge of a change of government, while the CSIR passed another milestone in its existence. Now two years old and employing more than 250 people, it had begun to acquire some degree of newsworthiness amongst the local press, who displayed an encouraging interest in the fascinating world of state-sponsored science into which the country was just beginning to venture. Schonland, though no showman, was very conscious of the need to communicate some of the achievements of the CSIR to the populace at large and he struck up a ready, if slightly formal, rapport with the journalists. He was also in demand as a speaker to the various service clubs and women's groups around the country and he seized these opportunities to appeal for an awareness of science and engineering amongst his fellow citizens. Speaking to the Pretoria Women's Club, he described rather colourfully the scientists working under him as 'the new pioneers of South Africa, travelling with an electron-microscope instead of an ox-wagon'. Such analogies presumably raised a laugh from some of the citizens of the city within whose limits stood the imposing granite memorial to the Voortrekkers, for whom the ox-wagon had almost a religious symbolism. His point, though, was quite subtle for he was very conscious of the stifling effect that a government driven by ideology would have on science and especially on scientists. 'Scientists', he said, 'must be left alone in the right atmosphere and under the right conditions; the public service conditions were quite unsuitable for this work'. Scientists must be given 'the freedom and the elasticity they needed for their research and they should be left alone for a number of years before the public asked for an accounting of what they had done'. Whether he sensed the looming demise of the Smuts government or just chose to stake out his ground for any government to see is less important than the simple fact that he spoke out so strongly in defence of unfettered freedom for research. Most newspapers consigned the speech to an inside page but one or two took up Schonland's call with enthusiasm. 'Patience for the Scientists' was how one editorial responded and then went on to remind both the people and the State, that they had a duty to heed the words of the man in charge of South Africa's 'back-room boys'. Schonland's pleas, said the editor, must not fall on deaf ears [23].

Whatever the view of South Africa's scientific potential within the country's borders, there was certainly interest from beyond them in a by-product of its gold mining industry. It had recently been realized that there were vast reserves of uranium, albeit of low grade, to be found within the tailings of the gold refining process that was a daily activity carried out on a huge scale across the Reef. South Africa was now seen as a key source of the raw material required both for weapons and ultimately for power stations. In some quarters it was even believed that South Africa possessed the most important deposits of uranium in the world, but its economic exploitation was still some years away. No government would miss the opportunity to exploit such a windfall and Smuts did so to the full, in fending off attempts by Clement Attlee's government in Britain to obtain preferential rights to it. The United States, too, had a great interest in this new, untapped reserve and so, inevitably, South Africa became a player in the international political arena. Smuts, for all his fundamental allegiance to Britain, said only that he would ensure that the material did not fall into 'hands that might abuse' this 'new source of frightfulness'. All his options were to be kept open. Of course, the man who would play the leading part in initiating the process of refining South Africa's uranium was Basil Schonland, since it would be the CSIR that had the ultimate responsibility for the underlying science. He was well aware of this and had already reached an agreement, during his visit to England the previous year, for the construction to commence of a pilot-scale purification plant in South Africa [24].

No meaningful discussions in England on the subject of atomic energy and all its implications could take place without the active, though frequently silent, participation of John Cockcroft. The war had hardly ended when he began to look for a site for the British nuclear research establishment that would make the country independent of the Americans and, ultimately, allow it to develop its own technology and, if that was the requirement, its own atomic weapons. Amongst his scientific colleagues, Chadwick, Oliphant and Peierls, with whom he had remained in constant touch while director of the Chalk River Laboratory in Canada, it was taken almost for granted that Cockcroft would become director of the new British facility to be constructed at Harwell [25]. He was therefore the man to whom Schonland turned, as he had done so often before, for advice and well-informed comment on the matter of the uranium that evidently existed in abundance in South Africa. The first formal communication between them on the subject had been a letter from Schonland, written in September 1945, in which he informed Cockcroft that 'we are looking for uranium with a Geiger counter' but 'we are quite uninformed [and] our resources in trained men are negligible'. Essentially, only Schonland and Meiring Naudé, soon to take charge of the National Physical Laboratory, could claim any background at all in the subject. Both Schonland's letter and his search for uranium were exploratory and tentative: 'we are not likely to push ahead unless the search is important' which, as it soon transpired, it certainly was [26]. That very month George Bain, whose collection of South African gold-bearing ore had first sparked the interest in 1941, and Dr C F Davidson of the UK Geological Survey, visited South Africa to conduct preliminary radiometric surveys in the Rand and Orange Free State gold fields. Their findings indicated the presence of extensive reserves of uranium and their report was greeted with considerable relief by the Combined Development Trust set up in terms of the 1943 Quebec Agreement between Churchill and Roosevelt to control all matters to do with the exploitation and use of atomic energy. Until now the primary source of uranium had been the Belgian Congo; the news of a second one was a godsend [27].

Cockcroft reacted swiftly. Collaboration with South Africa was obviously to be of great importance to the British nuclear programme and his very close relationship with Basil Schonland enabled him to pursue the matter both formally and informally, as the occasion demanded. By the middle of 1947 he had informed Roger Makins, the Foreign Office official who chaired Britain's Atomic Energy Official Committee (sic), that Schonland would value 'a fairly free discussion on this subject'. In July, Cockcroft told Schonland about the availability of declassified British documents and he also offered any help, if required, in establishing a nuclear power programme in South Africa. And he went even further by offering to assist in getting information from the Americans, which might assist whatever programme Schonland had in mind [28]. Such was the influence that John Cockcroft was able to bring to bear, and such was the esteem in which he was held in the USA, that he felt quite confident in making such an offer to Schonland. However, in this Cockcroft was surely being rather naïve. For all her close ties with the countries that had been her Dominions, Britain's future as an atomic power—and she was determined to have one—depended critically on the maintenance of the closest possible cooperation with the United States in all matters to do with atomic energy. Since the Americans vigorously opposed the divulgence of information to anyone other than Britain herself and, to some extent, Canada because of her involvement in the Manhattan project almost since its inception, collaboration with the wider Commonwealth was most unlikely at this stage.* There was, though, the matter of South Africa's uranium, that trump card in Smuts's pack, which was surely vital to all Britain's atomic aspirations.

Within some quarters of the British Government, there had even been a suggestion regarding the possible location of an atomic energy plant in South Africa, but that was an idea well ahead of its time. On 15 July 1947 Cockcroft wrote to Sir Henry Tizard, now scientific adviser to the recently formed Ministry of Defence and also chairman of the Defence Research Policy Committee that was intimately involved in all matters concerned with Britain's own atomic energy programme. To Cockcroft, informed as he was by Schonland, the idea of a South African atomic facility was fraught with difficulties and these he spelt out to Tizard. There was a shortage of engineers and drawing office staff in the country; there was no suitable inspection organization, particularly in the chemical field; and, finally, there were no local sources of the specialized engineering components that go to make up such a plant. For these reasons he recommended, if plans for such a plant were to go ahead, that it would have to be designed in England and many of the engineering components manufactured there as well. If these shortcomings could be overcome, then the type of plant best suited to South Africa would be either air- or gas-cooled.† In any event, Cockcroft suggested that it would be most advantageous to discuss the matter with Schonland before proceeding any further. Tizard readily agreed and there the matter seems to have rested, at least for the time being [29].

On 3 May 1948 Schonland flew to England for what he told the press was merely a 'routine' scientific visit, but from there he went to Ottawa and then on to Washington [30]. As reported in the South African newspapers, the routine matters to be discussed covered such subjects as atomic energy, terrestrial magnetism, atmospherics and what they called 'wave motion'; all, presumably, of interest to South Africa and certainly to Schonland himself in one form or another. But atomic energy was the dominant, though decidedly hush-hush, theme. He was

---

* The Americans were certainly not averse to making their own, sometimes bizarre, inquiries in South Africa. Schonland once asked a visiting American seismologist at the BPI why, during field trips, he was buying quite so much cheese from towns around South Africa. The American declined to reply but Schonland said later that they believed it was possible to determine the uranium content in the soil by analysing cheese produced by cows in the area in question! [Anton Hales: letter to the author 27 September 1998].

† A similar suggestion had been made in the summer of 1947 that Britain might build a water-cooled atomic pile in Rhodesia with possible sponsorship by South Africa. This 'curious suggestion was (however) abandoned'. See Gowing ref. [6] p 148.

in England for a month and much of that time was spent in the company of Cockcroft at Harwell, but he also visited the National Physical Laboratory at Teddington, as well as other research establishments around the country before departing for North America.

Whilst he was away, South Africa was shaken to its very core. On 26 May the Smuts government was defeated in a General Election that brought Dr D F Malan's Nationalists to power by the narrowest of margins. To professed liberal thinkers in the country a Nationalist government was a tragedy with ill omens for the future. To many Afrikaners it was the glorious event that broke the yoke that had tied them to England and had dragged them into 'England's war'. There were those such as E G Malherbe, who was Director of Military Intelligence during the war and then Principal and Vice-Chancellor of the University of Natal immediately afterwards, for whom it was a black day. He expressed his views most directly in a letter to Smuts. 'If this lot who are now on top remain in power, it is going to ruin our country', he wrote [31].

The results of the South African general election, if not actually emblazoned across the front pages of the British newspapers, had at least caught the attention of those who still thought of South Africa and Field Marshal Smuts as being almost synonymous. Now, just like Churchill three years before, he was out of power. For more details, Schonland turned to his former colleagues at South Africa House in London, and for her personal views he waited anxiously for a letter from Ismay. It eventually reached him in the United States but, other than the bald statistical details of the vote Ismay could do little more than speculate. In reply Basil agreed with her that all they could do was 'to wait and see'. But of immediate interest was his own future as the President of the CSIR and his personal relationships with the ministers of the new government, who would influence its policies and the future direction it would take. It also remained to be seen whether he would still have direct access to the Prime Minister. Of some immediate concern to both Basil and Ismay was the appointment of Eric Louw to the portfolios of Economic Affairs and Mines. Louw, an outspoken admirer of the Axis powers during the war years, would clearly be in a position of considerable influence as far as the CSIR was concerned and this fact alone was disturbing.

In accepting the Presidency of the CSIR, Schonland had viewed it very much as a duty to Smuts as well as to South Africa. His expressed wish was really to return, ultimately, to the BPI and to his personal research, and his dual appointments at the CSIR and the BPI ensured that this would occur at some appropriate time. With the agreement of Smuts in 1945 the period of the CSIR appointment was just for three years. What happened after that would depend on many factors, of which one would undoubtedly be the political hue and philosophy of

South Africa's government at the time. Basil's reply to Ismay's letter is therefore most revealing on this matter [32]:

> I don't want to be impetuous but you may bet about 3 : 1 that we go
> back to the B.P.I. at the end of next year, which is not really so far
> away. After all it was 50/50 before [the election] and so it is not
> unrealistic to put the odds up.

He clearly intended to keep his options wide open and soon they widened even further with the arrival in August of a letter from Cockcroft. Now Sir John, in recognition of the massive part he had played in establishing Harwell as the largest research establishment anywhere outside of the United States, Cockcroft was a major figure in British and, indeed, in world science. Any hint of pretentiousness, though, was totally foreign to the man and his letter to Schonland was, as they always were, concise to the point of a footnote and almost illegible. With the agreement of the British Council, whose function was to promote Britain abroad in every sphere, Cockcroft was to deliver a series of lectures to South African universities on the subject of atomic energy and he now sought Schonland's views on the most appropriate time for the visit. But it was the message contained between the tightly compressed lines of that letter that indicated that there was more to his trip than just the lectures. South Africa's uranium was crucial to Harwell and Cockcroft wished to discuss terms for obtaining it with the South African government, but there was also another purpose. Cockcroft's antennae were acutely tuned and he knew that the result of the recent election had changed many things in the country, not least of which were the Schonlands' own perceptions of the future. His closing lines touched on the matter [33]:

> I hope the political events have not affected your world materially.
> You will wish to keep your Institute, & this will no doubt provide
> for any uncertainties of the future. There is however always a need
> for someone of your prowess over here if ever you want a change.

If ever he wanted a change? Change had always been part of Schonland's life, precipitated as much by his own ambition as by events. Both would now conspire to determine his next move and here was Cockcroft easing open the door. Other invitations to cross similar thresholds would follow soon.

*   *   *

The BPI's annual report for 1947, written by its Director, struck a very positive note and indicated that the Institute could certainly be said to have recovered from the effects of the war. All remaining staff positions had been filled during the year and, even though the delivery of equipment from overseas was still slow, most of the requirements were being met from the BPI's own workshop, under the competent direction of Jock Keiller. Schonland had sounded a particularly bullish note by asserting that 'the achievements recorded in this report [are] an index of still better things to come ...'. Once again he reminded his readers of the pre-eminent position that the BPI occupied both nationally and internationally: 'It has become one of the important centres of research in Southern Africa and holds a unique position in the world in regard to some of its activities' [34]. Certainly in terms of its research publications it had no rivals in the country and, in the lightning area, even in the world. The research was buoyant and the atmosphere of open and free discussion, that Schonland took great pains to encourage, saw new ideas percolating through all the time. As always, he was only too aware of the BPI's physical isolation from the international centres of science and, to counteract its effects, he encouraged his staff to keep abreast of progress by reading the scientific literature.

Occasionally, research was reported in those journals that really fired the Director's imagination and this was never more so than if the author happened to be one of his many acquaintances from the Cavendish. In 1947 Patrick Blackett published, in *Nature*, a speculative paper*, in which he proposed that a fundamental property of any rotating body was the generation of a magnetic field proportional to its angular momentum. The earth being just such a body was a good example. Actually the idea was not new; it had first been proposed by Arthur Schuster at Manchester some years before but Blackett now applied his considerable intellectual powers to understanding what he considered to be a phenomenon of fundamental importance. Others, too, were intrigued by Blackett's suggestion and speculated upon both the mechanism at work and the characteristics to be expected of such a magnetic field. Another of Schonland's former Cambridge colleagues and now Professor of Physics at the University of Toronto, E C Bullard, who was no stranger to the BPI, suggested in correspondence with Schonland that the gold mines on the Rand provided just the site to test the theory. Bullard also drew Schonland's attention to the fact that, while Blackett's theory predicted a decrease in magnetic field strength with depth, the so-called core theory of the earth's magnetic field showed that it increased with increasing depth. A deep mine might well be the ideal place for mounting an experiment to determine which of these competing theories was correct.

During September, Schonland arranged that Dr Anton Hales, with the assistance of Mr D I Gough from the CSIR's National Physical Laboratory, should spend four days at the Blyvooruitzicht Gold Mine on the West Rand making measurements of the earth's magnetic field at depths as great as 1600 m below the surface. Their results, which were

---

* 'The magnetic field of massive rotating bodies' *Nature* 17 May 1947 **159** 658–666.

published later that year in *Nature*, fell somewhere between the two competing theories and so were inconclusive. Since these were the first experimental results obtained by any group in the world, Schonland was especially keen to capitalize on the head start the BPI had obtained, but in order to do so would require an improved instrument with which to measure the field variation with depth. Immediately, all the Institute's resources were thrown into designing and building a considerably more sensitive variometer, which was then sent to the Carnegie Institution in Washington, for any comments and advice that they might care to offer about its design and performance.

Equipped with this new instrument, Hales and Gough continued their measurement campaign at Blyvooruitzicht. While they obtained some data that tended to agree with Blackett's hypothesis these were not convincing, and they speculated that the presence of large, magnetic ore bodies in the vicinity of the test site were interfering with their measurements. However, by now Blackett was himself rapidly coming to the conclusion that his theory was incorrect and that any measured results tending to confirm it were themselves just spurious. The matter was finally laid to rest when, in the very best traditions of scientific accountability, Blackett published a monumental paper in 1952 entitled 'A negative experiment relating to magnetism and the earth's rotation'.[*] Though erroneous, Blackett's theory was by no means the cause of a wild goose chase, either at the BPI or anywhere else where the idea had caused more than a little interest. Certainly, the hypothesis itself was flawed but the research that followed from it yielded much of value in the field of palaeomagnetism, plate tectonics and about the very evolution of the earth's crust itself. It could well have been the stimulus for an idea then beginning to cross Schonland's mind and which would take shape in just a few years' time, when he proposed the formation of a section at the BPI with the specific aim of investigating Precambrian geology, geochemistry and tectonics [35].

Another idea of Schonland's, though almost certainly attributable to C T R Wilson's early work in the area which was later to bear abundant fruit in a most unexpected quarter, also saw the light of day early in 1948. Schonland was convinced that, by periodically screening, from an external electrostatic field, an isolated plate connected to a sensitive amplifier, it would be possible to enhance their ability at the BPI to make very precise measurements of the electric fields generated by thunderstorms. Until now the preferred method of doing this was the technique developed by Appleton, Watson Watt and Herd as long ago as 1926, which used what was essentially a monopole antenna and an oscilloscope to determine field intensity and polarity. However, that

[*] *Phil. Trans. Roy. Soc. Lond. A* 1952 **245** 309 *et seq.*

required exceptional insulation properties in the mounting of the antenna to obtain the long time-constant necessary for accurate measurement, and it was therefore a somewhat inadequate tool for detailed research purposes, especially in the rain. By alternately exposing and then shielding their sensing element to the field Schonland and Dawie Malan, who was mainly responsible for developing the device, converted what was almost a DC effect into a much higher frequency that made it easier to amplify electronically. This yielded considerably increased sensitivity. The instrument they constructed was actually a field mill but soon it became known, rather more evocatively, as the Spinning Jimmy. It was in that guise many years later that it came to the attention of the Russians when they were planning to launch Sputnik III, by far the largest of their orbiting satellites, and the first such vehicle to carry an array of scientific instruments into space. In 1958 the local press proudly announced that South Africa had an important part to play in the Russian space programme as a result of the Spinning Jimmy aboard Sputnik III. *Die Transvaler*, the frequently paranoid Afrikaans language newspaper whose founding editor just before the outbreak of the Second World War was Hendrik Verwoerd, ultimately to become Prime Minister, assured its readers that the Russians had not stolen any secrets from the BPI but had acquired the knowledge to construct their own instrument from the scientific journal that carried Malan's and Schonland's original paper in 1950 [36].

*   *   *

1948 was more than just a year of political upheaval in South Africa. It was, as we have seen, also the year that the country suffered three grievous blows with the deaths of J H Hofmeyr, H J van der Bijl and Bernard Price. In their very different ways these three men were giants on a small stage. Hofmeyr, the child with a precocious intellect that saw him sweep the boards at the South African College and then again, when hardly out of short trousers at Oxford, become, at the age of 24, the Principal of Wits. As Smuts's closest political ally he was the liberal conscience and standard-bearer of South Africa and was often Acting Prime Minister when the 'Ou Baas' was walking the wider international stage during the war. Reviled by the Nationalists for his outspokenly liberal views, 'Hennie' Hofmeyr, as he was known to Ismay Schonland's side of the Hofmeyr family, was lost to South Africa at the very moment in the country's history when his unrivalled intellectual powers was most needed. Hendrik van der Bijl was the South African-born scientist whom Smuts had called home from an outstanding research career in the United States to turn a country fast becoming too dependent on just its gold and diamonds into the industrial power-house of Africa. ESCOM, the electricity supplier that combined the best elements of state control and private entrepreneurial flair, was his first major venture. That was followed soon after by ISCOR, the iron and steel corporation, whose output was used to such massive effect by the local factories that produced military hardware for the Allies during the war. Then, as Director General of War Supplies, van der Bijl's marshalling of South Africa's resources during that critical time was outstanding and earned him great accolades. He was one of the few South Africans to have been elected to the distinguished ranks of Fellows of the Royal Society and on his death, at the age of 60, with many projects still unfinished, Basil Schonland wrote the biographical memoir to commemorate a remarkable scientist and patriot [37].

The death of Bernard Price touched Schonland most closely of all. Price, electrical engineer, benefactor and philanthropist, was a big man in every sense of that word. His presence alone in any gathering guaranteed him attention but it was his support for scientific pursuits far removed from the world of electrical power, in which he had made both his name and his fortune, that drew him apart from the world of business and industry. It was his recognition of the almost unique position of the Witwatersrand as a natural laboratory of lightning and, in view of the richness of its palaeontological deposits, as the cradle of civilization that led Wits to establish two Institutes that bear his name. The first, of course, was to become the launching pad of Basil Schonland's own illustrious career while the other, the Bernard Price Foundation became, after his death, the Bernard Price Institute for Palaeontological Research. Both are memorials to a remarkable man without whose benefaction scientific research in South Africa would have been immeasurably the poorer [38].

The uncertainties of the political events of the year and the passing of some of its most familiar and important public figures must have left many South Africans with a feeling of numbness and not a little uncertainty as to what the future held for them. For Basil Schonland his own future at the CSIR was assured, at least in the immediate short term, by his re-appointment as its President for a further period of one year as from 5 October 1948 [39]. The new Prime Minister saw no sense nor political capital in losing the services of the man who had done so much for South African science, while Schonland himself, though more and more attracted by the scientific challenges at the BPI, was prepared to see how the new government's policies, both towards science and in a broader sense, affected him personally before opting to return to research or, indeed, to explore other pastures.

Two pressing matters required his immediate attention. The first was a geophysical conference to be hosted by the BPI and held at Wits. It had long been Schonland's aim to hold such a meeting in Johannesburg as a showcase of his Institute's work and this was now the ideal time to do it. The conference was scheduled for the end of July 1949 and it was the intention that it should be an event of international proportions, attracting speakers from abroad as well as from all South African institutions with an involvement in geophysics and the related sciences. The other matter which fell under his CSIR hat was to follow soon after in October. It was the outcome of discussions first held in London three years before at the Empire Scientific Conference and the ensuing resolution that recognized the growing need for a long-term and highly specialized research programme dedicated to the problems of Africa. Needless to say, Schonland had been the prime mover behind it, but its actual genesis could probably be traced even farther back to the decidedly unabashed speech he had made in 1942 before Sir William Bragg's committee on Imperial Scientific Collaboration on the subject of 'Lost Chances and Science in Africa'. Now it was all to come to pass and much remained to be done.

Then, like seeds borne on the wind, there arrived a personal and confidential letter addressed to Brigadier B F J Schonland CBE FRS and bearing the signature of Sir Edward Appleton. Wheels that had started turning in 1915, when Schonland and Appleton were subalterns preparing for war, had revolved some considerable way. Each was now a man of consequence and their paths had crossed frequently since those days at Bletchley. Now Appleton, about to leave government service and return to academia, had a proposal to make to his South African friend and colleague.

CHAPTER 17

'COME OVER TO MACEDONIA AND HELP US'

The tenancy of some of the most senior posts in the British scientific estab-lishment fell vacant in 1949 and men of stature to fill them were being sought in rather an uncoordinated fashion. The present incumbents were men whose names, almost without exception, had been linked with Basil Schonland's since the days when they had all graced the hallowed halls of Cambridge and, especially, the Cavendish laboratory. Though it was their scientific prowess that had first gained them attention, it was the demands of war that saw all of them thrust into positions of considerable administrative responsibility. Sir Edward Appleton, ionospheric physicist and Nobel laureate, was Secretary of the DSIR throughout the war years and the man ultimately responsible for the Tube Alloys project that led to the concept of the atomic bomb. In May 1949 he would return to acade-mia as Principal and Vice-Chancellor of Edinburgh University. Sir Charles Darwin, mathematician and one of Schonland's first scientific collabora-tors, spent the early years of the war as Director of the National Physical Laboratory and then became the first Director of the British Central Scien-tific Office in Washington responsible for all scientific liaison between the two countries, including the atomic bomb. After a year he was appointed the first Scientific Adviser to the War Office and then returned to the NPL in 1943. He was due to retire from that post in March 1949. Sir Ben Lockspeiser, whose career within the walls of the Civil Service shielded him from much of the wider world, had, since 1946, been Chief Scientist at the Ministry of Supply, that multi-tentacled organization responsible for the provision of research, development and equipment for the Ministry of Defence, and he was now about to succeed Appleton at the DSIR. There were thus two large gaps in Britain's scientific edifice that had to be filled.

Early in January 1949, Appleton wrote a personal letter to Schonland in Pretoria [1]:

10/1/49

My dear Schonland,

I am sending this strictly personal approach to find out whether you would be prepared to accept the post of Director, National Physical Laboratory, Teddington, if it were offered to you.

Darwin is retiring in March of this year, as I expect you know. The salary is £2500 per annum without residence or £2250 with free accommodation in Bushy House (also free heat, light and power). The latter alternative is the more attractive.

There is also a fairly substantial entertainment allowance from the Petavel Bequest.

However I won't go into detail at this stage but you can count on me to let you have information on any point on your request.

If your reply is, as I hope personally, a favourable one it would help me if you could give me some idea of date of release from your present post — assuming the Lord President offered you the N.P.L.

You will, of course, understand that the appointment is not in my hand, but in the hand of the Minister!

> With warm regards
> Yours ever,
> E.V. Appleton

Schonland's immediate reaction on receiving Appleton's letter was to do what he had done so often before, when in need of an informed and considered opinion: he wrote to John Cockcroft. 'I write for your advice — though I expect you will laugh', he began. After telling his old colleague of Appleton's tentative approach to him, he explained that it was his intention anyway to leave his post as President of the CSIR 'at the end of the year' and to return to the BPI. His decision to do so, he said, had been taken for two reasons. The first was the very attractive financial offer made to him by Wits, while the second was his 'unsatisfied longing' to get back to some research of his own.

Now, out of the blue, had come Appleton's exploratory letter and, though tempted, Schonland also wavered. His next move would require careful thought and consideration of the many issues this raised. As always the financial implications were uppermost in his mind: British rates of income tax and the fact that he was still supporting his daughter, Ann, at university. On the matter of his suitability for the post he put his cards on Cockcroft's table and showed that side of his character that relished the challenge once any initial doubts had been convincingly assuaged: 'I am not an engineer & the NPL is very engineering — but if you think it would be fun to give it a go, I will. The BPI is after all a very small place, with limited resources, though it is attractive because of its freedom'. His enthusiasm then became almost palpable as it occurred to him that Cockcroft, whom he felt sure was actually behind the offer, might even be about to take over the running of the DSIR from Appleton. 'Please tell me', he asked 'if that is so because it would make a big difference to my decision if the offer gets really definite' [2].

Cockcroft's name was indeed one of those on the merry-go-round then spinning in Whitehall as key posts came up to be filled, but he had been sounded out, not for the DSIR, but for Darwin's job at the NPL, as well as for Lockspeiser's old position at the Ministry of Supply. The man charged with making the recommendations to government was Sir Henry Tizard who, as chairman of the committees dealing with both civil and military research, knew better than most who were the men of real consequence. But Cockcroft turned down both offers. On the scientific level the experimental programme at Harwell had reached a critical stage and he was intent on seeing it through. Personally, the upheaval, the need to move home and the financial demands of still supporting a young family were, for him and for Elizabeth his wife, factors also to be considered and when taken together they weighed heavily against any move [3]. The result was that two of the key positions in British science, the Director of the NPL and the Chief Scientists at the Ministry of Supply, were essentially in direct competition with each other for men of real calibre and competence and so the net was now cast abroad, in various directions and by various people, simultaneously. Basil Schonland soon found himself to be a highly desirable catch twice if not three times over.

Schonland's subsequent reply to Appleton, written after giving the matter much thought, indicated that he would certainly be prepared to consider the offer of the NPL were it to be made to him, but before committing himself there were some issues on which he sought clarification. Of prime importance was whether the responsibilities of the Directorship would be so onerous as to prevent him from pursuing his own line of personal research. Here he struck a chord, for Appleton himself had, in whichever position he found himself, always kept his research hand in and so his response was encouraging, though tinged with reality [4]:

> We should of course be very glad indeed if the Director could continue some research, though things would work best if that research were on the Laboratory's programme or at any rate closely related to it. I could well imagine, for example, your taking an interest in some lightning or spark discharge problem with, say, your own Assistant so the work could go on when you were otherwise engaged. I have, again, talked this over with Darwin and his feeling is that, welcome though such a proposal sounds, you may find other things including the outside activities which tend to come the way of the Director, N.P.L., crowding you out. My own comment there is that one must just see that things like this do not crowd a little research out entirely. As you know, I've always kept something going myself since coming to the Department and I think I would say that I have experienced more pleasure at having something simmering than displeasure because it is not always boiling!

Early in February John Cockcroft wrote again to Schonland. He now had a proposition of his own: 'I ought to have said to you in my letter of last week that if you didn't want the N.P.L. or if for other reasons it fell through & if you would still consider work in England I have a very great need of a Deputy to share my work' [5]. *Après nous le déluge!* Within the space of just a month Schonland found himself on the receiving end of two tantalizing offers from colleagues with whom he had cut his teeth as a scientist. One, seemingly clear-cut and almost official from Appleton; the other, couched in the style of John Cockcroft: certainly brief and decidedly informal. And there was yet another to follow.

Cockcroft then went on to explain that the job at Harwell had expanded almost like wildfire to encompass fields as diverse as internal politics, mining, intelligence and even biology, in addition to the nuclear physics that was its bedrock, and he was now finding it increasingly difficult to keep pace with it. To compound the problem, his right-hand man and head of the General Physics Division, H W B Skinner, had been offered the position just vacated by Chadwick at Liverpool University and, though vacillating, it seemed he would accept it. Cockcroft therefore saw himself in need of a deputy to whom he could delegate a great deal of the administrative burden and Schonland's ability in that regard was legendary.

Sir Charles Darwin now joined the list of those penning letters to Schonland. Naturally his was in connection with the NPL position that he was about to vacate and he was forthright in expressing his support for Schonland as his successor. 'Appleton told me a day or two ago that there was a chance you might accept the invitation to come here, about which I was highly delighted. I am therefore writing in the hope of persuading you . . .'. With such support now forming up behind him, it seemed as if Schonland's position was unassailable and all that really remained was his agreement to come to England to explore the situation on the ground. To help him further understand the details of the job, Darwin provided a complete break-down of his duties and responsibilities. Administration, he admitted, was not his strong point but the Laboratory was well-served by a secretary who took care of the details and understood the inner workings of the Civil Service. The position of Director entailed an involvement in various activities beyond the confines of the NPL itself and this, he warned, would make inroads into any plans Schonland had to pursue some research of his own, but it would not be impossible to do. In fact, it was actively to be encouraged and he cited Appleton's own ongoing research at the DSIR as a good case in point. His own days of making a contribution to research into atomic theory were now over but he still dabbled in ' "potty" mathematical things' that, though they might not shake the world, certainly kept his hand in.

To Schonland this would have come as encouraging news, for at least his yearning to do some research was an urge that he shared with his old Cavendish colleagues.

Darwin seemed so certain that the NPL position was Schonland's for the asking that he even elaborated in some detail on Bushy House, the very grand stately home and once the abode of Queen Adelaide, that essentially came with the job, should the incumbent Director so desire. Its luxuriant gardens would prove no burden in their upkeep because they were well catered for in terms of the Petavel Bequest—a generous amount intended to ensure that the Director lived and entertained at a level appropriate to his status. Darwin's letter concluded: 'I do very much hope you will come and, though I want to get away, I should be tempted to stay on a bit to keep it warm, because past experience has shown what a bad thing an interregnum is' [6].

A veritable rash of interregnums seemed about to paralyse the British scientific establishment in 1949 as, one after another, directors, superintendents and other custodians of the nation's research institutions either died or decided to move on from positions many had held during the war and for the few years since, as the country righted itself after those years of turmoil. As well as the gaps that were now evident in the realm of the physical sciences, both the Medical and Agricultural Research Councils were deprived of their heads and the Lord President of the Council, Herbert Morrison, under whom such matters fell, found himself hard-pressed either to stem the tide or to muster the necessary manpower to take over. He decided therefore, through Sir Archibald Rowlands, Permanent Secretary of the Ministry of Supply, to take active steps to set the ball rolling by, at least, finding Lockspeiser's successor as Chief Scientist at the Ministry. As soon as Cockcroft had ruled himself out of contention he was asked to suggest other possible candidates and immediately put forward Schonland's name even though he was hoping that Schonland might be tempted to join him at Harwell. Such magnanimity was characteristic of the man; but Cockcroft was also very acute and no doubt reasoned that if Schonland agreed to come to England, in whichever capacity, it would in the long run benefit British science and so all options should be placed before him. He therefore wrote yet again to South Africa to let Schonland know of 'the hole in the Ministry of Supply' which he didn't want to fill himself and then, in his customary frugal style with words, enquired: 'Would you consider it? Salary £3000 p.a.!' [7].

A rather more formal request reached Basil Schonland on 7 March. It came from Sir Archibald Rowlands in the form of an Army Message, classified 'Secret Immediate', and sent in cipher to the Military Adviser of the British High Commissioner in Pretoria. It read:

GT 559. SECRET. PERSONAL FROM SIR ARCHIBALD
ROWLANDS. ON ASSUMPTION THAT YOU ARE VACATING
YOUR POST IN SOUTH AFRICA MINISTER WOULD BE GLAD
TO KNOW WHETHER YOU WOULD CONSIDER THE
APPOINTMENT OF CHIEF SCIENTIST TO MINISTRY OF
SUPPLY IN SUCCESSION TO SIR BEN LOCKSPEISER WHO AS
YOU KNOW IS SUCCEEDING SIR EDWARD APPLETON AT
D.S.I.R. IF SO WOULD BE GRATEFUL IF YOU WOULD SAY
WHEN YOU COULD BE AVAILABLE TO TAKE UP
APPOINTMENT. SIR EDWARD APPLETON IS AWARE OF
THIS APPROACH TO YOU. WOULD BE GRATEFUL FOR
EARLY REPLY. ENDS.

So, not only was Cockcroft fully aware of the numerous offers now
landing on Schonland's desk, in various styles and formats, but so was
Appleton who was himself responsible for the first of them. If their collective involvement was not guaranteed to influence Schonland one way or
the other then yet another of their wartime colleagues, of even more
elevated stature, now took up his pen and wrote to Schonland from the
Ministry of Defence on 8 March 1949. Sir Henry Tizard was even more
keenly aware of the goings-on in both the civilian and military scientific
sectors than anyone else. 'I thought you might possibly like a letter
from me' he wrote 'as you will have to make a difficult choice soon'.
An understatement if ever the was one, yet Tizard knew only of the
NPL and Ministry of Supply offers made to Schonland but seemingly
not of Cockcroft's regarding Harwell, though he did surmise that there
might be a reluctance in South Africa to letting Schonland leave his
present post as President of the CSIR. Difficult choices indeed. Tizard continued [8]:

> My opinion is that you would enjoy yourself at the National
> Physical Laboratory. You would have the kind of scientific freedom
> there to which you rightly attach importance, and you would be
> able to do a lot of good because the place badly needs a good
> experimenter as a leader.
>
> On the other hand you would have the opportunity of doing
> far more good for the British Commonwealth at the Ministry of
> Supply. There are very few men available for that difficult job, and
> it is difficult. I am not going to hide that problem for a moment.
> You would not enjoy it anything like so much as you would enjoy
> the National Physical Laboratory.
>
> Perhaps that is enough for me to say at this stage. Why not get
> into an aeroplane and fly over here? Send me a cable to say that
> you are willing to do this, and I will make arrangements about your
> travelling expenses. I don't believe you can arrive at a satisfactory
> answer to these invitations and suggestions unless you come here
> and talk to people on the subject.

> Anyway it is time we saw you again and showed you some of the latest work in order to get your advice.
>
> Yours sincerely,
> H.T. Tizard

This was as straight from the shoulder as Schonland was likely to get from anyone and its frankness left him in little doubt about the difficulties that lay ahead.

Together Basil and Ismay wrestled with the choices now facing them. Three offers from England seemed almost like a glut, though Cockcroft's Harwell proposal, made almost as an aside, seemed only to be testing the wind. The other two were much more formal and deliberate and required considered responses with some immediacy. Ismay would go where Basil went and England was certainly dear to both their hearts. Only Ann, their younger daughter, was still dependent on them with some years of university education ahead of her. Leaving the CSIR was not an issue because it had never been Basil's intention to spend more than three years as its founding President. Only the Prime Minister's request that he should extend his period of office by an extra year while the new government found its feet had caused him to stay on.

Letters continued to arrive from England and they contained no shortage of advice. John Cockcroft, even though he had more than a little interest in the outcome of the selection process, adopted a completely independent stance and proffered his carefully considered opinions about both the NPL and the Ministry of Supply. He wrote twice in quick succession. The first letter, of 9 March from the Atomic Energy Research Establishment at Harwell, jumped to an unjustified conclusion, with Cockcroft expressing his pleasure that Schonland had agreed to take the NPL post. 'It will be a very good thing for this country since I am sure that the N.P.L. can be directed towards helping industry much more than it has done in the past. I expect too that you will be able to help us by putting more life into the Metallurgy department there.' He generously indicated to Schonland that the Deputy that he sought for Harwell 'would not have made so full use of your powers but it will be good to have you in the country so that we can push together for the things that want doing.' Skinner had finally made up his mind and was going to Liverpool so Cockcroft was resigned to the fact that he would have to 'try to steal someone else' [9]. For all his canny understanding of the workings of the scientific establishment, Cockcroft was, though, clearly unaware of all the machinations then taking place around Whitehall and elsewhere.

The NPL appointment was by no means cut and dried, as events soon to unfold would make abundantly clear. Many hands were in the pot and the brew was becoming a concoction. Ten days later, after word had reached him that matters were by no means settled, and that great efforts were being mounted to attract Schonland to the Ministry of Supply, Cockcroft wrote again [10]: 'Rowlands told me yesterday that he had written to you to try to persuade you to take on the Chief Scientist's job in the Ministry.' At issue were numerous matters to do with the very structure and running of that vast organization, as well as the importance of the various scientific programmes that were so vital to Britain maintaining her position as a major military power. The Permanent Secretary was well aware that the heavy administrative responsibility attached to the Chief Scientist's post made it unattractive to most scientists, and had indicated that he was intent on changing this by ensuring that the Chief Scientist would be responsible for research in the various establishments, with development becoming much more the function of industry. Cockcroft had his doubts about how feasible this would be but, as he told Schonland, he hoped that the channels of negotiation would remain open. He had no doubts, though, about the size of the job and the efforts that would be needed to coordinate the work on radar and guided missiles so as to give 'the Establishment* a clear lead'. To do this the Chief Scientist would have to be in a position and have the stature to influence decisions at the very highest levels of government, while also having 'unimpeded access to Establishments with [the] power to discuss problems with anyone'. In Cockcroft's opinion he knew who the right man was for that position: 'It is an important, but very difficult job & there are very few people who might make it go & you are the only person I know who will'.

This was not only powerful support from Cockcroft but it was also a ringing endorsement of Basil Schonland from someone who knew him and his qualities better than anyone else. As well as dispensing the wisdom of a man as able as any to judge the politico-scientific world in England, Cockcroft was lending all his weight and prestige to his old colleague, at the time when Schonland needed them most. And then, as letters shuttled between them, he wrote yet another setting out the pros and cons of the Ministry of Supply post and stating in even more detail his own reasons for turning it down. 'After all this, you might well ask whether you should consider the post at all. The only reason for doing so is that our research & development policy is not in good shape and really requires a concerted examination & a firm co-ordinating hand.' Guided weapons, improved radars, an electronic predictor for anti-aircraft gunnery to replace the aged mechanical system then still under development and the huge, multi-engined Brabazon aeroplane, built at enormous cost but of doubtful commercial viability, were all projects

* TRE: the Telecommunications Research Establishment, one of many similar establishments within the Ministry of Supply.

straining Britain's scientific and engineering resources. Their future, as well as many others', were just some of the many problems awaiting the Chief Scientist. Once again, Cockcroft was unequivocal in his choice of the man to provide that firm coordinating hand. 'As I said in my last letter, you are about the only person available who could do this job'.

By comparison, the directorship of the NPL seemed so very straightforward. 'The N.P.L. will be easier' he wrote '& you would be more sure of success—certain I ought to say'. Cockcroft, always so taciturn, even amongst his family, and renowned as a man who kept his own counsel in any company, had excelled himself in these recent letters to Schonland. He wrote, almost expansively, both as a friend and as an admirer of Schonland the man, the scientist and the soldier he knew so well. He closed with the simple wish: 'I very much hope you will come over, to one job or the other. I must leave the choice to you'.

By the time that letter arrived in Pretoria Schonland had already made his decision and his reply, which he too had classified 'Secret' was already in Rowlands's hands in London. To get it there so expeditiously he used the same cipher facility at the British Military Mission in Pretoria by which Rowlands' had reached him. His reply was brief and to the point [11]:

> For Sir Archibald Rowlands from Dr Schonland stop Your GT 559 refers stop I deeply appreciate Minister's offer but must respectfully decline stop Have agreed to accept Directorship of the National Physical Laboratory if finally offered to me stop Failing that I will return to academic research post for some years.

But even this did not close the matter. It was now apparent in London that Schonland was the man most sought after for the position as Chief Scientist of the Ministry of Supply. His recent rejection of it served only to make it imperative that he be invited to England, where it was hoped they would be able to persuade him to change his mind, once all the facts had been presented to him and he had been able to raise all issues of interest and concern to himself. His visit would, no doubt, also provide a useful opportunity to discuss the other matter at the NPL. So, on 9 April, yet another officer from the United Kingdom Military Liaison Staff based at General Headquarters in Pretoria made his way across the city to the old Mint and the CSIR. There, he delivered a letter to its President [12]. It was from Sir Evelyn Baring, British High Commissioner to South Africa, and contained three important messages. The first was from the High Commissioner himself who was attending the Parliamentary session in Cape Town. He informed Schonland that he had been instructed by his government to invite him to travel to London 'where they can give you good reasons for acceptance of post offered'. If that were not possible, the High Commissioner indicated that he had been told to speak to Schonland personally, either in Cape Town or in Pretoria—whichever might suit Schonland better. The other two were no less compelling. Rowlands acknowledged receipt of Schonland's earlier missive and extended to him the same invitation to come to London post-haste. He was confident, he said, that were Schonland to do so he would be able to persuade him 'that there are special reasons for your acceptance of the appointment'. The last came from Cockcroft and, typically, was the briefest of all. 'Have talked to Rowland [sic]. I think you should come for discussions. John.'

Sir Evelyn, well aware of the passage of secret messages across Pretoria and the unease in Nationalist circles about all manner of relations with Britain, then broached the sensitive issue of protocol, were Schonland suddenly to fly off to London. 'I should be grateful for your advice on what I am to say to Union Government in case you decide to go to United Kingdom. I am not quite sure how exactly you stand with them at present.' But Schonland knew exactly where he stood. On matters of scientific policy and his consequent responsibilities to the Government, he had no qualms but, when the fundamental tenets of Afrikaner Nationalism began to intrude, and the signs were that they would to an increasing extent, he was implacably opposed to them. For all that, he was very careful to keep his Prime Minister fully informed on matters of protocol, especially in view of events ahead.

April 1949 was the occasion of the Commonwealth Prime Ministers' Conference in London and Dr D F Malan would be attending such a gathering for the first time. It would clearly have been extremely bad form for him to have discovered whilst there that his scientific adviser was being tempted to accept not just one but any one of three offers of senior positions in the British scientific hierarchy. Schonland was especially aware of that and had therefore chosen himself, even before receiving the High Commissioner's message, to appraise Malan of the position, thereby relieving Sir Evelyn Baring of the need for delicate footwork in his dealings with the Union government. On 4 April, just before Malan left for England, Schonland wrote to him in confidence [13].

> I would not be bothering you with a personal letter just now were it not that certain proposals have been made from England which perhaps may be mentioned to you while you are there. Some months ago I was asked whether I would consider taking up the position of Director of the National Physical Laboratory in Britain if it were offered to me; more recently I have been asked, in confidence, to take the post of Chief Scientist of the Ministry of Supply. The second offer I have refused, but the first I have replied that I would be willing to take the Directorship of the National Physical Laboratory if the post is offered to me after the end of this year.

He continued:

> It may come as a surprise to you that I intend to ask you to let
> me give up the helm of the Council for Scientific and Industrial
> Research. When the Government first honoured me with this duty I
> undertook to hold the post for three years only. In October of this
> year I shall have been in it for five years, the first of which was a
> year in preparation. With five years of war service I shall have been
> ten years away from actual personal scientific work and I feel it is
> time that I returned to it from administration. The C.S.I.R. is well
> established as a national undertaking and I believe that a change in
> direction may well do it good.
>
> It was my intention to go back to the Geophysical Institute in
> Johannesburg which I left at the beginning of the war. What is
> attractive about the National Physical Laboratory in Britain is that it
> offers larger scope to me as a physicist and that the administrative
> side of it is not overwhelming. The Ministry of Supply post I regard
> as purely administrative and I should not wish to leave the service
> of my own country for it, though it is being urged on me that in the
> present uneasy state of the world it is of special importance.

Lest his Prime Minister think that he intended to abandon previously
carefully laid plans for the attractions now on offer in England—or
indeed for those at the BPI—Schonland explained that it was not his
wish to resign until 'about the beginning of next year', by which time
he would have taken full responsibility for the African Regional Research
Conference that had been planned to take place in Johannesburg during
October 1949. The Department of External Affairs replied almost immediately on the PM's behalf. Dr Malan's reaction, it said, was one of sincere
regret that Dr Schonland intended to leave his post at the CSIR. As soon as
he returned from the Commonwealth Conference in London, the spokes-
man assured him, the Prime Minister would be in contact with him [14].

Basil Schonland flew to England late in April. It was clear to him
from the urgency of the messages received from Britain's High Commissioner, as well as the gentle nudge from Cockcroft, that every effort would
be made in England to try and persuade him to accept the position as
Chief Scientist at the Ministry of Supply. But he had made up his mind
that he did not want it and so, once disposed of, the only purpose of
the visit was to discuss the tentative offer of the Directorship of the
NPL made to him a good while before by Appleton .

What he did not expect was the news that reached him on his arrival,
because neither Appleton, Darwin nor even Cockcroft had ever told him
that his was not the only name soon to appear before the Privy Council
for its decision. For reasons known only to themselves, Appleton and
Lockspeiser managed, during their change-over at the DSIR, to overlook
the key part played by the Royal Society in the appointment of the man

to run the NPL. What had transpired was directly in line with the agreed procedures for the appointment of its Director, but there were some in senior positions who were not as familiar with its complexities as they should have been. The NPL had, since 1943, reported to two masters: the DSIR, on matters of financial management, and the Royal Society on issues to do with its scientific programme and policy [15]. When the need arose for the appointment of a new Director, the procedure, though logical, soon became tortuous. First of all the NPL's own Executive Committee, of which the Secretary of the DSIR was vice-chairman, considered a list of possible candidates submitted by its members. In this instance 26 such names appeared. These were eventually whittled down to eight and submitted to the Royal Society in a recommended order. At its meeting on 16 December 1948 the Royal Society's Council pruned that list to just four, whom they ranked and then submitted it to the Committee of the Privy Council for Scientific and Industrial Research for its consideration. The names that went forward that day were: Sir John Cockcroft, Professor B F J Schonland, Professor E C Bullard and Dr W G Penney [16]. Both Cockcroft and Penney were fully committed elsewhere to programmes of work of the very highest priority in the field of atomic energy—Cockcroft at Harwell and Penney, first at Fort Halstead and then later at Aldermaston where he was the scientist in charge of the British atomic weapons programme. Both withdrew from further consideration, leaving just Schonland and Bullard in contention for the NPL position. Appleton must surely have known this but never mentioned it to Schonland while Lockspeiser, now in Appleton's shoes, seemed somewhat oblivious of the whole procedure!

So, to Schonland's great surprise and not a little consternation he discovered that Bullard, the same man who had spent some months in 1938 at the BPI working on the thermal conductivity of the rock in the Witwatersrand mines, and who was now head of the Department of Physics at Toronto, was also in London and was referring to himself as a candidate for the NPL position. Whereas Appleton had made it clear that it was not in his gift to offer the post to anyone he had, equally, not made it clear that anyone other than Schonland was being considered for it. Similarly, both Darwin and Cockcroft, who were so fully supportive of Schonland in their various letters, were either unaware of the appointment's procedure, which seems most unlikely, or had just assumed that the post was Schonland's for the asking. It certainly wasn't, as Bullard's very ebullient presence now made abundantly clear. With some urgency and not a little irritation, Schonland called on Lockspeiser for an explanation. Was it not 'tactless' of Bullard, he asked, to refer to himself as a candidate in view of his own correspondence with Appleton? Sir Ben was clearly embarrassed. He admitted that he had been unaware of the part played by the Royal Society in the appointment of the NPL's

Director, but knew that Bullard's name had also been put forward, and when he heard that Schonland was on his way to England for interview, he then took it upon himself to invite Bullard to come over from Canada for the same purpose. Now he found himself face-to-face with the man who was not only being pursued with some vigour to succeed him as Chief Scientist at the Ministry of Supply but he was also the same man who had assumed, since nobody had said otherwise, that he was the only contender for Darwin's former post at the NPL!

If not high drama the situation was certainly bordering on farce. Here was Schonland, invited from South Africa, under the mistaken impression that the NPL, if not quite his for the taking was at least there for the asking, whereas that was by no means the case. Now, a very annoyed Basil Schonland required not only an explanation but suggested that it would be best if he withdrew altogether since he had presumed that Appleton's offer had been made to him and him alone.

Schonland's withdrawal under such circumstances of confusion and crossed wires would, at the very least, have been embarrassing while, at worst, it could lead to unpleasant repercussions at higher levels. So Lockspeiser, conscious of his own waywardness in the matter, took a conciliatory line and persuaded Schonland not to withdraw. Schonland, for his part, clearly wanted the job but he stressed that he had not come to England to participate in any process of selection. He therefore asked Lockspeiser to inform the Royal Society that if the post were offered to him he would accept it without reservation but he was adamant that he was not a 'candidate'. Lockspeiser agreed to write to the President of the Royal Society to explain the situation and said that he personally would take steps to change what appeared to be a very anomalous procedure in the process of selection. No doubt heartened by all this, Schonland assured Lockspeiser that, if appointed, he would work loyally under him. He added, too, that he had heard that Lockspeiser intended to make some arrangement whereby the Director of the NPL assisted the Secretary of the DSIR in industrial research and, if this was so, he would be very glad to do so [17].

But all this was just a side-show. The real reason, so it soon transpired, for asking Schonland to come to London was to try and persuade him to take the post of Chief Scientist at the Ministry of Supply. That same evening, after his forthright encounter with Lockspeiser, Schonland found himself at a cocktail party as the guest of Herbert Morrison, Lord President of the Privy Council and the man ultimately responsible for making the appointment to the Ministry. There too was the Permanent Secretary, Sir Archibald Rowlands, whose spate of secret telegrams and messages had precipitated Schonland's hurried trip to England in the first place and, in supporting roles to add the necessary ballast, were various heads of sections and other officials in the Ministry. This array of Whitehall barony was clearly intended to provide a degree of persuasive gravitas to a convivial occasion, but Basil Schonland was a resolute man and was not in any way deflected from his original intention. With his resolve undoubtedly stiffened by the events of the morning that he presumably felt had gone his way, he did not shift his ground at all. He informed the Minister that he was not accepting the offer of Chief Scientist for the reasons he had given in his previous correspondence, and that was that. But Morrison and Rowlands, too, were persistent men and were not about to lose the opportunity to ply Schonland with persuasive argument. 'Was he not anxious to leave South Africa?' Rowlands asked. Word had clearly got about of Schonland's misgivings about longer term developments back home, and the source of these would almost certainly have been John Cockcroft, with whom Schonland's correspondence was decidedly more frank that it was with anyone else. If that were the case, said Rowlands, they should surely talk again and a further meeting was arranged for the next day.

But Schonland had not reached his decision lightly. Neither the Minister nor the Permanent Secretary, for all their collective coaxing when they all met again, succeeded where Cockcroft before them had failed. Schonland's mind was made up and he was only prepared to entertain the post as Director of the NPL. If such an offer were not forthcoming then he intended to return to the BPI in Johannesburg and to his own research career that, for so long, had been in abeyance. They then played the last card they had in reserve. If Cockcroft could be talked into taking over as Chief Scientist at the Ministry of Supply would Schonland be prepared to take charge at Harwell? This was not quite the bombshell it should have been because Cockcroft had, of course, already floated a not unrelated proposal when Schonland's name first appeared on the Whitehall agenda. Mildly amused that it should now come from this most unexpected quarter, as the pressure to lure him increased, Schonland countered only by saying that even if Cockcroft could be tempted he wouldn't stay long because he, too, still hankered after academic life and had made clear his desire to return there some day.

Rowlands, now recognizing well enough that he was beaten, confided in Schonland that the blame for the NPL episode must be laid at Appleton's door. Lockspeiser, he said, was particularly annoyed that Appleton had not consulted him before making his 'firm offer' to Schonland when, all along, it seemed that Bullard was Lockspeiser's first choice for the post. Schonland had thus been the unwitting victim of ham-handedness at the DSIR. Appleton had never told his successor of his intentions regarding the NPL and, after taking over, Lockspeiser had not read his predecessor's correspondence on the matter [18]. If Schonland still entertained any hope that the NPL might come his way, this inside information would surely have dashed it. All, of course, would still depend on the Lord President and the deliberations that would undoubtedly soon follow, but the portents were now certainly not good.

Basil Schonland returned home immediately after this series of encounters with British officialdom and the complexities of its scientific civil service. Matters were out of his hands and all that remained was to await the decision to be taken by the Lord President, Mr Herbert Morrison. It took some time in coming, since matters to do with a forth-coming General Election inevitably had first call on a politician's time and Morrison still needed to consult his advisers. Eventually, early in June, he made his announcement. If Schonland still harboured any hopes at all for the NPL position after the disclosures by Rowlands they were soon dashed when a letter arrived from Sir Ben Lockspeiser. Within a week it was followed by another from the Lord President himself [19]. Their contents were similar. 'The Lord President has decided', wrote Lockspeiser, 'that Professor E. C. Bullard should be appointed to the Directorship of the N.P.L.'. Lockspeiser had got his man; now he tried to placate Schonland. He continued:

> This is not an easy letter for me to write. I and others have spent much time and thought on this appointment and I am not a little distressed at the personal aspects of the matter. The generous attitude you took, however, when we talked in London, convinces me that this decision will not affect our friendship or lessen the opportunities we may have for future collaboration.
>
> I am hoping it may be possible for me to visit South Africa in the near future as it is clear, from such preliminary examination as I have been able to make, there is much we can do together.

The Lord President, who had himself become enmeshed in the farrago, was no less diplomatic.

> You know something of the complications about this post and the considerations which had to be weighed up in filling it so as to produce the most balanced team in civil research for the Government. It is a matter of personal regret to me that I shall not have the pleasure of seeing you added to the Government's team of scientists in this way, but I would like to assure you that all of us are very sensible of your qualities and appreciate the spirit in which you have assisted us in exploring how to fill two important posts for which the Government has been considering you.

Between them, Lockspeiser and Morrison hoped to have extricated them-selves from the sequence of events that covered no-one in England in any glory. Schonland was too big a prize to risk losing in this way and the ineptness in certain quarters could have cost them dear. Someone else within Whitehall who was acutely aware of that was Sir Henry Tizard, who still sought a Chief Scientist to take charge of the myriad scientific problems looming on the military horizon. Until now he had taken a back seat as the inducement charade unfolded, but he chose to make one last appeal to the man for whom he had had the highest regard since their first encounter in 1942.

Early in August he wrote to Schonland and inquired whether he could be persuaded to change his mind about the Ministry of Supply. 'Is your refusal of the M. of S. post absolutely final? Please reconsider it and let me know if I can do anything to make the conditions right.' If the request in 1944 to become scientific adviser to Montgomery was Schonland's moment of glory in his career to date, Tizard's almost impassioned appeal for his services now must have, at least, helped to counterbalance the sense of dismay and disillusionment that had descended upon him after the recent debacle. That Britain's top defence scientist should try, yet again, after all others had failed was a sure indication of Basil Schonland's standing amongst his contemporaries. The timing of the call itself could not have been more crucial, for within just three weeks of the Tizard letter the Russians exploded their first atomic bomb and the world was once more teetering on the brink of war. Tizard's closing words in the letter, for all their biblical resonance, were stark: 'Come over to Macedonia and help us!'* [20].

---

* Tizard's slight modification of Acts 16, 9.

CHAPTER 18

COCKCROFT'S MAN

In August 1949 Sir John Cockcroft arrived in South Africa as a guest of the CSIR. His visit had been on the cards for some time but the date was very much dependent upon the presence in the country of Basil Schonland and recent events had required him to be elsewhere. Ostensibly, Cockcroft was there to lecture to South African universities on the general subject of atomic energy, but the real reason was related much more fundamentally to South Africa's reserves of uranium. The stimulus for the visit had come from a decision of the British and American governments, made public in a press release from Shell Mex House in London on 11 June. Its very terseness indicated its importance: the USA and the UK were interested in the production of uranium from South African gold-bearing ore; no more, no less [1]. Cockcroft was the obvious man to discuss the subject with the South African government because his path to them would be carefully cleared by Schonland.

There was also another reason for Cockcroft's visit, but that was not announced at all. As Basil Schonland's closest confidant he was the one man with whom Schonland would discuss openly his own plans and intentions for the future, and Cockcroft was now fully aware of the unhappy sequence of events that had recently occurred in London. Those plans of Schonland's may have altered radically since then, and John Cockcroft was very interested to find out if they had and what they were. The opportunities to do so were many, because all the hospitality of the Schonland home in Pretoria was extended throughout his stay to their British friend and visitor. Included amongst them was a trip by car to the Kruger National Park to see the animals, which John Cockcroft presumably did, though the deep discussion with Schonland, seated alongside him on the capacious rear seat, suggested that their minds were far removed from such distractions [2]. Of immediate concern was Tizard's letter and his earnest call to Schonland to reconsider the offer of Chief Scientist at the Ministry of Supply. Before replying, Basil not only wanted to discuss every aspect of this most demanding of positions with Cockcroft, but he wished, as well, to inform him of developments much closer to home.

Shortly after returning from England, Schonland had submitted his resignation as President of the CSIR to the Prime Minister, Dr D F Malan. In his letter he told Malan that he would be returning to the BPI with the express intention of becoming, once more, actively engaged in research. Malan refused to accept the resignation; he wanted Schonland to remain at the CSIR. He now realized that, not only was South Africa's pre-eminent scientist much in demand elsewhere, but that his own wish was to shed the administrative burden that had been his lot for so long, in order to turn his mind to problems of geophysics and especially lightning. Malan therefore realized that his only chance of retaining Schonland's services depended upon being able to guarantee his scientific adviser both the facilities and the time to conduct his own research, and he said he was prepared to do both. Once again Schonland was faced with a dilemma in the form of Tizard's *cri de coeur* seeking his services and Malan's stated wish to retain them. Both would soon require a reaction. While all this was unfolding, there, in the background, was the comforting presence of the BPI and the Chair of Geophysics that Wits had for so long held open awaiting his eventual return.

These matters and all their various ramifications consumed many hours of discussion with John Cockcroft and Ismay, when the three of them were together. The influence that both had on Basil Schonland was profound. In Ismay he had not only a wife but a friend whose strength of purpose was every bit the equal of his own. No decision that affected them both was ever reached without her incisive examination of the details and the implications it might have for their future lives and livelihood. Equally, in John Cockcroft, he had not only a friend but also a sage, whose wisdom Schonland found compelling and whose views he seemingly never questioned [3]. As Cockcroft prepared to deliver his lecture* at the University of the Witwatersrand, Schonland wrote his delayed reply to Tizard [4].

21 Aug 1949

Dear Sir Henry,

I am sorry to have delayed a week before answering your letter, but I wanted to have a word with Cockcroft before replying. I am sorry that I must finally refuse the Ministry of Supply post.

---

* The rather involved title Cockcroft chose for this was 'A general lecture on atomic energy covering the whole field'. It was delivered on 23 August 1949 in the University's Great Hall to a full house of students and members of the Associated Scientific and Technical Societies of South Africa. Schonland, as chairman of the Nuffield Foundation Liaison Committee in South Africa which brought Cockcroft to the country, seconded the vote of thanks. Less than a week later, on 29 August, the Russians detonated their first atomic bomb to the dismay of the western world.

He then explained how his Prime Minister had refused to accept his resignation from the CSIR and had indicated his intention to accommodate Schonland's desire to devote considerably more time to his own research. 'I am now considering whether I can accept this generous offer', he wrote. But those recent unhappy deliberations in England continued to gnaw away in his mind, surfacing now and then either as minor irritations of what might have been, or as interludes that were best ignored. Tizard, who knew Schonland well, was of another school compared with some whose reputations were forged when England was facing fire. They were similar men, born to serve as well as to lead and neither tolerated, nor even understood, what sometimes passed for finesse around them. Schonland could therefore readily confide in Sir Henry:

> I feel that I should say to you, as an old friend, that had the position
> on my arrival in England recently been what I understood it to be,
> that I could have either the NPL or the Ministry of Supply post,
> I should have found it most difficult, perhaps impossible, not to
> yield to arguments that I was more needed at the Ministry position.
> For then I should have accepted the fact that I had removed my
> roots from S. Africa & become an Englishman and as such it was
> my duty to do what my new country required of me.

This was the Schonland they all knew; the man who had responded to Cockcroft's call in 1941 to take on the responsibilities of the ADRDE(ORG) and who, three years later, with even greater trepidation, had accepted the appointment as Scientific Adviser to a Commander-in-Chief. Then, once again, as the war drew to its close, his sense of duty to another Field Marshal saw him return home to found the CSIR and to lead it through its critical, formative years. Now, at the end of that decade, he found himself torn between a burning desire to return to research and the ever-pressing calls for him to render yet more unstinting service to a greater cause. But this time that route to duty had been much impeded by, at best, bureaucracy and bungling. Schonland's choices now seemed to lie entirely in South Africa. He could remain at the CSIR, but under changed contractual conditions, if Malan was true to his word, or he could return to Wits. A major decision would therefore be required within the next few months.

*   *   *

As if to bring science and scientific research once more into sharp focus, the BPI held its first conference 'for scientists in Southern Africa interested in the geophysical sciences' over three days at the end of July 1949 [5]. The organizing committee was chaired by Schonland but his frequent absences of late meant that the details fell to others to sort out. This they did admirably and a programme of 39 papers covering a most diverse field of geophysical activity was prepared. Approximately 80 delegates attended, with some travelling from the Rhodesias and even as far afield as the Belgian Congo, while a real international dimension was added by the participation of two scientists from the Carnegie Institution in Washington. Schonland's occasional boast, that the BPI was one of the leading laboratories of its kind in the world, was probably not too far off the mark. The sheer diversity of the research it had done in the space of a decade was commendable. South African participation extended well beyond the walls of the BPI and included contingents from the CSIR, the Union Observatory, the Geological Survey, the Meteorological Office, as well as from other universities across the land. Amongst them was Schonland's pre-war collaborator and wartime SSS colleague from Durban, Professor David Hodges, who had resurrected his earlier interest in the ionospheric reflection of signals generated by lightning.

The staff of the TRL naturally turned out in force as well. Trevor Wadley used the occasion to make public for the first time the results of his research into the feasibility of radio communication underground in mines, while Frank Hewitt described and demonstrated his cloud radar that had so excited Schonland as a new tool for the study of lightning. Philip Gane, from the BPI, discussed the seismology of the Witwatersrand earth tremors, while Dawie Malan described his recent advances in lightning photography. He also demonstrated the measurement of electric field change, using the instrument he called the 'BPI electrostatic flux-meter' before it became known more evocatively as the 'Spinning Jimmy', and showed, too, the lightning-triggered switches for daylight photography. Anton Hales and Ian Gough, a BPI and NPL collaboration, discussed their gravimetric survey of the country, using pendulums on loan from Cambridge, while Gough also reported on their measurements of the variation of the geomagnetic field with depth and their attempts to reconcile their results with Blackett's postulations. Temperature and radioactivity logging in boreholes, the crustal structure of the Western Transvaal, measurements of the ionosphere and the prediction of its characteristics, the electrical resistivity of the earth and the artificial stimulation of rainfall were amongst many other topics aired during those three days, when Schonland's brainchild put its scientific wares on public display for the first time.

The local press reported on this gathering at Wits and followed its deliberations with interest. None was more enthusiastic than the editor of the *Pretoria News*, who wrote of the pride that South Africans should feel in its scientists: 'This [conference] emphasises the emergence of South African science from colonial to national status.... [It] is truly a sign that the Union, first among African territories, is becoming a fully

Figure 22. The first three Presidents of the CSIR: B F J Schonland (1945–1950), S M Naudé (1952–1971) and P J du Toit (1950–1952). (Reproduced by kind permission of Dr Mary Davidson.)

fledged nation. It increasingly contributes to that common pool of human knowledge on which it has drawn for so long and from which it must always draw' [6]. Later, when writing his Director's Report on the BPI for the year 1949, Schonland was happy to record, perhaps a little ironically too, that the conference was very successful and that its single social function, a cocktail party, had been hosted by the President of the Chamber of Mines himself [7].

Within a month it was the atomic bomb that grabbed the headlines again, following Russia's detonation of a device that shook the West because none seriously believed they had the technology to do it. The result was a dramatic change in the balance of power and this immediately caused certain Western nations to show more than a passing interest in uranium and especially in South Africa's, as yet, untapped stockpile of the stuff. In 1948 the South African Parliament passed an Act establishing the Atomic Energy Control Board. The CSIR, through Schonland, was crucially involved in the *modus operandi* of this new body and particularly in the importation and use of radio isotopes for a variety of purposes, as well as for providing advice on research into fissionable materials [8]. With such matters much on his mind, Meiring Naudé, the Director of the NPL, had visited both Britain and the United States in 1946, where he toured their atomic research establishments and arranged for South African physicists to be sent for training at Harwell. Then, more recently, as his laboratory began to take shape, Naudé decided to set up a programme of research on deuteron accelerators and, through Schonland, made contact with Cockcroft at Harwell [9]. About this time Naudé proposed that the CSIR should build its own small cyclotron for research purposes and for the production of radio isotopes for medical applications. Though only a stripling in the research field, South Africa was undoubtedly a major source of the most important raw material, uranium, and this made her a key player on the nuclear stage. However, on coming to power, Malan's Nationalist government had other priorities and the contract negotiations with Britain, which had been started by Smuts, soon foundered [10]. Sir John Cockcroft had therefore been prevailed upon to visit South Africa and to use his special relationship with Basil Schonland to resuscitate the process.

Quite naturally Cockcroft and Schonland spent some time discussing South Africa's own research programme in the field of atomic physics, but excluded Meiring Naudé from their meetings. The subject of the cyclotron was of particular importance and, according to Naudé's subsequent writings on the subject, Cockcroft recommended against it in favour of a linear accelerator since, he said, that was the route being followed by the Medical Research Council in Britain and he believed it was the most effective and cheapest option for South Africa to follow as well [11]. Schonland readily acted on this advice and piloted the proposal through the meeting of the CSIR's Council in September. Subsequently, according to Naudé again, Schonland changed his mind after receiving the views of Dr Harold Himsworth, Secretary of the British Medical Research Council, who evidently actually favoured the cyclotron! The CSIR's Council, though presumably rather confused by the conflicting advice it was receiving from its President, acquiesced and reinstated Naudé's original proposal that a cyclotron was indeed the better choice.[*] This puzzling sequence of events is not satisfactorily explained, even by Cockcroft's letter to Schonland, written a few months later in which he makes the statement in typically brief and unelaborate style that it was the 'right decision to build the cyclotron' [12]. Had Cockcroft changed *his* mind or was Naudé's understanding and recollection of what went on somewhat wide of the mark? His exclusion from the meetings between Schonland and Cockcroft that effectively mapped out the country's nuclear future would not have helped [13].

On 5 October 1949 the Office of the Prime Minister announced that the period of office of the President of the CSIR, Dr B F J Schonland, had been extended by a further year 'at the same salary'. Though the letter did not say so explicitly, Malan had agreed to Schonland's request that the

---

[*] A 16 MeV cyclotron wholly designed by the CSIR and built mainly in South Africa by local industry was eventually commissioned in January 1956 at a cost of £49 500.

post should allow him the time to carry out his own research [14]. What had commenced as a three-year commission from Smuts in 1945, to found and launch South Africa's national research facility, had grown incrementally by two twelve-month extensions that should now see Schonland in harness beyond the turn of the decade. Dr Malan had certainly got what he wanted and could now turn his attention to matters of greater importance to himself and his party – those of embedding the Nationalists in power and of implementing the philosophy that would soon become known as 'apartheid'. Whether Schonland was really satisfied is less certain. He was though extremely busy, for the much-heralded African Regional Scientific Conference was due to commence at Wits on the 17th of the month and he was its convenor and leader of the South African delegation.

The Conference was a direct result of a resolution passed in 1946 at the British Commonwealth Scientific Conference, in which the growing need for the development of long-term and highly specialized research on African problems was officially recognized. Now it was all to come to pass, under the auspices of the CSIR, with Schonland's right-hand man in the planning and organization of its activities being his Deputy President, P J du Toit. In fact, du Toit's part in the run-up to the conference had been considerable ever since his retirement in 1948 as Director of the Veterinary Research Institute at Onderstepoort. The programme was largely his doing, as were the intricate details involved in arranging this first event of its kind in sub-Saharan Africa.

On Monday, 17 October 1949, Basil Schonland offered his arm to a slightly infirm 75 year-old, Dr D F Malan, whose health had never quite recovered following the momentary blackout he had suffered while addressing Parliament some months before [15]. Together they ascended the steps to the Great Hall of the University in Johannesburg that was bedecked with the flags of Wits, the Union of South Africa and the United Kingdom. There, Malan was to perform the opening ceremony of the most prestigious gathering of scientists to have assembled in South Africa since Sir Ernest Rutherford and his Cavendish entourage visited the country in 1929. It was a short but symbolic journey that they took together up those steps beneath the colonnades of the Great Hall. Two men of Cape stock: one, from its western reaches, a former *predikant** turned politician of marked anti-British sentiment; the other, a 53-year-old Anglophile born in the very heart of British South Africa in the eastern Cape, with liberal views and a personal philosophy of scientific rationalism. Their closeness as they walked together that day was a massive contradiction. Schonland, the visionary who saw his country as the pilot of Africa, able to offer it the means to a better future through

---

* *Predikant*: parson or preacher.

Figure 23. Schonland assisting a slightly infirm Prime Minister, Dr D F Malan, up the steps to the Great Hall of the University of the Witwatersrand when Malan performed the opening ceremony of the African Regional Scientific Conference on 17 October 1949. Obscured on Malan's right is Ismay Schonland, while Humphrey Raikes, Vice-Chancellor of Wits, is on Schonland's left. (Reproduced by kind permission of Dr Mary Davidson.)

scientific cooperation; and Malan, a bigot whose view was of a South Africa turning in on itself and excluding, by law and by force, if needs be, all 'men of colour' from the world of the white man.

But that day each was doing his duty and the occasion was accorded considerable attention by the South African press: 'Science plans to dispel Africa's darkness'; 'Dr Malan says problems of Africa set science a great challenge'; 'Planned development of continent urged'; 'Scientific Council proposed for Africa' and so on, ran the headlines [16]. The Conference elected as its president P J du Toit. He was proposed by none other than Sir Ben Lockspeiser, the leader of the British delegation, and was seconded by his counterpart from France, Professor J Millot. All told, more than a hundred scientists representing the major European powers, as well as the United States and Australia, and the scientific arms of the United Nations, were joined by others, all expatriates, from 25 territories south of the Sahara including the Belgian Congo, Northern and Southern Rhodesia, Madagascar, Mozambique and collective group-ings referred to as East and West Africa. Never before had South Africa played host to such a cosmopolitan group of men and women whose sole objective was to apply their collective scientific wisdom to the solu-tion of the problems of the 'dark continent'.

Basil Schonland, speaking in both English and French, introduced South Africa's Prime Minister to the audience. Malan, in English alone—his French roots long since submerged by his new Afrikaner identity—welcomed the visitors to his country and informed them that 'the scientific problems facing the countries of Africa are in many ways greater and more urgent and more challenging than those to be found anywhere else in the world'. 'Africa's problems', he continued, 'were greater not only because of the unique climatic conditions of the continent, the unique nature of its indigenous people and the special character of the carriers of disease encountered, but also because so little had hitherto been done to solve them.' Indeed, as the *Pretoria News* put it so starkly, Africa was not only the dark continent, it was the backward continent that had remained almost impervious to scientific and social advancement even 400 years after its shores were first touched by the ships of early navigators. 'The West does not even know now how to turn it into a lasting asset to mankind'; this 'least attractive and promising bloc of undeveloped land left in the world', it wrote [17].

Such pessimistic thoughts though were not the stuff of Schonland's conference. His own views, put so trenchantly in 1942, that the West had failed Africa and had allowed the Africans to stew in the own juice had shaken some but had almost certainly caused many to reflect a little more deeply. Now, a mere few years on, the forces of science were being marshalled to address the issues to which many before them had merely turned their blind or simply covetous, monocled eyes. Together, du Toit and Schonland had identified the key subjects on which the conference would concentrate during the next two weeks of its deliberations. Six sections were formed, each under a chairman carefully chosen for reasons of both his scientific esteem and his national affiliation. They were to address problems under the banners of the physical environment, soils and plants, zoology and animal husbandry, health and medical research, social research and, finally, technology. Those chairmen represented the flags of Europe and its colonial offspring that had been planted across the African continent: Portugal, Belgium, England, France and their host South Africa. A German presence, for all its earlier acquisitions in Africa, was missing—the dust of recent events was still very much in the air.

By its end two weeks later the conference had covered a vast territory of subject matter and had at least, with the assistance of simultaneous translation between English and French, been able to breach some of the language barriers that confronted the continent. Its final recommendation to their various governments watching the deliberations from the side, and which ultimately would decide what, if any, action would follow, was that a Scientific Council for Africa South of the Sahara should be formed.* Its purpose would be to ensure that 'the available scientific resources were used to the best advantage in the development of the human and material resources of the continent' [18]. The need for such an international body had been foreseen by Basil Schonland even before the conference had got under way. Without some means of coordination and communication between the few islands of scientific endeavour that existed within the huge sub-continental sea, their efforts would just dwindle and drain away. At the closing plenary session the Conference President, whose *joie de vivre* alone had ensured that the proceedings never faltered, was elected as the first chairman of the Council. In accepting the honour and responsibility, du Toit was unstinting in his praise of Schonland whom he described as the 'father of the scheme', a fact readily accepted by all the delegates who had seen Schonland at work, mostly behind the scenes, promoting the idea and arguing strongly for its acceptance. When Sir Ben Lockspeiser spoke he reflected the views of everyone when he said that Schonland's efforts had opened a 'new era of co-operation on the scientific problems of Africa.' He then continued: 'The great difficulty has always been for scientists from different countries with African territories to get to know each other and to talk to each other.' He might well have added that such difficulties even prevailed much closer to home, given the recent sorry tale of events to which he and Schonland had been party, in England!

*   *   *

For all the pressures that the year seemed to bring to bear, 1949 was also the year, by way almost of recompense, in which further honours were bestowed on Schonland, both at home and abroad. In March, the University of Natal, South Africa's youngest such institution, became an independent institution able to confer its own degrees and it chose, on the occasion of its first congregation, to confer honorary degrees on four men of considerable international stature. One of these was Schonland.† The main address was delivered by General Smuts who was the principal guest in his guise as Chancellor of both Cambridge and Cape Town universities, and the university's public orator, when warming to his

---

* The Scientific Council for Africa South of the Sahara (CSA) duly came into being in 1950. Its first meeting took place in November at South Africa House in London under P J du Toit's chairmanship. Schonland represented South Africa. Thereafter the Council met in various African cities and held numerous regional conferences across the continent. Though significant progress was made on the scientific front throughout the next decade, rising costs and particularly rising nationalism began to thwart it. Soon the political map of Africa changed out of all recognition as every colonial link was broken. The CSA was seen by some to be a vestige of that colonial past and so it was dismantled in 1965.
† The others were all from Great Britain: Lord Eustace Perry, Sir Raymond Priestley and Professor Dover Wilson.

task, was quick to point out the debt that the country owed to both Smuts and Schonland for the many parts they had played in the service of South Africa. 'Mr Chancellor, I present to you, in the presence of "The General", Basil Ferdinand Jamieson Schonland, that scholarship and education, science and technology, law and statesmanship may march together into the new era of light and promise that lies before the University of Natal' [19]. All who were there will no doubt have hoped, but probably with somewhat less conviction, that such a new era lay ahead of the country as well.

Later, in July, the Royal Society of Arts awarded Schonland its silver medal emblazoned with the head of the young Princess Elizabeth, who had become its President just two years before. This Society, with a long and distinguished lineage dating back to 1753, filled a rather unique place in British cultural and scientific life in that it addressed itself to 'the encouragement of Arts, Manufactures and Commerce' — activities frequently not seen to be either kith or kin by their respective proponents. Schonland's medal was awarded for the paper read the previous year on his behalf by Eric Boden, scientific counsellor at South Africa House, at a meeting of the Dominions and Colonies Section. That Schonland's paper had dealt with 'Recent Scientific Developments in South Africa' was now particularly germane since it was those developments, not just in South Africa but in regions to its north as well, that had so captured his imagination and had inspired the international conference recently concluded at Wits.

As the new year dawned yet another letter bearing a foreign stamp arrived for the personal attention of the President of the CSIR. It came from Canada and was written by the President of the University of Toronto, Sidney Smith [20]. The Chair of Physics at Toronto, Smith wrote, would soon become vacant as a result of the resignation of Professor E C Bullard, who was returning to England to become Director of the National Physical Laboratory, and Smith offered the post to Schonland! This was indeed a strange irony since Schonland's recent encounter with Bullard had left behind a lingering taste of some bitterness, but it was surely just coincidental for it is most unlikely that Smith would have known anything of Schonland's own involvement in the NPL saga. The offer to Schonland came with some solid backing in the persons of men who knew him well, and whose opinions Smith valued greatly. Both Professor C J Mackenzie, President of the National Research Council of Canada, and Dr O M Solandt, now Chairman of the Defence Research Board of Canada, were unstinting in their support of Basil Schonland, but Schonland's reaction was not long in coming. He declined the offer and informed Smith in his reply that both he and his wife believed themselves to be too old to consider starting a new life in Canada and, of necessity, as he always saw it under the circumstances, of becoming Canadians. Actually, there was more behind that decision than mere advancing years and a reluctance to change his national allegiance. Schonland by now knew his worth within the international world of physics and it is questionable whether a Chair in Toronto would have had anything like the allure for him of an equivalent position in some British universities or similar institutions. By declining, of course, he was merely leaving open all his other options.

Having taken one major decision presumably made it easier to take another. On 26 January, Schonland wrote to the Prime Minister's private secretary regarding the 'generous offer of the Prime Minister's that I should remain head of the CSIR with sufficient staff and leisure to enable me to carry on some of my own work. I have reluctantly decided [that] I wish to leave'. However generous the offer, the basic idea was flawed and Schonland had soon come to realize this. The responsibilities attached to such an onerous position would surely make it well nigh impossible to carry out any meaningful research, while any pretence at doing so would be contrary to everything he believed in. 'The country', he said, 'has a right to expect the full time of the head of such an important organisation . . . ' and he no longer felt able to devote his full time to it. The question of his successor should be addressed soon, he suggested, and the man he would personally recommend to take over the reins was P J du Toit. However, du Toit was now 62 and Schonland suggested that, if appointed, he would probably wish to serve for no more than two or three years. 'After that you will need a younger man with more energy and less wide interests and commitments' [21]. Though he never said so, the man he had been grooming all along to succeed him was Meiring Naudé.

Word of major developments within South African scientific circles filtered abroad and soon reached the ears of Cockcroft. On 22 January he wrote again to Schonland [22]. By now Cockcroft was fully aware of the outcome of the NPL episode and Schonland's refusal of the Ministry of Supply post, yet his curiosity about Basil's intentions had been aroused as a result of a chance meeting he had had at the Royal Society with Sir Harold Spencer Jones, formerly H M Astronomer at the Cape and a long-time friend of Schonland. According to Sir Harold, who had recently returned to England, it appeared as if Schonland was not too certain as to his future plans. Cockcroft immediately seized the opportunity he saw before him to make clear his own intentions and, particularly, to sound Schonland out, once again, about Harwell.

> Skinner is leaving at the end of April & I have not yet taken any
> steps to fill his post. I have [been] asked whether I will agree to
> succeed Tizard in the D.R.P. [Defence Research Policy] aspects of
> his job, but not affecting the Civil side. I would not want to move
> till Autumn 1952 & there are still a number of points to be cleared

up so that the move is by no means certain. The point about this is that if by any chance you were interested in coming here it would have to be on the understanding that you had the succession.

This was as plain as it could be for 'the succession' meant but one thing: the succession to Cockcroft as Director of Harwell! Schonland replied immediately and thanked his old friend for remembering him but denied being the source of Spencer Jones's information. He did not deny, however, that he was restive and suggested that it might well have been General Smuts who had informed Sir Harold. Of course, Smuts, as Leader of the Opposition in Parliament, would naturally have been informed by the Prime Minister on matters of national interest, and the future direction of the CSIR was certainly one of those. Schonland now made his position clear [23]:

> I am definitely and finally leaving the C.S.I.R. in about six months' time and returning to the B.P.I. I find it impossible to run such a complex organisation as the C.S.I.R. and keep up with my own research and reading without getting worried and tired.

The proposal to join his old colleague at Harwell he found flattering, he said, but this time, unlike his reaction to the Canadian offer of just a while before, it was not in any way impeded by suggestions of old age or issues of citizenship. 'The position with regard to England is quite different', he wrote, 'for we have lived there long and love the country and its people'. However, his memory of other recent events too was fresh and he was not prepared to tolerate a repeat performance : 'We were upset by the way it was handled', said it all. He then put his cards four-square on the table:

> So I suggest that I go back to my little laboratory and if I am really considered suitable to succeed you at Harwell, someone must *give me an official offer a year before the post is vacant,*[*] to enable me to brush up the subject and to come over six months before I take over. It would be as well if I were told sooner than that, that it is likely to happen so that I could direct my reading to nuclear physics instead of entirely to geophysics. This "likelihood" statement would not commit anyone and I would have no grievance if the official offer did not follow. I would, of course, accept the official offer, unless I had by then become bedridden!

The old Cockcroft magic had worked yet again. For all his memories of events, ham-fisted or otherwise,in the recent past, there was one man in whom Schonland seemed always to have implicit trust and that was John Cockcroft. Harwell was indeed tantalizing but the implications of

---

[*] Schonland's emphasis.

the Cockcroft invitation raised numerous questions. The letter therefore continued:

> There are just two points more. The first is that I should like you to have my assurance that I will more gladly accept the offer, if it comes, if I knew that you were yourself still concerned with policy direction in this field.
>
> The second is that I should like to know what the government's policy is about Harwell, and what is the likelihood that it may be maintained only on a diminishing scale. Is its primary policy one of military significance or is it a national laboratory like one of the D.S.I.R. laboratories, placed under the M.o.S. for reasons of security and because it *has* a profound military significance in addition? What is going to happen to it in the future? I realise you can't answer all this directly but perhaps you could do so through a top secret letter via Sir Evelyn Baring, the High Commissioner, which could be read and destroyed in his presence!

Schonland, as ever, had made his case perfectly clear, but what was equally clear was that he considered Cockcroft's continued part in influencing British nuclear policy to be of major importance if he was to accept the invitation. Whatever happened next would depend very much on Cockcroft's ability to provide those answers. Timing, though, was all important and so he added a footnote. 'I have omitted to say that my suggestions above may prove unacceptable to you because you need a successor-deputy in April. If that is the case, you must forget about me.'

On 17 February, Schonland wrote directly to the Prime Minister and stated that his 'sole reason for leaving [the CSIR] is to return to scientific research work'. He then reminded Dr Malan that his 'year of office' was due to end on 5 October and he suggested that his successor should take over from him about two months before that so that he could devote his attention to preparing 'the Council's estimates for the coming year' [24]. From then on the machinery of Government went into action, with an immediate reply to Schonland from the Department of External Affairs* [25]. It informed him that the Prime Minister had accepted his resignation with regret and would announce the fact formally on 6 October. What followed was a list of rather bureaucratic demands. Schonland was required to submit a letter containing the following:

(a) that he had accepted the position of President of the CSIR on the understanding that he had the right to leave after three years but had, in fact, remained in post for six;

---

* The Prime Minister also held the portfolio of Minister of External Affairs, as indeed Smuts had done before him.

(b)   that he was leaving at his own request, though the Government had wished him to stay; and

(c)   that he was leaving so that he could devote his whole time to research.

Presumably this somewhat heavy-handed reaction from Government was intended to ensure that any political and Press-driven backwash resulting from the resignation of the Prime Minister's scientific adviser and President of the CSIR could be countered decisively. The recent dismissal of W Marshall Clark as General Manager of the South African Railways to make way for a man of more acceptable 'Suid-Afrikaanse kleur' had stirred up much acrimony, both in Parliament and in the newspapers, and the Nationalists clearly wished to avoid it happening again [26].

When it came in October, the official announcement of Schonland's departure stressed each of these points, along with the accompanying expressions of appreciation of Dr Basil Schonland's considerable contribution to the furtherance of scientific research in South Africa. 'He was', said the Prime Minister in the House, 'a person who is internationally known as an outstanding scientist. South Africa can be proud of him' [27].

And so ended an era. It was an era that had brought together two men, whose close association over a quarter of a century, had not only changed the face of science in South Africa, but it was an era that had seen both elevated to international status, higher than that achieved by any of their countrymen before them. It ended not only because Schonland stood down at the CSIR but because Field Marshal Jan Smuts had died on 11 September 1950, and he was surely a colossus. Basil Schonland, the son of Smuts's long-time botanical acquaintance whose herbarium in Grahamstown was the oldest repository of the country's flora, had become scientific adviser to the 'Ou Baas'. Together they had brought into being the CSIR and it was surely their monument. Now, the country was in different hands; its political course changed to a heading that seemed predestined to those at the helm while others, with more breadth of vision, saw much rough water ahead. Many South Africans were concerned, the Schonlands amongst them. What had before been practised as a policy of 'enlightened segregation' under Smuts now became a rigid ideology of enforced separation of men for reasons of skin colour alone. Apartheid was no longer just a syntactical aberration, it became a cancer within the country.

CHAPTER 19

SCIENTIA ET LABORE

At its meeting on 1 March 1950 the Board of the BPI recorded 'with much gratification' that Dr B F J Schonland had accepted the full-time appointment as Director of the Institute from the beginning of October 1950, or 'possibly a few months earlier' [1]. However, no such early release from harness at the CSIR occurred. For all P J du Toit's willingness and enthusiasm to step into the President's shoes, there were still numerous matters that required the personal attention of the man who had initiated so much, and so Schonland only drew his six-year term of office to a close when he chaired his last meeting of the CSIR's Council at the end of September.

Three months before, in July, his attention had been very much on a series of meetings that had recently taken place in England. None concerned himself directly, but all dealt with a matter particularly close to his heart: a piece of equipment invented by one of those highly talented individuals whom he had begun to draw together in 1941 to form the nucleus of the SSS. Trevor Wadley, whom we have encountered already, was the man concerned. An engineer of exceptional talent and inventiveness, Wadley's skills had blossomed after the war in the excitingly free and challenging atmosphere that Schonland and Frank Hewitt encouraged at the Telecommunications Research Laboratory of the CSIR and it was his genius, in 1946, that was to change the whole face of high-performance radio receiver design [2]. The continuous tuning capability made possible by the triple-loop concept, soon to be embodied in what was always referred to as 'the Wadley receiver', was quite unique and the CSIR wished to exploit it to the full.

Schonland had been so convinced of the importance of Wadley's receiver, that he had described it to Cockcroft during his visit to South Africa the year before and Cockcroft elected to make some initial inquiries on his return home. This he duly did and not only brought this novel receiver to the attention of various people in England but he also mentioned it to H B G Casimir, the Director of Research at the Philips Company in Holland, whom he had met in the autumn at the first post-war Solvay Physics Conference — always a gathering of some significance.

Cockcroft then wrote to Schonland in June and informed him that Philips were working on a receiver evidently based on a similar principle to Wadley's and he suggested that Schonland should make contact with Casimir [3]. Schonland was not aware of the Dutch activity but knew that a British patent had been issued in 1948 to a French company describing an instrument similar to that developed by Wadley. All indications were that this company did not have the know-how to actually produce an effective receiver, but it would not be long before some company in Europe did. If the CSIR wished to exploit this revolutionary idea they would have to demonstrate it as soon as possible, so Schonland instructed Wadley to take his prototype to England [4].

At its meeting in July, Schonland reported to the CSIR's Council on the outcome of that visit and the demonstrations of the Wadley receiver that followed. A provisional patent had been registered in South Africa but none was possible in England because of the prior claim* registered there by the French. British reaction had been mixed. Marconi, Redifon and the Ministry of Supply though 'initially extremely cold and somewhat superior [are] now showing more interest', but Mullard, Schonland told the Council, thought the set 'a good one'. Its only deficiency was some lack of pre-selectivity, which Wadley had indicated presented no problems to resolve. Outside of the military sphere there was also interest from the BBC who wished to purchase six sets as soon as they appeared on the market [5]. This was indeed a moment of triumph, for it demonstrated so pertinently just how the CSIR should function, by fostering a brilliant idea, exploiting its application and then transferring the technology to an industrial body for eventual manufacture. That the Wadley receiver, conceived during the earliest years of Schonland's CSIR, should have been the inspiration that ultimately launched an electronics industry in South Africa was fitting indeed. Its eventual manufacture followed another series of demonstrations, carried out this time by Hewitt, with the assistance of another former SSS man, T V Peter, who was then on secondment from the TRL in Johannesburg to the TRE in Malvern. The Royal Navy had a specific requirement for a new HF receiver and the Wadley design more than met its specifications and so, in 1954, the Racal Company, in England, commenced production under licence to the CSIR of its famous RA17 receiver based precisely on the Wadley principle.

At its last meeting under Schonland's chairmanship, on 20 September 1950, the CSIR's Council took leave of its first President and resolved, as well, to send a message of sympathy to the family of General Smuts

---

* British Patent 606910 'Improvements in or relating to Frequency Transposing Devices' issued to the *Société Indépendante de Télégraphie sans Fil* on 23 August 1948.

who had died just a few days before. It was indeed poignant that the passing of Smuts, the man whose holistic philosophy had given life to the CSIR, should coincide with the departure of Schonland, its founding President. P J du Toit, soon to take over as the caretaker-President for two years, was as gracious and warm-hearted as ever when he spoke for all his colleagues and said how great a loss Schonland's leaving was to them all. The CSIR's very existence and its obvious viability, he added, were indeed lasting tributes to a man of 'extreme efficiency' [6]. This was a view that few would dispute.

What Schonland left behind was an organization consisting of five national research institutes, populated by a research staff that he described as being 'of a quality and keenness second to none in the world' [7]. What he had not done was to give the CSIR a permanent home. It still occupied the same site in the Old Mint in Pretoria where it had commenced operations immediately after the war. Buildings had not been Schonland's priority, for he was the first to admit that large machines and the structures to house them were things he had sought to avoid. His scientific ambitions had all been achieved with relatively simple apparatus, though the BPI was surely testament to the fact that, when buildings of a very particular type were required, he was anyone's equal in providing them. But in general he was much more like a Rutherford than a Cockcroft, for whom bricks and mortar had a strange fascination. Stories were legion of John Cockcroft's obsession with such details, even to the extent of choosing the very stone from which his laboratories were to be constructed [8].

For all that, Schonland had not ignored a permanent home for the CSIR; indeed the subject had been a matter of much discussion from as early as 1948, when four possible sites in the vicinity of Pretoria were being considered [9]. Of these, two were given particular attention. The first lay south of the city, within easy reach of Johannesburg, near the imposing granite monument that commemorated the Voortrekkers and their Great Trek into South Africa's hinterland in 1836 to escape British rule in the Cape. The second was a considerably larger area to the east of the city in rural land owned by the University of Pretoria and recently offered as a gift to the CSIR. Of the two, Schonland favoured the more accessible monument-site, since he was firmly of the view that industrialists from Johannesburg, the country's largest city and the focal point of its mining and industrial activity, must be encouraged to embrace the CSIR and its research facilities. But in this he was vigorously opposed by Meiring Naudé, who would succeed du Toit as President in 1952; and Naudé was absolutely right, for Schonland's view on this occasion was undeniably short-sighted. He would have been the first to admit that the CSIR, in order to remain viable in the midst of an ever-expanding economy, must itself expand and this required the physical space to do so. Only the site to the east of Pretoria could accommodate that. Naudé's views therefore prevailed and the Council agreed that the CSIR would build its laboratories and supporting facilities in that park-like setting amongst rolling hills soon to be known as *Scientia.*[*]

*   *   *

Professor Basil Schonland's imminent return to the BPI was received with enthusiasm by his old colleagues, who had become rather used to associating him just with tea-times on Fridays. He now intended to revive his personal research into the mechanisms of lightning as quickly as possible while continuing to collaborate with Dawie Malan, who had kept the lightning flag flying in the years when seismology was rapidly becoming the dominant subject in the tea-room. Their renewed contact soon bore fruit and together they published two scientific papers that year. One described the field mill while the other was an invited review of current knowledge of the distribution of electricity in thunderclouds [11]. Such knowledge of the electrical structure of thunderclouds included much from Schonland's own work during the previous two decades, which had begun with his reversal in 1931 of the model of the thunder-cloud that had been so staunchly defended by his old protagonist Sir George Simpson. Now, in one of those delightful ironies that abound in science, Schonland found himself having to call upon Simpson's more recent work to explain the trigger mechanism that initiates the first stroke of a flash. Simpson had shown by careful measurement, during the intervening years, that there were indeed pockets of positive charge embedded within the predominantly negatively charged cloud-base and it was these, Schonland now averred, that provided that trigger which began a cascade of pyrotechnic events.

Lightning was now back in Schonland's blood and the urge to write about it had returned as well. Since producing his monograph on 'Atmospheric Electricity' in 1932, he had long wanted to do the same about lightning for the intelligent lay reader and had even intended starting work on it while sailing to England in 1941, but the distractions aboard ship had made the task difficult. Then, events of an even more urgent nature, after his arrival in England, soon put paid to the whole idea. But the embers still glowed and he revived them once more while at the CSIR, and worked on the book whenever opportunities presented themselves. Entitled *The Flight of Thunderbolts*, a title inspired by Gilbert and Sullivan's operetta *The Mikado*, it was published in 1950 and

---

[*] When the new CSIR site was ready for occupation many of the staff favoured naming it 'Schonland Park' but this view did not receive Government support and *Scientia* was chosen instead [10].

dedicated to his friends who, as he put it 'shared the weakness'.* The writing was Schonland at his most lucid, as he set out not only to demystify the intricacies of lightning but also to describe the research that had helped unravel this most complex of natural phenomena. In what one reviewer described as a 'genial narrative' he mixed folklore with physics in a way that caught the imagination of his readers. Another, more verbosely, described the book as 'a most stimulating exercise in the classical application of the scientific method to a natural phenomenon'. But not everyone saw it that way. The august columns of the journal *Nature* carried a review that suggested that Schonland may have lowered his sights a little too far. 'The book is clearly intended primarily to appeal to the general reader and is most successful in this respect. From the point of view of the specialist however it has the defect that the treatment is descriptive only and that few references are given to the published sources.... Nevertheless, it will be read and enjoyed by all specialists who like to have an elementary but authoritative account of the subject' [12]. For all his own initial enthusiasm, Schonland himself became less convinced that its publication had served his reputation all that well. Even before the first reviews appeared he wrote to John Cockcroft and enjoined him not to fall into the trap of writing anything similar. 'Don't ever write a "popular" book. It is easy to dash it off and horrifying to see it in print, with all one's standards lowered!'. Cockcroft had been tempted but his 'General Account of Army Radar', written in 1945, remains unpublished even though it is a most readable, though not the most reliable, monograph on a subject on which both he and Schonland had spent a fair bit of their time [13].

In May, Schonland was informed by the Franklin Institute in Pennsylvania that he had been awarded its Elliott Cresson Gold Medal for 1950. The accompanying citation read:

> In consideration of his extended work in the field of Atmospheric
> Electricity and of his experimental investigations into Lightning,
> and of his conclusions regarding the mechanism of the Lightning
> Discharge as it occurs in nature.

This was recognition indeed from the institution that bore the name of the greatest of all lightning experimenters, and Schonland joined a distinguished group of former recipients, such as Tesla and Roentgen in the previous century and the Curies, Alexander Graham Bell and E O Lawrence in the twentieth—all honoured for scientific achievements of the very greatest merit. As he was just about to resume duty as the

---

* Those similarly afflicted were James Craib (Schonland's father-in-law who had assisted with his very earliest experiments at Somerset East), D J Malan, D B Hodges, H Collens, J S Elder, J A Keiller, E C Halliday and Bernard Price.

full-time Director of the BPI, he wrote to Humphrey Raikes, Wits' long-serving and long-suffering Principal, to inform him of this honour just bestowed upon him but he emphasized that he saw the award not just as recognition of his contributions in this field of research but equally as an honour for the BPI itself [14]. Sadly, Bernard Price, the man whose generosity and far-sightedness had made it all possible, had not lived to see Schonland and his Institute honoured in this way.

But Price had not forgotten them. In his will an amount of £40 000 was earmarked for augmenting the Revenue Endowment Fund of the Geophysical Research Institute that bore his name. In addition, he left a further £60 000 to the Bernard Price Foundation for Palaeontological Research set up in 1945 as a result of Robert Broom's discoveries at Sterkfontein and Kromdraai that had so fired Price's imagination about the origins of mankind and South Africa's key part in the story. These had followed Professor Raymond Dart's remarkable identification of a species of pre-man in 1928. The University was now not only financially more comfortable but it also enjoyed an international reputation as a result of the work, both in the laboratory and beyond, of Basil Schonland, Raymond Dart and Robert Broom and their many disciples [15].

On 5 October 1950 Schonland returned to his posts as full-time Director of the BPI and as Professor of Geophysics in the University. Notwithstanding the Bernard Price bequest, the matter of most immediate importance requiring his attention was the longer term financial viability of the Institute and, as he had reminded the Board at their meeting in March, the original intention on the founding of the BPI was that it was a venture with but a limited life, to be determined by the rate at which it used up its capital and the interest accruing from it. Originally, it was thought that once the problems associated with lightning had been 'cleared up', as he put it, the Institute would probably have served its purpose. On the other hand, if new problems had arisen in the meantime that it might tackle, it would be necessary to find additional resources. Such was very much the case, with a host of problems in seismology seemingly about to displace lightning as the BPI's major field of research. Schonland believed that an amount of £10 000 would be required per year to maintain the present level of activity, but the capital sum available for investment was barely enough to provide £7500. It would, therefore, be necessary to seek support elsewhere and the Board resolved to make the strongest case possible to a number of likely benefactors [16].

Prime amongst them was the Chamber of Mines, who would be requested to increase their annual grant to the BPI, as well as to provide an additional grant to cover the cost of seismic investigations then being planned on the goldfields of the Orange Free State. This new mining area, some 200 km to the south of Johannesburg, saw its first tentative boreholes sunk in 1939 but then drilling was suspended when the country went to war. Soon after the resumption of drilling in 1946, this region in the northern Free State was being heralded as South Africa's Klondike, and within five years the first mine was in production with the number soon to rise to eleven. But gold was not the only product of interest. In November 1949 highly secret negotiations had taken place between the Chamber of Mines and a mission from what was called the Combined Development Agency (CDA) (the joint purchasing arm of the United States and Britain) charged with securing supplies of uranium for the West's atomic bomb programme then gaining momentum as the Cold War began to bite. From South Africa's point of view, the timing of these negotiations was certainly fortuitous, even though the discovery of the new goldfields with their uranium deposits was certainly not. Whereas a combination of accident and good luck had so often led to the uncovering of a mining bonanza, this was by no means true of the Free State goldfields. Their discovery was the result of dedicated scientific effort by geologists who, after careful study of the reef on the Witwatersrand, had pin-pointed the sites of the boreholes which produced the evidence [17].

Geophysical science clearly had a most important part to play in determining the country's future, yet there were still many within its most important industry who took some convincing of this fact. This was remarkable because the University of the Witwatersrand, since its inception, had enjoyed considerable financial support from the mining industry. Indeed, the University's very existence was due to the needs of the mines for trained engineers who, previously, had come from Wits' predecessors, the South African School of Mines in Kimberley and the Transvaal Technical Institute, that had opened in Johannesburg in 1904. As early as 1911 the first major research scholarship was provided by the Chamber of Mines, when the industry recognized that there were many problems that could only be solved by a scientific approach [18]. Notwithstanding this occasional burst of enlightenment, it seemed as if the mining industry viewed the University in the purely utilitarian role of training engineers. Activities of a more speculative nature, such as those at the BPI, were mostly thought of as being research very much for its own sake with no obvious practical benefits likely to accrue from them. This was the misconception that Schonland strove so hard to overturn, but it was by no means easy.

And so Schonland turned, just as he had done in 1934, to the Carnegie Corporation of New York for a further subvention to add to that which he had obtained those many years before, when the idea of a lightning research institute in Johannesburg had first been mooted. But, as generous as the Corporation was, it was also very careful as to how it disbursed its funds; therefore any suggestion that the BPI's research programme might have become somewhat stagnant would be

401

damning and so it was most important to be able to indicate promising new areas of research into which the Institute was moving. Schonland had already given the matter some thought and he believed that the earth sciences, in general, offered a most fruitful field of study. 'We should move into more direct contact with geology', he wrote, and he proposed the setting up of a 'battery of techniques' to look into such phenomena as gravity anomalies; heat flow; the local variation of crustal thickness; magnetic anomalies such as dykes and other intrusions; and the determination of the ages of rocks and their radioactive constituents. It was an ambitious plan and one that would most certainly be beyond the capabilities of the BPI to tackle on its own. However, the opportunities now existed for collaboration between institutions such as the departments of physics and geology at other universities, the Geological Survey and, of course, the NPL at the CSIR. The coordination of such a programme lay very much within the domain of the CSIR, and Schonland proposed the establishment of a national committee along these lines; but the responsibility for initiating the scheme rested, he believed, with the BPI. Additional financial support should be sought from the Royal Society in London, as well as from the local electrical manufacturing companies, whose expertise and equipment would play an important part in the research programmes.

Schonland's Board at the BPI was most receptive to these ideas. The return of their Director had once again injected into the laboratory the spirit of enterprise and enthusiasm that was so reminiscent of the SSS at the very beginning of the war. The Board duly authorized the setting up of a new section to work in the fields of Precambrian geology, geochemistry and plate tectonics, and approved the expenditure of £1500 for this purpose, with the assumption that further support would be sought from the various bodies to whom this work should be of interest. It also granted leave to the Director to accept an invitation from the Carnegie Corporation to visit its headquarters in New York for further consultations on all these matters [19].

*   *   *

Though he had only been back at the BPI for a few months when he wrote it, the Institute's annual report for 1950 bore all the hallmarks of the Schonland of old. It opened with the Director's most buoyant introduction [20]:

1. <u>General</u>.

> The past year has been perhaps the most successful and interesting in the Institute's short history. As will be seen from the more detailed account which follows, the work of the sections dealing with thunderstorm electrification and with earth-tremor

seismology has in each case been carried to a stage where new information of considerable value has been obtained. These studies have opened up fresh avenues of research of much interest and importance. The other smaller sections of the Institute have also made considerable progress during the year.

The lightning programme, for all the suggestions that it might well have run its course, was continuing to produce results of great importance. The new electric fluxmeter had shown itself to be a most useful instrument with which Malan had amassed a large amount of data during the three previous lightning seasons. This would produce a much better understanding of the process of the electrification of thunderclouds during the intervals between the individual strokes to ground. Schonland and Malan had set themselves the task of analysing these data, in order to try to explain some interesting effects.

The Director's enthusiasm was almost palpable. For years he had cherished the belief that he would return, one day, to active research and now it had happened. He soon took a decisive lead and advanced a theoretical explanation for the effects that Malan and his assistants had observed. It was apparent, Schonland conjectured, that the individual strokes did not arise, as had been previously thought, from separate charge-centres widely separated within a thundercloud, but were, in fact, the manifestations of the intermittent discharge of a single column of charge that extended vertically upwards from near the base of the cloud to a region that may be as much as 9 km above the ground. This was a radical proposal because it implied that there were no horizontal components of this column of charge, and this seemed rather surprising given the marked horizontal extent of any thunderstorm, but Schonland stuck to this view because the quasi-electrostatic model that followed from it produced results which seemed to agree closely with the observations. Malan, though not entirely convinced of the validity of Schonland's model, could offer no alternative and so they used it to decide which of the various competing mechanisms then under discussion in the scientific literature was actually responsible for the lightning discharge. The one they plumped for again saw them in agreement with Simpson, who maintained that the process of charge generation occurred in the region of the cloud where ice and snow abound, in other words, at great altitude. The fact that this so-called freezing level existed at around 4 km above the ground in South Africa, while the inter-stroke processes occurred above 5 km, in general, added the necessary weight to their theory. These important findings duly appeared in the *Proceedings of the Royal Society* and not only proffered a new theory but also introduced the BPI's fluxmeter or field mill to the wider scientific community. This paper, published in April 1951, and another that followed soon after, on the distribution of electricity in thunderclouds, were soon to stimulate a great deal of research and not a little controversy that rumbled on for many years, particularly in the United States [21].[*]

The seismic investigations made at the BPI during the years 1948–49, also described in detail in Schonland's report, were particularly significant, in that they contributed measured data about the thickness and characteristics of the earth's crust along the line of the Reef to the west of Johannesburg — a region previously uncharted. The method used was based on the very frequent occurrence of earth tremors in the vicinity of the deep gold mines, which were easily detectable with suitable instruments at distances of up to 500 km from their epicentres. The study was carried out over an eighteen week period in 1948 and 1949 by Anton Hales and Philip Gane from the BPI and a visitor to the Institute, Dr P L Willmore of the Department of Geodesy and Geophysics at Cambridge.[†] Using equipment designed by Gane, with features that made it superior to any other similar system in use elsewhere, they obtained extremely good data, from which they were able to calculate that the depth of the crustal layer to the Mohorovicic discontinuity in this region was about 36 km. Though unremarkable in itself, since that was fairly typical of the earth as a whole, their results indicated that the crustal layer was not uniform, but consisted either of several layers of different constitution or of material which changed in some manner with depth. Since the question of the uniformity of the crust was a matter of considerable debate in international geological circles at the time, these BPI data provided valuable experimental evidence on which to base further study. Of immediate interest to the gold mining industry was a finding from this and other related work that 85% of the first movements of earth tremors recorded were downwards. This implied that they were caused by collapsing cavities underground and their focal points implied typical depths of between 1 and 2.5 km. Since the tremors all originated in a narrow strip running through Johannesburg, their source was undoubtedly the collapse of old mine workings, which lay at an average depth of 2 km below the city. A similar investigation was then carried out in the vicinity of the new goldfields being developed in the Orange Free State and, though 42 tremors were observed during the period under observation, they all originated from Johannesburg and none was detected from the new goldfields. This was surely a finding of some importance, since it indicated the stability of such workings over a

---

[*] In 1969 T Ogawa and M Brook from the New Mexico Institute of Mining and Technology concluded, on the basis of a very extensive study, that 'the "nearly vertical" aspect of the charge distribution [of Malan and Schonland] has been over-estimated'.
[†] Before coming to South Africa Willmore had just completed a study of the seismic effects of the great explosion on the island of Heligoland that destroyed the German underground fortifications there in 1947.

period of time, and a report was sent to the Tremor Committee of the Chamber of Mines.

Schonland's return to his 'little laboratory' and to active research, had not only been like a breath of fresh air to all his colleagues; the University, too, was pleased to have him in its midst, for he was now a man of vast experience on a much bigger stage and his views and counsel were sought on many issues. With the death just a few years before of Dr Bernard Price, Wits had lost much more than just a benefactor, for he was in some ways its elder statesman as well. To preserve his memory it was only fitting, therefore, that the University that had benefitted so directly from his generosity should collaborate with the South African Institute of Electrical Engineers in hosting, each year, the Bernard Price Memorial Lecture. The first of them, equally fittingly, was delivered by Basil Schonland on 26 July 1951.

The lecture was the central theme of what was a rather grand occasion. Schonland chose to talk about the work and achievements of the BPI since, as he said, he had to 'confess to an enthusiasm' for geophysics, as the subject was 'second only in interest and importance to the problem of the atomic nucleus'. His enthusiasm was certainly infectious. Soon his audience of academics and engineers alike were gripped as if by the excitement of the chase as he led them through nearly fourteen years of scientific discovery and pioneering wartime developments that took place within the Institute that bore Bernard Price's name. In describing its work, from lightning to seismology and from gravity to radar (a diversion so secret that not even Price himself could be told what they were doing), he brought his audience to its feet in a standing ovation when it ended. This was vintage Schonland now released from the burden of bureaucracy that had been his lot for more years than he cared to remember [22].

It was almost inevitable that further recognition of Schonland's stature in South Africa and in the world should follow the pattern already set elsewhere. Next to honour him was Grahamstown and its now fully fledged university. In March 1951, as part of the ceremonies to celebrate the transfiguration of Rhodes University College into Rhodes University, Schonland's *alma mater* conferred on him the honorary Doctorate of Laws, *honoris causa.*[*] What was more, he was also elected the new university's first Chancellor. This was indeed due and just reward for the man who had never lost contact with the Eastern Cape, even though his international reputation had soared since that day in March 1915 when he

---

[*] Honoured with him were Sir Arthur Trueman, Chairman of the United Kingdom University Grants Commission, Sir Philip Morris, Vice-Chancellor of Bristol University and Professor (later Sir) Keith Hancock, soon to write the biography of Smuts.

sailed from Port Elizabeth for England. Now he had returned to receive the acclaim of Grahamstown's citizens and academics alike.

In October, wearing the robes of his new office for the first time, he delivered his address to the Congregation of Rhodes University in the City Hall of Grahamstown [23]. It was stirring stuff. Whether they needed to be reminded or not, he reminded them all of their educational heritage. There, in St Andrew's College, at Gill College in Somerset East, the Grey Institute in Port Elizabeth and the Graaf Reinet College, were to be found the seeds of a tradition of education second to none in the country. It was those institutions that had laid the foundation for the University over which Basil Schonland now presided and for which he had the very deepest respect and the greatest appreciation for what it had done for him. He regaled his audience with the names of men who had passed through those places and whose names were now indelibly recorded in southern African history. There was Henry Burton, a member of the first Cabinet of the Union of South Africa; H U Moffat, the second Prime Minister of Rhodesia; and Sir James Rose-Innes, Chief Justice of the Union, amongst many others. Men of science abounded too, with possibly the greatest being Henri Fourcade, the multi-talented botanist, forester, mathematician and surveyor who developed the stereoscopic method of photographic surveying that was now of such importance throughout the world. Lest anyone think that men alone had contributed so handsomely to South Africa's rich heritage, he reminded his audience of one whose literary work 'would be remembered for centuries to come'. In Olive Schreiner, the English-speakers of South Africa's eastern frontier had not only produced a great literary figure but, he said, 'We have produced one of the greatest in the world'.

He was warm in his tributes to Cecil John Rhodes, in whose honour the university was named, for it was Rhodes, with his vision and his unswerving belief in the value of education, who had left his immense fortune 'to give this privilege to others, and in particular to South Africans'. The Rhodes Scholarship was his lasting memorial and the university that bore his name could surely hold its head up high. Schonland then took his audience on a guided tour through South Africa's often rocky past from colonial outpost to, as he put it, 'the stage of an industrial revolution'. That revolution had been fuelled by the mineral wealth that lay beneath the veld of Kimberley and within the dolomite of the Witwatersrand and Free State goldfields. Fortunes were made and lost there and the country fought a bitter war when others cast covetous eyes upon it. Then, when peace was restored and the Union was born, it was South Africa's universities that had produced so many of the pioneers who developed its industries and tended to the health and educational needs of the nation. 'High ideals, great self-sacrifice, generous financial support and the achievements of our graduates have brought us to a position of some importance in this land', he said. With an even greater revolution now ahead of them, as the world emerged from the misery of an even greater war, there were those who saw a future only in large universities based around medical, dental, agricultural and engineering schools. Schonland disagreed and told his audience that the fostering of science, in its fundamental forms, was vital to the prosperity of South Africa now entering this new stage of its development and Rhodes University, particularly, had a proud history of achievement in this area. He allowed himself to make a prediction based on the lofty view of the South African scene he had had as President of the CSIR.

> This country and the Rhodesias during the next five years will reach a position where not less than 1500 posts will be occupied by or available for professionally trained chemists, physicists, mathematicians, biologists, geologists and applied psychologists in fields ranging from the purely academic to the applied. These people will have to have a five-year university training. Anything less will not be acceptable to their employers.

Every incentive was there for the University, now robust in its independence, to take the lead, unencumbered by the presence of the large professional faculties, in order to meet that demand. He then addressed himself specifically to the students whose representatives were assembled in front of him.

> If a country gets the Government it deserves, students get, by and large, the instruction they deserve. Their influence upon their instructors is subtle but strong. If they are content with the minimum of effort, their professors and lecturers find it hard to maintain the standards of teaching and to exert the inspiring influence they would like to. If the students are determined to work hard and to learn all they can, the effect upon the staff is well-nigh miraculous. I would ask the students to realize this. I would ask them, 'Could you not forget for the next few years that you are South Africans and work at least as hard here as students do in Europe?'. You would be doing a service to this University the magnitude of which you can hardly realize. You yourselves could change its whole future immediately.

Schonland's words were met with acclamation, even by those who apparently struck their Chancellor as being a mite less industrious than their colleagues in more temperate climes. The student newspaper *The Rhodian* warmly congratulated and welcomed him back to Grahamstown, to which Schonland replied with further encouragement and a thought-provoking remark [24]:

...these are the happiest days of your life. *Enjoy* them to the full—
and *work*—and *think*. South Africa's future rests with those who
come from Rhodes in greater measure than with any other group in
this country.

What he had in mind might well have been linked with his very next duty
as Chancellor. Near the little town of Alice, about 100 km north-east of
Grahamstown, was the South African Native College of Fort Hare, the
only residential centre for higher education of South Africa's black
population. This Methodist college, known everywhere simply as Fort
Hare, had been founded in 1916 on the site of what had been the largest
British fort on the Eastern Cape frontier a century before. It had an
enviable reputation throughout Africa as an institute of higher learning
steeped in the traditions of England, and it boasted amongst its staff
and graduates some of the most influential of South Africa's black
leaders.* When Rhodes University separated from the University of
South Africa and became a degree-awarding body in its own right it
also became the responsible institution that accredited the degrees
awarded by Fort Hare. Schonland was now also Chancellor of Fort
Hare and he well understood the magnitude of that responsibility.

On 26 October, just a day after delivering his address at Rhodes,
Basil and Ismay travelled to Fort Hare, where Schonland addressed the
other congregation of Rhodes University [25]. He regarded it as a great
privilege, he told them, to be their Chancellor and it was one which he
much appreciated for he had 'a profound belief in the value and in the
possibilities of this College. Its importance to South Africa and to its
future cannot be overestimated. I shall very gladly do whatever I can to
help Fort Hare in its development'. In his mind and, no doubt, in the
minds of all his like-thinking contemporaries was the belief that
development depended upon the African students embracing to the full
the ideals of western civilization. Precisely what constituted this 'western
civilization' therefore formed the crux of his address.

If what he had said the day before in Grahamstown were the words
of Schonland the scientist, those that followed them at Fort Hare were the
words of Schonland the philosopher. In Grahamstown, his eye was on the
future of South Africa as an emerging industrial power; at Fort Hare, he
looked deeply into the heart and soul of Africa. It was a much more
demanding task.

He began by introducing himself to his audience, for most of them
would not have known him, but he surely felt that he knew them.
Grahamstown, at the turn of the century, when he was a little boy
living in Oatlands Road, had been a simple place. Its peoples, though

---

* Foremost amongst them is undoubtedly Nelson Mandela but Professors Z K Matthews and
D D T Jabavu were also notable names from an earlier era.

torn by bitter frontier wars that had lasted many decades, had come together in a sort of loose-knit community where black and white farmers rubbed shoulders on the lands that surrounded what would soon be known as 'the city of saints', such was the proliferation of churches and church schools in the place. The church and education had brought civilization—western civilization—to the Eastern Cape. Now, at Fort Hare, Basil Schonland, scientist, soldier and western man asked the question 'What is western civilization?'.

> We are given to accepting the phrase 'western civilisation' without much thought. It is a cliché, a catchword. It is confused, and often identified with 'western democracy' as if democracy has a compass direction. And it is often identified with Christianity, which is a creed and not a civilisation. If, as I think, it is none of these things, it is worth while our enquiring what western civilisation really is and how it has developed.

He identified two features that, he said, distinguished western civilization from all other civilizations that had preceded it or which were contemporaneous with it. They were: 'its creed, which is Christianity, and its enormous achievements in science and technology. No other civilization has had this religion and no other civilization has developed this fantastically powerful technique of scientific enquiry'. Until about the year 1600, he told his audience, the principles of Christianity alone had supported and developed western civilization, on the basis that all men were free; though the curse of slavery, in one guise or another, had shown it to be forever an ideal. Then, a remarkable change had taken place. A new doctrine had appeared that extended the original Christian doctrine which preached that the individual should be free in body, as long as he did not injure others, to one that demanded that he should also be free in mind while doing no harm to others. Of course, it had come from an even earlier time, from the Greeks in fact, but now its fundamental tenets found fertile soil amongst the nations of western Europe and 'it created a profound and revolutionary change.' The consequences of this need for free enquiry, even within enlightened communities, were both fear and scorn from those who felt threatened by it, but its momentum was unstoppable and, though its early protagonists, like Galileo, were subjected to authoritarian treatment, neither he nor those who followed him with equally revolutionary ideas were persecuted for their convictions. 'Thenceforth', said the Chancellor, 'by embracing the new doctrine of the absolute freedom of thought and inquiry, our civilization became unconquerable and enduring'.

In many ways these words must have contained little, if anything, that his audience of educated and inquiring young men and women did not already know. But maybe, in the context of South Africa at that time, it was a very significant warning to all who heard them. In Europe, just a few years before, it was Fascism that had sought, at best, to subjugate and, at worst, to annihilate those who dared to think in ways that did not submit to the absolute authority of 'The State'. What Schonland feared now, and what he wished to warn his congregation of academics and students about, was Communism—yet another totalitarian scourge that had enslaved millions yet was beguiling to those who saw themselves as the exploited victims of a tired, discredited, capitalist Western society. He went on:

> I think it is important for us to realise that it is not western civilisation which is tired and worn-out and effete, it is communism. It is not western civilisation which is unable to accept new ideas and alter its social structure, it is communism. And it is not western civilisation which has to exclude religion and to force its scientists and philosophers to follow the 'party line'. For communism is, in my opinion, the ideology of tired men and tired minds, and western civilisation is the ideology of fresh men and fresh minds.
>
> The thing that tired minds dislike most of all is change. And change, as I have said, is the essence of western civilisation. Its social structure is never static; first it was monastic, then it was feudal then capitalistic and now to a large extent it is becoming social-democratic. The truth is not final nor is it simple. It does not lie at the bottom of a well, it is many-sided. Western civilisation, like its religion, represents a constant striving towards an ideal which it never reaches. There is always another peak to climb, another social injustice to put right.

This was stirring stuff and it was Schonland speaking to Africans who knew full well what he meant; for many of them felt beguiled by an ideology that promised to break their bonds, and they all knew about social injustice.

That social injustice in South Africa had come to a head during the years 1950–51. The Nationalist government of Dr D F Malan had piloted a number of Acts through Parliament which made apartheid not just a matter of misguided social convention but established it as a system that was underpinned by the full weight of the law, and its implementation was forced on the country with much Nationalist zeal and fervour. To oppose it was seen by many in government-supporting circles as being tantamount to supporting the aims and philosophies of Communism. This branding of any outspoken critic of apartheid with the Communist tag was a dastardly tactic that unsettled many South Africans of no political persuasion at all, who simply wished to see justice and fair play in all aspects of life in their country. That 'McCarthyism' was then at its height in the United States served only to add even more fuel to the vehemence of the attacks levelled at their opponents by South Africa's government. What made it even more invidious was the belief of the Afrikaner Nationalists that theirs was a mission sanctioned by the Almighty, for, as many a *dominee* would have them believe, they were His chosen people [26]. And so the threat, or at least the allure, of Communism was very real in South Africa, particularly amongst those who now found themselves dispossessed of their birthright simply because of the colour of their skin [27].

The Schonlands were liberal-minded, English-speaking South Africans. Basil had always been a staunch supporter of Smuts, but of Smuts the man and of Smuts the philosopher rather than of Smuts the politician. There was no dichotomy in this; it was simply that politics played no part at all in his make-up. He was patriotic and placed huge store by doing what he saw as his duty, and neither patriotism nor duty required the making of any political choice. Ismay, by contrast, held firm political views that now placed her squarely in the socialist camp [28]. She too, though, both admired and supported Smuts as a man and as a statesman. Together, the Schonlands were decidedly uneasy about the direction in which their present Government clearly wished to steer the country, for it was certainly not the direction determined by western civilization.

In his closing words to the congregation at Fort Hare, Schonland touched on these sensitive matters so close to home. 'The social history of western civilization', he said, 'has been marked by a continuous series of surrenders of power and wealth by those who possess them'. But people did not do this readily, regardless of the injustices which may have bestowed the privileges on them in the first place, and they always had reasons to justify that reluctance. 'They say that others are not deserving of it, that they have no enterprise, that they are not ready for it because they are not educated, that they will waste these precious gifts which have been accumulated with so much sacrifice'. It was, he said, a familiar cry and it was so often based on prejudice and fear. This was the struggle of western civilization and it had been the same since the days of Magna Carta. But, he averred, such views did contain a certain truth for they provided a safeguard 'against over-hasty giving, [and] against giving before the recipient is ready to receive the gift'. By implication, of course, the recipient would ultimately reach that stage of readiness, however it was defined or measured, and then the next step must follow automatically, for this was but the natural course of western civilization.

In but a few years Schonland would meet some of the African leaders who were, even then, pressing their demands not just for racial equality but for much more besides. That meeting, in conditions rather different to the formalities of Fort Hare, would bring home to him quite how far removed his cautious words were from the aspirations of the mass of his black fellow countrymen.

*   *   *

The political motives of South Africa's Nationalist government were watched with eagle-eyed intent from the Opposition benches in Parliament. Any move that may have smacked of political manoeuvring for its own ends provoked a speedy response, and one that was seized upon was the change that had recently taken place at the level of the presidency of the CSIR. Schonland's resignation as President was raised in the House when the Opposition asked the Prime Minister to comment. Dr Malan assured the Members that though he had wished that 'Dr Schonland would become a permanent official' [of the CSIR] he had expressed the desire to return to research, even though he had been made 'most important offers in connection with a similar post in Britain'. But, said Malan, 'he [Schonland] said he was born in South Africa, that he is a South African, that he lives in South Africa and that he is going to offer our country his further services' [29].

This was rather overblown political rhetoric but it was perfectly true, and those further services were soon to fall under yet another of Schonland's many hats when he became Adviser on Defence Research to the CSIR and Commander* of the South African Corps of Scientists of the UDF. He was confirmed in both posts immediately he resigned from the CSIR and took the associated responsibilities very seriously. The UDF counted itself fortunate in being able to secure his services, even on an occasional basis, when so many other men who had made significant wartime contributions were only too pleased to shed their uniforms forever once hostilities were over. Though even further removed from the war that had recently broken out in Indo-China than she had been from Hitler's war 'up north', South Africa watched the events unfolding there with growing concern and would soon be offering the services of its forces to the United Nations. The UDF, though, were once again ill-equipped and under-manned and many of its most senior officers with wartime service had been summarily retired when the new government came to power. Schonland therefore found himself in the interesting position of having to advise a General Staff who numbered amongst their ranks some whose careers had involved little in the way of active service. By contrast, their scientific adviser had been in the thick of it under no less a figure than the man who had commanded the massed armies that landed on the beaches of Normandy on D-Day.

---

* Soon to be called the 'Director'.

As always, the quietly influential figure of John Cockcroft, though all of a continent away in England, continued to influence the thoughts and actions of Basil Schonland from time to time, and it happened yet again as the Korean War reminded everyone that a new threat, in the form of global Communism, was now looming. It was in March 1945 that Cockcroft had first suggested that Schonland should write of his experiences with Montgomery but Schonland declined and nothing further transpired. The thought, though, that he might eventually do so never left him and, now that he had more time for his own affairs and interests, and given the increasing need for close scientific collaboration with the military, the opportunity to pick up his pen was ripe.

In February 1951, while recovering from a bout of 'flu, Schonland wrote an essay which he titled 'On being a Scientific Adviser to a Commander-in-Chief' [30]. It was, he said, his attempt to put on paper what 'the Army might call the "doctrine" of the use of a scientific adviser' and, though he was the first to concede that what he presented may well have been out of date, and that he was somewhat hamstrung by not having access to relevant war diaries and other papers, his thoughts would at least be on paper. If nothing else, he wrote, his views might amuse anyone who had to fill such a post in the future!

This document was unique in many ways because it was very likely the first such account ever written. It was a contribution from a man who was not a professional soldier but who, as a scientist with a particular bent and aptitude for soldiering, had been able to present a view of the scientific issues of modern warfare hitherto unknown or at least unpublished. Its bias was naturally towards 21 Army Group and the campaign in North West Europe in 1944 and 1945 but its author's observations, comments, suggestions and recommendations were as applicable to that theatre of war as they would have been to most others involving vast armies in the field. As has been recounted already, Schonland had enjoyed a remarkably easy relationship with the Generals and Brigadiers with whom he shared the Generals' Mess during that campaign. Such a relationship was clearly vital to the success of his mission, which could easily have been perceived as a mixture of fish and fowl. In his own words, 'The S.A.'s [Scientific Adviser's] nickname in the Mess will probably be "The Professor", but it should be a term of affection invented by his friends and not a form of sarcastic wit'.

He believed that it was most important that the SA should have had previous military service, especially in an operational research role and ideally in a campaign, otherwise he would 'simply be a dressed-up civilian'. The alternative option of attaching an experienced staff officer to the SA to act as an 'interpreter' would simply not work in the heat of battle, when the Commander and his staff required quick, decisive and informed answers to their immediate questions. He also had definite

413

views on where, within the HQ structure, the SA should operate. Montgomery's system of three headquarters, Tactical, Main and Rear, was clearly thought out, with each having a very specific function within his order of battle. It was at Main HQ that the Chief of Staff to the C-in-C operated, with Montgomery himself always forward at his Tac HQ, in virtual isolation except for his small personal staff. At Main HQ were the various Brigadiers' General Staffs (BGS) responsible for such matters as Intelligence, Operations, Staff Duties and Plans and therefore, if the SA were to have any useful function at all, he too had to be at Main HQ. Schonland was equally clear on what rank he believed the SA should carry. 'Main H.Q. consisted of four Major-Generals and about fifteen brigadiers, with their staffs. The principal direct advisers to the C-in-C and Chief of Staff, and, therefore, the men with whom the SA would have the most frequent contact, were BGS(I), BGS(Ops), BGS(SD) and BGS(Plans). It would have been difficult for the SA to be on easy terms with these last four if he had himself been a Major-General. 'Unless the arrangements in a future H.Q. are different', he wrote, 'I do not think he should rank higher than a Brigadier. On the other hand, a lower rank than Brigadier would make his work almost impossible'.

He stated what he believed to be the Scientific Adviser's primary function, as well as that of the supporting Operational Research Section (ORS) that operated indirectly under his command. The chief use of operational research, he said, was to study what happened in a military operation and to advise on any changes in the tactics of future operations. Its function was not to solve technical problems; the Army's own specialist Corps such as the Engineers, Signals and REME* were there for that purpose. The Scientific Adviser himself should not become embroiled in detailed technical matters such as the study of special enemy weapons like 'the V1 and V2 and other V monstrosities we overran in France'. If, as he put it, 'the S.A. has time for military archaeology he is unlikely to be doing enough proper work for his Commander.'

There was, though, a subsidiary but important function of a Scientific Adviser and that was to solve what he called 'conundrums'. These were usually put to him by the staff officers and they often encompassed the whole gamut of scientific warfare. As we have already seen, one of some urgency, when the massive sea-borne invasion was imminent was whether the Germans might manage to 'electrify the sea' by running cables into it from a nearby power station. Schonland managed to assuage his colleagues about that possibility. Another asked:

> If we knew of a panzer leaguer and turned Bomber Command onto
> it, what damage would result to both hard- and soft-skinned
> vehicles?

* REME: Royal Electrical and Mechanical Engineers.

414

How he handled that would have been interesting to know, given the facts about bombing accuracy that were then just becoming available. A third query had a very direct bearing on the well-being of the C-in-C himself and dealt with the possibility of the enemy applying radio direction-finding techniques to pinpoint the radio transmissions emanating from Montgomery's Tac HQ. Since Monty continued with the practice, one must assume that its likelihood was slight! And there were many other conundrums just like them to exercise the SA's mind.

Schonland completed this monograph in double-quick time and then decided that, to serve any useful purpose at all, other than as a collection of his reminiscences with his views included for good measure, the document had to be circulated widely amongst both his military and his scientific colleagues who had lived through these most testing of times with him. So he immediately sent copies to Cockcroft, to Sir Henry Tizard and to Omond Solandt in Canada, amongst others. He also sent one to Major General de Guingand, Montgomery's most admirable Chief of Staff who had recently retired from the Army. Now he awaited their reactions with interest and they were not long in coming.

'Freddie' de Guingand received the document enthusiastically, added a few gracious comments of his own and suggested that Schonland send a copy to 'Monty' himself. Tizard, as Chairman of the Defence Research Policy Committee, responded enthusiastically. 'Let me say at once that I think it is an extremely interesting record of your views and I find that most people are in general agreement with you. The War Office are anxious to keep a copy...'. Solandt, too, wished to circulate it widely amongst Canada's military chiefs, especially as Schonland was the man, he said, who was 'a pioneer in the task of being a scientific adviser to the commander of a modern army in the field'. Given the commander in question, that was no mean feat, and Schonland, remembering Montgomery's withering reaction when asked if he would like to have a scientific adviser join his headquarters, now awaited his response to that adviser's views with more than a little interest. In the meantime, he amended the document in line with the many comments and suggestions that had been made and further copies went off, as requested, to the Deputy Chief of the General Staff in London and to South Africa's own CGS, Major General C L de W du Toit DSO, in Pretoria [31].

The reaction of Field Marshal Sir Bernard Montgomery duly came in a letter to Schonland from de Guingand and 'Freddie' quoted it verbatim [32]: 'Thank you for the paper by Schonland. I have read it with very great interest and consider it excellent. Please tell him so from me.' Praise indeed from the man who was now Deputy Supreme Commander of NATO and as renowned in Brussels as he was wherever he encountered mere mortals for his demolition of the verbose and the inarticulate! To have obtained such encouraging comments from Montgomery himself was fair praise indeed.

* * *

South African military matters now had a permanent place in Schonland's week and, in an office alongside his own at the BPI, he set up the 'Defence Research Section of the CSIR' with Miss P Murray, on attachment from Pretoria, as his assistant. The Prime Minister and the Minister of Defence, acting on their Scientific Adviser's advice, had approved the setting up of a Committee on Defence Research under Schonland's chairmanship with a membership that was selected carefully to ensure that it represented all sectors of the country's scientific base. Humphrey Raikes, Principal and Vice-Chancellor of Wits, represented the universities; P J du Toit, the CSIR; Dr E Taberner, the Associated Scientific and Technical Societies; and Dr F J de Villiers the *Suid-Afrikaanse Akademie vir Wetenskap en Kuns*.* This committee met at the BPI for the first time on 23 April 1951 [33]. In front of them was a document that set out the structure of the special body known as the South African Corps of Scientists, which Schonland had been appointed to command by the then CGS, General Sir Pierre van Ryneveld, in 1947. Under him were eight scientists who had all been commissioned as officers in the Active Citizen Force of the UDF, but there were to be no uniforms, nor would they attend any form of peace-time training. However, they would bear military ranks: in this company Schonland himself was still 'the Brigadier'.

He then explained the thinking behind this rather peculiar military unit which it was his lot to command. Concern had been mounting in defence circles about the deteriorating international situation due to the increasing Soviet threat and the possibility, however remote, that war might once again envelop the world. Should that happen, South Africa had not only to be militarily prepared, which was the CGS's responsibility, but the nation's armed forces must be assured of adequate scientific support, which was Schonland's. Together, the CGS and Schonland had agreed to form a Corps of Scientists who would be 'interwoven with the Defence Research Section of the CSIR'. Since no specific defence research section actually existed at that time, the idea, if not actually stillborn, lay dormant. However, in Schonland's mind, at least, he could readily identify the elements that would rapidly constitute such a section should the need for it arise in a hurry. Many of the members of his famous Special Signals Service, the SSS, now worked at the CSIR's Telecommunications Research Laboratory (TRL) and some had even been sent on secondment to the Telecommunications Research Establishment (TRE)

---

* lit. 'South African Academy for Science and Arts'.

in England for special training. They, at least, were competent to be pitched into the scientific battle, should the need arise.

By early 1951, the nucleus of this rather unusual military outfit that had officers but no 'other ranks' was firmly established. No second-in-command to Schonland had as yet been appointed but the establishment table catered for such a person in the rank of Colonel or naval Captain. The psychologist, Dr Simon Biesheuvel, whose wartime work on air-crew selection had so impressed Ismay Schonland that she had earmarked him for Basil's attention as the man to lead the CSIR's Institute for Personnel Research, found himself as the Staff Officer Personnel Research in the rank of Commandant.* Ranked immediately below him as Majors were Eric Boden and Frank Hewitt, both SSS men of considerable wartime experience, who looked after general duties and telecommunications respectively. Another member of the SSS recently returned from TRE, Captain P Meerholz, assisted Hewitt, while Boden could call on the services of Lieutenant J P de Witt. The breadth of responsibility of this new 'Corps' was certainly wide. Naturally, given Schonland's recent past, it included operational research under Captain H M Cooke, while the realities of the modern, post-war world made it essential to have people capable of addressing the two fearsome areas of atomic and chemical warfare. The officers charged with those responsibilities were Captains G H Stafford and D E A Rivett, who were soon on their way to England for special training. With Cockcroft's assistance, Schonland had arranged for South African scientists to work at Harwell and at other research establishments and so become proficient in areas well beyond South Africa's reach at that time. Not only would the Corps of Scientists benefit but so would basic science in South Africa.

The objectives of the Corps as Schonland agreed them were quite simple. They were 'To assist the armed forces of the Union by means of scientific advice, by research and development on technical methods of warfare, offensive and defensive, and by the provision of an operational research group'. They were not there to perform routine technical work for any of the services, nor to provide, as the document spelt out, 'operating personnel for chemical warfare or radar or engineer units.' Those were the responsibilities of the armed services themselves and the Director was emphatic that any additional personnel, provided either by the CSIR or by any university charged with defence-related work, would probably be better served were they not commissioned officers in the Corps but remained as civilians and so able to discuss scientific

---

* 'Commandant' was originally a Boer War rank roughly equivalent to Lieutenant Colonel that was substituted for the 'English rank' soon after the Nationalist Government came to power in 1948.

and technical matters with senior officers without the difficulties connected with rank. Whether this would necessarily work in practice had often been an issue in England during the war, when scientists interacted with the military. Some, like Schonland, had both the scientific stature and the necessary military rank to carry it off; others, lacking both, did occasionally run into difficulties.

Soon, both the scope of the Corps' activities and the number of scientists who were commissioned to serve in it increased. Philip Gane was appointed to take charge of what were called 'special devices of a physical character'. Similarly, T V Peter, ex-SSS, became an officer, as did F P Anderson of the NPL, who found himself responsible for both the equipment and the operational research associated with coastal and submarine defence. The laboratory facilities of one of South Africa's coastal universities (Cape Town, Stellenbosch or Durban) were suggested as being most appropriate for this task and Humphrey Raikes was asked to pursue the matter with his fellow university Principals.

Raikes, though no stranger to the military, having served with much distinction in the Royal Flying Corps during the First World War, and more recently as a lieutenant colonel in command of the University Training Corps at Wits during the next conflict, soon proved to be less than happy with some of Schonland's ideas, particularly that associated with the gathering of scientific intelligence. Evidently Schonland had proposed that academics be asked to pass to the Corps of Scientists any information they had picked up in the course of their normal research which may have a bearing on military matters behind the Iron Curtain. Without actually saying so, Raikes clearly considered this a little naïve. Some academics, he suggested, with no Communist sympathies whatsoever, would see this as an affront to their academic impartiality and would 'shut up shop like an oyster', while others might well be tempted to play a practical joke by offering some spurious information which could then lead to pandemonium. He left Schonland in no doubt as to his views: 'No this sort of thing won't work in this country, and an attempt to try it would lead to a lot of trouble' [34].

Despite reservations such as these, the Corps of Scientists soon found itself employed in various ways and Schonland, especially, found himself kept busy administering it and advising both the senior defence staff and the Minister of Defence on matters of scientific importance. He also maintained close ties with his many friends and colleagues in England and in Canada who now occupied positions of similar influence with their Ministries of Defence, and this served South Africa very well.

Most prominent amongst these colleagues, of course, was Sir John Cockcroft who, in November 1951, had been awarded the Nobel Prize for Physics, which he shared with E T S Walton for their research at the

Cavendish that led to the splitting of the atom in 1932. As always, Cockcroft's services in England were much in demand and, as well as having full responsibility for the Atomic Energy Research Establishment at Harwell, he was also chairman of the Atomic Energy Sub-Committee of Britain's Defence Research Policy Committee (DRPC). With Sir Henry Tizard about to retire as its chairman, Cockcroft was tipped as his successor, but even he, for all his peripatetic capacity, doubted whether holding two such important appointments (the one at Harwell and the other in London) was either feasible or advisable. By July of that year he had made up his mind to accept the full-time chairmanship of the DRPC and suggested to the British Government that Basil Schonland should be appointed as his successor at Harwell and feelers went out to South Africa. Whereas the rumoured departure of such a prominent scientist as Cockcroft from one important senior post to another always generated comment in the British Press, the name of his suggested successor was not made public. However, some scientists close to Cockcroft, such as his former deputy at Harwell, Herbert Skinner, now Professor of Physics at Liverpool, were told and regarded the suggestion as something of a bombshell for, as Skinner wrote to Cockcroft, 'Schonland does not know the job' [35]. But Cockcroft was not deterred for he knew Schonland and presumably felt that what he had to offer by way of all round acumen more than offset any shortcomings he may have had in the details of atomic energy.

As had happened just two years before, when his name was being touted for the Ministry of Supply and NPL positions, Schonland felt a moral obligation to inform the man to whom he was ultimately responsible that his services were being sought elsewhere. He immediately informed his Vice-Chancellor, Humphrey Raikes, of the communication he had received from Cockcroft and promised to keep him informed of any further developments. Raikes, by now a tired man following years of financial stringency, staff cut-backs, political pressure and the not altogether whole-hearted support of his Council, will no doubt have regarded the loss of Schonland as a bitter blow, for Schonland was the University's scientist *par excellence* and Raikes had long sought to develop Wits into a institution of some standing. Schonland's presence was crucial if he was to achieve it. With such pressures upon him it came as no surprise when Raikes himself informed the University of his wish to retire after more than twenty years' service at its helm [36].

While Schonland awaited further developments, the political situation in England exercised the controlling hand. Clement Attlee's Labour Government had been re-elected in February the previous year with but a six seat majority. In September 1951 Parliament was dissolved and the country faced a General Election, which Churchill's Conservatives won, and this saw the old wartime leader returned to Downing

Street with many of his former colleagues, such as Lord Cherwell his scientific adviser, back in the Cabinet with him. Cherwell, though Paymaster General, was also Churchill's adviser on atomic energy and related matters and he wanted Cockcroft to remain at Harwell, and he duly agreed. However, Cockcroft was ever conscious of the mounting pressures on his time and so he still hoped to attract Schonland to join him. But Schonland now weighed up his own options and, after consulting Ismay, they reached a conclusion which Schonland communicated to Raikes in December [37]:

12.12.51

Dear Mr Raikes,

It is with some relief that I write to let you know that the question of my going to Harwell has been settled by the decision of Mr Churchill's government to retain Sir John Cockcroft there. Although there is some talk of my going as a second-in-command or 'fifth wheel' in the general set-up, I shall not accept any such offer if it is made.

And so the possible loss by the BPI of its Director at such an important stage of its post-war revival was forestalled and Schonland's plans to expand the Institute's activities into the field of geochemistry could proceed. To do this would require a significant injection of capital and the invitation that he had received a while before from the Carnegie Corporation to pay a visit to New York presented the ideal opportunity once again to seek its support. He now saw that the time was ripe and if he could also use it to explore matters of common interest in the defence field with his colleagues in Canada and England, then so much the better. He therefore put all this to Raikes:

The offer of the Carnegie Corporation to pay for a visit to the United States by my wife and myself has been received and if you are agreeable I should like to pay such a visit to the United Kingdom and the USA between March 1st and July 1st next year. Will you please let me know if this arrangement has your approval?

It did, and Schonland then sought the approval of the Chief of the General Staff in Pretoria to discuss defence issues of a scientific nature with his counterparts in those countries. This too was approved and the South African Military Attachés in London and Washington were advised of Brigadier Schonland's forthcoming visit [38]. At home Basil and Ismay began to make their arrangements for an extended visit to the northern hemisphere the following year.

It was now apparent that Schonland saw the emphasis of the BPI swinging away from problems to do with lightning and atmospheric electricity in general since, as he said, these 'are likely to have been cleared up [in a few years' time] and work on this subject will be curtailed'. However, what remained to be completed was the problem of the distribution of electric charge within clouds and that of the space-charge in the air below them. The first of these now lay squarely in the court of Dawie Malan, while the second was a new problem to which the Director himself would be turning his mind. Before doing that though he again took up his pen and completed a second edition of his little book on 'Atmospheric Electricity' that had made its first appearance in 1932. He dedicated the new edition to his former mentor at the Cavendish, C T R Wilson, who had been so instrumental in launching his lightning research career those many years before and Wilson, now long since in retirement in Scotland, responded warmly to Schonland's request for permission to do so [39].

*   *   *

On 26 February 1952 the Schonlands flew to the United States as the honoured guests of the Carnegie Corporation. It was to be the first leg of a mission with many purposes and none was more significant than Basil's return visit to the Corporation's headquarters in New York. There, he set off along the path that he had first walked almost twenty years before when, in 1934, he presented himself to the Director, the astute Dr Frederick M Keppel, and convinced him of his intention to make a success of the geophysical research institute planned for Johannesburg. Now, with the BPI as an institution of some renown behind him, as well as a personal career studded with significant achievements, Schonland felt somewhat more confident when making his new appeal. His request this time was for financial assistance to equip a laboratory in the field of geochemistry, and to assist with the training of suitable scientists to run it. Once again his efforts were rewarded. The Carnegie Corporation announced soon afterwards that it was to award a grant to the BPI of $102 000, over a five-year period, and designated it for disbursement in three specific categories: $56 000 for work on the history and structure of the earth's crust; $40 000 for setting up a geochemistry section charged specifically with work on the Precambrian formations, and $6000 for the establishment of a special fund to assist a post-graduate student from amongst the Bantu peoples of South Africa.

This most generous offer could not have come at a more important time in the evolution of the BPI and Schonland was intent on capitalizing on the opportunity that the visit gave him to see some American institutions working in those areas. It was, therefore, no hardship at all to agree to the Corporation's request that he deliver a series of lectures on the work of the BPI at prestigious universities and research laboratories across the United States. He relished the opportunity to see such monuments to

American scientific progress as the Massachusetts Institute of Technology, Harvard University, the Universities of California and Chicago, the National Bureau of Standards in Washington and the School of Mines in New Mexico [40].

And they wished to hear him elsewhere, too. That year, 1952, happened also to be the bi-centenary of the famous 'Philadelphia Experiments' on lightning conducted by Benjamin Franklin in which he identified the similarity between the electric spark and lightning, and from which he devised the lightning rod — a discovery of great consequence. It was only fitting, therefore, that Basil Schonland, whom many saw as Franklin's natural successor, should be invited to play his part in the commemorations planned in Franklin's honour. Some time before he had accepted an invitation from the Institute to write a scientific paper on the work of Franklin and it was published in the special commemorative issue of the *Journal of the Franklin Institute*. To this day it remains a fitting tribute to the pioneer who, according to his successor 'found electricity a curiosity and left it a science' [41].

After a week's stay in Canada, where Basil visited Omond Solandt and renewed an old acquaintanceship, the Schonlands flew on to England. There his immediate intention was to seek assistance from the Nuffield Foundation to staff and run the research unit in geochemistry which, as he told them, had as its specific purpose the study the Precambrian rocks throughout Africa south of the Sahara. The extent of the study northwards across the continent was precisely in line with the objectives of the Scientific Council for Africa South of the Sahara (the CSA) that had been set up at his instigation just a few years before, and this guaranteed him a sympathetic hearing. In due course he was informed that the BPI would receive an amount of £13 500 over a five-year period in support of what Schonland proposed to call the Nuffield Geochemical Unit. The grant would be used for the specific task of doing research 'on the history of rocks (by mass-spectrographic and other methods now being developed), on the determination of the radioactive content of rocks, and on their general geochemistry' [42]. If, after that time, progress was satisfactory, the Foundation informed him, consideration would be given to continuing the grant.

Of course, a visit to the Cockcrofts was obligatory and Basil and Ismay therefore spent some time with John and Elizabeth at Harwell. The possibility that Schonland might still join his old friend there as his deputy was never far from Cockcroft's mind, particularly as he had resigned himself to the fact that he had been unsuccessful in his wish to become Master of his old College at Cambridge and so had accepted that his own future, at least for the next few years, lay at Harwell. For Schonland it was comforting to know that his services were much in demand, both at home and in England, but he was cautious about both the salary and the conditions of service then on offer at Harwell. Besides, he had more than a full plate to deal with in South Africa. Cockcroft, too, appreciated that Schonland's continued presence in South Africa had obvious benefits for Britain, especially as it was now apparent that the Dutch government had offered South Africa assistance in the construction of a 6 MW atomic pile and in the training of its scientists and technicians in that field. To counter this he had even written to the Under Secretary in the Atomic Energy Division of the Ministry of Supply, with the suggestion that the United Kingdom should invite young South African scientists to spend some time at either Harwell or in British universities, and the immediate reply was in full agreement [43]. Basil Schonland's part in all of this was clearly crucial if Britain was to retain any influence at all in South Africa's atomic affairs.

*   *   *

The South African Association for the Advancement of Science (the S_2A_3) had a long and proud history by the time it elected Basil Schonland its President for 1952, its golden jubilee year. His own involvement with the Association went back at least as far as the famous meeting of 1929 attended by Lord Rutherford and his colleagues from the Cavendish and it had continued ever since. In 1935 Schonland had served as President of Section A: Mathematical and Physical Sciences, but now his colleagues had done him the supreme honour and elected him President of the Association itself. Very soon after arriving home in July 1952 he delivered his presidential address in Cape Town [44]. Its central theme had been in his mind ever since he had spoken to the Rhodes University Congregation, but it had really gelled when he saw the power of basic science at work in the United States.

As was appropriate on such a special occasion, Schonland commenced by reviewing the Association's achievements and the scientific highpoints in its history. From its inception immediately after the end of the Boer War this body, whose purpose was rather quaintly described as being for the 'presentation and discussion of general reports and surveys', had grown rapidly such that by 1905 it boasted more than 1300 members countrywide—this in a country of but 40 professors and lecturers at the time. In 1912, the 'great Onderstepoort Institute' of veterinary medicine came into being under the legendary Sir Arnold Theiler who saw pure science as a possible way of solving South Africa's problems. Certainly, on the agricultural front, these were dominated by cattle disease and by drought-stricken pastures, and only radical new measures based on science could save them. He recalled J H Hofmeyr speaking in 1929 when he appealed for the increased development of 'Science in Africa and of Africa by Science'. Though modesty prevented him from doing so, Schonland might well have reminded his audience that he, for one, had taken Hofmeyr's words to heart, with the recently-formed CSA bearing grand testimony to his initiative.

He then moved into more sensitive territory—the relationships between South Africa's peoples of different cultures, races and hues. Other Presidents of the S_2A_3 before him had gone there and some had made much of the part that science could play in resolving what Sir Carruthers Beattie, in 1928, called 'Africa's greatest problem, the contact of non-homogeneous races'. Hofmeyr again, a year later, described Africa as 'the world's most significant laboratory for human relationships' and he believed that science alone could find a proper solution to its problems. And so did many others who followed them. But Schonland took issue with this and dissected its logic: 'I cannot agree with the view that problems involving social and racial relations can be *solved*[*] by the scientific method alone. But I am sure that they cannot be solved without it'. He then offered his fellow-countrymen, scientist and layman, politician and *predikant*, some thoughts of his own:

> In this field of race relations the scientist can advise only, but his advice may be very precious, for without knowledge of facts and probabilities, action may have to rely entirely on prejudice and guesswork. We spend millions on research on atomic bombs; race relations are at least as explosive.

Behind South Africa's troubles as Schonland saw them were, undoubtedly, the recent laws of the Nationalist government, the Group Areas Act and the Population Registration Act of 1950, that defined where one might live on the basis of the colour of one's skin and to which statutorily-defined racial group each and every South African belonged. That same government was now hell-bent on forcing South Africa down the road of racial segregation, enshrined in law and bolstered by the forces of the state. It presented a bleak view to anyone who cared to peer into the future.

But Basil Schonland was ever the optimist, as well as a realist, and he moved away from his review of the past, and comments on present uncertainties, to concentrate his thoughts on the future of the Association and on the future of scientific and technical education in the country. In 1950, South Africa had nine universities and its scientists and engineers were served by no fewer than 22 scientific and technical societies. On the surface this proliferation of learned institutions was surely healthy by indicating, as it seemed to do, that these were now thriving communities from which the country at large could only benefit. The President, though, sounded a word of warning:

[*] His emphasis.

> Our scientific workers and engineers need not only something more
> for themselves than narrow specialization and specialized societies,
> they should also *give* more than these societies and their own work
> require of them. If this is all the scientific community life they have,
> science will suffer and the country will suffer. Once a year, or at
> least once every few years, the scientific expert should mix with
> others of a totally different outlook and expertise and by personal
> contact and discussion with them help to give new life to the
> scientific community as a whole.

Of all those bodies in the land allied in the pursuit of scientific excellence,
their association, he suggested, was surely the one most able to take the
lead in achieving 'that mixing of the scientist and the applied scientist and
the engineer and the interested layman ... without which many things can
go quite wrong in the future'. Another vehicle for such cross-fertilization
were the publications of the association and here he stressed the need for
articles of breadth rather than depth. 'We exist primarily to create and
foster a scientific fraternity in South Africa, not to publish original
work. We exist to provide a common meeting ground for South African
scientists and a forum for general discussion of the problems of this
country from a scientific angle'. He reminded his audience that the
Afrikaans language, as one of the two official tongues in the country,
had just as much part to play as English in the dissemination of scientific
information to the nation. It was the S_2A_3 that should be encouraging both
Afrikaans lectures and articles in order to widen their reach as much as
possible. He also said that he saw a need for publications of a '*semi-popular*
character' whose purpose was to review and survey the new ideas and
discoveries of science and so bridge the gap between 'those who teach
and do advanced scientific work and those who pay for it'. If practising
scientists were not prepared to communicate their thoughts and findings
in this way to the lay public and the politicians, then they must bear the
consequences of any subsequent financial decisions that did serious
damage to that very research. Science (and scientists too) had a responsibility to those who supported their activities but did not understand
them.

This was all rather controversial but Schonland more than anyone in
the country could venture to say such things because he was speaking
almost as the elder-statesman of science in South Africa and, in many
ways, that is precisely what he was: a figurehead, and a father-figure
too, to a generation just striking out on scientific careers of their own.
But he was also speaking to those who determined scientific research
policy and who, through the schools across the land, were setting out to
attract the next generation of students to university. His views on the
teaching of science at school to enable the pupils 'to understand the
world and to be useful citizens of it' will have taken many by surprise:

> It would, I think, be a good thing for us to enquire how suitable is
> the training in science in our schools to give girls as well as boys
> this necessary understanding. I hope that the humanists, for whose
> point of view the scientists have much greater sympathy than they
> perhaps realize, will not misunderstand the purpose of such a
> proposal. It may well be that we need less school training in
> science, not more, and that what we need is to change the ideas
> behind science teaching. It may well be that that what is needed is
> to treat science in our schools more historically and humanistically
> and less as a technical subject, to teach at the school stage more of
> Newton and his times and less of Newton's laws of motion and
> their applications.

If such views represented a heresy then Schonland was merely following
a well-worn tradition established by his predecessors in this office.
During his presidential address the previous year, Professor E G
Malherbe, Vice-Chancellor of the University of Natal, had delivered his
broadside by expressing his dislike of the 'huge laboratories and institutions directed by national councils' and called for their inclusion within
the universities. He was surely referring to the CSIR and its laboratories
and institutions and this was as radical a suggestion as only the President
could make without inviting immediate ridicule. Schonland, as the
greatest protagonist of those national laboratories and institutions, used
the occasion of his address not to chide his old colleague but to suggest
gently that such a move would not be in the best interests of the universities or of the scholars within them. 'In fact', he said 'it might destroy
them'. Then he gave free rein to his own views, which echoed a sentiment
from his address at Rhodes just a while before, but which now were given
much wider exposure:

> What they [the universities] need is more staff and facilities for
> advanced post-graduate work in the fundamental sciences, and not
> in the applied sciences; in physics, not in engineering; in chemistry,
> not chemical engineering, and in botany, not agricultural research.
> The difficulty about such developments is that there are not enough
> men of high quality in this country — or available from elsewhere —
> to staff such advanced research and teaching departments
> adequately. Nor is there enough money to provide the expensive
> facilities they need, if more than one or two universities are to
> enlarge themselves in this way.

The problem he had identified was not unknown. In Australia the radical
solution had been taken to set up, in Canberra, a national university
intended only for advanced study, but Schonland saw this as a drastic
measure likely to lead to the migration of all the best post-graduate
students from other universities to the one national facility. The ideal
solution, he believed, already existed in South Africa. By creating a

426

number of specialized institutes attached to and forming part of particular universities, as was the case with the Bolus Herbarium at UCT, the *Instituut vir Plantfisologie* at the University of Pretoria and, of course, the Bernard Price Institutes of Geophysics and Palaeontology at Wits, he believed that an ideal solution could be achieved and he left it to his audience to consider how such a scheme might be extended to include other areas of special interest.

For an hour Schonland had held the attention of as illustrious an audience of South African and visiting scientists yet assembled in the country. In closing he reminded them of the call, made half a century before by the founding President of their association, Her Majesty's Astronomer at the Cape, Sir David Gill, for a movement whereby science could serve South Africa. Now, said Schonland, with many great achievements to their credit, 'and a future that is very great indeed', the need was to show how scientists *themselves* could serve South Africa. 'I believe that for this we need a new outlook, an outlook of service to the community, as well as to science itself'.

* * *

Schonland's circle of friends was wide and he had many former colleagues dotted about the academic institutions of Britain and her Commonwealth. One former colleague was Solly Zuckerman, who was now Professor of Anatomy at Birmingham. Though he and Schonland shared South African roots, Zuckerman had long since severed all links with his birthplace and was a rather grand figure in the British scientific establishment. Operational research during the war had occasionally seen them sharing common ground but never in collaboration. The 'Zuckermen' designed by Frank Nabarro and christened by Schonland was about as close as they came to working together during the war! Then a curious incident occurred in 1952 which suggested that Zuckerman held Schonland in much higher regard than he might ever have guessed from their previous encounters. It also involved Professor Rudolf Peierls, who was well-known to Schonland as one of the discoverers of the critical mass of uranium necessary to cause a chain reaction. Of course, what had followed from that had now changed the world forever.

In June 1951 the University of Birmingham began the process of finding a successor to Sir Raymond Priestley, who was due to retire as its Vice-Chancellor. By May the following year the list of suggested candidates had been reduced to one and an offer of appointment was made. In October, the nominee withdrew, thus leaving the University in the embarrassing position of soon being without a Vice-Chancellor. A second round of selection was therefore imminent but the whole process had, in the eyes of Zuckerman, Peierls and others on the University's Senate, been a rather lacklustre affair and this prompted the two of them to take action. Peierls immediately sent a telegram to Schonland in Johannesburg informing him of Birmingham's need and inquiring whether he might be prepared to take on the job. Schonland indicated his interest and said he would consider it, providing the conditions were right, and so Peierls put Schonland's name forward for consideration.

That the matter was taken no further was no reflection on Schonland but simply served to confirm just how desultory the procedure of finding a Vice-Chancellor had become in Birmingham. Zuckerman was incensed and wrote immediately to Schonland in explanation and expiation [45].

13.11.52

Personal & Confidential

My dear Schonland,

I could not be more depressed than I feel at the unsuccessful outcome to Peierls' last-minute intervention with your name in the selection of a Vice-Chancellor for Birmingham. And I kick myself for having failed to pay attention to the matter till he got hold of me last week. I was fully under the impression that the selection committee was ambling along gently, and that fair warning would be given to members of the Senate before they committed themselves. But that was not so. The die appears to have been cast even when Peierls cabled you and the meeting at which the Senate was "consulted" was in effect a rubber-stamping of a foregone conclusion. That happened yesterday. The excuse given about your name was the assumption you were unobtainable.

So now you will be spared the unenviable task of herding a flock of sheep in Birmingham—for the V.C. is president of the Senate—and people like myself will go back to sleep in university administrative measures [sic], secure in the belief that a proper, safe, and mediocre decision will have provided the University with far less than it really deserves. I was so excited when Peierls told me that you might be available if the conditions were right. Under your direction Birmingham could have realised itself—and it would have been fun to have worked with you.

Yours ever,

Solly Zuckerman

While his treatment in 1949 at the hands of some rather inept British bureaucracy, and not a little double-dealing, left Schonland feeling somewhat bruised, this latest attempt to attract him to another senior position in the country he so admired served only to indicate to him that he was held in much regard a long way from South Africa's shores. However, Peierl's telegram had been very much a bolt from the blue and so the eventual outcome of this rather unsatisfactory affair troubled him little.

The BPI was about to move into a new era and the Director had his hands full. As a direct result of his successful appeals for support from the Carnegie and Nuffield Foundations, Schonland was able to build four new laboratories and two new offices on the top floor of the building. He also set about recruiting additional members of the research staff to work in the area of geochemistry. Though lightning was rapidly assuming a lower priority in the order of things, it was by no means defunct and Schonland himself was pursuing a particular avenue with his usual enthusiasm. From the many measurements that he and Dawie Malan had made of the electrostatic fields that accompanied lightning, it struck him as strange that the field associated with the stepped leader exhibited no significant pulsations, even though the variations in light intensity were so obvious and beyond doubt. It was surely reasonable to expect that the one process was associated with the other, but why were their effects so different? To explain this he fell back on an idea that he had first put forward in 1937 in his Halley Lecture at Oxford. This was the pilot streamer, then a decidedly novel idea, but evidence now existed from laboratory measurements on the mechanism of the long electric spark which seemed to support his hypothesis.

In what ultimately turned out to be his last formal research paper on lightning, and essentially the last piece of dedicated research that he himself conducted, Schonland developed an elegant theoretical model of the ethereal javelin that guided the stepped leader to ground. Since measurements had shown no variation in field strength, it was obvious, he maintained, that the charge producing the pilot streamer and, hence, the electric field that he and Malan had measured, must be descending in a continuous manner ahead of the visible bursts of light of the stepped leader, which the Boys camera had so faithfully recorded. That the camera had shown no evidence of the pilot itself just indicated how weak it was, yet its function was vital to the whole lightning process for it cleared the path along which the subsequent stepped leader, and the massive return stroke, would travel. Schonland's calculations were impressive and his argument to support them was persuasive. Since the pilot and the stepped leader were so intimately linked, he proposed, they would carry the same current and have the same distribution of charge, and their streamers would be of the same radius. Where they differed so markedly was in their speed of advance: the pilot moved forward tentatively, feeling its way, in previously uncharted air, leaving behind an ionized trail along which the leader could follow in rapid spurts, emitting light as it did so. Once it reached the tip of the pilot streamer the leader, on finding no charged path ahead, would halt while the slower, but doggedly persistent pilot pressed ever onward, to be followed within microseconds by its energetic companion until they reached the ground and were overwhelmed by the massive and dramatic return stroke that rushed back along the route they had chartered for it. The image of the tortoise and the hare in Aesop's fable was too tempting for Schonland to ignore!

Experimental evidence to support this theory existed in the form of a letter published by T E Allibone in *Nature* in 1948.* There, Schonland's old Cavendish colleague and collaborator of 1934 described his discovery in the laboratory of a pilot streamer that preceded the stepped leader. Schonland was now convinced of the direct link between lightning and the long electric spark. His conclusion, in what would be his swansong in the *Proceedings of the Royal Society*, said it all [46]: 'The pilot streamer processes in lightning and the long spark have nearly identical velocities, current densities, channel field strengths and ratios of length to radius; it is concluded that they are produced by the same mechanism'. The subject was by no means closed and was to remain controversial for years to come. But, on the basis of the evidence and the limited computational techniques available at the time, Schonland's conjecture was as good as any on offer.

* 1948 *Nature* **161** 970.

CHAPTER 20

'COMETH THE HOUR'

In common with all the universities in South Africa at that time, Wits experienced a period of rapid growth in the immediate post-war years. The influx of returning ex-volunteers after demobilization saw the number of students increase dramatically from 1946, and by the end of the decade nearly three thousand former servicemen and women had been admitted to the University. Staff numbers had to increase just to cope and, in fact, the growth in the academic staff establishment during that period actually exceeded that of the student body, with a doubling in their numbers. The main impetus behind this expansion came from the Vice-Chancellor, Humphrey Raikes, who saw it as his and the University's duty to those of South Africa's citizens who had so willingly sacrificed their careers for war service and who now, on their return, wished to equip themselves with professional and academic qualifications. What Raikes may have lacked in dynamism he made up for in dogged determination, of which he needed much in the face of a University Council dominated by an autocratic chairman, with very narrow views of the purpose of Johannesburg's university, and by a Senate populated by some strong-minded professors who alternately challenged and ignored him in equal measure.

The Faculties of Engineering and Science, too, underwent massive changes during those years. From being rather old-fashioned in outlook and poorly-equipped in their laboratories, they now faced a rather different world of post-war expansion and industrial development, and they had to adjust accordingly. South Africa sat poised on the edge of a revolution brought about by the creation of vast technological complexes such as Sasol in petrochemicals, Iscor, the iron and steel corporation, and Escom, the electricity supply commission. All, of course, now drew on the skills and expertise of Basil Schonland's own CSIR and of the universities. The country's need for engineers and scientists seemed insatiable, and Wits, with its annual output of more than a hundred graduate engineers each year, in eight branches, exceeded that from any other South African university. Equally, as the demand for scientists across the spectrum began to spiral, the Faculty of Science produced in September 1951, even before Schonland made public his views on the subject, a visionary report on the role that science could, and indeed should, play in the development of South Africa. Of critical importance, if the University was to meet this challenge, were the leaders who had to take their departments forward by developing appropriate courses and by initiating ground-breaking research. Finding and nurturing them, if needs be, soon assumed the very highest priority within both the University's Council and Senate and this unity of purpose was due as much to the low-keyed persistence of Humphrey Raikes as anything else. His wish, on retiring in 1954, that Wits should become a leader amongst South Africa's universities in the areas of research and postgraduate education was soon to be fulfilled [1].

Of course, the foundations for some of this had already been well and truly laid under Schonland's inspired leadership at the BPI. But the BPI's primary purpose was research and not teaching and, though the Director did some teaching it was more by way of a service to other departments than as a means of fostering postgraduate work in his own. Schonland did, however, take a close interest in the quality and calibre of those whom the University was considering for the various chairs in the Faculty of Science, and he was primarily responsible for an appointment that changed the face of Physics at Wits.

He had long been particularly aware of the rather uninspiring ambience in the Department of Physics ever since his first visit there with his apparatus for detecting 'the penetrating radiation' in the early thirties. Then, on his return to South Africa from England in 1945, he discovered that nothing much had changed except that the Head of Department, Professor H H Paine, who had held the post since 1920, was about to retire. Something should now be done. Armed with his own recent experience of British physicists, and their superb performance in a multiplicity of roles during the war, he was certain that the right man to pull Wits out of the doldrums could be found from amongst that pool and he advised Raikes accordingly. Raikes favoured making some tentative inquiries and inevitably it was to John Cockcroft that Schonland turned for suggestions. In his reply Cockcroft provided no fewer than thirteen names for Schonland's and possibly the University's consideration [2].

But no appointment was made. The new Head of the Department of Physics, Professor G T R Evans, merely moved from one office in the department to another and, almost predictably, the state of much lassitude continued. It would, though, be unfair to blame the staff too much for this. Their lot in life at that time consisted of little else but the teaching of physics to all and sundry in vast numbers, including engineers, medical students and anyone else whom the University believed required some vestige of knowledge in the subject. Time for scholarship, let alone research, was severely restricted by that regime.

What was needed was strong and inspirational leadership. Time passed, students increased in number and the burden in teaching them did too, but research remained a rarity. Then Schonland returned to the BPI after having launched the CSIR and he once again found the Chair of Physics vacant, following the sudden death of Professor Evans. He immediately offered his services to the Vice-Chancellor to attempt to fill it with a man of the highest calibre—and he had someone in mind.

The Schonlands' extended visit to England and North America in 1952 provided the perfect opportunity to talk to Frank Nabarro. 'Nab', as he had been known to all in the AORG, where he had made a significant contribution across a broad front, was a theoretical physicist and a disciple of Nevill Mott. Like virtually all the young scientists who found themselves under Schonland's command at Petersham and then at Roehampton, Nabarro thought very highly of this South African Fellow of The Royal Society who suddenly appeared in their midst in 1941 in the uniform of a lieutenant colonel. Schonland's natural authority, his rapid grasp of detail and his obvious rapport with the military undoubtedly made the AORG the success it was and Nabarro both admired and liked him.

The interview for the Chair of Physics at Wits took place in London in the spring of 1952. Its timing was perfect in that Nabarro was on the look-out for a professorial appointment, having just been unsuccessful in his application for one in Theoretical Physics at King's College, London. He and Schonland met and, in an atmosphere so reminiscent of the AORG, reminded each other of events now slowly fading into history. Their reunion was made all the more congenial by the fact that Schonland's personal assistant at the AORG became Frank Nabarro's wife after the war and this news added to the conviviality of the occasion. However, their meeting had a serious purpose and almost imperceptibly the conversation changed tack as Schonland became his Vice-Chancellor's representative as he sounded out and assessed Nabarro as the man who might inject life into a moribund Department of Physics. Nabarro's own memory of the interview was of Schonland playing the part of the impartial 'honest broker': friendly yet detached; aware of his young colleague's worth from times past but probing and testing his worth and standing in the physics of today [3].

Was Nabarro the man that Wits wanted and needed? Schonland believed he was and wrote immediately to Raikes telling him so. Nabarro was duly invited to apply for the chair, but he still had to convince himself that his career would be best served by moving from England, where physics was all focused on Harwell, to South Africa where only the presence of Basil Schonland at Wits and R W James at UCT provided any scientific attraction at all. In addition though, there was the spectre of apartheid to confront, with South Africa now under an intensely

433

nationalistic and increasingly isolationist government. This presented both Frank and Margaret Nabarro with yet another dilemma. However, encouragement was soon forthcoming from one of Nabarro's own academic colleagues, Professor A H Cottrell,* then visiting Wits from Birmingham, where he was Professor of Physical Metallurgy. Cottrell, on hearing of the offer to Nabarro, wrote to him in August with his views that Wits operated no academic colour bar and that it was a university essentially 'liberal in outlook' with clear potential for development in the field of physics [4]. More was to follow with a letter from Schonland himself written at the end of November. In it Schonland expressed the wish that he would soon be welcoming Nabarro to Johannesburg. He elaborated a little on the University and on the need to stiffen its academic spine with men of real stature, particularly in the light of a recent appointment of one Arthur Bleksley to the Chair of Applied Mathematics.

Bleksley, a graduate of Stellenbosch University, had been on the staff of the Department of Applied Mathematics since 1931 and had built up an admirable reputation as a lecturer of the very highest quality whose manner, personality and clarity of instruction had captivated the generation of students who had passed through his courses. His own research, in the exciting new field for Wits of theoretical astronomy, had petered out before the war as he concentrated more and more on the teaching of applied mathematics in his inspirational way. In addition, Bleksley became the popularizer of science in South Africa. With his encyclopaedic knowledge and easy-going manner he, more than any other scientist in the country, had captured the public imagination for the wonders of science and the mysteries of nature. In doing this so very effectively he had achieved just what Basil Schonland had called for in his address to the S_2A_3 only a short while before — 'closing the gap of ignorance, the iron curtain between science and the public'. Then, Schonland had spoken out strongly in support of the scientist who did this, especially as there were so many who shied away from it with the excuse that the 'semi-popular exposition [of science] was neither his interest nor his métier'. Now, in his letter to Nabarro, he took the contrary view and strongly opposed Bleksley who, he declared, had 'no pretensions to scholarship'! [5].

Schonland's rather dismissive view of Arthur Bleksley's worth was more than a little surprising. Before the war Bleksley had made a significant contribution to astrophysics by his analysis of the characteristics of a group of stars known as the pulsation variables. The fact, though, that he then decided to abandon research in order to concentrate on his lecturing and the popularization of science seemed to count very much against him

---

* Sir Alan Cottrell FRS was Vice-Chancellor of Cambridge from 1977 to 1979.

in Schonland's eyes. However, history is always the judge in these matters and in this case Schonland's condemnation was surely too severe. Bleksley, during his more than thirty years on the Wits staff, probably did more than any scientist or academic in the country to close that gap of ignorance that always seemed to separate science from the public. He was also the very epitome of the Schonland ideal: the lucid scientist.

Frank Nabarro arrived in South Africa to take the Chair in Physics at Wits early in 1953. With Schonland backing him he believed that he could throw himself into the task of turning what had been a department almost trapped in a backwater into a centre of excellence in solid-state physics. He intended to build up a programme of postgraduate research and to forge strong links with industry. His dynamism was obvious, his approach direct and his impact on his new colleagues was little short of cataclysmic. Their reaction, when it came, was one bordering on mutiny and soon Nabarro found himself without his most powerful friend and ally, for events had moved on and Schonland himself was now in demand elsewhere.

In January 1953 Basil Schonland received a letter from Professor Mark Oliphant, an old colleague from the Cavendish, who was now Director of the Research School of Physical Science at the Australian National University (ANU) in Canberra that had recently been founded with the view of becoming a prestigious, research-only, institution. In his letter Oliphant invited Schonland to attend a conference at the ANU, but Schonland declined. He was, though, much taken by Oliphant's next proposal [6].

Oliphant was a product of the school of forthright expression; he was also a considerable physicist with a great reputation of achievement behind him. It was in his laboratory at the University of Birmingham, early in 1940, that two of the most remarkable scientific developments of the war had taken place. At almost the same time, two pairs of scientists working in adjacent rooms, but because of wartime secrecy completely unaware of what the others were doing, produced both the cavity magnetron which revolutionized radar, and the basis of what they called a 'super-bomb'. J T Randall and H A H Boot, watched over closely by Oliphant, designed the device that, for the very first time, made possible the generation of large amounts of radio frequency energy at wavelengths of 10 cm and less. The effect of this device, soon to be called the cavity magnetron, on radar and on the outcome of the war was climactic. Next door, Otto Frisch and Rudolph Peierls performed the calculations that led them to write a short memorandum 'On the construction of a "super-bomb"; based on a nuclear chain reaction in uranium' [7]. The fates of the cities of Hiroshima and Nagasaki, and the immediate end of the Second World War, were the postscripts to that paper. Oliphant himself soon became one of the first physicists from Britain to join the

highly secret Manhattan Project set up in the United States to manufacture the atomic bomb. Soon after the end of the war he returned to his
native Australia to take up the post offered to him at the ANU where,
in 1952, he was its acting Vice-Chancellor. Of peripheral but not unrelated
interest was the fact that Oliphant was also chairman of the Australian
atomic energy committee [8].

The Cockcroft–Oliphant relationship was equally as strong as that
which existed between Cockcroft and Schonland and, once again, it was
that all-pervasive influence of John Cockcroft on his former colleagues
from the Cavendish that had served to bring Schonland's name to Mark
Oliphant's attention. Cockcroft had spent the months of August and
September 1952 in Australia dealing with matters on behalf of the Defence
Research Policy Committee. Whilst there, in the presence of the Australian
Prime Minister, R G Menzies, and the founding Vice-Chancellor of the
ANU, Sir Douglas Copland, he performed the opening ceremony of the
University's first permanent buildings. Copland was soon due to retire
and the University was now seeking his successor. Not surprisingly,
given both the ANU's strong scientific bias and the personality and
influence of Oliphant, the name of Cockcroft featured prominently in
early discussions as Copland's likely successor, but Cockcroft could not
be persuaded to take the post, even by Menzies, and so the search
continued. It was Cockcroft himself who then suggested to Oliphant that
Basil Schonland might well be the man they were looking for. And so it
was that Oliphant, in his letter of January 1953, put it to Schonland that
he might consider becoming the next Vice-Chancellor of the Australian
National University.

The prospect of assuming the senior position within a university,
particularly one based on such visionary ideals for the advancement of
science, produced a rather different reaction from Schonland than had
been the case when the Canadian chair was offered to him two years
before. Then, his rejection of the Chair of Physics at Toronto was due as
much to the rather lowly position it afforded him within the hierarchy of
the university as to the Schonlands' stated reluctance to pull up their
roots in one country and transplant them elsewhere. Now, the attraction
and the challenge of the top post at Australia's newest university was a
very different proposition and he replied to Oliphant indicating his interest.

Oliphant set out immediately to marshal support for Schonland
from amongst the members of the University Council. However, the
appointment of any Vice-Chancellor is never far removed from some
political consideration or another and this one was no different. But
Oliphant had something of the political animal in him and between himself and Roland Wilson, a government-appointed member of the ANU's
Council, they marshalled support for Schonland from amongst the
academic community. The only other contender for the post was Leslie

Melville, an Australian who was Executive Director of the International Monetary Fund in Washington. That gave him considerable status, but Melville had no experience of university management at all. Support within the selection committee was soon polarized between those who favoured the well-known and well-connected Australian and the others who believed strongly in having an academic at the helm, even though he was known personally to very few of them and just by reputation to some. There was also another interested party, for in the background was the Federal Government of Robert Menzies, which had its own firm ideas on the matter. An impasse looked likely. At the suggestion of Dr H C Coombs, the chairman of the University's Council, who was about to leave for England, it was agreed to ask Schonland to fly to London where they would be joined by the ANU's Chancellor Lord Bruce for a searching interview. Oliphant immediately sent a telegram to this effect to Schonland in Johannesburg [9].

In all haste Schonland made arrangements to travel to England where he duly met the two Australians. The interview took place over lunch and was a resounding success. Lord Bruce, a former Prime Minister of Australia and at one time his country's High Commissioner in London, was much taken with the South African, pronouncing him 'a gift from Heaven'. Even Coombs, who initially had been rather sceptical about appointing a complete outsider of whom so little was known, saw in Schonland all the qualities that they were seeking in their new Vice-Chancellor—an easy-going, pleasant personality likely to fit in well, broad interests that extended beyond the natural sciences and genuine enthusiasm for the concept of the ANU. The outcome appeared to be assured, but between that lunch-time meeting and sending their joint report to Canberra Coombs, who drafted it, began to waver. Schonland, he believed, would be very good but when all things were considered he now expressed a slight preference for Melville. Lord Bruce, by contrast, felt strongly that Schonland was the man they were looking for but, presumably at Coombs's urging, was prepared to concede that Melville might well be equally good. Such equivocal advice, when it was received, served only to polarize the selection committee in Canberra still further. To resolve the impasse would require that Melville be flown to London from Washington and the whole process repeated, which it duly was. Again the ANU's Chancellor and its Chairman of Council found themselves backing opposing horses but before they prepared their report Lord Bruce received an urgent cable from his Prime Minister. Menzies wrote:

> I feel strongly that the Institution's [the ANU's] interests would best
> be served by Melville particularly having in mind some opposition to
> National University expenditure in Cabinet circles. I need hardly add
> that a unanimous recommendation from yourself and Coombs would
> make Melville selection much more easy for the University Council.

Bruce, well-versed in the art of political wiles himself, declined to take the hint and cabled Menzies for clarification, but none came. It was quite clear that the Federal Government had set its mind on an Australian Vice-Chancellor for its National University and was prepared to use financial leverage, if needs be, to ensure that they got the man they wanted. It was, as Lord Bruce later confided, a clear case of 'No Melville, no money' [10].

Sir Leslie Melville was duly appointed Vice-Chancellor and served the ANU well until his retirement 1960. If Schonland knew of the political intrigue that had thwarted him he kept his views to himself. Mark Oliphant probably did not!

*   *   *

If the events of 1949 in England and these most recent machinations in Australia were anything to go by then Basil Schonland seemed destined to be the eternal bridesmaid, at least when it came to significant appointments beyond South Africa's shores. His appetite, though, for new challenges had certainly been whetted by what he had heard on that flying visit to London and one can be absolutely certain that whilst there he and John Cockcroft would have conferred on other matters too. But there were yet more political issues on Schonland's mind and these were taking place in South Africa's own backyard.

1952 was a year of great upheaval in the country. Since coming to power four years earlier, the Nationalist Government of Dr D F Malan had set about systematically to restructure the very soul of the country. Unlike Smuts's policy of 'pragmatic segregation', which was hardly different from that followed in many Western countries at that time [11], the Nationalists' was to be one of rigid separation of the races enforced by a series of laws that soon bordered on the draconian. The Group Areas Act, the Separate Representation of Voters Act, the Bantu Authorities Act, the Riotous Assemblies Act and the Pass Laws had all been introduced to circumscribe the lives of the majority of South Africa's population and to curtail any protest. Any hint at opposition, from whichever quarter it came, could so easily be attributed to the pernicious influence of Communism which the Nationalists portrayed as an international conspiracy bent on overthrowing the South African State by violence. To counter it they passed the Suppression of Communism Act in 1950. But opposition there was indeed, and the African National Congress (ANC), founded in 1912 to 'develop a strategy to defend the rights and privileges of blacks' [12], called on the government to repeal those laws by 29 February 1952. If, as was expected, this did not happen, the ANC would then instigate across the country a system of mass action and civil disobedience to be known as the Defiance Campaign.

The first of these was planned to coincide with an event of some significance in South Africa's history, as least as far as its white population was concerned. Three hundred years before, on 6 April 1652, three ships of the Dutch East India Company under the command of Jan van Riebeeck anchored in what would later be called Table Bay. On board were the first white settlers at the southern tip of Africa. Commemorating their arrival three centuries later was therefore a matter of great significance to the government of the day, some of whom could trace a direct link to the early settlers aboard those ships. Any action, therefore, that was counter to those celebrations would be seen by the Government as highly confrontational. But the ANC were unmoved and the Defiance Campaign went ahead with protests in the heart of Cape Town. This action on the foreshore of South Africa's Mother City was to be the well-spring of organized resistance against apartheid and 1952 was, according to Albert Luthuli, the newly-elected president of the ANC, 'a turning-point in the struggle', the struggle against white rule that was to last for many, many years [13].

Basil Schonland's direct contact with these events was remote, like that of most of his white fellow-countrymen. By one of those strange quirks of fate, however, there was a remarkable coincidental connection between him and the man who was now driving this pernicious legislation through Parliament. Hendrik Frensch Verwoerd, the political ideologue whose policies made race and racial differences the defining elements of South African life, was the Minister of Native Affairs. He was the same H F Verwoerd who, in 1929, had followed Basil Schonland to the podium in Cape Town to read his paper at the conference of the British Association for the Advancement of Science, in the company of so many illustrious men of science. Embittered as a youth by the Britishness that he found around him, both in South Africa and in Rhodesia where he grew up, this Dutch-born son of a missionary became an ardent supporter of National Socialism while a postdoctoral student in Germany in the 1930s. On his return to South Africa the young Verwoerd abandoned academic life to become, in 1937, the founding editor of *Die Transvaaler*, the Afrikaans language newspaper that was the mouth-piece of the then Hertzog government. During the war, with Smuts once more in power, Verwoerd's editorials were as unashamedly pro-Nazi as they were consistently anti-Semitic and he was every inch the hard-line Nationalist who was soon to become the architect of rigid, inflexible apartheid [14]. Now, a decade later, along with his colleagues in a Cabinet that believed sincerely that the hand of Providence had given them charge of 'their country' forever, he set about implementing a programme of grand apartheid that was based, he said, on the soundest of scientific *and* Christian principles.

The ANC's Defiance Campaign was well planned and was intended to be non-violent. Its purpose was publicly to disobey the laws they perceived to be unjust and, if needs be, to suffer arrest and possibly even assault in the process. There were to be three specific phases — the first by small groups in the cities would involve the deliberate flouting of discriminatory laws which prevented access by 'Non-Europeans' to facilities such a railway compartments, public toilets and Post Offices reserved for their 'European' co-residents; the second would extend those activities to the country towns; the third phase would involve mass action such as strikes and industrial actions across the country. Those participating in the Campaign were all to be volunteers under nominated leaders and all would be trained in the tactics to be employed. On every occasion when a protest was to take place the police were to be informed [15]. Prominent among the ANC leadership who planned and co-ordinated these activities was a young Johannesburg-based lawyer, Nelson Mandela, who was then president of the ANC's Youth League. Beginning in earnest on 26 June 1952, with displays of mass disobedience in most of the major centres of South Africa, the Defiance Campaign ran for nearly five months and involved 8500 people. Predictably, the Government saw it not only as a confrontation and a challenge to their authority but were adamant that Communists were behind it.

For all the good intentions of its organizers that the protests be peaceful and as orderly as possible, it was in the interests of some that the Campaign should be seen to precipitate riots, violence and, almost inevitably, deaths. During the latter part of its activities, rioting broke out in the Eastern Cape and more than forty people were killed, many being innocent bystanders. Tension ran high throughout South Africa and the flames were fanned by the press on all sides, with the Government-supporting newspapers reacting almost hysterically. All too easily the actions of the ANC were seen by many white people as the beginnings of turmoil in the country and there were calls for strong measures to stamp it out. Not surprisingly the Government reacted with vigour and the ring-leaders were rounded up and charged under the Suppression of Communism Act. The ANC called for a judicial inquiry into the real cause of the violence, suspected by them to be *agents provocateurs* operating in the wings. This was refused, while the reaction of the Minister of Justice, Mr C R Swart, was blunt: 'violence would be met with violence' [16].

Relations between South Africa's various racial groups were severely strained and peace in the country balanced on a knife-edge. Basil and Ismay Schonland, along with thousands of their fellow South Africans whose politics were of a more even-handed kind, viewed all these developments first with dismay and then with mounting concern. While Schonland himself had maintained the strictest political neutrality in all his dealings with Government, regardless of its hue, his views in private were unequivocally those of a liberal; Ismay's were outspokenly so. They gave expression to their feelings and beliefs in the company of friends, of whom they had many amongst South Africa's liberal thinkers and opinion formers. Schonland himself served as a member of the board of directors of the *Forum*,* a 'weekly independent review', as it referred to itself, which first appeared in 1938 with the stated intention of presenting to its readers a 'broad kind of South Africanism', in marked contrast to the myopic view or, worse, the tunnel-vision, of the Nationalists then lining up behind Dr Daniel Francois Malan. Throughout the duration of the war Ismay had sent copies to Basil so that he could keep abreast of developments on the political and intellectual scenes back home. In Theo Haarhoff, a former Rhodes scholar and one of the *Forum*'s editor-directors, they had a close personal friend who not only wrote thought-provoking pieces for the journal but had followed Schonland's progress through the elevated echelons of the Army with growing interest and much pride. During the war any thoughts that he may have had to give Basil Schonland and his exploits some publicity 'at any time and in any way whatever' were quickly quashed by Ismay's intervention in 1941. 'Basil', she said, 'would never speak to him again if this happened' and this little admonishment served to keep that Schonland story out of the public gaze. Haarhoff focused on other issues closer to home [17].

The *Forum*'s treatment of South Africa's problems, especially those involving the various peoples within its borders, was both refreshing and enlightened. These were best summed up by Alan Paton, a regular contributor to its pages, in his admirable biography of Jan Hofmeyr [18]. The journal's views stood for freedom of thought and speech, he wrote, and 'the fearless expression of opinions by others', with denial of hospitality only to 'the intolerant and ill-mannered'. There was to be one loyalty, namely 'South Africa undivided'. Of the franchise, the *Forum* reflected a view of gradualism when it said that 'Rome built up a state by gradually extending the franchise. This too is a parable for us, particularly in regard to our native population. We are not here discussing the Bantu, but we believe that if the Afrikaner were released from his fear for his culture, there would be more chance of a constructive contribution to the Native question'.†

Such views accorded well with the Schonlands' idea of liberal democracy but, as was their way, Basil and Ismay were not just passive

---

* Its first editor was R J Kingston Russell, the managing editor John Cope, with S H Frankel an editor-director. The Board of Directors included Sir Robert Kotze, Sir Carruthers Beattie, Sir Brian Robertson DSO MC, H R Raikes, Professors R B Young, J Y T Greig, I D Macrone, C M van den Heever, R A Dart, E G Malherbe and R Currey.
† *The Forum* 4 April 1938.

observers of events. They actively supported the South African Institute of Race Relations,* the body that tried to prick the collective conscience of the country. Since 1947 this institute had been under the energetic leadership of Quintin Whyte, a Scot who, after being invalided out of the Indian Civil Service, came to South Africa, obtained an education diploma at Rhodes and then found himself employment as a school-master at Healdtown and Lovedale Missionary Institutions at Fort Beaufort in the Eastern Cape, before moving to Johannesburg and the institute. Its supporters amongst South Africa's white population numbered many in positions of prominence within the community but it never counted itself amongst the most popular of organizations. Ismay assumed the prominent role in the Schonland family of encouraging Whyte in what was often a thankless task, as he strove to encourage better relations between black and white, often in the face of outright hostility from some and detached disinterest from so many. Basil provided quiet support in the background, wise counsel when needed and a little financial assistance, always from the 'Anonymous Donor'.

*   *   *

South Africa was, therefore, at yet another cross-roads in its history when Schonland wrote to Sir John Cockcroft in August 1953. It continued the correspondence that had been passing between them throughout the year, mainly on matters of South Africa's uranium, the possibility of Commonwealth collaboration in the field of atomic energy and the training of South African physicists at Harwell. This was a matter of particular importance because of the austerity budget just introduced in South Africa and its effect on the 'nuclear physics effort', which Schonland had been trying to encourage. Trained manpower, in the event of little support for research in the field, would at least provide some continuity for the future. All was not entirely gloomy, however, because the Government had reacted favourably to Schonland's suggestion regarding some form of collaboration with the United Kingdom in the nuclear field. To explore the possibilities, the Deputy Chairman of the South African Atomic Energy Board and Secretary of Mines, Mr V H Osborne, was to be sent to England for discussions. Osborne was insistent that Schonland should accompany him and Basil asked John Cockcroft to be in attendance at all those meetings [19]. But this was all relatively routine compared with what followed next.

---

* Founded in 1929, the South African Institute of Race Relations was based at Wits where its first Director, J D Rheinallt-Jones, strove to promote the 'scientific study of the Native'. In 1923 he persuaded the Chamber of Mines to provide the financial support that led to the establishment at the University of the Department of Bantu Studies.

442

Cockcroft never wavered in his wish for Schonland to join him as his deputy at Harwell and he raised the matter yet again during that fleeting visit to London. Whereas Schonland had declined on every other occasion, this time he had strong cause to reflect on the future: his, the BPI's and the country's. To Cockcroft he wrote briefly on his return home: 'I won't reply yet about the Deputy post. This is an advantage since I have need of time to form an opinion about the way things will go in this country in the future.'

The pace of events on the political front in South Africa and the increasing forthrightness of the demands being made by the ANC, had brought into very sharp focus the glaring divide that now separated the black and white peoples of the country, and they stressed the need for urgent action. The government had made it clear how they intended to react, while the Defiance Campaign had shown that the ANC were organized, capable and determined. The threat of violent confrontation loomed large and the only way to avoid it was surely to talk. The Nationalists were both obdurate and truculent—their way of talking meant only castigation and threats. Their opposition in Parliament, the United Party that Smuts had led throughout the war, was rapidly losing its way and its response at this critical time was ineffectual. The churches divided almost along political lines, with those sympathetic to the ANC cause being torn between their black and white constituencies. Many individual clerics voiced their outrage at discriminatory legislation and stood firm in the face of Nationalist threats and government edict. Only the newly-formed Liberal Party, unique in the country with a membership not determined by the colour of one's skin, took a united and principled stand and sought a meeting with the ANC Executive headed by Chief Luthuli. Amongst those it invited to participate was Basil Schonland.

Soon afterwards Schonland wrote to Cockcroft. The meeting, and particularly its participants, had had a profound effect on him [20]:

> The meeting with the native [African] leaders was a great surprise
> to me. I had no idea at all that they were so well educated and such
> fine men, nor that they were determined to seize power in this
> country in 10–20 years' time! On this they were quite
> uncompromising. They would not listen to any proposals for
> 'gradualness' and cited India and West Africa as examples of
> countries where education was not required as 'a qualification for
> full citizenship'.*

---

* In his autobiography *Long Walk to Freedom* (Little, Brown: 1994) Nelson Mandela mentions, on p 147, a meeting held early in 1953 'with a group of whites who were in the process of forming the Liberal Party'. He was not present himself.

It had taken rioting and bloodshed, fuelled by years of pent up anger, to bring together these men who knew so little of each other's thoughts, ideas and aspirations. To sophisticated, liberal men whose own philosophies embraced the widest reach of western thought, the realization that their African kinsmen were their equal in every way had come both as a revelation and as a great surprise. This was not the Africa as it appeared and as it had been presented to generations of white South Africans, however well-intentioned they may have been. These were no serfs or mere children whose intellectual progress depended on the dispensing of education at a rate set without consultation or consideration by those who 'knew best'; these were men who were determined to set the pace themselves and whose objectives were clear-cut and unequivocal. South Africa's future lay ultimately in their hands.

There was now much for the Schonlands to ponder and to think about. Had the time now come for Basil to accept the offer from Cockcroft and leave both his beloved BPI and the country that meant so much to him for the enormous challenge offered by Harwell and the world of nuclear physics? He was almost 58 and not far off retirement. What other opportunities lay ahead in science? Certainly fewer where he was than at the very heart of the most exciting scientific and technological challenges then confronting his British colleagues. Cockcroft had indicated his great desire to end his career in the university where Rutherford had stunned the world in the opening decades of the century and where, in the 1930s, Cockcroft himself had played such a significant part during the most glorious decade of British physics. And there was already talk of creating a 'new technology university', possibly centred on Cambridge with Cockcroft one of its most enthusiastic supporters [21].

What did South Africa have to offer Schonland; what did he have to offer South Africa? As things stood then, the answers to both questions were probably the same. Little. Schonland had more than paid his dues to his country as scientist and soldier while South Africa, a laboratory unequalled in its geophysical riches, was now beginning to fracture, not under electric or seismic stresses, but as its disparate peoples turned in on themselves and were separated by the fault-lines of apartheid.

* * *

In April 1954 Basil Schonland visited England, again on official business. 'Nothing mysterious, in spite of the papers', he wrote to his brother Felix, 'just part of my job as scientific adviser to the Dept. of Defence'. The visit lasted all of a month and, though much hard work on defence matters consumed a portion of the time, he also spent some of it at Harwell with Cockcroft. While there he agreed to accept the appointment as Deputy-Director of the Atomic Energy Research Establishment. 'It is a great opportunity for a man of my age and training and it will be very pleasant for Ismay to live half-an-hour from Oxford' he told his brother [22]. Schonland's appointment as Cockcroft's deputy was duly announced in the British press on 7 May.

That Cockcroft was at last to have a deputy was greeted with relief in those quarters where his manifold responsibilities and peculiarly wayward style of management continued to sow no end of confusion amongst his colleagues and subordinates. However, the name of his deputy met with some surprise, for Schonland was, at least within Harwell, something of an unknown quantity. For all his Cavendish pedigree and research under Rutherford, he had long since abandoned nuclear physics for lightning and had played no part in the wartime developments that had spawned Britain's atomic research activities. But the bond that he had with John Cockcroft out-weighed everything else and Cockcroft's word carried great weight in high places. In South Africa the news was met with a mixture of pride and resignation. To the country at large and its newspapers this was an honour bestowed on one of South Africa's greatest sons: 'one of the select few who are the builders of a new age opening to mankind', one wrote. Among others who had worked with him in various capacities over many years, there was pride too, but also just a twinge of resentfulness that 'Schonland never stayed long in any job' — there was always the bigger challenge, the greater glory awaiting him elsewhere. But such, rather small-minded, remarks were few. More generous commentators saw in his appointment great honour for South Africa: 'It was inevitable that Dr Schonland should have been marked for the highest honours in his profession, in a wider sphere than South Africa could offer him. Yet it was in the land of his birth, with its relatively modest opportunities, that he was able to initiate much of the work which made him famous and which is inspiring a later generation of scientists in this country'.

In his brother Felix he had no greater admirer nor compiler of family achievements and, in his letter of congratulations, Felix saw the appointment as deputy to Cockcroft as just the first rung on the ladder, but Basil was quick to disabuse him of this: [23]

> I fear you are going to be disappointed, for this is my last fling. I do
> not want to be the Director. We are looking forward to going
> though in many ways it is a somewhat adventurous and risky step.
> I shall hope to get a good Director of the B.P.I. in my place.

He soon did. Anton Hales, who had been appointed to the chair in Applied Mathematics at UCT in 1949, accepted Schonland's offer to return to the BPI to succeed him as Director and he duly arrived in September. The Schonlands were booked to sail for England from Cape Town on 1 October, but before then there were a number of loose ends to be tied up and the intervening few months would be very busy. One of these was the celebration of the fiftieth anniversary of the founding of Rhodes University at which the Chancellor was to speak. There was also an invited monograph on the lightning discharge to be written for the German *Handbuch der Physik*, and the BPI was holding its second conference on geophysics in July – an occasion which promised to be an even greater success than the first that was held in 1949.

There was also a visitor to entertain in the person of Dr Ernest Marsden, who was in South Africa for the express purpose of delivering the Royal Society's Rutherford Memorial Lecture of 1954. Marsden was continuing the tradition where a distinguished scientist, who had also been one of Rutherford's students, would present the memorial lecture a number of times in one of the countries of the Commonwealth and 1954 was the turn of southern Africa. When in Johannesburg, Marsden stayed with the Schonlands and visited the 'Beep' where South African radar had started as a result of that memorable shipboard meeting that he had had with Schonland in September 1939. That wartime conclave during their three days at sea must surely rank as one of the defining moments in South Africa's scientific history. From it sprang not only the SSS and the defensive chain of radars that eventually ringed the country's coastline, but it provided the inspiration that soon led Schonland to England, paired him again with Cockcroft and then set in train the series of events that changed the course of South African science forever. Now, on meeting again, the two men compared careers that rivalled each other's remarkably in their similarity and in the stature of their appointments. Marsden, about to retire and soon to be knighted, was New Zealand's most eminent scientist; Schonland was undoubtedly his South African *alter ego*.

Those remaining few months passed quickly. The geophysics conference took on a special significance, now that Schonland's colleagues at the BPI knew that he was leaving them. They decided to use the occasion to honour him and his work in the field of geophysics in South Africa; the fact that such geophysics luminaries as John Jaeger from Australia and Francis Birch from Harvard were there just added extra lustre to the four-day meeting held at the end of July. The BPI Board duly recorded its grateful appreciation of the work of the first Director of the Institute and wished him well on his new appointment.

Schonland himself remained dedicated to the last and, as well as working on his promised monograph, he collaborated with Dawie Malan in writing a paper on an intriguing phenomenon. First reported in 1939, the recent photographs of lightning striking upwards from the Empire State Building in New York provided much food for thought and the BPI was well placed to comment. Schonland and Malan offered a cogent argument, which disputed the conclusions reached by their

American colleagues. The polarity of that first upward-moving leader was, they asserted, negative and not positive as had been presumed, and it was on this note that Basil Schonland brought to a close his scientific career in South Africa [24].

He also resigned his commission in the Union Defence Force and with it his appointment as Director of its Corps of Scientists. Thus ended Schonland's life in uniform that had begun as a youthful private in the First Eastern Rifles in Grahamstown in 1913 and had progressed through two World Wars to its culmination as a brigadier and scientific adviser to a Field Marshal and to a Prime Minister. Always a volunteer, he never baulked at the need to serve in the colours for King, country and for those highest ideals with which he so readily identified throughout his life. An inspirational leader, a disciplinarian if required, but one who was always seen by those who served with him as a man of exceptional ability, who had that rare talent of being able to combine the best of soldiering with the best of science. The achievements alone of both the SSS and the AORG would forever stand testament to Basil Schonland as the scientist *and* soldier.

It was most fitting that the last duty he performed in South Africa before taking up his appointment in England was in Grahamstown, his birthplace. Together, he and Ismay left Johannesburg for the last time late in September. Their route to Cape Town and the ship to England was not direct, but went via the Eastern Cape. On 24 September 1954, in the Grahamstown City Hall, Dr Basil Schonland once again addressed the assembled congregation of Rhodes University on the golden jubilee of the university's foundation. In a speech that radiated warmth for his *alma mater*, its Chancellor reminisced about those days, long past, when as a boy of nine he had almost felt the energy and enthusiasm displayed by the university's founding fathers as they turned a former military barracks into a place of learning and scholarship. Of course he could, because his own father was one of them. It was with great pride that he then reminded them of the university's rich academic heritage, but he stressed that he did not propose to talk about people but rather about purpose—and particularly the purpose of a university. Whereas its immediate purpose was the provision of instruction necessary for gaining a higher qualification, its ultimate purpose was very much more and that, he said, was to be found in the answers to three short questions: What is the nature of man himself, of his inner motives, his wants and frustrations, self-denials, joys and sorrows? What is the nature of man's relationship with others, of his existence as a social being? What is the nature of the external world in which he exists?

To try to answer them he offered the audience his own belief of what constituted that ultimate purpose of a university [25]:

> All science, all history, philosophy, art, music and literature is the
> record of man's attempts to become more civilised by finding better
> answers to these three fundamental questions. When a university
> conveys them to its students it provides a spiritual aid to living
> more important than the degrees which it confers. The ultimate
> purpose of a university is thus neither practical or materialistic; it is
> not even rational. It is imaginative, aesthetic and indeed ethical.
> And its final justification, to use an old-fashioned phrase which
> once clearly defined the purpose of every university, is no less and
> no more than 'to glorify God'.

And so their Chancellor left Grahamstown to travel much farther afield.
Soon he would watch over the fortunes of his university from a very
different vantage point—a former wartime RAF aerodrome situated
some few miles from Oxford and bordering on a village known as
Harwell.

CHAPTER 21

HARWELL

On 15 October 1954 Basil Schonland joined the payroll of the Atomic Energy Research Establishment at Harwell [1]. He and Ismay moved into a house at No 2 South Drive, next door to the Cockcrofts and within the perimeter fence of the establishment. The Schonlands thus became part, physically if not yet spiritually, of one of Britain's most illustrious research establishments.

Nine years before, the Research and Development Establishment, as it was then known, came into being by an Act of Parliament in one of the first actions of the newly-elected Labour government of Mr Clement Attlee. Its function, made known in grand understatement, was 'to deal with all aspects of the use of atomic energy', whether of a civil or military nature. The responsibility for this new and extremely sensitive facility within the structure of the state was given to the Ministry of Supply which, throughout the war, had marshalled Britain's scientific and technical resources in the development and production of sophisticated scientific devices and weaponry. Because of the very nature of atomic energy and its devastating effects, that were so chillingly demonstrated just a few months before, when the atomic bombs were dropped on Japan, it was vital that all further British research be conducted under the watchful eye of a Ministry already well versed in all the ramifications of defence. The man appointed by the Government to head this operation was the former Chief of the Air Staff, Lord Portal, who became Controller of Production, Atomic Energy within the Ministry of Supply. Portal himself soon set up the Atomic Energy Council under his chairmanship with its members to be the directors of the various establishments about to come into being. Two of those directors had recently been appointed. The question as to who should be the scientist to lead the research establishment itself had been the cause of much discussion. The favoured candidate was Sir James Chadwick, Britain's most senior representative on the Manhattan Project in the United States during the war and the man who had the undoubted trust of the Americans — a vitally important factor at this critical time in the exploitation of the power bound up in the atomic nucleus, with all its military implications. But Chadwick declined, wanting instead to return to Liverpool University and his research, and he in turn suggested Professor John Cockcroft, who was then in overall charge of the joint British and Canadian heavy water and uranium programme at Chalk River near Montreal. Cockcroft vacillated, but after receiving certain assurances, he agreed and was duly appointed Director of the new establishment at Harwell in January 1946.

In Cockcroft, Chadwick had seen most if not all the qualities that he believed were necessary for the running of what would surely be a large and highly complex organization. If anyone had demonstrated the ability to immerse himself in a multitude of tasks, no matter how demanding and at one and the same time, it was John Cockcroft. Behind his quiet, purposeful demeanour lay an expansiveness and an enthusiasm for new ideas that was strangely at odds with his seeming inability to communicate his views to others. This almost mute of a man was unflappable, benign and much admired by all who worked with him, yet that abundance of equanimity was not entirely without cost, for behind it lurked a capacity to ignore matters which to Cockcroft himself, were but pedestrian or bureaucratic. His idiosyncratic running of the ADRDE, only a few years before, has already been recounted and its echoes still persisted in many quarters. To Chadwick and those within the very select circle of Attlee's Cabinet charged with setting up Britain's own atomic research programme, these foibles could be countered by ensuring that Cockcroft was well served 'by a suitable choice of assistants' [2]. However, in the beginning he was on his own.

Like Schonland, Cockcroft, too, had had every intention of returning to academic research once the war was over and so, before accepting the appointment to Harwell, he had obtained a written assurance from Sir Oliver Franks, Permanent Secretary in the Ministry of Supply, that 'subject to rules for safeguarding military secrets' the research undertaken at Harwell would be carried out without restriction [3]. The Cockcroft vision was that scientists at Harwell would be as free to exchange their ideas and publish their results as were their counterparts in the universities. Not only was this important to him personally but, on being made public, it served immediately to placate those in the wider scientific community who had viewed as ominous the whole exercise of atomic research and especially the appointment of Lord Portal, with his illustrious wartime career just behind him, as its controller. Harwell was immediately seen to be different from the other government research establishments because it would enjoy far greater freedom in its scientific pursuits. This was soon to be the source of much concern and not a little friction later on. The implications of such openness meant that many other facets of Harwell, including both its administration and the method by which its annual parliamentary grant of money was to be disbursed, were handled in a manner previously unknown to Whitehall. Even

Cockcroft's position was unusual. Until 1950 he reported not to Portal but to the Ministry of Supply and was given almost unfettered control of everything that happened at Harwell. This he exercised in his own characteristic and inimitable way. The Atomic Energy Research Establishment was certainly unique and soon its name became synonymous with Harwell. And Harwell soon became synonymous with Cockcroft.

In January 1946 another atomic organization came into being under the wing of the Ministry of Supply. It was initially called the Production Division, with the purpose of designing, constructing and operating the very large industrial facilities needed to produce the fissile material for the atomic energy programme, whatever its application. With its headquarters at Risley in Lancashire and other plants scattered across the north-west of England, the Production Division was the responsibility of an engineer with a formidable reputation. Christopher Hinton, at forty-four, was four years Cockcroft's junior in age; as an engineer he was second to none. Soon after the outbreak of war he had been loaned by his company ICI, where he was Chief Engineer of the Alkali Group, to the Ministry of Supply to take charge of the 'filling factories' responsible for the production of ordnance. From what had been almost a chaotic situation on his arrival, Hinton set up the Royal Filling Factories that would soon employ more than 100 000 people in the munitions industry and he turned it into a highly efficient operation. By 1942 he had become Deputy Director-General within the Ministry of Supply and was regarded not only as a man of immense intellectual capacity, but as a hugely gifted manager, who was also blessed with unusually creative skills as a designer. As a person Hinton was a complex character. Externally formidable, almost domineering, he towered both intellectually and physically over most others, but behind that façade was an austere and at times a shy and lonely man who displayed great concern for the welfare of all who worked for him even though he dominated them with a will of iron. Nicknamed 'Sir Christ' he was, in the opinion of many highly regarded men in the field, the only person ever to display the qualities of true genius that any of them had ever known [4].

The third side of the British atomic energy triangle was completed by the appointment, soon after, of Dr William Penney, the youngest of the trio and a mathematician who, at just 27, had achieved professorial rank at Imperial College. During the war, Penney had done important work for the Admiralty on the Mulberry harbours for the D-Day landings and on the effects of high explosives, before joining the British scientific team at Los Alamos, where the first atomic bomb was being designed. His expertise lay in the blast and shock effects of the bomb and he was regarded as one of the few people to fully appreciate the operational aspects of the weapon. Penney witnessed the dropping of the atomic bomb on Nagasaki on 9 August 1945 and, soon after, visited the ruins of that city and Hiroshima, that had suffered the same fate just three days before, to quantify the effects of the blasts. He so impressed the Americans that they tried extremely hard after the war to encourage him to stay in the United States but, once again, it was Chadwick whose influence on the British Government prevailed and Penney was appointed Chief Superintendent of Armament Research at the Ministry of Supply in the spring of 1947. Though initially responsible for all armaments research, his task, ultimately, was to produce Britain's own atomic bomb.

These three men would dominate nuclear science in Britain over the next decade and, such was their eminence, that they too were a triple diadem to follow Cockcroft's creation that Schonland had encountered at Petersham some years before. Whereas Hinton always seemed well turned-out, Penney, like Cockcroft, displayed a lack of sartorial elegance. A highly gifted mathematician with a flair for developing elegant and simple solutions to problems of great complexity, Penney also had Cockcroft's unpretentious manner and got on well with the military (on both sides of the Atlantic) and this made him a highly-prized asset in his line of business. But when he took up his appointment as Chief Superintendent, CSAR in abbreviated form, at the Ministry of Supply's Armament Research Department at Fort Halstead in Kent, his arrival, unaccompanied by any fanfare, did little to arrest the rapid decline in morale in an establishment which was itself rapidly running down in the uncertain post-war period. However, it was soon evident to all the staff that this rather dishevelled, schoolboy-like figure, was more than just a mathematician. As CSAR he was soon to become Caesar! Penney's informality was inspirational and his enthusiasm rapidly regenerated that waning commodity amongst his colleagues. Though at times blunt in his manner of speaking, he was oblivious to rank or station and addressed all in his company as his equals, be they generals or general factotums. Where he differed from Cockcroft was in his tendency to withdraw from situations and groups that failed to hold his interest or attention. He was not the natural mixer that Cockcroft was and lacked his older colleague's ability to endure the tedium of forced social contact with never a hint of boredom. Penney's eyes, by contrast, would soon glaze over [5].

It was into this circle that Basil Schonland soon found himself inducted, though his personal responsibilities as Cockcroft's deputy seemed right from the outset never to have been spelt out in clear and unequivocal terms. Was this 1941 all over again? Others within the establishment, such as the recently appointed Chief Engineer, G W Raby, reacted with some dismay at the arrival from South Africa of a virtual stranger, particularly as he had been catapulted into such a senior position. Raby had enjoyed a long and fruitful association with Cockcroft at the ADRDE during the war, where he had shown himself to be a man of conviction who got things done. He actually knew Schonland too from those wartime days, and had met him again during his own visit to South Africa in 1948 on important Harwell business, when the issue of South African uranium was receiving particular attention in England and elsewhere [6]. When Cockcroft failed in his first attempt to persuade Schonland to join him as his deputy in 1951, he had turned to Raby and had offered him the post, but Raby had recently accepted an appointment in the Sudan and so was not available. That African sojourn over, he returned to England in 1954 and took up the new appointment as Chief Engineer at Harwell which he assumed (though Cockcroft had almost certainly never said so) made him Cockcroft's deputy in everything but name. His consternation was almost palpable when Schonland duly put in his appearance and assumed the mantle of Deputy Director, and it was not long before Raby made his feelings clear to Schonland. Since both knew the Cockcroft style of management well, they sought a diplomatic solution from the Technical Secretary of the Establishment, Mr D R Willson, who, with wisdom rivalling Solomon's, decided that Raby should henceforth be called Deputy Director (Engineering). A crisis was averted and Schonland and Raby remained on amicable terms [7]. Cockcroft was probably oblivious of the fuss.

*   *   *

Before continuing with Schonland's introduction to the business and personalities of Harwell, it is necessary, at this point, to examine in some detail the structure of the British atomic research effort and particularly to see it in relation to the strategic balance existing in the world in the immediate post-war years.

Harwell was the site of a former RAF aerodrome, from which the paratroop-carrying gliders destined for Arnhem had been dragged into the air by their 'tugs', the Stirlings and Halifaxes of the RAF's Troop Carrier Command. Now, a decade later, it had all the trappings of a university but without any sign of students nor any colonnades and architectural finery. The old hangars had been refurbished and pressed into service as laboratories and offices to accommodate the scientific community that now inhabited the place. Its perimeter was ringed by a security fence but there were almost no restrictions on the mental freedom that Cockcroft so enthusiastically encouraged amongst everyone who worked there. Research, even for its own sake just as long as it was interesting and productive, was all that mattered. If Schonland's terms of reference were vague, that applied equally to those of Harwell as well. In fact, it had been implicitly agreed during the very early discussions in 1946 that such formalities would be undesirable 'having regard to the unsure problems with which we have to deal and the fact that there is no dividing line between atomic energy research and nuclear physics research' [8]. It was left to Cockcroft to decide what role Harwell would play and indeed how it would play it. And so it was that this enigma of a man had almost complete control of the work done in the establishment that was to be so fundamental to the British atomic energy programme. Though intensely loyal to the objectives of the overall project, Cockcroft's firmly held belief in unshackled research, plus his generous and ever cooperative nature and his great propensity for encouraging good ideas in others, meant that Harwell soon became the very centre of curiosity-driven research that knew few bounds. Soon, there flourished within its perimeter a vibrant community of independent-minded and, in some cases, highly original, young scientists. Amongst them, not surprisingly, were some who nurtured particular ambitions of their own.

Tangible research results were soon forthcoming. From 1949, just three years after Cockcroft's appointment, the scientific literature began to carry papers written by Harwell scientists in fields as diverse as structure analysis using neutron diffraction, the emission spectroscopy of transuranic elements and the plasticity of hexagonal metals [9]. Papers covering an even wider range appeared with increasing frequency and Harwell was soon recognized as one of the leading research establishments of its kind in the world. Certainly nowhere else in Europe was there anything even remotely comparable with it. The presence of Cockcroft was the undoubted drawcard that had attracted so many first-class scientists to Harwell, and he encouraged and supported these men of ideas by his philosophy that was driven, not by how much it would cost, but by whether a proposal had scientific merit. As reluctant as he was to discourage good ideas, Cockcroft was equally unhappy about shutting down those projects that had begun to lose their way and soon Harwell became the very epitome of untrammelled research, come what may [10].

For all Cockcroft's apparent encouragement of the *laissez-faire*, he was true to Harwell's fundamental brief, which was to carry out research into all matters to do with atomic energy. Its overriding, though as yet unstated, purpose was to provide the basic scientific know-how necessary for the development of a British atomic weapon—once official sanction for its development had been given by government. Naturally intertwined with this, because of the very nature of the processes involved, was the generation of electric power, but its exploitation was assigned a very much lower priority. The dedication of Cockcroft, Hinton and Penney saw Britain explode its first atomic bomb in 1952, whereas it would not be until October 1956 that the first large-scale atomic power station in the world, at Calder Hall in Cumberland, would come into service and even then its primary purpose, and that of a similar one at Chapelcross on the Scottish border commissioned in 1959, was to produce plutonium primarily for military applications rather than to supply electricity to the national grid [11].

Such an objective naturally meant secrecy and only a select few within Attlee's Cabinet were privy to the details. The nation at large, and most of their political representatives, were unaware of all this since they were much more preoccupied with the travails of a country faced with the enormous task of re-building an economy shattered by nearly six years of war. However, matters of great strategic importance and not a little national interest, were certainly on the minds of those in the upper echelons of government. Amongst the scientific community, as usual, feelings were mixed. Most vocal of those in their opposition to any talk of re-armament and, especially, the development of atomic weapons, was Patrick Blackett. His powerfully voiced opinions, however, met with little sympathy in Whitehall, because bigger political issues were at stake and they certainly cut no ice with either Attlee or the Chiefs of Staff, ever conscious of the looming threat of Communism. And so the rigorously argued case by one of the world's greatest scientists was dismissed quite peremptorily by them 'as the views of a layman' [12].

It was within this climate of national need that work began in earnest in 1946. Though Cockcroft's restraints seemed few, both Hinton at Risley and Penney, first at Fort Halstead and then at Aldermaston, had clear, unambiguous objectives—to produce the material and to design Britain's own atomic bomb. Penney needed plutonium and Hinton was to produce it for him at the various factories soon to be set up for this purpose at Springfields near Preston, at Windscale* on the Cumbrian coast and at Capenhurst in Cheshire. Such a diffuse distribution of facilities was deliberate—security alone demanded it—but it did add to the general complexity of the project. Even more so were the technical requirements. The fundamental process, whether to make a bomb or to generate heat from which to produce electric power, involved nuclear fission: the splitting of an atomic nucleus of a fissile material such as uranium or plutonium by bombarding it with neutrons, with the consequent release of considerable amounts of energy. Under the right conditions of critical mass, the process is self-sustaining by what is known as a chain-reaction mechanism which sees every such collision release about 2.5 additional neutrons. Uncontrolled, the effect is catastrophic—an atomic bomb—whereas the introduction of a material that absorbs neutrons will either regulate it, for the controlled generation of energy, or stop it all together. In general, the nuclear reactor (or atomic pile, as it was originally known) involves a slow fission process of neutron

---

* Windscale was originally called Sellafield but the name was changed to avoid confusion with the plant at Springfields. Subsequently, after the 1957 fire, it became Sellafield once again and embraced the adjacent Calder Hall reactors as well.

bombardment under so-called thermal conditions.* By contrast, the atomic bomb requires fast fission and uses either enriched uranium (^{235}U) or plutonium as its 'fuel'.

A basic understanding of the physics involved is important to what follows. The only naturally occurring element that will undergo fission is uranium, and then only in its enriched form as the isotope ^{235}U, which makes up just 0.7 per cent of the uranium that occurs fairly abundantly in various parts of the world, the rest being ^{238}U. Research progress during the war had been rapid. It was found that ^{238}U, if subjected to neutron bombardment, will itself be transmuted into what was referred to as the mysterious 'element 94' when first announced in an unguarded moment in an American scientific journal in June 1940 [13]. This material was soon called plutonium, ^{239}Pu, a material of considerable importance in the events about to unfold. The fact that plutonium is a by-product of uranium under neutron bombardment was especially important, because the very process within a nuclear pile intended to produce energy would itself produce the plutonium required for other purposes, such as the production of an atomic bomb. Cockcroft's and Harwell's task at the outset was to understand the underlying science and, in collaboration with Hinton, to design the reactors required to produce plutonium. The decision to commence construction of a single atomic pile for the production in Britain of plutonium was taken on 18 December 1945.

Any suggestion that the British research effort would be augmented by American assistance was short-lived. The intense scientific collaboration between the two nations that had been ignited by the Tizard mission to the United States late in 1940 was now a thing of the past. The United States was rapidly becoming a very different place from the country whose military might and gum-chewing soldiers had been such familiar features of British life in the closing years of the war. Attitudes towards Britain amongst American politicians had hardened noticeably since those days and they would do so even more as a result of events soon to follow. America was distancing itself from Europe and, in the process, it clutched to its chest what some Americans believed to be the fruits of their scientific expertise and theirs alone. Nowhere was this more evident than in the field of atomic energy. Then, as if to seal their fate, it was announced in March 1946 that a British scientist, Dr Alan Nunn May, who had been working at the Chalk River facility in Canada, and had made regular visits to American atomic plants, had been arrested on

---

* The exception is the fast breeder reactor (the first of which came into operation at Dounreay in the far north of Scotland in 1959). By encasing a very dense core of ^{235}U in a blanket of natural uranium the fast fission reaction converts ^{238}U to plutonium at a greater rate than the ^{235}U is consumed in the core. The reactor thus produces more fissile material than it consumes and hence is known as a fast breeder reactor.

suspicion of passing atomic secrets to the Soviets. The news infuriated the Americans who saw it as a massive betrayal by their former allies and it contributed greatly to the mounting hysteria within the United States Senate against any form of future collaboration with the British. Cockcroft and his colleagues were stunned, for though Nunn May was known to have been a Communist sympathizer while a student at Cambridge in the 1930s, such views at a time of mounting Fascism in Europe were hardly surprising. The British security services had seemingly agreed because, after completing his PhD under Chadwick and a period at King's College, Nunn May was sent to Canada in 1943 to join Cockcroft's team at Chalk River. It was while he was there that he was recruited by a Communist espionage cell and passed to them information gleaned during his visits to various sites of the Manhattan Project in the States. His final act of treachery was to hand over to his Soviet handler microscopic samples of ^{233}U and ^{235}U, two isotopes of uranium that were crucial to the bomb project.

But even more was to come. In 1950 Klaus Fuchs, a refugee from the Nazis who had escaped to Britain in 1933 and by 1946 had become head of the theoretical physics division at Harwell, was sentenced in London to fourteen years imprisonment after pleading guilty to a string of charges relating to his duplicity. Whereas Nunn May's treachery had been serious, Fuchs's was positively devastating, for he was, by then, more deeply involved in the weapons programme than anyone at Harwell. Between 1943, when he worked at Los Alamos, and his arrest six years later Fuchs had passed to the Russians virtually all the details of the plutonium bomb tested in New Mexico in July 1945, as well as those of the processes involved in its manufacture and those of the so-called H-bomb, then under discussion in the United States. To further embarrass the British, just six months later another Harwell scientist and arguably the most brilliant of the three, Bruno Pontecorvo, defected to Russia. Like Nunn May both Fuchs and Pontecorvo had worked under Cockcroft at Chalk River and, though they had all been subjected to the various vetting procedures then used by Britain's security services, all had managed to evade detection for a number of years [14].

Whereas both the climate of increasing isolationism in the United States and the perceived lapses in British security had a hand in precipitating it, it was the McMahon Act, passed into law in the United States in August 1946, that sealed Britain's fate officially. This piece of legislation was intended to safeguard the American atomic energy programme, but in the eyes of the American public-at-large—and thus in the eyes of their national representatives—atomic energy meant only one thing, the atomic bomb, and this led almost to national paranoia about 'atomic secrets'. As a result, an essentially sensible piece of legislation designed to protect the United States against the illegal dissemination of its secrets was successively amended until, eventually, it stated that there was to be no exchange of such information between the United States and any other nation even for industrial purposes. This law, which ran counter to three solemn commitments made between 1943 and 1945 by the United States Government to both the British and the Canadians, in regard to co-operation in the field of atomic energy, effectively sounded the death knell for any such collaboration in future. There were many in England who, with some cynicism, saw this legislation as little more than the Americans setting out to maintain their atomic monopoly. Whichever way it was, the British (with their Dominions) were, as in 1939, once again on their own [15].

In January 1947 the British government took the momentous decision to produce an atomic bomb. It was a decision to be shrouded in secrecy, so much so that the great majority of those recruited to work in the various establishments set up for this purpose were to be kept in the dark as to the real purpose of their work. Of particular importance, it was deemed, was the need to allow no connection to be made, by anyone not privy to the nation's greatest secret, between Penney's activities at Fort Halstead and those at either Harwell or Risley. As the various programmes of work got under way, such watertight secrecy began to show all the signs of springing a leak. It was clearly impracticable to prevent the great majority of scientists, engineers and others involved in this huge enterprise from knowing what its ultimate objective was. In addition, as the year was drawing to its close, it was obvious that the stringent procedures put in place to maintain such secrecy were proving a real impediment to progress. This was readily appreciated at the Ministry of Supply and the solution agreed upon at the highest level was for the Government to let it be known almost incidentally that Britain was developing an atomic bomb. Further speculation and all probing likely to follow by the press could then be curtailed by issuing a standard D-notice.* The existence of the British atomic bomb project was therefore made public in the House of Commons on 12 May 1948 by way of a short question from the floor of the House to the Minister of Defence. The answer confirming that all types of weapons, including atomic weapons, were being developed was equally brief [16].

Research at Harwell, though driven by the military imperative, extended over many fields. The decision to use plutonium rather than

---

* D-notices are a system of self-denying ordinances operated by the press itself in which certain information of military importance will not be published. The request for D-notices is made by a Press Committee of press and Service department representatives, and editors are free to accept, reject or ask for amendments of any proposal. Though backed by no legal sanction the D-notice system was widely supported by the press because it applied only to publication of information about the Armed Services and could not be used to impose a security ban on other government or industrial activities.

uranium in the bomb accelerated research into the types of reactor needed to produce it and this, in turn, had implications over a very wide area from reactor physics itself, to chemical processes and to the metallurgical factors involving containment vessels and the effects of radiation on these, plus many more. The mood at Harwell was ebullient. Staff had been recruited from amongst the British scientists returning from the Manhattan Project, from Chalk River and from amongst others who had spent the war years working on radar at TRE. There was also a number of young graduates fresh out of university. In September 1947 the British Association held its annual meeting in Dundee and it featured on its programme a symposium on the 'Peaceful Applications of Nuclear Fission'. Cockcroft, naturally, was the main speaker and his contribution was eagerly awaited, for, as Sir Edward Appleton said in his presidential address, 'no scientific subject had ever aroused quite the same mixture of hopes and fears [as atomic energy]'. The Director of Harwell used the opportunity to discuss both the research potential and some applications of the Britain's first low-power graphite pile, known as GLEEP,* that had gone critical just a month before. It was, Cockcroft said, a much more powerful source of neutrons than the largest cyclotron then in existence, and this would enable Harwell to conduct a range of new experiments in nuclear physics—though he divulged few details except that GLEEP would be the main source of radioactive isotopes for medical purposes and their production would begin immediately. What many wanted to hear about was the use of atomic energy for the generation of electrical power, but here Cockcroft was much more cautious. It appeared, he said, that 'the pile is a potential, though not certain, new source of power for the world' and any large-scale development in that direction would have to await the results of considerable research. On one point, though, he was certain; the safe disposal of the radioactive waste materials that were by-products of the power generation process would pose a serious problem that could not be ignored [17].

Britain's first atomic pile was a purely experimental machine, designed at Chalk River and built at Harwell for the specific purpose of gaining familiarity with the technique and for performing some fuel acceptance tests. A pile capable of producing a stronger neutron flux was required for other experimental work and, particularly, for producing radioisotopes for medical and other applications, and this implied an output of about 6000 kW. Once again the basic design was done by the British scientists then in Canada and construction took place in

---

* GLEEP (Graphite Low Energy Experimental Pile) was designed at Chalk River and constructed at Harwell under the dynamic leadership of Charles Watson-Munro who had played a significant part in New Zealand's wartime radar developments following the release of the RDF information to the Dominions in 1939.

Harwell. In 1948 this second pile, known as BEPO,* went critical. It was to form the basis, though on a considerably smaller scale, of the first atomic reactors to actually go into service at Windscale in October 1950. Though no British scientist had been allowed access to the top-secret American site at Hanford in Washington state, which produced the plutonium for the American weapons, enough was known about the process to indicate that those piles were water-cooled and fuelled by natural rather than enriched uranium. That in itself meant that the process of neutron bombardment had to be slowed down or moderated in order to increase the probability of fission and so the choice of the most appropriate material for the moderator was crucial. A light element is required, such that collisions between the neutrons and the moderator will lead to energy absorption and, hence, a slowing down of the neutrons themselves. Water acts as an effective moderator but also absorbs neutrons, which would bring the fission process to a halt. By contrast, so-called heavy water (deuterium oxide) absorbs far fewer neutrons but is expensive to produce, is corrosive and might boil away at elevated temperatures. Research had shown that carbon, especially in the form of pure graphite, was well-suited to the task and, since it is also plentiful, it was selected. A pile of 28 000 blocks, all individually machined in the Harwell workshops and occupying a cube 8 m on a side, was built. Horizontal channels through the graphite housed the rods of uranium encased in specially designed aluminium cans to protect against corrosion and to prevent the escape of radioactive products formed during fission. The process was regulated or controlled by the insertion, as required, of boron-loaded control rods to absorb neutrons very rapidly. The whole structure was then encased in a reinforced concrete shield 2 m thick for the protection of all those in its vicinity.

Thoughts now turned to the full-scale production facility charged with producing plutonium [18]. Whereas BEPO was not the pilot plant for the piles under construction at Windscale, there was certainly a link between them, particularly in the type of cooling system eventually employed. Water cooling, as used by the Americans, soon proved to be a major problem for the British designers. Not only would vast quantities of water of exceptional purity be required but there was always the risk that its supply to the reactor might be interrupted, thereby causing a rapid and possibly disastrous rise in temperature of the pile before the control rods could shut it down. Even though it was believed that the chances of this happening were small, the American site at Hanford had been specifically selected with just such an eventuality in mind. It was in a very remote and sparsely populated region of the United States with a special, four-lane, 40 km long highway leading from it, for

---

* BEPO: British Experimental Pile.

the specific purpose of evacuating the site in the event of such a disaster. No such possibilities existed in an island with the size and population of Great Britain, except in the remotest reaches of Scotland and Wales, and there were good reasons for rejecting both. Air cooling was therefore the only practicable option and it was agreed that the piles should be at least 65 km from any large centres of population to afford the greatest degree of protection in the event of a catastrophic failure. The site at Windscale on the coast of Cumbria was selected with these criteria in mind. A significant factor in the design of the air-cooled system was the use of fins on the aluminium cans, carrying the uranium rods. By increasing the surface area of the cans, the fins improved the efficiency of the cooling process markedly, while the increased neutron absorption caused by the additional aluminium was kept to a minimum by reducing their thickness.

The issue of the moderator to be used in the Windscale piles was to be highly significant. A phenomenon known as the 'Wigner effect' occurs in graphite when it is subjected to neutron irradiation. In essence there are actually two effects. The first is a displacement of the atoms in the crystal lattice which results in a change in its dimensions and, consequently, of those of the graphite itself. This could conceivably cause structural changes to occur in the pile, such as a narrowing of the channels within which the aluminium-encased uranium rods and the control rods have to move freely. In addition, such constriction would reduce the quantity of air flowing through the pile, with obvious adverse effects on its cooling. The second effect is potentially even more serious. It is possible for Wigner energy to be stored within the graphite, with the eventual build-up of heat reaching such proportions that the uranium rods themselves could ignite. This possibility had been foreseen by a Harwell physicist, W G Marley, in the early days of the programme and was raised again in 1948, when the first signs of a thaw were beginning to occur in Anglo-American scientific relations. While on a visit to Harwell the noted American nuclear physicist Edward Teller drew attention to the danger of a release of Wigner energy leading to a possible fire in a fuel rod [19]. Both warnings were to be proved portentous in a few years' time when Pile No 1 at Windscale caught fire; particularly as no one, apparently, at Risley was aware of the dangers!

* * *

On 29 August 1949, the Russians detonated an atomic bomb and the political fallout was immense. Within the very highest echelons of the British Government, for only they knew any details at all about the British work on atomic weapons, there had been the unchallenged view that only the United States and Britain had the capability of developing such a weapon. Now their former ally, with intentions unknown, had leapt

461

into the big league with the United States. From that moment the balance of power within the world shifted irrevocably; no longer could Britain consider herself to be a 'Great Power' alongside the United States, but, on the strength of the evidence just presented, the Russians certainly could. The first British atomic bomb test was still three years away.

There followed soon after an agonising reappraisal by the Chiefs of Staff as to whether Britain should continue with her atomic weapon programme. Sir Henry Tizard, as chief adviser to the government on defence research policy, was by now a vehement opponent of British efforts to produce an atomic bomb. To do so to the exclusion of other equally pressing needs such as the means of delivering the weapon and the accurate guidance of those aircraft when so doing would, he earnestly believed, be foolhardy. In this the Chiefs of Staff agreed with him but to abandon the effort all together would signal the end of any possible resumption of collaboration with the Americans in the field of atomic research, particularly as the balance of power in the world had now shifted so demonstrably and there was now a real likelihood of American isolationism being reviewed, notwithstanding the very obvious evidence of Klaus Fuchs's massive treachery which had presented the Russian's with the very blueprints from which they had manufactured their bomb. The arguments, both pro and con, waged for months. Eventually, in April 1950, the Chiefs of Staff reported to the Defence Committee that they recommended continuation of the weapon development programme, but not to the exclusion of other vital projects, and they expressed the hope that some agreement would soon be reached with the Americans so that ultimately their joint resources could be pooled [20].

Tizard's concerns extended well beyond the realms of the military uses of atomic energy; he was to prove just as sceptical about its use for the generation of electrical power. To him the whole idea of atomic energy as a source of heat and power was much exaggerated, especially in a country like Britain that had an abundance of coal. However, he was also the first to appreciate the need for Britain not to lag behind the Americans in acquiring the necessary know-how and, however reluctantly, he supported continued research on both aspects at Harwell and the design, at Risley, of appropriate reactors for the generation of electrical power. By contrast, his former wartime colleague Lord Portal saw any diversion of effort away from the military objectives as unacceptable and only allowed Hinton's Industrial Group (formerly known as the Production Division) to conduct theoretical studies, whereas Hinton himself had had every intention of using one of the Windscale piles to produce power as well as plutonium. Cockcroft's view of the possibilities of electrical power were naturally conservative. In June 1950, at a lecture in Oxford, he stated that at least another decade of research would be necessary before there was any certainty about the viability of nuclear-generated power. The first flush of post-war enthusiasm had been somewhat tempered by the experience gained in the five years that had followed [21].

Within a year, though, it was clear that the initial objectives set for Harwell had been met. The fundamental research effort, no matter how unrestrained it had been, had borne considerable fruit. Both reactors at Windscale were now in operation, as was the uranium metal plant at Springfields, and Cockcroft believed that he had achieved what he personally had set out to do. His services had been requested elsewhere as the successor to Sir Henry Tizard as chairman of the Defence Research Policy Committee and it was then that Cockcroft first saw Basil Schonland as his own successor at Harwell. But the results of the General Election in October 1951 changed everything. The Conservatives, under Churchill, were returned to power and Cockcroft remained at Harwell. There was now a strong move afoot to separate the atomic research establishments from the rigid policies and especially the inflexible salary structure of the Civil Service. Whereas Harwell had attracted exceptionally talented scientists and engineers, due undoubtedly to the presence there of Cockcroft[*] himself, much dissatisfaction had been expressed with the low salaries then on offer, compared with those in universities and in industry. It was evident that the atomic energy organization should be removed from the control of the Civil Service and should have its governing body empowered, amongst other things, to pay competitive salaries. The powerful driving force behind these deliberations was none other than Churchill's wartime scientific adviser Lord Cherwell, himself now a member of the new Cabinet as Paymaster-General.

More than 10 000 people were employed in the various laboratories, factories and operating sites involved in all aspects of atomic energy. Harwell alone accounted for 3500 of them by 1952 [22]. The birth of the new Atomic Energy Board to replace Portal's Council (Portal having resigned earlier over his dissatisfaction with intransigence of the Ministry of Supply) was not without considerable labour pains. Cherwell, who believed strongly that the atomic energy programme should reside in its entirety within the Ministry of Defence, had to accept the fact that this was untenable for very many reasons, not least of which was the warlike flavour that it would impart to the whole enterprise, but also because the MOD was, at that time, only a small coordinating ministry. In April 1952 the government decided that the Paymaster General would take responsibility for the board and would, henceforth, advise the Prime Minister on all matters to do with atomic energy. Its links

---

[*] In November 1951 Sir John Cockcroft and Ernest Walton were awarded the Nobel Prize for Physics for their experimental work nearly 20 years before at the Cavendish on 'the transmutation of atomic nuclei'.

with the Ministry of Supply, however, would still remain, though these led to frequent clashes between Cherwell and the responsible Minister, Duncan Sandys, since little love was lost between these two powerful characters. The membership of the board would consist of Cockcroft, Hinton and Penney with extra appointments from the Ministry. In Cherwell the board had as a chairman a scientist who understood precisely what the problems were; more importantly, he was a man of immense resolve who was seldom thwarted and, most important of all, he had the ear of Churchill, with whom he had stood so resolutely throughout the war. But the arrangement was beset with problems from the start and it was soon evident that it was far from satisfactory; in fact it was questionable whether it offered any advantages at all over the original structure put in place by the previous government, and in many respects was identifiably worse. Cherwell was unrelenting in his attempts to see the whole atomic energy programme, both military and civil, under the independent control of an Atomic Energy Authority, because he could see the direction in which atomic energy was going in the United Kingdom. What had previously been predominantly an experimental programme, intended to develop the techniques and the facilities for producing plutonium and for designing and testing atomic bombs, would now also have to produce weapons for stockpiling. In addition, there was the distinct possibility, he believed, of using atomic energy to produce electrical power. Neither of these vast undertakings could be contemplated without participation by industry. Henceforth there would have to be industry participation right down at the levels of development and conceptual design. None of this was possible if the atomic energy programme remained under the control of a government department [23].

On 3 October 1952 at Monte Bello, a group of uninhabited and barren islands some 80 km north west of Australia, the first British atomic device, codenamed *Hurricane*, was exploded.* It was the thirty-third such detonation since the war: the Americans had already tested 29 and the Russians three. The cloud that ascended from the cauldron in the sea that just an instant before had been but a speck on the map marked the entry of Britain into the 'Big Power' league. It was the culmination of nearly seven years of research as well as a massive construction effort at Harwell, Risley, Springfields and Aldermaston. The reactors at Windscale, which Hinton was later to dub 'monuments to our initial ignorance', had produced the small sphere of plutonium at the very core of the bomb. At first, to those who witnessed the blast, it seemed almost an anti-climax but the shock-wave that followed within seconds was not.

---

*One of the official observers at this momentous event was Dr Omond Solandt, then chairman of the Canadian Defence Research Board.

But the effect in London was not immediate, at least as it affected the future of the atomic energy programme. Cherwell's Cabinet paper setting out his proposal for an Atomic Energy Authority beyond the reaches of the Civil Service had been timed to coincide with the Monte Bello test and, like its shock-wave, the reaction to it was delayed. When it came there was strong opposition from the Ministry of Supply and even Churchill himself was lukewarm. Cherwell considered resignation but then, in the words of his biographer, 'stooped to lobbying' by calling for support from several Ministers whose views he knew coincided with his. This had the desired effect; the cabinet committee responsible was won over and, by January 1953, a committee of experts was set up under Lord Waverley to work out the mechanism for transferring responsibility for atomic energy from the Ministry of Supply to a suitable 'non-departmental organization'. But the Ministry fought doggedly and, noting the general mood, set its sights on retaining control of the atomic weapons programme at Aldermaston. Its case hinged on the security implications of allowing such sensitive matters to be handled beyond the immediate confines of a government department, but Cherwell was immovable. He pointed out to Waverley that, if weapons development were divorced from atomic energy in general, the inevitable result would be that power production would dominate and the weapons programme would suffer 'disastrous results' [24]. Waverley's committee debated these matters long and hard but eventually concluded that it would be a 'grave blunder' to separate the military aspects from the rest of the atomic energy programme. In July the Waverley Committee report was signed and after the usual perambulations through Parliament the Atomic Energy Authority formally came into being a year later and assumed full responsibility for the project on 1 August 1954. Lord Cherwell had achieved, in the face of enormous opposition at times, what to him was one of the most important politico-scientific objectives of his, at times, turbulent career. He was to serve as a member of the Authority until his death three years later.

CHAPTER 22

FISSION BUT NO FUSION

Basil Schonland came to Harwell, as had so many others before him, because Cockcroft was there. But Cockcroft had one, final, unfulfilled ambition which would take him elsewhere and then, despite his protestations to the contrary to his brother Felix, Schonland knew that he would be Cockcroft's heir apparent because his old friend had told him so.

To all who worked there Harwell *was* 'The Establishment'. Since its transformation almost nine years before from aerodrome to a scientific encampment, this site of die-straight tarmacadam and unprepossessing brick and concrete hangars, some 20 km south of Oxford, had undergone huge changes. Some of the largest aircraft hangars had been turned into laboratories, others housed scientists, engineers and support staff and the runways became roads. The array of research machines and reactors that were now the establishment's stock in trade was impressive. At one end of hangar 7 was the 43 cm synchrocyclotron sunk into a pit, while GLEEP and a 5 MV van de Graf accelerator occupied hangar 8. BEPO, the big experimental reactor, stood alone in hangar 10.

After 1952, the research and development branch at Risley took over responsibility for the technical support of the production reactors at Windscale, while Harwell switched much of its attention to the physics, chemistry and metallurgy associated with the power reactors then gaining ground within the thinking of Cockcroft and his colleagues. Defence-related work, though still important, was decreasing at Harwell as Aldermaston, under Penney, came into its own. Most recently techniques for detecting nuclear explosions had been moved there and, by April 1953, defence research at Harwell amounted to only about 12 per cent of the research effort of the establishment, as opposed to nearly 30 per cent for work on both nuclear materials and reactor technologies [1]. While Cockcroft had initially recruited staff from amongst his former colleagues at Chalk River, more recently there had been an influx of scientists and engineers from the Telecommunications Research Establishment at Malvern. Their expertise in radar and electronics in general was to hold them in good stead when it came to developing both the exotic high-energy machines, such as a 4 MV linear accelerator in 1949 and a much larger 13 MV machine operating with much shorter pulses in 1952. The basic power source for these machines was the klystron, which was fundamental to all radar transmitters of the time [2].

By 1953, the plans for Britain's first reactor designed specifically to produce plutonium, but with electric power as an important by-product, were well advanced. It was the work of a small Harwell design team under the leadership of B L Goodlet, formerly professor of electrical engineering at the University of Cape Town, whom Cockcroft had enticed to Harwell in 1949, and it came to fruition as PIPPA, the Pile for Industrial Production of Plutonium and Power. Its route to production was by no means either straight or level. Tensions existed between Harwell and Risley — and that meant between Cockcroft and Hinton — as to the precise role each would play in the development of a nuclear power programme and, particularly, where PIPPA would be built. In addition, the chiefs of staff were pressing hard for the increased production of plutonium and insisted that it take priority over any power programme. Harwell, at least in Cockcroft's eyes, was primarily there to conduct fundamental research, but its expertise was required across the whole spectrum of activity and Cockcroft was also insistent that it should remain so. Such a breadth of interests, from nuclear reactors to radioisotopes, from particle accelerators to health physics, from power systems to electronic instrumentation meant that the establishment had, in the space of less than a decade, become everything that Sir John Cockcroft had wished for. Not surprisingly, it was recognized throughout the world as a place of scientific distinction. The Cockcroft vision of a university-type research establishment had surely been fulfilled, but Harwell's other function, though never formally defined, of meeting Hinton's and Penney's more pragmatic needs was vital to Britain's security interests. It soon became clear that success in serving so many masters was more evident in some laboratories than in others. While chemistry and metallurgy had achieved it, that reconciliation had not really been evident in physics. There, the pursuit of international excellence in high-energy physics had occurred at the expense of the industrial programme and this had caused grave disquiet at Risley. It had even led to questions being asked as to whether Harwell really served Risley's interests adequately. A clear definition of purpose, and particularly one setting out the relationship between Harwell and the production organization, was required but none was forthcoming as the dust from Monte Bello began to settle [3].

PIPPA was a thermal reactor using natural uranium with a graphite moderator. Where it differed markedly from the two reactors already in service at Windscale was that it was cooled by carbon dioxide rather than air, and the carbon dioxide was forced through the pile at high pressure, thus necessitating that the whole arrangement be contained within a pressure vessel. Another new feature was the use of an alloy of magnesium, beryllium and calcium, developed at Harwell and known as Magnox, instead of the aluminium previously used for the cans that housed the uranium fuel elements. Magnox had been shown by exhaustive tests to have both excellent mechanical and heat transfer properties, coupled with a lower neutron absorption cross-section, so allowing the use of larger cooling fins that were crucial in improving the efficiency of the energy conversion process. There were, however, competing technologies, with a heavy water reactor (HIPPO) and a fast reactor (ZEPHYR) both serious contenders and so, in 1952, it was by no means certain that PIPPA was the favoured approach. However, it was a question with grave military implications that sealed it. With the Korean war at its height and the threat of nuclear escalation looking ever more likely, the chiefs of staff called for a statement as to what would be required to double Britain's output of plutonium. There were only two contenders: an expanded Windscale programme or the construction of one or more PIPPAs. The answer was soon forthcoming in a report from Harwell that showed that PIPPA could meet that need at 60 per cent of the cost of the Windscale reactors and, by the end of that year, the foundation of Britain's nuclear power industry had been laid as a by-product of the demand for military plutonium [4].

By the end of 1954, soon after Schonland's arrival, the world's first atomic power station was already under construction at Calder Hall just alongside Windscale on the Cumberland coast. The initial feasibility studies and conceptual design had been done at Harwell, but then Hinton's Industrial Group took over and began to turn concept into reality. A major research problem then confronting Cockcroft and Harwell was the type of reactor that should eventually become the workhorse of a nuclear power industry—should on-going research confirm both its scientific feasibility and its economic viability. Since Britain had by this time established a clear lead over any other nation in power-generation technology, there were no ready-made solutions to follow and Harwell would have to determine the optimum path for itself, although the first signs of cooperation across the Atlantic were just appearing. Since 1948 and especially after Monte Bello, the American's blanket refusal to collaborate with anyone in the nuclear field had slowly become less obdurate. Persistent British efforts, in which Cockcroft had been prominent, to encourage their trans-Atlantic cousins to cooperate led eventually to a relaxation in attitudes, and what became known as the *modus vivendi* made a degree of scientific interchange possible once more, though it by no means led to a flood of information. Such was the scene that confronted Basil Schonland at Harwell in October 1954.

*   *   *

Schonland's reaction on seeing the vastness of the site at Harwell and the scale of its operations was not entirely one of awe but of some surprise as well. He had expected to find an applied research establishment; instead he noted how closely its atmosphere approached that of a university. To Anton Hales, his successor at the BPI, he wrote: 'The Harwell establishment as you can imagine is a somewhat frightening organisation to enter with its enormous staff and varieties of activities but everyone has been extremely kind and I'm beginning to know my way around' [5]. His introduction to the scientific staff was handled in a typically low-key manner by Cockcroft. No formality, no ceremony, just an unannounced arrival in a laboratory by the Director and his recently-appointed Deputy, followed by a few words from Cockcroft and hand-shakes all round. Schonland's first opportunity to introduce himself rather more generally came at Christmas when the *Harlequin*, the 'leisure and social magazine' of the Establishment, offered him its foreword to do so. He wrote [6]:

> Any newcomer to Harwell is struck by the amazing range and
> complexity of its activities. That it all works smoothly and
> effectively is probably due to some secret discovery in scientific and
> human relations. No doubt the kindly goodwill of the British people
> and that now declassified invention of theirs, the Committee
> System, play some part in the mechanism, but these alone would
> not fully explain it.

That it did seemingly function so well owed as much to Cockcroft's mystique as to his reputation and also, of course, to the quality of those whom he had attracted to work with him. But the cracks that were now beginning to appear in the edifice could be traced to that other Cockcroft characteristic: his almost wanton disregard for the formalities of management upon which the smooth running of any large and complex organization so depends. Schonland's appointment was recognition, at least on Cockcroft's part, that the day-to-day running of his Establishment was beginning to pass him by and he knew of no better man than Basil Schonland to sort it out.

The Cockcrofts and the Schonlands lived next door to each other. The Director and his wife Elizabeth occupied what had been the RAF Station Commander's house, while his deputy and Ismay were comfortably accommodated next door. The two men frequently walked together across the grounds to their offices that faced each other across a corridor in the administration building. Schonland's was decidedly spacious, having been rapidly converted from the three that had previously served the needs of clerical staff into one commodious office for the Deputy Director. Once he had settled in, he began the process of familiarizing himself with each and every detail of the disparate activities then under way across the length and breadth of the establishment. Just finding his way around initially proved troublesome because one former runway that had become a road looked just like another. To add to the confusion, he discovered that the many buildings scattered around this former aerodrome carried different numbers above the doors that served each end. To the readers of the *Harlequin* he remarked wryly that he presumed it all had a purpose and that one day it would be explained 'for the benefit of future historians'. To assist him in his navigation the roads were obligingly given appropriate names such as Fermi Avenue, Curie Avenue and so on! [7].

One of Schonland's first responsibilities, and one which Cockcroft must have gladly deputed to him, was to organize and host the many visits to Harwell by dignitaries. In December the dignitary was none other than the Prime Minister himself and, on the 17th, Sir Winston Churchill, accompanied by Lord Cherwell, arrived and was escorted through selected laboratories and around various exhibits by Schonland, with Cockcroft quietly bringing up the rear. That visit followed a special

Figure 24. The visit of the Prime Minister, Sir Winston Churchill, and Lord Cherwell to Harwell on 17 December 1954. From left to right: Dr H Seligman, R F Jackson, Dr B F J Schonland, H Arnold (security officer), Churchill (seated), L F Potter (deputy security officer), the PM's bodyguard, Cherwell (seated), Sir John Cockcroft, D R Willson, Mr Christopher Soames MP, Sir Edwin Plowden. (Reproduced by kind permission of the UKAEA.)

briefing in September given to Churchill by Cockcroft, Penny and Sir Edwin (later Lord) Plowden, the first chairman of the Atomic Energy Authority, on the military aspects of atomic energy. Two months later, in the twilight of his career, Churchill approved the decision that Britain should develop the hydrogen bomb and, soon after, in April the following year, he retired from active political life, ending a career of stunning importance to the British nation.

December 1954 had seen the publication of the first annual report of the new Atomic Energy Authority and Schonland played a part in preparing the first draft. When it was sent to Plowden for his comments he indicated that its introductory chapter—the crucial one in his view—was not quite what he had in mind and asked for certain changes to be made. Schonland, with echoes resounding of the AORG and his insistence on the need to 'know your reader and what he requires', readily took it upon himself to rewrite it in a form more suitable for ministerial consumption [8]. Plowden must have approved for it was soon published and Schonland's literary stamp thus appeared on the document that was to help set in train the future of atomic energy in Britain.

That stimulus had first come from what was known as the Trend Report, the work of a committee under Burke (later Lord) Trend, a Treasury official soon to become Cabinet Secretary. Its data came from a study initiated by Hinton at Risley and, though cautious in tone, Trend saw nuclear power providing at least a bulwark against spiralling oil prices and diminishing coal reserves. However, the government seized the opportunity to trumpet in no uncertain terms the potential of this exciting new source of energy in a white paper published in February 1955 [9]. Its tone was intentionally optimistic—some said unrealistically so—but its purpose was clear. Britain was totally dependent for its energy needs on its finite resources of coal, calculated to be sufficient for only another 200 years. The prospect of nuclear power was, therefore, a bonanza waiting to be tapped and to some of the country's politicians it represented an opportunity for publicizing good news that must not be squandered. Twelve nuclear power stations were to be built over the next decade with a total capacity of almost 2000 MW and at a capital cost of some £300 million. Initially, the technology would be that of the Calder Hall station then under construction but this would be superseded by liquid-cooled thermal reactors then being developed at Harwell and Risley. Finally, the newest technology of all, the so-called fast breeder reactor, pioneered at Harwell by ZEPHYR,* was to be exploited. In this case, a core of enriched uranium containing about 46 per cent ^{235}U is

---

* ZEPHYR, the zero energy fast reactor, was designed at Harwell and a full-scale version was constructed at Risley and went critical in February 1954. Construction of the Dounreay fast reactor in Caithness began in March 1955.

encased in a 'blanket' of natural uranium, ^{238}U. Since no moderator is used, the fast neutrons convert the ^{238}U to plutonium at a rate greater than that at which the ^{235}U in the core is consumed. The reactor thus produces more fissile material than it consumes and was called the fast breeder reactor. With such a combination of technologies at hand it was believed, at least at Government level, that Britain would have built up an unassailable lead in the field of nuclear power engineering and so could exploit the world markets for this expertise and reactor hardware. Quite naturally industry had been invited to participate and four consortia were established based around the major electrical engineering companies in the country at the time (AEI, English-Electric, GEC and Parsons/Reyrolle), with each being augmented by companies with the necessary mechanical and civil engineering expertise. The white paper served another useful purpose. That same month had seen the publication of another announcing that Britain was to develop the H-bomb. Its implications were considerably more doleful and public attention (and criticism) needed to be deflected by every possible means: what better way than by a positive nuclear story to counteract a negative one? [10]

Harwell's scientists and engineers were much more cautious about the ultimate supremacy of nuclear power over that generated from the burning of fossil fuel. Brian Goodlet was quite clear about this when he wrote in 1953 that, while nuclear power was a certainty and economic nuclear power a possibility, cheap nuclear power seemed unlikely [11]. Not only was Goodlet's contribution to these early developments at Harwell very significant but, being a sensitive man, he had a great personal attachment to the project. When, in March 1953, the news broke that PIPPA was to be built at Risley and that all the engineers who had been engaged on the project at Harwell would be transferred there, he was dismayed. He foresaw this as 'the end of Harwell' if they did not build PIPPA themselves, for not only would it provide the plutonium for further research but it would also give Harwell's engineers considerable experience in the use of this newest of technologies for the generation of electrical power. PIPPA, he stressed, 'was absolutely central to Harwell and its future'. Others, though, like D R Willson, the establishment's secretary, took a much broader and rather less possessive view of the project and warned instead that Harwell would be finished if it tried to build everything [12].

By the mid-fifties it was apparent that Harwell's honeymoon was over. Cockcroft's ideal of unfettered research, with up to 30 per cent of the establishment's effort devoted to work of a fundamental nature was clearly unsustainable in the light of Treasury demands that funds be cut for any government-supported research unlikely to be 'financially productive' in what was deemed to be a relatively short period of time. In addition,

Cockcroft felt his establishment being pulled in three directions at once. There was the obvious need to support the nuclear power programme; there was also pressure to diversify well beyond that, into other fields entirely; and then there were those who saw Harwell as the nucleus of a British technological university [13]. The first was vital if nuclear power was to have any future in Britain at all; the second was diversionary and unlikely to command much support from Parliament; the third, though closest to Cockcroft's heart, would serve only to incur the wrath of Britain's existing universities, some of whom believed that Harwell was already receiving a disproportionately large share of the limited funds made available for research by the government [14]. In the view of Sir Edwin Plowden, to whom Cockcroft now had to report, the Director's 'indulgence towards long-term research [made him] a "soft touch to his younger colleagues"' who clearly found the atmosphere at Harwell and the attitude of its Director very amenable indeed. But Plowden had other ideas. When confronted by the multiplicity of schemes involving new research reactors such as LIDO, DIDO and PLUTO, all of which had their respective champions among Cockcroft's colleagues, he appointed the Authority's Member for External Relations and Commercial Policy, Mr William Strath, to carry out an assessment of the projects that would be most beneficial in the long term. Since Strath was in no position on his own to make recommendations on a scientific basis, all he could do was to rely on Cockcroft's guidance and then try to indicate the order of priorities as he saw them. The two men duly met in November 1955 and Cockcroft, for all his reputed inability to engage in argument and discussion, managed yet again to persuade his inquisitor that Harwell should continue to work on new reactor designs while, in addition, a new site should be established elsewhere for the study of 'advanced reactor concepts' [15]. That proposal was eventually to lead to the setting up of a new site at Winfrith Heath in Dorset, which will be described later. The fracturing of Harwell was therefore under way.

While Strath and Cockcroft discussed matters affecting Harwell's future policy and direction, Basil Schonland suddenly found himself pitched into an event of international significance. Britain and Russia had agreed to lift, if only by a corner, the veil of secrecy that covered their nuclear research activities. The first visit ever by a party of British nuclear scientists to Russia took place late in November, when Schonland led eight of his AEA colleagues on a six day visit to Russia [16]. What had prompted this sudden parting of the Iron Curtain that allowed this delegation from Harwell and Risley to meet some of their Russian counterparts, and to see the research facilities where Russia's own atomic energy programme was evolving, was the highly successful conference on the Peaceful Uses of Atomic Energy held just two months before in Geneva, under the auspices of the United Nations.

An idea first proposed by Isidore Rabi, the chairman of the General Advisory Committee of the US Atomic Energy Commission, and whole-heartedly supported by Cockcroft who helped him organize it, the conference was a resounding success attracting to the city's Palais des Nations nearly 4000 delegates from around the world. The opportunity to display their expertise in the field of nuclear power, and to advertise their reactor hardware before an international audience of prospective clients, overrode years of secrecy that had prevented even former allies from knowing what each other was doing. Schonland, in the company of his British colleagues, found the atmosphere truly electric when the Conference President, the Indian nuclear physicist Homi Bhabha,* dropped a bombshell when he discussed a topic that, until then, had been highly classified. Controlled thermonuclear fusion was well-known in theory but until then its name had never been breathed in public, such was the secrecy surrounding all the research then under way on both sides of that cold-war curtain. Now Bhabha chose to use this grand occasion in Geneva to predict that, within two decades, 'the energy problems of the world will truly have been solved for ever, for the fuel will be as plentiful as the heavy hydrogen in the oceans' [17]. Bhabha's dramatic announcement and the obvious progress made by the British, the Americans and the Russians in developing the technology for nuclear power indicated the futility of keeping secret any longer the possibilities of the most exciting source of energy known to man. In the almost euphoric mood that followed, not only were proposals made to declassify aspects of thermonuclear research, but it was evident that the Russians wished to improve their relations with the West, especially in the fields of science and technology, and an invitation was extended to John Cockcroft, particularly, to pay them a visit to view some of the Russian work in progress. The seed had already been sown during the conference, when Britain provided the air transport to bring to Harwell, on a one-day visit, representatives of all the countries in attendance. Russian curiosity to see what lay behind Harwell's perimeter fence and Britain's obvious willingness to display some of it surely prompted the reciprocal invitation that, in November, saw a party of British scientists boarding an aircraft at London Airport for the flight to Moscow. Their leader, though, was not Cockcroft but Schonland, for other more pressing commitments in England prevented the Director from making the trip and so he delegated the responsibility to his deputy.

That Schonland was able to enter Russia at all would have raised some eyebrows in South Africa, because Schonland was a South African and relations between Dr Malan's government and the Soviet Union were just about to be severed. But Basil Schonland travelled to Russia

---

* Bhabha himself had been at the Cavendish from 1927 until 1939.

Figure 25. Schonland and five members of his British delegation about to board an aircraft at London Airport for Russia in November 1955. From top to bottom: Dr A A Smales, D W Fry, Dr J V Dunworth, B L Goodlet, Dr Willis Jackson, Dr B F J Schonland. (Reproduced by kind permission of the British Library.)

not on a South African passport but on a British one, having relinquished his South African citizenship within just a few months of his arrival in England. When questioned about this some years later he refused to discuss it, saying simply that it was a personal matter. Whether personal or not, the sensitivity of the position he occupied at Harwell must have had everything to do with it. After the arrest of Klaus Fuchs and the defection of Bruno Pontecorvo, there was an almost hysterical reaction within certain quarters in the United States, and British security practices were subjected to very close scrutiny. This led to a general tightening up of procedures and anyone who was not British-born of British parents, and who was likely to have access to secret material, had their cases considered personally by the Permanent Secretary and the Minister responsible [18]. Schonland fell very much into both categories, but the speed with which his application for citizenship was processed (the usual requirement at the time being twelve months' residence in England for citizens of the Commonwealth) suggests that significant wheels had been set in motion as soon as had accepted Cockcroft's offer to join him at Harwell. And indeed they had, for within just months of arriving in England he was working in the closest possible collaboration with the most sensitive of all British nuclear installations: Penney's Atomic Weapons Research Establishment (AWRE) at Aldermaston. The AWRE was in urgent need of a wide range of nuclear data, especially in the critical region required for megaton weapons, and, although some classified US information was still reaching Penney through the usual intelligence channels, he knew that such sensitive data would not be supplied. The AWRE would, with Harwell's assistance, have to conduct the research themselves [19]. To assist in this Schonland set up, under his chairmanship, what was called the Ψ (Psi) Committee, a joint grouping of selected scientists from AERE and AWRE whose purpose was 'to review the physics research concerned with thermonuclear weapons (T.N.W.)' within the two establishments. Its first meeting took place in Schonland's office at Harwell on 20 December 1955 soon after his return from Russia [20].

The Harwell delegation that left London under Schonland's leadership consisted of Dr A A Smales, head of the Analytical Chemistry Group, Mr D W Fry, head of the General Physics Division, Dr J V Dunworth, director of the Reactor Physics Division, Mr B L Goodlet, head of the Engineering Research and Development Division, all from Harwell, as well as Mr J C C Stewart and Dr L Rotheram, from Risley, and Dr Willis Jackson* who represented the Royal Society and the recently formed British Nuclear Energy Conference [21]. Their hosts were the Soviet Academy of Sciences who spared nothing in their efforts to welcome their British visitors. A tour of an atomic power station under construction was the focal point of the visit, at which Schonland was able to return the favour by presenting his hosts with a film showing some of the equipment being installed in Britain's first such station soon to be opened at Calder Hall. Then followed carefully choreographed visits to the Lebedev Institute and the Academy's own Institute of Geochemistry, when informal contact between scientists was enthusiastically encouraged. The Russians clearly intended to capitalize on the mood of increased international scientific collaboration that had sprung from the Geneva conference, but Schonland's party were under no illusions about what they would really be allowed to see and hear.

Their very presence in Moscow was an opportunity to be exploited to maximum effect, while their own curiosity about Russian scientific progress would be reined in by a blanket of security. Throughout the

---

* Subsequently Baron Jackson of Burnley.

six days of the visit Schonland felt that their every move was monitored and their every word probably recorded. Great vigilance was necessary and he took it upon himself to brief his colleagues in detail about what topics they might discuss and even the types of questions they might ask. Of particular concern to him, prompted no doubt by the briefings he would have received from Harwell's security officer, was the possibility that the Russians might, for whatever reason, use the opportunity to confront the British delegation with their former Harwell colleague, the defector Bruno Pontecorvo. Schonland therefore studied numerous photographs of Pontecorvo and then circulated them amongst the group so that everyone would be sure to recognize him should he suddenly appear. Under no circumstances, he told them, would he allow any contact between them and the man who had disappeared so mysteriously whilst on a family holiday in Italy just a few years before. All were therefore to keep a sharp lookout for Pontecorvo wherever they went.

The visit turned out to be as intriguing as it was fascinating. As expected, Russian security was both tight and all-embracing. A walk through a Moscow park, ostensibly for a breath of fresh air but actually to allow the British scientists to converse away from hidden microphones, soon attracted the attention of one or two men whose interest in them, though superficially friendly, seemed to extend well beyond the pleasures of an afternoon stroll. Quite naturally some members of the party found the experience unnerving and one even went to the trouble of wedging his bedroom door shut each night in case of unwelcome visitors! But, for all their slightly paranoid precautions, Schonland's concerns about Pontecorvo seemed well-founded. Whilst visiting a particular laboratory, an ever vigilant member of his party caught sight of a man who bore an uncanny resemblance to their erstwhile colleague and Schonland was alerted immediately. Very deliberately the Deputy Director of Harwell marshalled his colleagues and led the British party out of the laboratory in question. No diplomatic incident followed, for their hosts seemed almost to expect this and, after a brief flurry of activity, invited their British guests to return to a room from which 'Pontecorvo' had disappeared! [22].

After this ground-breaking British trip to Russia with all its cloak-and-dagger interludes, the reciprocal visit to Britain by the Russians in April the following year was a very different affair. It was conducted with all the ceremony of state, since the Russian scientific delegation were but members of the party led by the Premier of the USSR, Nicolai Bulganin, who was accompanied by the first secretary of the Communist Party, Nikita Khruschev. Leading the Russian scientists was Dr Igor Kurchatov, rumoured to be the father of the Russian atomic bomb and one of his ports of call was to be Harwell. To return the favour there for the hospitality he and his colleagues had received just a year before was Basil Schonland. But this time, or at least for part of it, Sir John Cockcroft was to be in attendance and once again the inimitable way in which Cockcroft won the confidence of like-minded men came to the fore. His unassuming manner, owlish appearance and seeming ability to communicate by mere physical presence rather than vocal effort established an immediate rapport with the Russian. They met first at the Athenaeum in London, where Cockcroft hosted a lunch for the visiting scientific delegation. After an animated discussion on the staircase about the Russian reactor programme, Cockcroft was delighted by his visitor's offer to deliver a lecture at Harwell [23]. Such apparent openness from the Russians was truly remarkable at a time when Britain knew next to nothing of what their closest allies, the Americans, were doing and it was an occasion not to be missed. But Cockcroft would miss it because he had an important meeting in the United States where the very subject of removing the secrecy surrounding thermonuclear research was the main point on the agenda. His absence from the lecture and the events that followed it would soon have profound consequences for both himself and for Harwell.

*   *   *

As early as 1945 there was evidence to show that a new energy source existed that was based on the thermonuclear fusion of deuterium (so-called heavy hydrogen) ions. In principle, if either deuterium (^{2}H) alone, or deuterium and tritium (^{3}H), were heated sufficiently to cause them to be completely ionized, and so become a plasma, then the massively energetic collisions between the ions would be sufficient to overcome the repulsive Coulomb force between them and so lead to fusion of the deuterium, or of the deuterium and the tritium. Hence, in accordance with Einstein's relationship between mass and energy, the resulting change of mass that had now occurred must be accompanied by the release of energy as well as the generation of protons and neutrons — the latter being of especial significance, since they would be unaffected by the confining effect of any magnetic field and so would readily escape, thus providing an indication of the processes at work. Since deuterium occurs naturally in sea water the point is often made that the energy released by fusing the deuterium in just one litre of water is a hundred times that produced by burning a litre of petrol. In addition, and so different from nuclear fission, no harmful waste products are created in the process. Thus the promise of an almost limitless supply of energy was the glittering prize: the difficulty on the way to winning it was how to contain within some suitable reactor a plasma at a temperature in excess of 10^8 K, much greater even than that of the sun itself. By 1956, various methods of doing this were under investigation and all involved the use of suitably configured magnetic fields to trap the reacting plasma within so-called magnetic walls for long enough to allow fusion to occur. The problem, however, was far from simple and its history is most relevant to the saga about to unfold at Harwell.

By the end of the war the fission mechanism was not only well understood but had been used to devastating effect against two cities in Japan. Then, as peace returned, its potential as an energy source was exploited, as we have already seen. By contrast, the opposite process of fusing together two light nuclei, with the accompanying release of even greater amounts of energy, was still only a theoretical concept, even though it had been recognized at least twenty years earlier as the prime source of energy in the sun and the stars. But neither the fundamental difficulty of producing the incredibly high plasma temperatures nor that of containing the reaction in some suitable way had yet been solved. Then, the advent of the atomic bomb provided a solution to the former, while rendering lengthy containment quite unnecessary if the requirement was an even more devastating explosion. Nuclear fission could be used to produce nuclear fusion. Once this had been recognized it was only a matter of time before the first fission-initiated fusion bomb was detonated by the United States. It took place on the Pacific atoll of Eniwetok under the code-name of 'Ivy Mike' on 1 November 1952. The so-called hydrogen or H-bomb had arrived. Within less than a year the Russians had detonated theirs, while British efforts to do the same were being actively pursued under Sir William Penney at Aldermaston, but at the time of the Russians' visit to Harwell that first British thermo-nuclear explosion was still a year away. Obviously, neither group of scientists breathed a word about bombs, but the possibility of generating electrical power by thermonuclear means was a subject of enormous interest. However, it was the physics of the process and not just the ultimate applications that held Kurchatov's audience in fascinated and rapt attention on 25 April 1956.

Academician Kurchatov spoke in the Cockcroft Hall at Harwell and his words were carefully translated by an interpreter provided by the Foreign Office. The title of his address was 'On the possibility of producing thermonuclear reactions in a gas discharge'. It was an occasion of quite stunning significance and not a little irony; indeed it truly overflowed with both irony and contradiction. For here, in the very heart of Harwell, and enmeshed in its security, was the leader of the Russian atomic research programme speaking on a topic that was almost unmentionable in any conclave between the two closest allies in the western world. 'Thermonuclear reactions' were words that just must not be taken in vain! In the audience were scientists from Harwell with a predominant interest in the fundamental science and the possible exploitation of nuclear fusion for the generation of electric power, while alongside them were their colleagues from the Atomic Weapons Research Establishment at Aldermaston, whose sole interests in the matter were the possible military applications of thermonuclear fusion. There had long between tension between them, for many at Harwell strongly favoured declassification, while Aldermaston, with the support of London, insisted on its rigorous retention. But not present to hear their Russian visitor was the man who, on one hand, believed so passionately in the free exchange of the fruits of scientific progress but, on the other, feared that Britain might lose her presumed position of eminence if too much was disclosed too soon. That man, Sir John Cockcroft, had left just the day before for Washington to try to induce the Americans to declassify thermonuclear research.

Regardless of the motives of the Russians in allowing Kurchatov to speak, whether magnanimous, duplicitous or whatever, there was no relaxation in the absolute rigidity of Harwell's policy: they would listen but their talk would be resolutely guarded. To ensure that this was so, Schonland, once again responsible for all the arrangements of the visit, called a special meeting beforehand of all those attending Kurchatov's

Figure 26. Schonland with the Soviet Academician Dr Igor Kurchatov at Harwell in April 1956 prior to Kurchatov's remarkable lecture in which he revealed details of the Russian research on thermonuclear reactions in a gas discharge. (Reproduced by kind permission of Dr Mary Davidson.)

480

lecture. Just as he had done in Russia the year before so he insisted that nothing must be said during the subsequent discussion from which any details of the Harwell research might even be inferred by their Soviet guests, and a detailed list of such topics was issued to everyone [24]. Whether Schonland succeeded in this somewhat sublime expectation will probably never be known, but Kurchatov's lecture, certainly, was enthralling. To Dr R S Pease, one of the Harwell team of physicists who sat and listened while at times almost pinching themselves in disbelief that such an event was actually happening, it was a lecture the likes of which would have been unimaginable but a short time before. It was about the quest for one of the great research prizes of its time: the creation of thermonuclear fusion in high current discharges in deuterium and it was being delivered at Harwell by a Russian who was no recent émigré, political fugitive or defector but the director of the Atomic Energy Institute of the Academy of Sciences and a deputy to the Supreme Soviet of the USSR! [25].

Kurchatov described how currents of up to $2\,MA^*$, with durations of about 10 µs had been produced in straight gas discharge tubes. Accompanying the current were bursts of neutrons, which initially were thought to indicate that thermonuclear fusion had occurred. However, as Kurchatov explained, it was soon realized that the neutron reaction rates, their dependence on the current and other factors showed that this hypothesis was flawed, and that other explanations had to be sought in order to identify the actual source of those neutrons [26]. This point was subtle but absolutely crucial to the successful generation of a thermonuclear reaction and Kurchatov's frankness in admitting it was indeed a salutary lesson in the interpretation of experimental results. But it was not one that was grasped immediately by all who heard it.

Once the Russian party had left Harwell, Schonland convened a special de-briefing session for all the scientists who had attended the lecture. Its original purpose was to consider whether any disclosures possibly made by Kurchatov might affect the work already under way at Harwell. As it turned out it served a rather more immediate purpose for it became clear that Schonland, amongst others, had not grasped the crucial point that the neutrons detected by the Russians were not, in themselves, indicative of a thermonuclear reaction [27]. Once that vitally important point had been clarified there was general consensus that little had been said that was not already well known. The straight gas discharge tubes used by the Russians, with electrodes at either end to initiate the process, could never be scaled up in size to yield a useful source of power because those electrodes are always in contact with the gas and so would conduct away the heat that is fundamental to the

* MA mega amperes (10^6 A).

whole process. By contrast, if the ionized gas were contained within a toroidal configuration, as used at Harwell, and so behaved like the secondary winding of a transformer, it could, if suitably energized, reach the required temperature with no electrode conduction losses at all. Since research in Britain had concentrated on the toroidal geometry for the better part of a decade, the lecture would, in that respect at least, prompt no dramatic re-appraisal of Harwell's research programme. What it did emphasize so emphatically was that the generation of neutrons alone was not proof of a thermonuclear reaction, because they appeared at a temperature well below that at which the necessary thermal agitation of the nucleus takes place. A significant neutron flux occurs at temperatures of about 10^6 K, whereas very little power is produced until the temperature is at least an order of magnitude higher than that.

When the reports, released by Harwell, of the Kurchatov lecture appeared subsequently in *The Times* they were sober and accurate [28]. No great break-through was reported, for none had been achieved, except possibly in the field of personal relations between the scientists at Harwell and their counterparts beyond the Iron Curtain. The absence, though, of Sir John Cockcroft from both the lecture and the de-briefing that followed it was doubly unfortunate, particularly in the light of events soon to follow.

*   *   *

Though Schonland had left his native shores and, according to his passport at least, his alliances were now firmly rooted in Britain, his name and reputation were still engraved on many a South African memory and his advice and good counsel were much sought after in the land of his birth.

Early in December 1955 he received a letter from Frank Nabarro, now installed as Head of the Department of Physics at Wits, informing him that a chair in the field of nuclear physics, to be subvented by the South African mining industry, had just been established. This was indeed welcome news. Nabarro was seeking Schonland's help in attracting someone of appropriate calibre to the post and he included a list of likely candidates as well as his own comments about their suitability. Schonland's views, he said, would be regarded most highly when the time came to make an appointment. Amongst those names was that of the 25-year-old Friedel Sellschop, then just completing a PhD at the Cavendish, whom Schonland remembered well as a young man of considerable potential who had spent two years at the BPI after finishing his MSc at Stellenbosch University in 1952. It had always been Sellschop's intention to do his PhD at Cambridge and then to follow this with some postdoctoral work in the United States. He had seen the post at Wits advertised on the notice board in St John's College, Cambridge but had done nothing about it. But what he did not know at the time was that he was soon to have a powerful ally who would mount a vigorous case in support of his candidature.

Schonland wrote back immediately to Nabarro with the comment that Sellschop was far better than anyone else on that speculative list and that he would support him were Sellschop to apply. This was a bold move and it would mark the start of a protracted correspondence between 'Nab' and his former boss at the AORG on the matter of the professor of nuclear physics at Wits and on the very viability of such research in South Africa.

It was soon abundantly clear to both of them that Sellschop must be encouraged to apply for the chair and, in view of his relative youth and lack of experience, the University must be assuaged of any doubts it may have had about his competence to fill it. Nabarro believed that Schonland was the only person who could do both and, to start the ball rolling, he prevailed upon him to write to Sellschop at Cambridge asking him to consider applying for the chair. Schonland duly wrote, but rather obliquely [29]:

> I have had a note from someone in the University of the
> Witwatersrand suggesting you may be too modest to apply for the
> Chair of Nuclear Physics there. I hope that this is not the case and
> that you will try your luck with it for although your experience is
> perhaps not very great you would have a good chance to build up
> an interesting department there and I certainly would support your
> claims for consideration.

That letter was followed soon after by an invitation to Sellschop to visit Harwell on its next Open Day and, whilst there, to accept the Schonlands' hospitality at their home on South Drive. Sellschop accepted both with much enthusiasm, though aware that this was no social visit, for the invitation almost had the ring of a Royal Command behind it! On arrival, he soon discovered that Schonland intended to do all in his power to assist Wits and South Africa to establish themselves in the field of nuclear physics, but both depended upon the right man being appointed to the newly endowed chair. Schonland was in no doubt that Sellschop was that man even if it meant foregoing his own postdoctoral ambitions, at least for the immediate future. With such advocacy behind him Sellschop felt greatly encouraged and submitted his application to Wits for their consideration and cited Dr B F J Schonland CBE FRS as a referee [30].

Once notified that Sellschop's application had indeed arrived, Nabarro set out in yet another letter to Schonland his concerns about how to convince the university in which he himself was but a recent incumbent in a chair—and a none too comfortable one at that—that they should even contemplate making a senior appointment to someone about whom they knew so little. But he believed that Schonland's influence at Wits would carry enormous weight and so prevailed upon him to communicate with the university, reminding him in characteristically forthright manner that: 'It will obviously take a very firm recommendation on your part if we are to appoint someone who has nothing so far published and is still a PhD student in nuclear physics'.

Indeed it would, but Nabarro, like Sellschop had great faith in the power that Basil Schonland could still command on the Witwatersrand and so, on 29 May 1956, Schonland wrote to Professor I D Macrone, the Acting Principal at Wits, in support of Sellschop's application. He backed up his own strong recommendation with the comments of the director of the Cavendish, none other than Professor N F Mott FRS. Mott and Schonland, though sometimes at cross purposes during Schonland's early days at Petersham, had evidently reconciled whatever differences they may have had and were now occasional colleagues at the Royal Society. Mott confirmed that Sellschop was indeed a man of undoubted potential, whose work at the Cavendish had been of an experimental nature investigating 'some deuteron reactions' that had proved to be quite demanding. In tackling this he had left his supervisor Dr Shire in no doubt that he was a man of ability and drive. Wits clearly needed both and Schonland endeavoured, in closing, to convince Macrone that Sellschop, for all his youth and inexperience, was up to the task. 'In spite of his lack of published material', he wrote, 'I am sure that Sellschop should be considered as a very good candidate for the chair and directorship of nuclear physics which the University is considering. He is a young man of very great promise indeed' [31].

But Macrone, the University Council and its Senate all wavered. For all Schonland's support the fact remained that, were Sellschop to be appointed to the chair he would be the youngest professor at the University, and so they held back from making the ultimate appointment. Instead they offered him the directorship of the Nuclear Physics Research Unit for four years and announced that the chair would be filled at the end of that period.* Such reluctance was quite understandable in the circumstances: anything else might have been seen as the height of academic folly by the mining industry whose own overlords, after all, were no mere youths; and Wits, by dint of what may have appeared a rash appointment, could certainly not risk losing the imprimatur of the

---

* Friedel Sellschop was duly appointed to the Chair of Nuclear Physics in 1959 becoming the youngest professor then in the University and the second youngest in its history. Only the remarkable J H ('Hennie') Hofmeyr, Ismay Schonland's distant cousin who was appointed Professor of Classics at 22, after a double first at Oxford, had been younger. The fact that Hofmeyr went on to become Principal of the University at 27, Administrator of the Transvaal two years later and both Cabinet Minister and Acting Prime Minister during Smuts's frequent absences abroad during the war established an almost insuperable standard.

country's largest industry and the university's greatest, though sometimes unpredictable, benefactor.

*   *   *

Schonland, for all the assistance he was giving to Wits to attract a man of distinction, once again found himself being pressed to reconsider an academic career. In December 1955, following the sudden death of its Vice-Chancellor Dr T B Davie, the Council of the University of Cape Town passed a resolution in terms of which the 'Heads of Universities' in the British Commonwealth, the United States and Holland, as well as a number of 'other individuals', were notified of the vacant post of Principal and Vice-Chancellor and were invited to mention the post to anyone whom they believed might be suitable for the position. The closing date for applications was 30 April the following year and members of Council were themselves asked to submit nominations. When it met four months later, the special committee appointed to deal with the matter found itself with a number of formal applications, as well as a list of names that had come in from here and there. In all, there were thirty six, and it was agreed to give due consideration to each of them. Soon the list was whittled down to eight, and amongst them* was the name of B F J Schonland, who had been suggested by various people within the broad constituency, though Schonland himself had not been consulted by anyone! Those names then went to the University's Senate with a request that a secret ballot be held to indicate their acceptability and, if possible, to provide an order of preference. The outcome was a recommendation that the post of Principal and Vice-Chancellor should be offered to Dr Schonland and a formal letter to this effect was dispatched to England.

Whether Schonland had got wind of these rather convoluted proceedings is not certain, but his reaction on receiving the letter was not one of outright rejection, for all the commitment he had just made to Harwell and to Cockcroft. Ismay, despite her reservations about South Africa's future, was prepared to support her husband in whatever decision he made, but Schonland himself was very uncertain and sought further views. There were no colleagues he felt he could turn to and John Cockcroft was hardly the man to consult in this instance. It so happened that Don Craib, Ismay's brother and Basil's old room-mate at Caius in 1919, was in England at the time and so Schonland approached him for his reaction. Craib, whose own career as a pioneering electrocardiologist and then as Professor of Medicine at Wits had been tempestuous, to say the least, was characteristically forthright. To leave Harwell, one of the

---

* The eight names on the short list were: T Alty, C W de Kiewiet, J P Duminy, J M Hyslop, L Marquard, Sir Arnold Plant, B F J Schonland and F G Young.

most prestigious institutions of its kind in the world, for the principalship of 'a little university in an unknown country like South Africa' would, he stated in typically unabashed style, be 'a very big come-down'! [32]. That was as about as direct an answer as Basil would get from anyone and, though couched in terms that the Schonlands would hardly have used, it was a blunt, though not entirely inaccurate, summary of the situation. Further deliberation soon convinced both Basil and Ismay that it would indeed be a retrograde step in his career were he to return to South Africa and so he cabled Cape Town that he was unable to accept the offer of the principalship [33].

The reaction at UCT was one of great regret. The university, along with all its English-speaking counterparts in the country, saw itself coming under increasing government pressure to restrict its entry only to white students. Though no legislation to this effect was yet on the statute books (it would come in 1959), there were already ominous sounds coming from various quarters that it was but a matter of time before this happened. Seeking strength in unity, Wits and UCT coordinated their various actions in attempting to dissuade the government from imposing university apartheid, but the Nationalists' assault on the country's English-language universities was unrestrained. It was now vital that they should receive strong and enlightened leadership from their Vice-Chancellors and it was with this very much in their minds that the UCT Senate and its Council offered the senior academic and administrative post in the university to Basil Schonland — a man acknowledged throughout South Africa as being blessed in abundance with such attributes. On receiving Schonland's telegram declining the invitation the chairman of the University Council replied immediately: [34]

> It was with much regret that I read your decision even though I think I can see something of the reasons which prompted it. The university, and indeed all South Africa, will be the poorer for it.

It was clear though that Schonland would not come and much careful reflection was now required before offering this vital post to someone else. He continued:

> ... the present intention is not to offer the appointment to anyone else but to spend the next three months exploring new possibilities.... The post is so vital to the University and to South Africa that it is desperately necessary to get the right man.

*   *   *

On 17 October 1956, Her Majesty the Queen conducted the opening ceremony and threw the switch that brought into operation the world's first atomic power station at Calder Hall and a justifiable feeling of pride swept across the British nation. Whereas the Americans had developed an experimental breeder reactor in 1951 and the Russians had operated their first experimental power station three years later, Calder Hall was certainly the world's first large-scale reactor, intended to deliver some 90 MW of power to the national electricity grid. Its other purpose was secret for it would join the two piles nearby at Windscale in producing the precious plutonium for Britain's nuclear deterrent. About the former, Sir Edwin Plowden wrote that 'a new fuel and a new source of power is put to the service of mankind'; about the latter he, naturally, said nothing [35].

Other nations too were soon to embark upon major developments in atomic energy and, not surprisingly given Homi Bhabha's eminence in the field, India was prominent amongst them. Since Britain had visions of exporting reactors to the developing world, developments elsewhere were watched very closely indeed both within the AEA and at Harwell. On 20 January 1957 India was to inaugurate her own Atomic Energy Establishment overlooking the harbour of Bombay, with the commissioning of a so-called 'swimming-pool reactor' based on the Harwell-designed LIDO. The Commonwealth Relations Office were particularly keen to see the United Kingdom represented at an appropriate scientific level and not by a diplomat. Cockcroft was naturally the man to go but, as President of the European Atomic Energy Society and a member of the United Nations Scientific Advisory Committee, his life had virtually been predetermined for him by a calendar of meetings and official gatherings, so Schonland was once again to be Britain's representative. Diplomatic niceties required that he should visit Pakistan as well and he duly fulfilled both these engagements with all the aplomb that came naturally to a man so schooled in the ways of the British Army, whose traditions were the very buttress of almost everything on the sub-continent. If Schonland had ever wondered what his particular function might be at Harwell — since Cockcroft was not in the habit of making such things too clear — it soon became apparent that he would be called on increasingly to represent Harwell's interests abroad. In March he was due to go to the United States to visit a number of research laboratories, universities and government establishments and then in July there was a trip planned to Africa, where Britain's rather more fundamental needs, in the important matter of uranium, could well be met.

As it turned out, he had to postpone the American visit, much to the consternation of the British Embassy in Washington, that had been working very hard to increase the level of contact between the two countries in the commercial nuclear field. Schonland was taken aback by this and wrote an apologetic letter to the Embassy explaining that 'he did not think of himself as sufficiently important for anybody to bother'! He then pointed out that there was a particular need for him to visit Africa on matters to do with the supply of uranium while, in addition, his presence was also required in England where 'things were just about to boil'. The things in question were a new National Institute for Nuclear Research and the opening up of a 'second Harwell' at Winfrith Heath in Dorset which required 'some pacification of the natives' both within and without the boundaries of Harwell itself. His American visit would have to follow in September [36].

In June the Schonlands left London for Africa. Ismay flew all the way south to join her family in the Cape while Basil headed for the Rhodesian Federation, paying a brief call on the way to East Africa. His purpose was to explore all possible sources of uranium and to promote some research in the Rhodesias into detecting any reserves of radioactive material that may have been swept down and then deposited upon the banks of the great rivers that drain the region. He arrived in South Africa on the 25th. It was his first trip to the land of his birth since taking up the Harwell appointment and the newspapers announced his home-coming with general acclamation. Others had been expecting him as well. The very next day he was honoured by the University of the Witwatersrand with the degree of Doctor of Science *honoris causa*. By comparison with the orations delivered on previous such occasions, that from Wits was rather bland. At UCT, so soon after the end of the war, his exploits were still shrouded in secrecy but it was good enough for them that 'he was in the confidence of the military chiefs'. Cambridge's, a year later, was magniloquent in its Latin phrasing, counting him amongst the ornaments of the Royal Society. More recently, both Natal and Rhodes excelled themselves in their laudations. Schonland's contributions to South Africa were seen in Durban as leading that university, and presumably the country, 'into the new era of light and promise' while, in Grahamstown, the Public Orator regaled those present with an account of their distinguished alumnus's 'thrilling contribution' to science and his prowess as a soldier. Wits, in stark contrast, provided a long and detailed catalogue of their most illustrious former professor's accomplishments with little real evidence of soul, theirs or his. But Basil Schonland himself told 'The Man on the Reef', who interviewed him for Johannesburg's evening newspaper, that of all the honours he had received this was probably the one he most appreciated [37]. For reasons of its links with lightning and the BPI one can well believe that to be true.

As well as being one of the world's major sources of uranium oxide, South Africa had embarked upon the production of heavy water in 1955 as a by-product of her endeavours to produce oil from coal. The South African Coal, Oil and Gas Corporation (SASOL) was founded in 1950 and by 1955 was producing its first synthetic oil using local coal by means of the German Fischer–Tropsch process, first licensed to a South African mining company in 1935. Hydrogen and its isotope deuterium were generated in quantity during this operation and it was readily possible to separate them from which deuterium oxide, or heavy water, could then be produced. There was considerable interest in South Africa in exporting the heavy water along with its uranium and, equally, of using both in its own reactors then being considered. Schonland had been kept abreast of these developments by the President of the CSIR, Meiring Naudé. With this additional string added to its nuclear bow the South African government was keen to come to some agreement with Britain and exploratory contact between them was made in 1956. By the following year, a bilateral agreement on the supply of uranium was being negotiated and it contained an appendix that provided for collaboration in research into methods of producing heavy water. Before leaving for South Africa, Schonland wrote to Sir Edwin Plowden to inform him of the current position and of the likely arrival at Harwell of two South African scientists to participate in the collaboration. International sensitivities were never far from the surface in such matters and Britain's existing agreements with both France and Israel had to be protected, but Schonland had been advised that no difficulties were likely to occur by opening the door to South Africa as well [38].

Naudé also sought Schonland's advice on the issue of the CSIR's own nuclear physics section, but Schonland replied that it was too large a subject to deal with comprehensively by letter. His views though, however abbreviated, made clear a number of fundamental points for Naudé to consider. South Africa would have to decide what it intended to do in the field of nuclear energy. If there were larger objectives than merely being a supplier of uranium oxide then these should be clearly thought out and their cost implications carefully assessed. He compared the country with India and Australia, both of which had plans to become nuclear powers. In India's case her intentions were to join the league already occupied by the major powers whereas Australia's intentions were considerably more modest. Even so, the Australians had budgeted to spend a million pounds a year on running costs and 'a good few millions in capital charges'. He cautioned Naudé against any decision to venture in that direction unless South Africa was well-served by the necessary scientific and engineering manpower to staff the laboratories on an appropriate scale. Given the risks associated with an, as yet, untested technology he warned against making any precipitate decisions:

> The fact that a nation possesses uranium does not make it essential
> that the same nation should develop expensive facilities for the
> development of industrial applications of atomic energy, any more
> than the possession of coal requires a nation to produce coal fired
> boiler plants itself.

For all his reservations when advising the President of the CSIR, Schonland displayed none when talking more generally to 'The Man on the Reef'. His comments about the future of nuclear energy were most optimistic and certainly represented the view prevailing at Harwell at that time. By 1970, he said, a third of Britain's industrial power would be provided by nuclear power stations, 'while at night, the whole country's electrical supplies will be nuclear' with the thermal power stations shutting down at 6 p.m. every day. Its cost would be two thirds of that from coal-fired stations and, as progress was made, the costs would decrease accordingly. 'It is easy to visualise', he continued, 'the effect on industry throughout Europe with this cheap power. It should mean another industrial revolution as the present cost and short- age of fuel is holding all these countries back'. As far as South Africa was concerned he could not foresee the time when either the Transvaal or Natal, with their abundant reserves of coal, would find nuclear power cheaper 'but the Cape might find a nuclear plant more economical'. In response to questions he even ventured views on the use of nuclear propulsion for ships and aircraft. Such ships, he said, were already on the drawing board 'and it would not be long before giant tankers bringing fuel to Europe would be rounding the Cape at high speeds powered by a nuclear pile'. Nuclear-powered aircraft, though feasible, would present problems were they to crash, he opined, and so were unlikely.

This was heady stuff and his South African readers relished it. Here was the man who was just one rung lower on the ladder than the famous Sir John Cockcroft himself, and between them they were steering Britain's nuclear power programme relentlessly forward. The benefits were almost tangible and the world was facing a future where seemingly limitless sources of energy were just waiting to be tapped. But the reality was, of course, rather different and the priorities very definitely so. Ahead lay many pitfalls, some quite unseen.

The visit to South Africa had helped to put many issues into perspective for both Basil and Ismay. Their home was now clearly in England though Gardiol, always their retreat near Somerset East, continued to exert an almost magnetic attraction. And then there was Grahamstown and Rhodes University. Schonland was still its Chancellor but, living in England as he did, he fulfilled that function now in a purely nominal way and it concerned him. Would the University not be better served by someone who was amongst them and not just a figurehead, especially now? With much regret he therefore decided to offer his resignation as Chancellor on the grounds that his manifold responsibil- ities in England made it almost impossible for him to serve his *alma mater* in any useful way. But his letter to the chairman of the University Council, written in July, caused something of a furore when it was tabled at their meeting in September. The cry went up that to lose Schonland would be nothing short of calamitous, not only for Rhodes but for Grahamstown, and certainly for South Africa. A unanimous resolution was then passed that called on the chairman to write immediately to ask their Chancellor to withdraw his resignation and S B Hobson wrote in longhand: 'We would indeed be grateful to you if you would do us the honour of remaining on as our Chancellor. We are getting along fine as we are; are in fact expanding in a most promising fashion. We need no special services from you at this stage, just your name as the head of our University & your kindly interest in us' [39].

It worked. No man could ignore such sincerity and Basil Schonland agreed to continue serving Rhodes to the best of his ability.

CHAPTER 23

ZETA AND WINDSCALE

The tempo of events now began to increase dramatically. For all the euphoria that greeted the commissioning of the fission reactors at Calder Hall, it was the promise of almost limitless power from thermonuclear fusion that had really transfixed some in the world after Homi Bhabha's dramatic announcement in Geneva in 1955. Then, within the UK Atomic Energy Authority, it was the Kurchatov lecture a year later that added further spice to an already exotic mix that had long been simmering.

What was ultimately to be called ZETA had begun to take shape in the Clarendon Laboratory at Oxford late in 1946, soon after the arrival there of a young Australian doctoral student by the name of Peter Thonemann. The idea of releasing energy by the fusion of light elements had already occurred to Sir George Thomson, professor of physics at Imperial College, London, Nobel Laureate and son of Sir J J Thomson of the Cavendish, the discoverer of the electron. Thomson's pedigree in the field was undeniable; Thonemann, by contrast, had none. However, such was the secrecy that surrounded this speculative quest that Thonemann knew nothing of Thomson's work and so was not dismayed by an illustrious competitor and was able to pursue his own ideas with much energy and great enthusiasm.[*]

Both Thonemann and Thomson set out to contain, within a toroidal vessel, deuterium gas at low pressure, which would then be ionized by an external source. The resulting plasma could be heated by means of an applied radio frequency current to a temperature sufficient to cause fusion and the release of energy, as well as a significant discharge of

[*] Probably the earliest work on thermonuclear neutron generation had been done at Liverpool University as a result of Sir James Chadwick's pivotal association with the American research during the war. In 1946 J M Meek was appointed Professor of Engineering (Electronics) at Liverpool and was joined soon after by J D Craggs. Together they continued the work started by Meek at Metro-Vick, primarily at Chadwick's suggestion, on the possibility of producing nuclear fusion by the discharge of an electric arc. Though they detected neutrons after discharging 300 kA sparks through deuterium they soon showed that they were not indicative of fusion and the work ceased [1, p 19].

neutrons. Though unaware of each other's existence for quite some time, their proposed schemes were remarkably similar. This is not too surprising when it is appreciated that their hypotheses were themselves based on conventional physics that was well understood at that time. However, the application was certainly novel, but not unique for there had been work done along somewhat similar lines before the war in America, Germany and Russia, while Thonemann himself had tested some preliminary ideas of his own while still a student in Sydney in 1939.

By 1946 Thomson's work had reached a sufficiently advanced stage that he sought to patent it, not, it should be noted, for reasons of financial gain nor even to establish priority of claim, but simply to place on record his own contribution to the state of knowledge. What prompted this somewhat unusual decision was a meeting he had at the beginning of March that year with an old colleague, Rudolf Peierls, Professor of Theoretical Physics at the University of Birmingham. Their association went back to the days when Thomson was chairman of the MAUD committee—the highly secret committee set up in April 1940 to consider the possibility of constructing a 'uranium bomb' [2]. It was, as we have already seen, Peierls and his fellow émigré Otto Frisch who had calculated the critical mass of uranium required to make such a bomb—a result of cardinal importance and one which altered the future direction of the war. Soon Peierls would join the British scientific contingent at Los Alamos where he would then be privy to much classified American information. So, when Thomson approached him nearly six years later to discuss his ideas about thermonuclear fusion, Peierls was concerned that his own knowledge of the American programme could easily prove to be an obstacle to open collaboration with the other members of Thomson's team at Imperial College. Presumably his sensitivity about such matters and his heightened concern had been precipitated by the newspaper headlines that followed the arrest in Canada on 4 March 1946 of Alan Nunn May on suspicion of passing secret information to the Russians. Thomson, too, was concerned by the security implications of his work, and had long maintained the position that no research of a classified nature would be done in his Department at Imperial College; so he readily appreciated Peierls's point of view. To protect both their interests Thomson sought there and then to define precisely what his own state of knowledge was by meeting with the Ministry of Supply's patent agent on 26 March to draw up the necessary patent application.

But the matter did not rest there. The path of progress became increasingly tortuous as the number of interested parties increased. What triggered this was one of Thomson's claims in the patent that his device could also be used to produce plutonium because of its substantial yield of neutrons, and plutonium, of course, was the very stuff of bombs.

493

That fact in itself was therefore sufficient to cause the patent to be classified secret. Now, not only could Thomson no longer continue the research in his own department, but Harwell, in the person of Sir John Cockcroft, was insistent that it should not be done there and he immediately offered Thomson facilities at Harwell itself. But Thomson rejected this offer because by now he knew of Thonemann's work at Oxford and, for all his stature in the world of physics, he feared that, as the Oxford work had already been taken under Cockcroft's wing, his own ideas would merely languish in the combined shadows of Cockcroft and Thonemann. He therefore proposed transferring his research to the laboratory of the commercial firm Associated Electrical Industries (AEI),* recently set up at Aldermaston Court where it would come under the direction of T E Allibone, the man with whom Basil Schonland had collaborated so successfully at Metropolitan-Vickers in Manchester in 1931 and again in 1934 when they re-created in the laboratory the conditions of the lightning flash. This proposal, however, met with a somewhat lukewarm reaction from Cockcroft, who felt it was premature to involve industrial collaborators. This would all change in August 1951 after Cockcroft and Lord Cherwell, once again installed in academic office as head of the Clarendon, had seen Thonemann's impressive results and together realized its possibilities. Soon after, the Imperial College research team were accommodated by AEI and their fusion research programme itself was funded in full by Harwell, thus ensuring absolute control by Lord Portal's organization, with all its security implications intact. Ironically, in view of all Peierls and Thomson's concerns about secrecy, and the official procedures in place to enforce it, Allibone, at Cockcroft's specific request, had as early as 1947 discussed Thomson's secret patent with Klaus Fuchs, head of Harwell's theoretical physics division—a discussion which will have been of undoubted benefit to Fuchs' Russian masters [3].

By 1948 Thonemann's research was being funded by Harwell. Just over two years later it had reached a sufficiently advanced stage that the resources of Cockcroft's laboratories were now needed to take it further. Thonemann himself and one of his assistants joined the Harwell payroll in October 1949, though they remained in their laboratory at the Clarendon until 1952 when they finally moved the short distance south to become part of a rapidly expanding team of experimentalists and theoreticians assigned to the controlled fusion project. Many problems had still to be solved. One, of critical importance on which Thonemann had already made a start, was the method of containing the high

---

* The AEI group were an amalgamation of British Thomson Houston and Metropolitan-Vickers whose laboratories at Aldermaston Court, south of Harwell, were opened in May 1947 for long-term, fundamental research.

temperature plasma within the torus so that heat loss by conduction to its walls was reduced to an absolute minimum. This was achieved by means of the 'pinch-effect' whereby an electric current flowing in a conductor generates forces that actually compress the conductor itself: a phenomenon frequently observed after lightning had struck a lightning conductor! In this case, that conductor was the plasma and the current within it was induced by conventional transformer-action from a suitably orientated primary winding. Radio frequency energy was used to cause the deuterium to ionize and the plasma so formed now behaved as a shorted secondary turn of a transformer. Magnetic induction from the primary winding then took over to produce a current that would have the combined effect of both pinching the plasma and of increasing its temperature. But problems still persisted. One was a tendency for the ions to drift towards the torus walls during the zero crossings of the applied field. This was solved by an ingenious use of a capacitor bank discharging into a pulse transformer to set up a unidirectional rather than an alternating field. Such techniques had been extensively researched at the TRE for radar applications during the war and their use at Harwell was due to the considerable part now being played by former radar scientists in a rather different field.

A stable ring of plasma aligned with the toroidal axis was the ideal for which Thonemann and his colleagues were striving. What was observed by means of high speed photographs, so reminiscent of those that Schonland and his South African colleagues had used when photographing lightning was, however, very different. Even the slightest lateral displacement of the discharge would cause the plasma to wriggle or kink with a resulting imbalance of the magnetic forces and an immediate increase in the kink itself. This 'kink instability', as it came to be known, was a major problem because it would grow until the plasma made contact with the surrounding walls and dissipated all its energy. To try to maintain absolute symmetry in the tenuous, low-density gas that made up the plasma, or even in the pulsed field that produced it, was impossible, so the solution when it came was a stroke of genius. If, by means of a solenoidal winding on the torus, a magnetic field was made to coincide with the axis of the plasma, then any tendency for it to kink was resisted by the additional forces of that toroidal field. This stabilizing effect, coupled with the naturally reactive forces generated by the equal and opposite current induced in the walls of the torus, were the mechanisms by which the intensely hot plasma was anchored to the axis of the torus and away from its heat-sapping walls.

*   *   *

By the end of 1954, when Basil Schonland was still familiarizing himself with the newly-named roads that intersected the Harwell landscape,

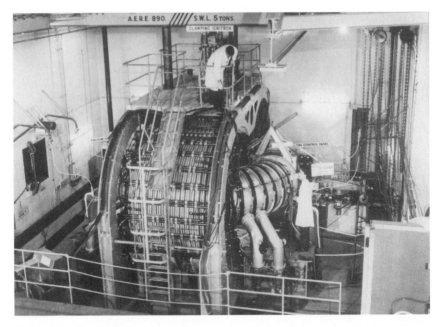

Figure 27. ZETA, the Zero Energy Thermonuclear Assembly, at Harwell in 1957. (Reproduced by kind permission of the UKAEA.)

the decision was taken to begin the detailed design and construction of a large-scale device within which controlled thermonuclear reactions might take place. Both Thonemann and Thomson's preliminary experimental work had reached the stage where such an increase in scale was justified. D W Fry, who had recently succeeded Skinner as the head of the General Physics Division at Harwell, backed Thonemann's proposal for a large transformer-driven toroidal device, while Thomson's ideas would come to fruition in the AEI laboratory, though on a much smaller scale.

As one of its first major decisions the newly-formed United Kingdom Atomic Energy Authority approved the expenditure for a 100 000 A experiment to be mounted at Harwell. In July 1955, at Fry's suggestion, the project was given the code-name ZETA (the Zero Energy Thermonuclear Assembly) and assigned the highest priority within Harwell. By now, informal cooperation with the Americans allowed some information on aspects of the work to be exchanged between the British and American research teams, but there was still no official confirmation that the Americans had a controlled thermonuclear programme of their own, though informed British opinion was in no doubt that they did. By the summer of 1955, the number of staff on the controlled fusion project at Harwell had doubled, with a further such increase planned over the next couple of years. Then, in August, Homi

With the compliments of CHRYS and the DAILY MAIL

DAILY MAIL, 25ᵀᴴ JANUARY 1958

" WITH 5-MILLION DEGREES OF 'EAT KNOCKIN' ABAHT INSIDE , YOU'D THINK THEY'D FIND AN EASIER WAY O' DOIN' THIS ! "

Figure 28. Schonland was much amused by this cartoon. The original was sent to him by its creator and it was framed and hung in the entrance to the Cockcroft Hall at Harwell. (Reproduced by kind permission of the UKAEA.)

Bhabha's confident prediction in Geneva, of the limitless source of power just waiting to be unleashed, served to fuel expectations even further.

ZETA was a significant piece of engineering. The aluminium torus was 3 m in diameter, with a wall thickness of 25 mm and a bore of a metre. It operated at a gas pressure of about 133 mPa*, through which a current discharge of 10^5 A was initiated by 3 ms pulses at 10 s intervals, which themselves were generated by 25 kV transients applied to the primary of the transformer. Though designed entirely at Harwell, its construction was somewhat beyond the capabilities of the establishment's well-equipped laboratories and workshops, but well within the bounds of Sir John Cockcroft's vision. Once again his instinctive engineering ability that had introduced Rutherford's laboratory to the era of the 'big machine', when all around him were still wedded to the string and sealing wax approach, came to the fore and his enthusiasm for this great venture was infectious. Basil Schonland, his deputy, who was rapidly assimilating all that went on around him, was by contrast a disciple of that Rutherford school and was having to adjust to a vastly different scale of activities. Soon, as he assumed more and more of Cockcroft's responsibilities, he would become foster-father to a very substantial collection of 'ironmongery' of the type that he usually

* mPa: milliPascals (10^{-3} Pa)

497

sought to avoid.* The construction contract for ZETA went to Metropolitan-Vickers—a name of much resonance to all those who knew of the sterling work done by the company during the war in constructing the transmitters for Sir Robert Watson-Watt's 'Chain Home' RDF system. Whereas Metro-Vicks' previous collaborations with Harwell had encountered many problems, by overrunning both their budgets and their schedules, ZETA was to be different thanks to exceptionally tight project management and close liaison between Harwell's scientists and the company's engineers.

On 12 August 1957 power was applied to ZETA for the first time and tests were run with hydrogen to give some idea of operating conditions. Then, the deuterium upon which all the predictions had been based was introduced. Within two weeks, the optimum conditions of gas pressure and axial magnetic field were found and the induced current was increased to 120 kA. Immediately the neutron detectors positioned around the torus began to register counts. By mid-September the current had been increased still further to 180 kA and up to a million neutrons were being recorded per discharge. But the scientists were cautious. Nobody was yet in a position to link those neutrons to the thermonuclear process they were all seeking. Neither the Russian experiments as described by Kurchatov nor the more recent American results—both of which made use of straight discharge tubes—had been conclusive in this respect and further evidence was sought. However, there was an increasingly optimistic mood in evidence at Harwell as word spread of these large neutron emissions.

Just a month before, on 7 August, Harold Macmillan, the new British Prime Minister, paid a visit to Harwell and, as usual, Schonland, operating assiduously in the background, arranged the programme. Macmillan was intent on restoring good relations with the United States after the Suez débâcle had brought down his predecessor Anthony Eden, and nuclear issues were high on his agenda since such weapons were seen as crucial elements in the western defence arsenal. He therefore showed particular interest when shown around the ZETA apparatus by Schonland, Cockcroft and R S Pease, a key member of its design team. The sagacious Macmillan took careful note of what he had seen and stored it away for subsequent use. Expectations were high on all fronts and Macmillan enthused, 'You will be pleased when you get this to work.

---

* To Nabarro, in Johannesburg, Schonland wrote in 1958: 'I have never been fond of large pieces of ironmongery' and remembering his own words to Anton Hales at UCT many years before, he cautioned his former AORG colleague and now Wits professor against succumbing to what he called the 'fatal weakness of South African physics: that it is always trying to keep up with the Joneses instead of striking out on its own'. A Harwell in South Africa was just not on the cards [4].

You should need a drink to celebrate', he said. 'Beer?', suggested Schonland. 'Good heavens no, champagne!' spluttered the PM [5]. Also visiting Harwell that year was Her Majesty the Queen, accompanied by HRH Prince Philip, and between them Cockcroft and Schonland shared the honours of showing them around. The establishment was indeed a jewel in the national crown and the Royal personages were treated to a display of Harwell's atomic wizardry.

But any thought of champagne or celebrations of any kind was premature until the neutrons had been shown unequivocally to be of thermonuclear origin, and Schonland was particularly cautious about leaping to such conclusions after his close encounter with calamity following the Kurchatov lecture. On 4 September, just before setting off on his delayed visit to the United States, he urged Cockcroft to play down the results. Then, after his first meeting with his American hosts within hours of his arrival the next day, Schonland cabled Cockcroft with the news that neutrons had also been observed in the American experiments and he stressed again the need for caution, particularly as American and British scientists were due to discuss their results at a meeting in Princeton in October. But that very day Cockcroft had informed Sir Edwin Plowden that, though he could not be certain of the thermonuclear origin of the neutrons, the probability of that being their cause was very high. Plowden immediately informed the Prime Minister and the cat was well and truly out of the bag.

Matters did not rest there. Just a week after the first significant numbers of neutrons were detected, two papers were presented at a meeting of the British Association in Dublin. Both had been written while ZETA was still under development and so made no mention of the results achieved within the previous week. Word had leaked out, however, and Sir George Thomson and J D Lawson, the two authors, found themselves bombarded by questions from an audience that included Homi Bhabha as well as the Irish Prime Minister. The secrecy surrounding the research, as well as both speakers' reluctance to be drawn on the implications of the leaked results, only added fuel to an already volatile situation [6]. On the one hand, American pressure was being brought to bear on Downing Street to make no disclosures at all, while the efforts of the press to generate increasingly graphic headlines was beginning to get out of hand. To further fuel the debate, there were those at Harwell, most notably Thonemann with Cockcroft's support, who believed that some sort of official announcement was called for, in order, once and for all, to lay to rest the wild speculation. And so a press release was drafted and sent immediately to the British Embassy in Washington.

The American response was both swift and unsupportive: no announcement should be made until definite evidence of the thermonuclear origin of the neutrons had been obtained on both sides of the

Atlantic. Harwell's reaction to this was one of some pique. This was surely just a delaying tactic on the part of the Americans while they did all in their power to catch up with the British work. Thonemann, who was particularly irritated by this, then suggested to Cockcroft that every effort should be made to confirm the thermonuclear source of the neutrons and then publish the result in the open scientific literature, with or without American approval. Cockcroft liked the idea and experiments were stepped up. On 19 September he informed the AEA that the probability that the neutrons were thermonuclear in origin was high, and Plowden sent his formal congratulations to the ZETA scientists. Immediately, Cockcroft asked the editor of *Nature* to prepare at short notice for the publication of a paper on the subject. But the Americans were obdurate: only simultaneous publication would be acceptable to them. For whatever reason, noble or otherwise, they held on to this view; and they were not alone, for even at Harwell there were those, most notably D W Fry, who cautioned against precipitate action until the evidence was watertight.

Then, as they so often do, events took charge. On 4 October 1957 the world leapt into the space age with the launch of the first artificial satellite. All attention swung away from Harwell and on to the Soviet Union and Sputnik 1, the 84 kg sphere then orbiting the earth every 96 minutes, its radio beacon communicating its presence and enabling scientists on earth to determine its orbit.* American consternation at being beaten into space by the Russians reached near apocalyptic proportions when their intended first satellite exploded on the launching pad a month later. Then to add to the consternation in the West, just six days after the Soviet triumph, a fire broke out in No 1 Pile at Windscale and the world's first nuclear accident grabbed the headlines.

The ramifications of all these events were enormous. The Americans were desperate for some scientific achievement that would help redeem a little of their national pride; the British nation, so convinced by both scientists and politicians of the mystique of nuclear energy, now had to come to terms with the possibility of a catastrophe. Only the predictions made about fusion and the promise shown by ZETA offered, if not a palliative, then at least a distraction. But ZETA remained an exclusively British affair while General Eisenhower, now US President, faced a nation unaccustomed to ignominy, who demanded an explanation.

---

* The first published orbital data for Sputnik 1 appeared on 21 December 1957 in *Nature* **180** 1413. It was based upon measurements and calculations made by J A Fejer, a member of the Schonland's SSS during the war, who was now working at the Telecommunications Research Laboratory (TRL) in Johannesburg. Fejer predicted an orbital lifetime for the satellite of 87 days, and 56 days for that of the rocket that launched it. The actual lifetimes turned out to be 93 and 58 days respectively. It was no mean achievement for the TRL, whose staff had cobbled together a tracking system in double-quick time.

Before the impact of Windscale struck home, the British press had tended to gloat at the American's discomfort as they themselves basked in the reflected glory from Harwell. But it was entirely without good cause for there was no proof that the neutrons emitted by ZETA were of thermonuclear origin. In fact, the latest results from Harwell were far from convincing, and the press release that had been hastily re-written now played down the significance of the neutrons and even wavered as to the accuracy of the temperature measurements upon which so much depended. But the clamour for an announcement was increasing and Cockcroft, Schonland and Thonemann, though aware of all the uncertainties, were smarting at what they considered to be American self-interest on the nuclear front riding roughshod over everything else.

On 11 November the announcement was duly made in the House of Commons. In reply to a question the Paymaster-General, Mr Reginald Maudling, said that ZETA was operating and that 'experiments were going on to identify the source of the accompanying neutron emission which probably arises from thermonuclear reactions but may possibly be due to other complex nuclear processes' [7]. This naturally triggered a massive response from the international press, with the more lurid accounts being little troubled by detail or accuracy as they heaped scorn on the Americans who had once again been pipped at the post. Washington's reaction was immediate and furious in the form of a telegram to Plowden from Admiral Lewis Strauss, Chairman of the US Atomic Energy Commission. A diplomatic incident was on the cards if not quickly defused, and it soon became clear to the American scientific community that publication of their own work was required sooner rather than later. However, it was not until 12 December, and only after a visit by a number of Americans to Harwell, that they finally agreed to declassify certain material and make it available for publication. Whereas scientific cooperation and even progress had long suffered under the heavy hand of secrecy, it was the boot of national self-interest that really stood in the way of progress. By now Cockcroft's keenness to announce that neutrons were being produced in profusion, and that there was some evidence to suggest a thermonuclear origin, could no longer be restrained, even by his more cautious colleagues. Strangely, he was far less enthusiastic about including mention of the similar success achieved by Allibone's AEI team, even though he and Schonland had witnessed the flux of neutrons for themselves that emanated from SCEPTRE III, the AEI torus, when they visited Aldermaston Court very early in 1958. Only when Thomson protested vehemently did Sir John agree to do so [8].

The events that followed within the last week of January 1958 were of seismic proportions. The earlier date agreed with *Nature* for the publication of the results was moved forward from 7 February to 25 January so as to pre-empt any disclosures that might be made at a conference in the US the following day. On the 24th, Harwell was to issue a press release which would state that 'there are good reasons to think that [the neutrons] come from thermonuclear reactions' but with the added caveat that this had 'not yet been definitely established'. And so it was that on 23 January the previously impenetrable security that surrounded Harwell was temporarily removed to allow in upwards of 300 journalists to attend the press conference. Just before it commenced, Cockcroft drew Thonemann aside to ask him whether he was really sure that the neutrons were thermonuclear in origin, but Thonemann wasn't at all sure. Cockcroft then spoke cautiously but optimistically and, on finishing his brief statement, agreed that he and his colleagues assembled on the platform would answer questions from the floor. The inevitable question came: 'How certain was Harwell that the neutrons had a thermonuclear origin?' None of the scientists seated alongside Cockcroft offered an answer and so Sir John, to the astonishment of his colleagues, agreed to do so and declared that he was '90 per cent certain' that at least some of the neutrons could be traced to a thermonuclear source [9]. His colleagues were stunned by such certainty and the press stampeded to get their copy to the newsdesks of the world.

Almost in keeping with the very make-up of the man was another announcement of Cockcroft's, issued in less public surroundings, that was but a mumble or a murmur by comparison. Not surprisingly it was drowned by the crescendo of activity as the press scrambled to write about Britain's remarkable achievement. On 22 January, it was made known that Sir John Cockcroft would be retiring as Director of Harwell later that year and his successor would be Basil Schonland.

In England the announcement that Cockcroft was stepping down appeared only on page 5 of *The Times*, whereas the news that a South African was to succeed him guaranteed banner headlines in that country: 'Surprise Downing Street announcement: Schonland to become Harwell Atom Chief; No reason given for take-over from Sir John Cockcroft' [10]. South Africa was fascinated but not really surprised, for the Schonland name had become almost a byword out there. That he had merely been associated with ZETA, this seemingly limitless source of energy now being trumpeted around the world, was sufficient to elevate him to heroic proportions in the eyes of many of his fellow-countrymen. 'Of course he would rise to the top; just a matter of time', was the oft-heard opinion.

Editorials in the South African press then elaborated on the appointment and sang the praises of the man who, in their eyes, had brought the country honour and prestige at a time when both were in rather short supply. 'Here is a man', ran one, 'whose brains and hard work have taken him to an outstanding position in a mysterious new science that was still virgin territory when he was a boy. Now, to crown a career already distinguished beyond the usual run, he becomes head of the world's leading nuclear research station at a moment when his colleagues have announced an achievement that is not only the greatest so far recorded in that particular field, but may well come to be reckoned as the most important triumph of mankind since the control of fire'. Cockcroft's precipitate announcement, made just the day before, had certainly ignited a fire of its own and it would run, not only through the corridors of Whitehall and Washington, but through the African veld too, as Basil Schonland's fellow countryman basked in their own reflected glory.

Once the congratulations had subsided the more staid editorials appeared that presented the background, as they knew it, to the changes soon to usher in a different structure within Britain's atomic energy establishments. But rumour and speculation abounded as well. Had Cockcroft been forced to step down over events in the immediate past? Was Windscale to blame? Had he oversold nuclear fusion? No-one knew. What was clear though was that Schonland would assume his new position on 17 February, when he would take over executive responsibility for the work of the establishment and relieve Cockcroft of his day-to-day duties, thus allowing him 'to devote more time to the formulation of policy and the broad oversight of the subjects constituting [his] own special field' [11].

The change-over from Cockcroft to Schonland was certainly not unusual in Cockcroft's mind because that is just what he had intended should happen. That it came when it did, however, fuelled many speculative fires. None of this made any difference at all to the very special relationship that existed between the men themselves.

* * *

The fire at Windscale was the world's first major nuclear accident. Basil Schonland was to serve as a member of the board of inquiry that investigated its causes and reported with great urgency to the Prime Minister. The actions that followed had ramifications that extended well beyond the north-western corner of England where Windscale's two reactors rose like primeval mushrooms from the Cumberland coastline.

The fire broke out on Thursday 10 October 1957 during the process of annealing the graphite moderator in Pile No 1 of the reactor, in order to release the Wigner energy that had built up within it. Until then this had been a fairly routine operation, but was not one that the designers of the Windscale piles had originally contemplated. As early as 1948, American scientists on a visit to Harwell had passed on information about the effect, described previously and first discovered by E P Wigner, a Hungarian working at Princeton, in which dimensional changes occur in

503

graphite when irradiated. Hurried redesign of certain features of the pile was necessary in order to allow for this 'Wigner growth' and research effort was immediately directed towards understanding the problem, both in Harwell and in Chalk River, Canada. It was, however, another phenomenon also attributed to Wigner that was to have very much greater consequences for the piles at Windscale. Such was its importance that the events soon to follow would have reverberations at the very highest levels in Whitehall. Remarkably, its existence had not reached the ears of the designers at Risley (though Harwell knew of it from the Americans in 1949) until it was too late to change the operating conditions of the piles to counteract it.

When graphite is subjected to intense neutron bombardment its crystal structure is disturbed such that atoms are displaced from their usual positions in the lattice. These atoms collide with others, causing them to be similarly ejected and this combination of displaced atoms and the resulting vacancies within the regular lattice leads to an increase in potential energy which, if unchecked, can result in considerable increases in temperature, the risk of catastrophic fire and the consequent ignition of the fuel rods and even the graphite itself. This stored 'Wigner energy' can be released in a controlled fashion by heating the graphite, to increase the thermal vibration of the crystal lattice and so cause the atoms to resume their normal positions, thus 'annealing' the radiation damage [12]. This Wigner energy phenomenon only came to light after the Windscale piles had gone critical in October 1950 and neutron irradiation of the graphite was well-advanced. Hinton, at Risley, had not been party to the discussions with the American physicists at Harwell in 1949 and seemingly no information was passed on to him about the danger of energy storage within the graphite. The problem first reared its head in 1952 when, in May and again in September, first Pile No 2 and then Pile No 1 experienced abnormal rises in temperature which, initially, could not be explained. Only after detailed investigations and consultation with the Americans was it apparent that spontaneous releases of Wigner energy had occurred, which then caused the subsequent increase in pile temperature. The first controlled anneal to release this energy was carried out in January 1953. Harwell scientists were by now involved in investigating the problem but, presumably because of American experience,[*] they treated it as a fairly routine matter and soon handed over responsibility to Hinton's staff at Windscale.

---

[*] American knowledge of Wigner energy was based on their experience with the graphite reactors at Hanford in Washington State. By increasing the operating temperature of the pile they reduced the amount of Wigner growth that restricted the vital flow of water, the essential coolant, through the pile. By so doing they also reduced the amount of stored energy which was the very essence of the problem at Windscale where the piles operated at a lower temperature.

Despite his most strenuous efforts, Hinton had fought a losing battle to convince Cockcroft at Harwell that more dedicated research effort was required to support the construction of the reactors at Windscale. To many of Harwell's physicists the graphite problems presented to them were seemingly mundane and more akin to 'high-school physics' and 'cookery' than the curiosity-driven research which Cockcroft so willingly encouraged [13]. Friction between the two men had a long history, stemming from the very earliest days in the development of the British nuclear programme when Cockcroft and most of his scientists were still in Canada, while Hinton was already confronted with real engineering problems in setting up the facility at Risley. His requirements were for support across the broad front of chemistry, metallurgy and physics and, in the face of Harwell's reluctance to provide this in any whole-hearted way, he sought, albeit reluctantly, to establish his own research and development organization; but he soon found himself over-stretched. Poor salaries and the unattractiveness of the isolated Windscale site, without adequate accommodation and schooling, worked against him, and his research facilities were always undermanned and, in many respects, his staff under-qualified. Given all this, the fact that the Industrial Group achieved what it did, virtually on time and within budget, is a remarkable tribute both to Hinton and to his staff who had well earned for themselves the sobriquet of the 'buccaneering corps d'élite'; but the pressures told [14]. Frequently at odds with Plowden, and disillusioned by developments within the Authority, Hinton left just two months before the fire to become the first chairman of the newly created Central Electricity Generating Board.

When the anneal of Pile No 1 at Windscale commenced in October that year, the operations staff there had already amassed a considerable amount of practical experience in carrying out such a procedure, since it was the sixteenth such anneal they had performed and the ninth on that particular pile. However, little or none of it was written down. The operation took the better part of a week to complete and, though well practised, it relied heavily on the operators' 'feel' of the pile rather than on a firm set of prescribed scientific instructions on which to base their actions and decisions. Success thus depended entirely on the adequacy of the instrumentation of the pile and on the operators' interpretation of the information it provided. Of course, since no-one had gone before them, they were the pioneers who were sailing towards a horizon beyond which may well have been an abyss.

Then, at 4.30 p.m. on 10 October, just over three days after the anneal had started, it was reported officially that the pile was on fire [15]. To follow the events that led up to it, it is necessary to consider the construction of Pile No 1 and the procedure used to release the Wigner energy by the process of annealing. These were amongst the mass of information that Schonland and his colleagues, charged with inquiring into the cause of the accident, had to assimilate in double-quick time.

Annealing was initiated by switching off the eight air blowers which, during normal pile operation, maintained the flow of cooling air through the multitude of horizontal channels that housed the aluminium-encased uranium fuel elements, or those of a lithium–magnesium alloy, from which the tritium so vital to Britain's hydrogen bomb programme was produced. Each of the Windscale piles contained 3440 such channels within the 2000 tons of precisely machined graphite blocks in an octagonal stack some 15 m in diameter and 7.6 m long. In addition to these horizontal channels there were four vertical channels used mainly for making measurements within the pile and to house special experimental apparatus. Control of the pile was achieved by the use of 24 horizontal control rods of boron steel chosen specifically for its high neutron-absorbing properties. In the event of an emergency the pile could be shut down by dropping into it, under gravity, 16 neutron-absorbing rods that would immediately shut down the reactor. The graphite pile was fitted with an array of sensors to monitor such things as neutron flux, pile temperature and the position of the various control rods. In addition, there were eight movable scanning devices, known as BCDGs (Burst Cartridge Detection Gear) to detect and locate any of the 70 000 fuel elements which might burst[*] at any stage during the operation of the pile. Such an eventuality would naturally be very serious since highly radioactive fission products would be released and the uranium would become oxidized, thus leading to the possibility of a fire. To protect the operators and all beyond the confines of the plant, this reticulated core with its rods and sensors was finally encased in a biological shield of reinforced concrete nearly 2 m thick.

When in operation, the hot air discharged from the rear of each pile eventually found its way to the skies above after passing through two chimney stacks some 125 m high. Though originally designed without filters, for these were not considered by Harwell to be necessary in view of the expected rapid diffusion of the minute combustion products from such a height, Risley's engineers found themselves faced with a major task of retro-fitting filters when construction of the stacks was already well under way. This had come about as a result of a visit to the US by Cockcroft in 1948 where he discovered that radioactive particles had been detected at ground level at the Oak Ridge reactor in Tennessee. Filters had immediately been fitted and Cockcroft accordingly reported this to Hinton. The reaction at Risley was understandably mixed and did little to strengthen the bond between Cockcroft and Hinton, or the relationship between Harwell and Risley. Filters designed in some

---

[*] The word 'burst' conjures up visions of catastrophic failure of a cartridge. In fact even a pinhole defect leading to an increased reading on the BCDG would constitute a 'burst'.

haste, with the assistance of the chemical defence research establishment at Porton Down, were fitted but could only be accommodated at the top of the two stacks, where conditions were not ideal from the points of view of either filter performance or maintenance. Though referred to at the time as 'Cockcroft's follies', their installation would soon be more than justified.

The purpose of the annealing process under way on Monday 7 October was to increase the temperature slowly towards the front of the lower part of the pile, where the stored Wigner energy was assumed to be greatest. This was done by the pile operators inching out the control rods in that area, while carefully monitoring the temperature readings from the appropriate thermocouples embedded within the pile. The intention was to increase the uranium temperature to 250°C in the first instance. In the early hours of Tuesday two uranium monitors were indicating this value while others showed the graphite to be in the range 50 to 80°C. However, one thermocouple reading graphite temperature had climbed to 210°C, which was taken to be an indication that a self-sustaining release of Wigner energy was taking place. Since this was the desirable condition the physicist in charge decided to run the control rods in again, thus shutting down the pile, in the expectation that the annealing process would be sustained by the energy now being released. However, within hours it was noticed that the process seemed to have stalled: temperatures, as the operators observed them, were remaining constant, or even falling, and there was clearly the risk of the energy release just petering out with much of the graphite being unannealed. So, based on experience of similar situations in the past, but with no 'operating manual' to consult, since none existed, the pile was run up again in order to initiate a second cycle of nuclear heating.

Though somewhat erratic in its behaviour, the pile seemed to be responding more or less as expected over the next 24 hours. However, on the Wednesday afternoon temperatures began to rise sharply and were approaching the danger levels beyond which the operators were obliged to implement the only written instructions that had actually been provided by the Pile Manager. These were soon put into effect: senior staff were informed, appropriate inspection ports and hatches were closed and the fans brought into action to drive cooling air through the pile. The temperature then dropped, but only briefly. Soon after midnight one thermocouple indicated a sharp rise. Dampers were opened but had no effect; they were opened again—longer this time—and some cooling was evident but within an hour the temperature began to rise yet again. Something was clearly seriously wrong.

Then the first indication of radioactivity within the stack was noted, and since the reactor was shut down this was clearly unusual. However, no action was taken because the view of the operators was that the

507

continual use of the fans had caused some radioactive dust to be blown up the stack. The situation was further confused when an increase in radio-activity was noted in the stack of Pile No 2. The immediate thought was that it may have come from a burst fuel cartridge, which would make it necessary to shut down that pile as well, but this was later discounted when a fault in the pile instrumentation was discovered. By now increased radioactivity readings were being recorded beyond the immediate confines of the piles themselves and, initially, these were thought to be coming from Pile No 2, but the erratic behaviour of the temperature of Pile No 1 was increasingly a cause for concern and all indications now were that something was seriously wrong. The Pile Manager was informed and took steps to use the BCDG to trace what he thought was a burst cartridge, but the gear was found to be jammed because of the greatly elevated temperature and it could not be moved. Until now the pile's own operating staff had dealt with the problem but it was obvious that the situation was very serious indeed and required attention at the highest levels.

Windscale's staff problems were soon to be cruelly exposed. Senior personnel were carrying considerably more responsibility than should have been the case, as a result of being unable to recruit suitably trained scientific staff. This immediately placed a considerable burden on those individuals performing multiple tasks, but they had shouldered them with equanimity if not enthusiasm and had pressed on. In all 52 of Windscale's 784 professional posts were vacant. To make matters worse the plant, along with much of the British nation, was in the grip of the 'Asian flu' epidemic, so staff numbers were even further depleted. The complement was barely able to meet the demands placed on them when all was running smoothly; now they were facing what could become a calamity and things had entered a realm where no one had been before. Glowing fuel elements were visible in four channels when viewed through the access holes on the face of the reactor and more were suspected. The Pile Manager instructed that fuel rods in the surrounding channels be forced out to form a fire break around the affected area. The immediate concern was that the graphite might reach what was felt to be its critical temperature of 1200°C, leading to an uncontrolled release of Wigner energy with the ensuing increase in pile temperature by 1000°C, followed by ignition of the graphite itself, an explosion and the release of radioactive material and gases into country-side around and well beyond. Temperatures continued to rise and passed the 1200°C mark but no spontaneous eruption occurred. Flames could, however, be seen flickering at the back of the pile. No one knew what might happen were it to reach 1500°C.

The pile had to be cooled. The intended method of temperature control by means of air at high pressure had failed to arrest what seemed, by now, to be an incessant rise. If anything, the air was only fanning the flames; it was certainly feeding the fire. Carbon dioxide was available nearby at Calder Hall but previous experience with it at Risley had shown it to be ineffective at such high temperatures. The only alternative was water, but water and hot graphite had the potential for forming a highly explosive mixture with the release of hydrogen and the generation of carbon monoxide. Could such a deadly combination be risked? Almost in desperation it was decided to try the carbon dioxide gas and, if that failed, then water would just have to be used. A tanker of CO_2 was brought in and the gas pumped into the pile but, as expected, it achieved nothing. And so all that remained was to call on the fire brigade, who had been standing by all along, to provide the water that might either douse the conflagration or precipitate an explosion with unimaginable consequences.

Sir Edwin Plowden, chairman of the AEA, had by now been informed of the situation, as had Sir Leonard Owen who had recently taken over from Hinton as managing director of the Industrial Group. Plowden wrote to the Prime Minister, who also held the sensitive portfolio as the Minister responsible for atomic energy, and set out the facts as he knew them in an attached memorandum. Holding the reins at Harwell, and presumably informed of the emergency at much the same time, was Basil Schonland. Cockcroft was then on his way by sea to the United States to attend the important meeting on fusion at Princeton. This sea journey was very much a consequence of the unyielding schedule of work that Cockcroft had endured over the past many months and his doctor had insisted that he slow down. News reached him of the Windscale fire while he was still at sea and it seemed as if this brief, leisurely interlude would be terminated abruptly because he was the Authority's Member for Scientific Research and much of what had happened fell within his brief. However, his immediate presence would serve no useful purpose, so he attended the Princeton meeting and then flew straight home, a much troubled, though outwardly impassive, man [16].

Emergency procedures at Windscale were now in place. The Chief Constable of Cumberland had been informed of a possible evacuation of the surrounding area and a fleet of buses was put on standby. Just before 9 a.m. on Friday 11 October the water was turned on under the direction of the deputy works general manager who crouched behind a bulkhead watching and listening for anything untoward. Nothing happened. The water pressure was increased but the flames continued unabated. It was then decided to cut the flow of air to the pile and so the fans were switched off. The effect was dramatic. Starved of oxygen, the fire died out and three hours later the emergency was over, at least as far as Windscale itself was concerned. How much radiation had been released into the air was unknown and what effects it would have on the surrounding population was yet to be determined. Of immediate concern was the threat to public health as a result of contamination of the milk produced on surrounding farms by iodine and strontium, fission products within the fall-out. Sensor-equipped monitoring vans had been touring the Windscale site from the Thursday afternoon and soon their area of coverage was extended to include the coastal routes both north and south of the station as determined by available meteorological information on wind direction throughout the period. It was evident, or so it was thought at the time, that the stack filters, 'Cockcroft's Follies' so belatedly installed, had worked most effectively by trapping the major part of the particulate material, for only iodine vapour had leaked through. Since this would readily find its way on to the grazing lands populated by dairy cows, particular attention was focused on the supplies of milk to the local population, but no firm guidelines existed on which to base any decisions. Calculations had therefore to be performed using reports in the scientific literature to establish some appropriate levels of tolerance, especially for children, for whom such figures were essentially non-existent. Numerous consultations by telephone took place and hastily arranged meetings were held with medical and health physics experts to agree on safe values. Milk was then collected from across the region and analysed both at Windscale and Harwell. The outcome was a total ban on the use of milk from all farms within an area of about 620 square kilometres around Windscale on 15 October, three days after Pile No 1 was cold. The ban remained in force until 23 November, by which time regular sampling of milk showed that the danger had passed.

It is not the place here to consider any further the health implications of this first nuclear accident. Many issues continue to attract considerable attention today and the case is by no means closed, even more than forty years later. The matter is very thoroughly handled in [17], as are a variety of other issues that followed from this near-disaster at Windscale. The inquiry into the cause of the fire followed with all speed.

* * *

Quite who should mount an inquiry in the event of an accident was a matter that had been given considerable thought by Hinton even before the AEA came into existence. In his original proposals, presented in 1952, he suggested that any incident involving loss of life or serious damage to property or health should be investigated by senior counsel, or even by a judge, sitting with a panel of experts. However, the practicalities of doing this at short notice ruled against it and the procedure was changed in 1956. It then became the responsibility of the Head of the Group concerned to set up the necessary inquiry while also notifying the Minister within whose bailiwick the Authority might reside at the time. In 1957 the responsible minister was the Prime

Minister, Harold Macmillan, and he had indeed taken a very great interest in the emergency at Windscale from the moment that the news first broke. There were undoubtedly issues at stake that extended well beyond the isolated reaches of Cumberland, and Macmillan issued instructions that he be kept closely informed of all developments. Since taking over as Prime Minister in January from the beleaguered Sir Anthony Eden, Macmillan had begun a campaign to restore Britain's relations with the United States that had stalled in the aftermath of the McMahon Act and had recently suffered even more as a result of the Suez invasion the year before, when America insisted at the United Nations that the British and French forces should withdraw. Macmillan's close association with President Eisenhower went back to 1942, when he was attached as the resident British minister in Eisenhower's command in north-west Africa. Now he intended to revive his strong personal relationship with the President in order to re-establish the bonds between the two wartime allies that had suffered many grievous blows during the post-war years. One of these, of course, involved the McMahon Act of 1946 that had put an end to all cooperation on matters of nuclear energy. The very last thing that the new Prime Minister could afford was an incident in Britain's own nuclear industry that might smack of incompetence. The inquiry, soon to follow, therefore carried many heavy responsibilities.

On the advice of the Authority's legal adviser, its chairman suggested to the Prime Minister's office that a board of inquiry, whose membership included a consultant engineer, would be able to ascertain the facts of the accident. Macmillan agreed and the inquiry was set up on 14 October, just four days after the PM had first been informed of the fire in Windscale's Pile No 1.

The man appointed as chairman of the board of inquiry was Sir William (later Lord) Penney, hastily extracted from Aldermaston. In the circumstances he was probably the only man who could have occupied this crucial position, though he himself was rather reluctant to take it on because one test of a nuclear weapon had just been concluded in Australia, while a vitally important H-bomb test was about to be carried out in November in the Pacific. However, as Sir Edwin Plowden informed him, there was no alternative. Cockcroft was too intimately involved with the decisions leading to the design of the Windscale piles, while his legendary terseness would surely have ruled him out of such an inquisitorial role. Sir Leonard Owen, Hinton's successor at Risley, was not yet an authority member but, even so, as the Director responsible for Windscale, his appointment would have been inappropriate. This would surely also have ruled out any suggestion of recalling Hinton for this task of the very greatest national importance, even though no-one had been closer to every detail of the design and construction of those piles, and his razor-sharp mind would have made him a superb, though formidable, chairman. So Penney, as the Board Member for Weapons Research and Development, and a man with an unparalleled reputation as the driving force behind Britain's entirely independent nuclear weapons programme, found himself plunged into very unfamiliar territory. Though he had no in-depth knowledge of nuclear reactors, he was unrivalled as an applied mathematician most deeply involved in the interdisciplinary problems of atomic energy. He was also a man of gentle disposition, a careful listener and a compassionate interrogator — all characteristics which he would use to the full when leading the questioning of those who had brought Windscale back from the brink of disaster.

Not one but two engineers were appointed, post-haste, to Penney's board. Both were consultants to the authority and both had previous service in Britain's nuclear industry. They were Professor Jack Diamond, Beyer Professor of Mechanical Engineering at Manchester University, and Professor J M Kay, Professor of Nuclear Power at the Imperial College of Science and Technology. As a member of the Royal Naval Scientific Service on loan to the Ministry of Supply between 1944 and 1953, Diamond had worked first in Montreal under Cockcroft and then followed him to Harwell, where he had played a significant part in the conceptual design of the air-cooled reactor, a scaled-up version of BEPO. Kay had been the Chief Technical Engineer at Risley from 1952, where he did ground-breaking work on the principle of fast reactors, and was a joint-author of an acclaimed report on a national strategy for civil nuclear power. In 1956 he was appointed to a chair in London. The fourth member of the board was Basil Schonland, then still deputy to Cockcroft at Harwell.

There was some disquiet, in certain quarters, both at the composition of the board and its terms of reference, which were 'To investigate the cause of the accident at Windscale No 1 Pile on 10th October, 1957, and the measures taken to deal with it and its consequences; and to report'. Some newspapers felt that the implications of such an accident demanded that an entirely independent group of experts, and not one so closely associated with the AEA, should investigate it. In the very peculiar circumstances that prevailed at the time in this most sensitive of areas one has to ask where such independent and expert arbiters were to be found, if not from amongst the members of that close-knit community. It was unthinkable that the British Government would countenance any foreign involvement in the matter, so the pool of available expertise was closely circumscribed. But there were still those who expressed their concerns, amongst them the Member of Parliament whose constituency included Windscale. He asked whether the full facts would emerge, and whether those Windscale employees directly involved in the operation to save the reactor would be adequately represented when called before Penney's board of inquiry. Plowden's brief allowed him only to comment on the second of these and he replied that no representation by either the trade unions involved or anyone else would be necessary, nor any allowed, because it was a fact-finding and not a disciplinary inquiry. But not everyone was mollified by this [18].

The board members assembled at Windscale and began work on 17 October. Within a week the Prime Minister was due in the United States with his ultimate mission being to attempt to persuade Eisenhower to have the McMahon Act repealed. On the very day that the Windscale fire was reported to him, Macmillan had written to the President urging him to consider a pooling of resources in order to meet the new Soviet threat that the launch of Sputnik had made so stark. A rocket capable of placing a satellite in orbit could just as easily deliver a nuclear warhead. Repealing the McMahon Act was, therefore, crucial if Britain and America were to throw in their lot together. The timing of the Windscale fire could not have been worse, for it could end up scuppering everything Macmillan had worked so hard to achieve. Across the Atlantic, the impact of the launch of Sputnik on the collective confidence of the American nation had been akin to that of the attack on Pearl Harbor. America was stunned and a witch hunt for scapegoats was soon in progress. Eisenhower was vulnerable and Macmillan knew he could offer just the sort of support he needed, but not at any price [19]. The Prime Minister knew too that any hint of British incompetence over Windscale would play right into the hands of those in the United States Senate who opposed any cooperation on nuclear matters. Windscale was therefore a highly sensitive issue.

Penney and his colleagues worked between ten and eleven hours a day for the nine days they spent at Windscale examining the evidence, reading reports, studying masses of data and interviewing numerous witnesses. 'We got what is generally agreed to be the right answer to the accident but it took some finding', Schonland wrote later to his brother Felix in Grahamstown [20]. The investigation was an impressive exercise in itself, and not at all unlike the operational research that Schonland had led so effectively at the AORG, and then when advising Montgomery's staff during the advance into north-west Europe in the final months of 1944. Evidence was scattered, often hidden, and had to be pieced together, analysed and critically assessed. In wartime, military strategy was most often the driver; now political considerations loomed large. To those unfamiliar with the man and his background, Schonland's appointment as Harwell's representative on Penney's board of inquiry may have come as a surprise. There were certainly some who had come to see him as an administrator of considerable ability but not as a front-line scientist. Since arriving at the AERE, his position had been that of John Cockcroft's foil, releasing the genius that Cockcroft displayed for

513

inspiring younger men to achieve great heights, while Schonland himself took on the increasing burden of running an establishment of immense complexity. In more recent years, financial and political constraints conspired to limit significantly the freedom that its scientists had had to work on whatever caught their fancy, something that they had just taken for granted under Cockcroft. While Cockcroft could always be persuaded by the sheer excitement of the scientific chase, Schonland had the unenviable task, and one that often made him decidedly unpopular, of reining in the free flow of money to support every good idea. He had frequently to say no when Cockcroft's singular utterances could be interpreted virtually the way one chose to hear them [21].

When Plowden and Owen came to appoint the members of Penney's commission they did so in close collaboration with Cockcroft, who ensured that Harwell was represented by someone of considerable weight. The choice fell naturally to Schonland, by virtue of his stature as a scientist and his considerable experience in the underworld of scientific intelligence. In addition, and to great advantage, he brought with him no previous baggage, having had no connection with the underlying philosophy of Windscale, nor was he party to any of the design decisions that brought it to fruition. What he did bring was an independence of mind and an ability to cut through the mire of any argument to get to the point. Cockcroft knew all that so well.

Penney and his three colleagues completed their report on Saturday 26 October. The Prime Minister was already in America, aware only that the events at Windscale were coming under the closest scrutiny. The document reporting the board of inquiry's findings had been written collectively by its members working alongside David Peirson, the Authority Secretary who had been appointed secretary to the inquiry. Peirson was an excellent draughtsman, Penney a clear and lucid writer of uncomplicated English, while Schonland, ever the scourge of the imprecise and the verbose, had the ability to produce vast texts with minimal re-drafting. Such facility with words served them all well. The testimony of 37 witnesses had been recorded, both on magnetic tape and by a stenographer, but the transcription of all that material would take weeks, so they relied very much on their own notes. They also examined 73 technical exhibits including Pile No 2, specially shut-down so that they could see at first hand the conditions prevailing in its counterpart when the process of annealing first began. Finally, they scrutinized the records of the monitoring surveys conducted by the Health Physics Centre at Windscale as the drama unfolded and then took a break of one day to collect their thoughts and to clear their minds before commencing their task.

In eight short chapters and 31 pages they concluded that the primary cause of the accident was the second nuclear heating that had commenced on Tuesday, 8 October [22]. Based on the evidence of the thermocouples, it appeared to the physicist in charge in the pile control room that, by early that morning, after some twelve hours of nuclear heating, 'the general tendency was for the graphite temperatures to be dropping rather than rising and it seemed probable that unless more nuclear heat was applied, the [Wigner energy] release would stop'. Nuclear heating was therefore resumed and this, according to the board of inquiry, set off the sequence of events that precipitated what so nearly became a catastrophe. The board studied closely all the chart recordings from the various thermocouples positioned around the pile. They agreed that some showed a general decrease in the graphite temperature, while others indicated little change, 'but a substantial number of the graphite thermocouple readings showed steady increases'. By either misinterpreting these readings, or by overlooking them, the pile physicist's decision to restart the heating produced what was essentially a cumulative effect. The temperature in a particular region of the pile began to rise alarmingly, eventually causing one or more of the uranium or lithium–magnesium fuel cartridges to burst—an event probably first detected by the radioactivity monitors on the roof of Windscale's Meteorological Station. As the radioactive materials oxidized, so more heat was produced, and this combined with that generated by the release of Wigner energy to cause the temperature to rise in the relentless spiral that eventually ignited the fire that so nearly raged out of control.

Having identified the cause, the report now sought to put its collective finger on any technical inadequacies of the pile and its instrumentation, or any shortcomings in the administrative procedures at Windscale that led to the accident. Its most severe criticism was levelled at the instrumentation of the pile, particularly the number and placement of the thermocouples. Whereas those intended to monitor the temperature of the uranium during normal operation were correctly positioned, this was certainly not the case during annealing operations. Calculations had shown that the control rod positions used during the anneal would cause the peak neutron flux to occur almost a metre closer to the front of the pile than under normal operating conditions. With the air flow minimized to encourage nuclear heating during an anneal, this peak neutron flux therefore corresponded with the position of maximum temperature. The effect of this, the report contended, was that the uranium temperature was actually some 40 per cent higher than that indicated by the nearest thermocouples.

Turning its attention to administrative matters, the report labelled the absence of an operating manual for Wigner releases as a serious defect. The brief instructions issued to the pile operators in the event of an untoward increase in temperature were clearly inadequate. Details of pile operations had had to be pieced together 'from committee minutes and from traditions'. All this appeared to the board to stem from a lack of any clear division of responsibility between the various branches of the Industrial Group and their technical advisers at Harwell. The channels of communication, if they existed at all, were vague, haphazard and tended to cause 'undue reliance [to be placed] on technical direction by committee'. The pile operators were not well-served by their managers, while the managers themselves were over-burdened by multiple responsibilities, none more so than the Works General Manager, whose brief included the reactors at Calder Hall and Chapelcross as well as those at Windscale. It was clearly an impossible task and should have been recognized as such long before now. Certainly Hinton had done so but his frequent warnings to London went unheeded.

The Penney report pulled no punches. It was both frank and forthright. The pace at which the many facets of the atomic energy programme had developed in Britain had spawned a highly complex industry and the infrastructure to support it was stretched at times to breaking point. Nowhere was this more evident than in the Industrial Group. Its staff and organization were overloaded; its objectives diverse. Most pressing of all was the need to provide the fissile material upon which the country's nuclear arsenal depended and yet, all the while, it had to design and sustain Britain's expanding cluster of nuclear power stations. No one knew this better than Penney himself and, in his leading of the questioning of the various witnesses, he exercised considerable constraint, compassion and tact. This was no disciplinary inquiry. Its purpose was to discover the facts behind the fire and the steps that had been taken to deal with it, and the multifarious issues that flowed therefrom. Of these none was more important than the possible health risk posed to the population around the site and, indeed, to those much farther afield as well. Though the responsibilities of the Chief Safety Officer, the Group Medical Officer and the Health Physics Manager had never been clearly defined, their joint action had been decisive. Penney's board agreed with their decision to concentrate all their efforts on the danger of possible ingestion, via the local milk supplies, of radioactive iodine 131. What the inquiry found alarming, however, was the fact that tolerance levels had never been satisfactorily determined and so they had had to be worked out, under great pressure, on the spot.

Penney apportioned no blame to anyone at Windscale. In fact, his report was fulsome in its praise of the 'considerable devotion to duty on the part of all concerned'. He was, though, most conscious of the heightened level of national concern following the accident and the need for a speedy report on its cause. Of especial importance were any likely consequences for public health. Such urgency had therefore made it impossible to carry out an in-depth technical assessment of all aspects of the matter. In addition, organizational changes were surely needed, but his board was not properly constituted to recommend these. The report, dated 26 October 1957, did make five far-reaching recommendations. The first called for the setting up of a Technical Evaluation Working Party to make a detailed study of all technical information to do with the accident; the second proposed that the AEA should conduct a detailed review of the organization of the Industrial Group, the relationship between the operational staff at Windscale and the other technical directorates, and of both the numbers and quality of its staff, who were entrusted with considerable responsibility. Clarification must be sought in regard to the authority's responsibilities in respect of health and safety, while steps should be taken to determine the maximum safe levels of exposure to radioactive substances. This, of course, was not solely the province of the AEA, but the Windscale experience had now made it of paramount importance. Finally, Penney's board of inquiry recommended that Pile No 2 at Windscale should not be restarted until its instrumentation was adequate and more was known about the factors involved in the controlled release of Wigner energy.

The accident at Windscale signalled the end of the first phase of Britain's affair with nuclear energy. The two piles adorned with Cockcroft's follies were never to operate again. Hinton had called them 'monuments to our initial ignorance'. Together they became stark reminders of what might have been.

*   *   *

The inquiry over, its members went their separate ways. The report went straight to the Prime Minister who had just returned from his most successful visit to the States, where he had signed the Eisenhower–Macmillan Declaration pledging, amongst other things, full cooperation between the United States and Great Britain in the fields of nuclear weapons and nuclear energy. The McMahon Act was effectively dead. Penney's report, however, filled the jubilant Macmillan with gloom and his immediate reaction was that, if released, it could so easily jeopardize the outcome of those most far-reaching negotiations with the Americans which still had to receive the blessing of Congress [23]. The Ministry of Defence recommended publication as did the Board of the AEA who, in addition, accepted collective responsibility and opted for full and frank disclosure of the shortcomings in the authority's own structure and the inadequacies in the instrumentation of the piles that had contributed to the accident. It was satisfied that the technical evaluation committee, including independent experts, soon to be chaired by Sir John Cockcroft, would uncover every relevant detail, while organizational matters would receive the attention of another under the chairmanship of Sir Alexander Fleck, Chairman of ICI. At issue, too, were the possible consequences of Wigner energy release for the Magnox reactors at Calder Hall and Chapelcross, as well as those in the civil nuclear power programme then well under way. Understandable public anxiety would only be assuaged by presenting the facts in a clear and forthright way, even though they would provide ammunition for those implacably opposed to nuclear energy in all its forms.

But the Prime Minister decreed otherwise. The Penney report would not be published; instead he would himself report to Parliament and a shorter, less technical document in the form of a government white paper would follow. It should never be forgotten that Windscale was first and foremost a defence installation there to provide the plutonium, polonium and tritium for nuclear bombs. Its other functions such as the production of radioisotopes for medical and scientific purposes, and for the testing of a variety of materials to assess the effects of radiation, were all secondary. When constituted, the Penney board of inquiry were given strict terms of reference, as we have seen, and it was decreed that its deliberations would not take place in public, nor would its report be published because of possible defence implications. The Prime Minister could have chosen to keep the report under wraps on defence grounds alone but there were none because the Ministry of Defence, having seen the report, cleared it for publication [24]. Macmillan, though, was driven by considerations that extended well beyond Britain's shores. Restoring American confidence in both British scientific ability and in its ability to protect the nuclear secret were paramount. Penney's committee was, therefore, hastily re-assembled and set about abridging their original document. As requested, they removed most technical details as well as their conclusions and recommendations. These then found their way into an introductory memorandum written by the Prime Minister himself.

The white paper [25], which included this abridged report amongst a group of six annexes, was the work of Downing Street, the Cabinet Office, the Atomic Energy Office and the AEA. It was presented to Parliament by Macmillan on 8 November [26]. The House, distracted by a leak of a different kind (of the bank rate) received it almost routinely. The reaction of the newspapers, other than some disquiet about the composition of Penney's commission and suggestions that the accident had been played down, was generally positive. The explanation offered was 'frank and satisfying' and public concern had seemingly been allayed. However, talk of 'faults of judgement by the operating staff', in the memorandum from the Chairman of the authority Sir Edwin Plowden, and the newspapers' interpretations of this slander, prompted a heated reaction from those concerned at Windscale and from their staff association. For all the assertions of the Board of Inquiry and the authority that the accident was a matter of collective responsibility, the inevitable political machinations and the hands of many scribes had managed to convey a different message. If blame was to be attached then it lay fairly and squarely within the chain of command and the lines of communication between Harwell, Risley and Windscale. In addition, the serious under-manning that existed within key scientific and engineering sectors, particularly the Operations Branch, had made entirely unreasonable demands of those who had held the safety of Windscale in their hands during those few days in October.

No-one was more outraged by the white paper than Sir Christopher Hinton. Now desk-bound in London at the CEGB, he fumed at the injustice done to the men at Windscale who had borne so much responsibility during the fire. For years he had warned the AEA against 'stampeding into a wildly expanded programme' as its objectives seemed to embrace an ever-widening arc from diffusion plants to reactors with pure science sandwiched in between. 'When any technological development becomes a totem of national prestige', he reflected much later when, no doubt, thinking of ZETA as well, 'common sense flies out of the window' [27].

Schonland, much caught up in that spiral, was soon in action again. It was now blatantly obvious that the behaviour of graphite under conditions of irradiation had been all but ignored at Harwell, while the resources at Windscale, where the study had been transferred, were not up to the task. It was therefore vital that an immediate and urgent study be mounted to understand it. He therefore worked with the research director at Risley and others, including R F Jackson, the Chief Engineer at Harwell, to devise a programme of research and this received the go-ahead from Cockcroft and Owen in mid-November. The pressure was immense because an interim report was required by 12 December [28]. In all, 37 scientists and engineers, split almost equally between Harwell and Windscale, worked on the research that was headed by Alan Cottrell, soon to leave Harwell for the Chair of Metallurgy at Cambridge. No project at the time had a higher priority and, given the findings of his recent inquiry, Penney was in no doubt that there was nothing more urgent for the AEA to do: on its outcome rested the very future of the UK's nuclear energy programme, both civil and military [29]. The effort involved was exhausting. Schonland made numerous trips from Harwell to the north and also to London for frequent meetings. The strain, too, was telling, particularly when the time came to take one of Harwell's research reactors, BEPO, through a controlled cycle of Wigner energy release. One had already been done successfully before the Windscale accident, now the question was asked whether to risk another or just to shut the reactor down. Recent events dominated everyone's thoughts; Jackson and his colleagues used the results of the most recent research to analyse the situation and he concluded that it would be safe. The responsibility for giving the order to start the anneal, though, fell to Schonland [30]. Some months later he described in a letter to his brother Felix how Harwell's heart almost stopped beating as he prepared to issue the order to begin [31]:

> After Windscale it was a bit of a worry and the place was thick
> with firemen and nitrogen tankers and experts from everywhere,
> including the USA, all night. I went to bed at 4 a.m. — a much
> relieved man, for the graphite people had had some rather hair-
> raising ideas in the course of their tests. None of these were borne
> out in practice.

The pace remained almost frenetic. Regardless of whether cooperation between Britain and the United States in the nuclear field had been sanctioned or not, the Americans took the very greatest of interest in the Windscale accident and in the outcome of the investigations that followed. Early in December, the US Atomic Energy Commission sent a team of scientists and engineers to England for three days of intensive meetings at Risley. Schonland joined Cockcroft and Owen and fifteen of their colleagues in discussions which peeled off every layer of the saga that had so nearly plunged Britain into disaster. Their deliberations were so sensitive they were classified 'Secret Atomic'. Now, for the first time, the Americans were able to reveal details of their own experiences with Wigner energy, fuel cartridges and operating procedures. Most revealing of all, since it had been so secret, was the information they now willingly imparted about the Hanford reactors, upon which the US depended for its supplies of weapons-grade plutonium [32].

*   *   *

As the year drew to its close, Basil and Ismay took themselves away from Harwell for a week's break in London. There they were joined by their son, David, who was then lecturing in mathematics at South-ampton, after completing his PhD at Birmingham. Schonland felt worn out. He was now nearly 62 and the events of the past few months had taken their toll. To Felix he wrote 'I have had a very heavy time up to Christmas owing to the Windscale accident ... then came a party of Americans to confer on graphite and another to confer on general incident matters. The net result was pretty grim for me!' Not recrimina-tion, just exhaustion. And it was by no means all over. Three more white papers landed on various desks in Whitehall and elsewhere. They were the outcome of the various inquiries set up immediately after Windscale under Sir Alexander Fleck. Their implications as regards Harwell and its future were far-reaching. A restructuring was now very much on the cards and its various shortcomings, for so long ignored, now had to be put right. The scientist in Schonland couldn't resist reminding his engineer-brother of the natural order of things in the world as he saw it. 'The Operations Branch', he wrote, 'has been run by engineers who don't know enough about science to avoid accidents—we shall [have] to draft some of our people North to keep them up to the mark!' [33].

CHAPTER 24

'CAPTAIN AND NOT THE FIRST OFFICER'

On 25 January 1958 the journal *Nature* published a group of papers by both British and American authors containing the latest results in the field of nuclear fusion. The editorial accompanying them was effusive: 'The announcement of the successful control of the thermonuclear reaction has now been made, and all the scientists and others concerned are to be congratulated on their magnificent achievement' [1]. The Harwell paper listed twelve authors, headed by Dr P C Thonemann. It acknowledged the encouragement and support given to the research by Sir John Cockcroft and the late Lord Cherwell.* Clearly, this paper was supposed to be the definitive statement on a most vexed subject and its publication, along with papers from both the AEI Laboratory at Aldermaston Court and from various American research institutions, was the outcome of protracted negotiations at the highest of levels. The mood in those British circles close to the fusion programme, over the preceding few months, had been one of exasperation at the intransigence of the Americans, whose own work was less advanced but whose real concerns were overwhelmingly related to security. There was, at least in some quarters of the US Administration, an almost paranoid fear that the massive flux of neutrons produced by a device like ZETA would enable small nations to 'quietly set about making atomic bombs without entering on the costly business of building reactors like Windscale or separation plants like Oak Ridge' [2].

Rather more fundamentally, though, there still remained one unanswered question. What was the source of those neutrons? Following Cockcroft's precipitate announcement in January the British press were in no doubt and were decidedly cock-a-hoop, but Thonemann's paper in *Nature* did not venture to answer it [3]:

> To identify a thermonuclear process it is necessary to show that
> random collisions in a gas between deuterium ions are responsible

---

* Somewhat belatedly, a week later, a note appeared in the next issue adding the name of Sir George Thomson to those whom they wished to acknowledge.

for the nuclear reactions. In principle, this can be done by
calculating the velocity distribution of the reacting deuterium ions
from an exact determination of both the energy and direction of
emission of the neutrons. The neutron flux so far obtained is
insufficient to attain the desired accuracy of measurement.

The matter was, therefore, by no means settled and even Cockcroft himself was beginning to moderate his views. As little as a week later he published a popular article in which he was careful not to ascribe the neutrons to a thermonuclear reaction [4] and followed this with an equally restrained performance at a special meeting convened at the Royal Society to discuss the whole subject. However, for all his sudden reticence, he still had an ally in the person of Sir George Thomson, who refused to accept that, purely by chance, there existed a non-thermal source that gave precisely the same neutron yield as that predicted for the temperatures they had observed. Since those temperatures were themselves hardly accurate, Thomson was on flimsy ground and Cockcroft's case was, at best, tenuous and he knew it [5].

And so, on 17 February 1958, with the air still thick with the aftermath of Windscale and the more recent repercussions from ZETA, Basil Schonland took over from Sir John Cockcroft as Director of the Atomic Research Establishment at Harwell. He was now also Director of the Research Group of the AEA but would report to Cockcroft who was the authority's Member for Scientific Research. It was by no means an easy transition, coming as it did at a time of great change within the AEA, at Harwell and in nuclear research in general in Britain. From the very beginning Schonland's authority was uncertain. The Cockcroft aura was still all-pervasive at Harwell and, to compound it, so was his physical presence because he chose to remain in residence at No 1 South Drive while the Schonlands continued to live next door! If subliminal effects mean anything, this one conveyed entirely the wrong message to the rest of the establishment. Though in his own words to Felix, Schonland declared himself pleased to be 'the Captain and not the First Officer' some will have wondered whether Cockcroft was not still the Admiral [6].

Where there was no doubt about Schonland's status was in South Africa. His assumption of the top job at Harwell again produced banner headlines in the local newspapers and soon letters and telegrams of congratulation were pouring in to the establishment. As ever, Schonland was meticulous in replying to each, many of which had come from complete strangers who just wished to express their pride in his appointment. One that pleased him particularly came from Dr Frank Hewitt, now director of the Telecommunications Research Laboratory in Johannesburg. Their close association in the SSS and subsequently in London during the war brought back pleasant memories to his former commanding officer. Schonland wrote:

> Many thanks to you and all my friends at TRL for their extremely
> kind cable of congratulation. I hope it will not be long before I see
> most of you over here on some visit or other. Meanwhile I can only
> say how much I admire the very beautiful work* which continues
> to come out of the little laboratory we formed long ago from a few
> people who had been in the famous S.S.S.

Schonland's elevation to high office had caught the eye of the Information
Attaché at South Africa House who wished to interview him for a special
article in *Panorama*, a glossy news magazine, but Schonland declined. 'I
am sorry to tell you', he wrote, 'that being a public figure is not at all in
my line and as in other cases I am not able to help you'. Even the
senior representative of the South African Press Association in London
had no luck in his attempt to persuade him to be interviewed 'for an
article in serious vein in which your name, new appointment and views
would be coupled with those of your and our compatriot Sir Solly
Zuckerman'. Schonland stood firm; Zuckerman was happy to oblige.

Letters arrived from many of Schonland's former associates in the
AORG and 21 Army Group. Generals Freddie de Guingand and David
Belchem wrote with their congratulations, as did Omond Solandt from
Canada. Closer to home was one from John Wrightson, now a partner
in a firm that was making headway in the rapidly expanding nuclear
construction industry. To Wrightson, a former infantry officer attached
to 21 Army Group Headquarters in Brussels in 1944, Schonland remained
the 'Brigadier' as he conveyed his best wishes and looked forward to their
next meeting. And there was even one from a former Lance Bombardier
who had worked as a clerk at Petersham. Schonland wrote back thanking
him for his kindness and said he remembered him well as the 'sole
member of our office staff who could work our ancient typewriter'.
And a letter also arrived from Rupert White-Cooper, Schonland's
school-friend at St Andrews and then fellow soldier in two world wars.
A reunion was certainly called for [7].

Such recollections of times past and of the men who were his closest
colleagues during those war-torn days caused Schonland to think of his
former chief who was now Deputy Supreme Commander of NATO.
Hosting visits to Harwell by dignitaries was, by now, very much his
stock-in-trade but they had all been there at someone else's invitation.
As his first official guest he invited Field Marshal Montgomery of
Alamein to visit the establishment on 9 April 1958 and to dine with him
afterwards. Monty readily agreed to visit, but declined the luncheon

---

* He was presumably referring to the 'Tellurometer', a microwave distance-measuring
instrument designed at the TRL by Trevor Wadley in 1954. The first production models
were produced in 1957 by a company in Cape Town managed by S H Jeffrey, at whose
artillery battery in East Africa Schonland and Hewitt had installed a JB radar in 1940.

Figure 29. The visit to Harwell in April 1958 by Field Marshal Montgomery of Alamein. Schonland, his host for the occasion, looks on approvingly as 'Monty' views an exhibit. (Reproduced by kind permission of the UKAEA.)

invitation: '. . . must get home. I fly to Paris on the 10th. Send me a map of the route'. It all seemed like yesterday — such orders and such places! The Field Marshal duly arrived and Schonland showed him around Harwell, with a pause on the way to explain to him the workings of ZETA, but Monty's main interest was to meet 'the men'. Amongst them were a number who had served under him in the 8th Army in North Africa and later in 21 Army Group and he obligingly agreed to be photographed with them all. His letter thanking Schonland for his hospitality followed soon after: [8]

10 April 1958

Dear Schonland,

I enjoyed every minute of my visit and learnt much that I did not know before. I was greatly impressed with the efficiency of the staff — and the happy relations that existed between master and men. You clearly have a first class show, of which you can be very proud.

Montgomery of Alamein

525

Figure 30. The Director, Dr B F J Schonland, discusses an exhibit with Her Majesty the Queen, accompanied by HRH the Duke of Edinburgh, when on a visit to Harwell. (Reproduced by kind permission of Dr Mary Davidson and the UKAEA.)

The new Director's first general communication with his staff, who now numbered more than 6000, was via the columns of the *Harlequin*. All were eager to hear how the land would now lie, especially as Sir John Cockcroft was still much in evidence. Diplomatic as ever, Schonland said that, along with all his colleagues, he was 'happy that [Cockcroft] will still be with us at Harwell, both because he will continue to live here and because he will continue to guide our policy and general progress, for that guidance remains his responsibility' [9]. Deep down he must have longed for just a little more distance between them, now that he had Harwell under his wing. It was those responsibilities that were uppermost in his mind and none was more pressing than piloting through a series of organizational changes within the Research Group — developments that would lead ultimately to the fracturing of Harwell itself.

It was the Fleck Reports following the recent events at Windscale that set in motion sweeping changes, the effects of which would long be felt throughout the AEA. The first of Sir Alexander Fleck's three committees to report did so in December 1957. It dealt mainly with the organization of the Industrial Group and, though full of praise for the vitality and efficiency of its staff, and for their significant achievements in bringing on stream the country's first reactors, the report also pin-pointed clear deficiencies that had to be addressed. Behind these lay a severe shortage of technical manpower, particularly at the senior level within the Operations Branch. Of course, the nuclear industry was not unique in this, for it just mirrored the national problem at that time when the universities were not producing in sufficient quantity the scientists and engineers for which industry was crying out. In addition, the pay structure within the authority, which was tied to that of the Civil Service, did little to help. Since the nuclear power programme required both maximum operational efficiency and the highest possible safety standards, all other activities, though of national significance, had to be secondary to these and this immediately had implications for the rest of the AEA establishments. In essence, the report called on the AEA to ensure that its development efforts were concentrated on fewer projects and, where necessary, to transfer staff from within the Authority to fill the key vacancies within the Industrial Group [10].

The second of Fleck's reports covered issues of health and safety within the authority, while the third evaluated the technical aspects of the Windscale piles and, particularly, the controlled release of Wigner energy from the graphite moderators. Schonland, as we have seen, was intimately involved with those Wigner trials, but there were also other working parties dealing with filters, instrumentation and cartridges and all their reports had implications for Harwell and its immediate research priorities. However, it was the expanded programme of research on new reactors that Cockcroft had initiated and the continuing work on fusion, that really taxed the new Director since, as history would show, the former was an extravagance that the country could neither afford nor justify while the latter was bedevilled by scientific and personal controversy.

As early as 1955, Cockcroft was of the view that it was Harwell's function to work on new designs of reactor, while an entirely new establishment, beyond its boundaries, would concentrate on what he called 'advanced reactor concepts' as well as on small experimental reactors. Already three such concepts were being studied and these proposals would see two or three more being taken to prototype stage. Industry was alarmed but the visionary in Cockcroft held sway and soon he, Schonland, Raby and D W Fry began to look for a suitable site for a 'second Harwell' [11]. After examining well over 70 localities throughout England and Wales the one they eventually selected was in Dorset. Planning permission was eventually granted in February 1957, after much negotiation and placating of local communities, for the building at Winfrith Heath, an isolated and somewhat desolate site some 130 km to the south of Harwell, of what would become the Atomic Energy Establishment or AEE. There, the newly-formed Development and Engineering Group of the Authority would be set up, under Fry, to study the basic physics of nuclear reactors with the purpose of reducing their capital cost. By this time the British nuclear power programme was well under way, with five power stations under construction and two more to follow. This massive effort was carried out by industry, on contract to the various electricity boards around the country, and their design was based on the original gas-cooled, graphite-moderated Calder Hall station.

The primary aim of the AEE's research was to increase reactor efficiency by operating them at higher temperatures. The intention, in what became known as the Advanced Gas-Cooled Reactor, was to use uranium oxide rather than uranium metal as the fuel, while the higher temperatures would require a new container material, such as beryllium or stainless steel, instead of the magnesium alloy in use at Calder Hall. Ultimately, to increase the temperature even more (to obtain yet higher efficiencies), would involve the use of special refractory materials then still under investigation. It was a vast project and the AEA itself was not in the financial position, on its own, to support the research and development effort involved. Cockcroft was undeterred. European interest in nuclear power had been aroused by the research undertaken in the States and in Britain, and in 1958 the European Nuclear Power Agency was formed. Within a year a collaborative agreement had been entered into with the AEA and an international DRAGON project team was set up at Winfrith to concentrate on this research [12]. But other issues were also in the air and soon this tranquil Dorset site would be the focus of much controversy, for reasons quite unconnected with fission reactors, no matter how conceptually advanced their designs or how economically strained their budgets. The successor to ZETA would be the cause and Schonland would be embroiled in it all.

*   *   *

In the meantime, research into nuclear fusion had developed a momentum all of its own and was proceeding apace. By the end of May it had been firmly established by a team of nuclear physicists working on the Harwell cyclotron that the neutrons observed by their colleagues in the fusion group were almost certainly non-thermonuclear in origin, and their findings were published in *Nature* in June. Yet another press conference was called, this time at the Authority's headquarters in London, and it was addressed by Basil Schonland himself in his new capacity as the Director [13]. In a carefully worded statement, that was based on this most recent research and a detailed re-examination of all previous work, he described the method of operation of ZETA and made it clear that the three milliseconds during which the hot deuterium gas was separated from the walls of the torus were insufficient to allow the plasma to reach thermal equilibrium—a necessary condition if the resulting fusion reactions were to be described as thermonuclear. This was a vitally important requirement and it not been appreciated at the time of Cockcroft's famous unguarded answer. The neutron flux upon which Cockcroft, at least, had placed so much hope was, therefore, spurious and no thermonuclear processes had been at work. This fact though, said Schonland, 'does *not* make the ZETA results less significant'. What was required were higher temperatures and longer times of isolation, and to achieve them a successor to ZETA was already being planned.

The reaction of the daily press was mixed. Most tried to be positive about what was still a rather bewildering subject, but there were some who saw this retraction by Harwell as a great blow to British prestige, just when it appeared that ZETA was a triumph on a par with Russia's successful launching of Sputnik. However, more thoughtful reflection revealed a deeper concern, and that was the effect that the blanket of secrecy imposed at Harwell (seemingly for competitive and not military reasons) had had on the free exchange of information between scientists. Such disappointments, opined the science correspondent of the *Manchester Guardian* perceptively on 17 June, could surely have been avoided had the small Harwell team working on ZETA been able to expose their ideas to 'sound analysis and interpretation' by rubbing shoulders with scientists of other specialities in the way that made universities 'so excellent in pure scientific research'. There were many at Harwell itself who had long espoused such views but the active participation in fusion research by Penney's group at Aldermaston— with rather different objectives in mind, none of which were commer- cial—had made such openness impossible. In addition, American reluctance to divulge anything at all had put Harwell in a doubly difficult position. However, the recent efforts of Harold Macmillan (his own act of censorship over the Windscale report notwithstanding) in convincing Eisenhower to repeal the McMahon Act were soon to bear fruit. Not only would the Americans now aim for complete declassification of their fusion work in time for the second Geneva Conference on the Peaceful Uses of Atomic Energy in September, but the mood had also spread to the Soviet Union, which similarly opted for full and open publication of its achievements in the field [14].

In December 1957, just a short while before Schonland took over from Cockcroft, the expectations for ZETA were running high at Harwell. Achieving controlled thermonuclear fusion in the laboratory seemed only a matter of time. Already there was talk of a bigger and better machine and budgetary approval had been sought for ZETA II, capable of sustaining a toroidal current of 2 MA for a pulse duration of one hundredth of a second. The energy storage was estimated to be of the order of 50 MJ, with a mean power consumption of 7 MW. Construction would probably take some three to four years at a cost of about £3 500 000. ZETA II thus represented a major engineering undertaking necessitating considerable outside involvement, but where this work would actually be done still had to be decided. A design specification was duly drawn up, in collaboration with Metropolitan-Vickers, and attention was now focused on expanding considerably the size of the Harwell fusion team to take on this most exciting development.

At the Authority's headquarters in London the mood was not quite so sanguine. Concern had been expressed about the size of Harwell, with the feeling that it was too big to be managed effectively. In addition, ZETA now seemed to have become as much an industrial enterprise as a research tool, and that dichotomy rather called into question its very existence at Harwell, which was still regarded as the focal point of fundamental research. Cockcroft, aware of his AEA colleagues' concerns, proposed that ZETA II should be sited at Winfrith—a suggestion first made to him by Schonland who was himself convinced that this was precisely the purpose for which Winfrith had been set up. Quite independently, Schonland had also made arrangements with Sir Leonard Owen that the Industrial Group at Risley would play a major part in the construction there of ZETA's successor. So the move of ZETA II to Winfrith was confirmed and with it would go a number of scientists, engineers and support staff from Harwell to make up what would become the Controlled Thermonuclear Reactions (CTR) Division. But when news of this impending move reached Harwell there was an immediate reaction against it. Leading the charge was Dr B H Flowers,[*] now Head of the Theoretical Physics Division. Flowers argued strongly to Cockcroft that, contrary to the impression held by both the AEA's and Harwell's senior management, ZETA II was every inch a fundamental experiment in physics. Its success, just as had been the case with ZETA, would require input from scientists who were not members of the CTR Division but who were very much part of Harwell. Moving ZETA II to Winfrith would be catastrophic, he said, because 'Winfrith was not a place to which the best research physicists will want to go, or should go, or are likely to go'! [15].

Flowers made a further point which was related directly to the moves afoot to slim down the Harwell monolith. He was greatly

---

[*] Later Baron Flowers of Queen's Gate in the City of Westminster. Flowers left Harwell in 1958 to take the Chair of Theoretical Physics at Manchester; in 1961 he became Langworthy Professor of Physics. From 1973 to 1985 he was Rector of Imperial College of Science and Technology, University of London, and between 1967 and 1973 was Chairman of the Science Research Council.

concerned that, since the AERE had already lost its reactor research programme to Winfrith, and was in the process of shedding even more of its functions to new establishments then springing up in its vicinity, it would soon be left without any role whatsoever. He had certainly touched a nerve, for 1957 had been the year in which plans to divest Harwell of much of its scientific infrastructure had taken root. All activities associated with the production of radioisotopes using the research reactor BEPO, as well as the two piles at Windscale, were soon to be transferred to Wantage about 11 km away. In addition, some of the Electronics and Engineering facilities were to be moved to Bracknell, while parts of the Analytical Chemistry Branch had already been accommodated at Woolwich and Chatham. But by far the most significant development was the decision, announced in February that year, to set up the National Institute for Research in Nuclear Science (NIRNS)* adjacent to Harwell. It had long been appreciated that universities were unable to afford the massive investment in equipment and facilities that were now very much part and parcel of modern nuclear physics. Gone were the days of glass tubes and rubber bungs; what was now needed were the accelerators and cyclotrons that were well beyond the reach of any single academic institution. By July 1957, when Cockcroft cut the first turf to mark the site for NIMROD, the Institute's 7 GeV proton synchrotron designed by the AERE's Accelerator Group, plans were well under way also to transfer the 50 MeV proton linear accelerator from Harwell to NIRNS. In addition, resident teams of support staff[†], to assist their university colleagues in the operation of the equipment and to conduct their own research, would be moved across from Harwell in the immediate future. What was to become known as the Rutherford High Energy Laboratory — a name suggested for it by Schonland himself [16] — would in years to come be renamed the Rutherford–Appleton Laboratory; both names of great resonance for Basil Schonland.

It was indeed an exodus. The vast empire that had been Harwell under the enigmatic inspiration of John Cockcroft was now to be passed to his successor in a rather different form. What Schonland actually inherited was a pedigreed specimen that was in the process of

---

* NIRNS was constituted on 12 March 1958 with 16 members incorporated by Royal Charter. Cockcroft, Schonland and Sir Donald Perrot were the three members representing the AEA.
† One of the first of these Group Leaders was Dr G H Stafford who had completed his BSc and MSc at UCT in 1939 and 1941 respectively. After the war Stafford joined the NPL (CSIR) under Schonland and was then seconded to Harwell for a period. On his return to the CSIR he served as a Staff Officer within the South African Corps of Scientists with responsibility for matters to do with 'Atomic Warfare'. In 1954 he moved to England and in 1958 became group leader for the proton linear accelerator at NIRNS, From 1969 to 1981 he was Director General of the Rutherford Laboratory and then Master of St Cross College, Oxford, from 1979 to 1987. He was elected a Fellow of the Royal Society in 1979.

Figure 31. The AERE Research Group in August 1958. Seated (L to R): Dr H M Finniston, J R V Dolphin, D W Fry, Dr B F J Schonland, Dr R Spence, D R Willson, Dr J V Dunworth. Standing (L to R): Dr P C Thonemann, Dr W P Grove, Dr W G Marley, E H Cooke-Yarborough, A S White, J F Jackson, Dr N F Goodway, W S Eastwood, R M Fishenden, Dr E Bretscher, T B Le Cren, Dr B H Flowers, Dr T G Pickavance, H J Grout. (Reproduced by kind permission of the UKAEA.)

being disembowelled by committee. With it he acquired a staff divided and, in some cases, disenchanted by what they saw as a wanton disregard for science, dominated by financial considerations imposed from outside and by indecisive management from within. No longer would Harwell flourish as the very epitome of the curiosity-driven research establishment that appeared so like a university to Basil Schonland on his arrival almost four years before.

The matter of a home for ZETA II was thus far from settled. Cockcroft was sympathetic to Flowers's views, while others such as D W Fry, under whom the whole project would fall were it to come to Winfrith, had vacillated all along. Initially opposed to its siting in Dorset, Fry then agreed with Cockcroft and Schonland that the move was necessary. However, after listening to Flowers he again changed his mind. In view of the obvious discord that was being stirred up, Cockcroft decided to put the matter before the Harwell Steering Committee, on which the scientists were strongly represented. The meeting actually took place in mid-January 1958 while many of those whose passions were aroused by the siting of ZETA II were also caught up in all the excitement of ZETA and the forthcoming publication of its first results. Passions had been aroused and argument was fierce on both sides. The case against the move was put most forcefully by Flowers. It rested on the fact that the fusion team, even in its enlarged form, must be kept together at Harwell which was, by far, the more suitable location. He recognized that the face of the establishment would change by this heavy emphasis on fusion research, but might that not be a good thing? By simply reorganizing the site and removing its security restrictions Harwell would become the centre for research into nuclear fusion and solid-state materials. In this way it would grow in stature and its international reputation would benefit immeasurably.

It was certainly a compelling argument but it was vigorously contested, particularly by the Research Group's two senior engineers, J R V Dolphin from Harwell and his opposite number at Winfrith, H J Grout. Both insisted that the development of ZETA II was an industrial project with clear commercial implications in the short term, thus making Winfrith the logical choice. Both arguments found support and the meeting was split between the perceived scientific excellence of the project and its possible commercial exploitation. Not surprisingly it failed to reach consensus. A policy decision would have to made at the very highest levels within the authority.

Schonland, as Director of Harwell and Director of the Research Group, had ultimate charge of Harwell as well as the new AEE site at Winfrith, plus the Radiochemical Centre at Amersham [17]. But he had no seat on the AEA Board, where matters of policy were still Cockcroft's responsibility as the AEA's Member for Research, and so Schonland found himself caught between the warring factions at Harwell, where the expectation was that he would take a firm lead, while he himself was tied to decisions that had been reached by a body, the authority's board, on which he had no direct say. It was therefore not an entirely happy position in which he found himself and it was soon made worse by a decision taken by Sir Edwin Plowden, Chairman of the authority, that there was to be no expansion in staff at Harwell. New areas of research could be started only if older ones were closed down and the purchase of new equipment was also to be severely constrained, but most important of all was the ceiling placed on the numbers of staff employed at Harwell itself. Plowden was insistent at a meeting he had with Cockcroft and Schonland early in May that, while he fully understood the merits of the case for keeping ZETA II at Harwell, there would be no increase in staff to work on it. This left little alternative but to go ahead with the plan to move the whole fusion project and, with it, the embryonic ideas for ZETA II, to Winfrith. The formal announcement to this effect was made by way of a press release in September [18].

On hearing this, Harwell's scientists were far from satisfied and Peter Thonemann, whose part in the fusion programme had been so pivotal from its very inception, was incensed. Even before this announcement had been made he had informed Fry that he would be most reluctant to move to Winfrith and warned that other key members of the team might drop out as well. But both Cockcroft and Schonland were insistent that there was no alternative to Winfrith and they emphasized that the decision to move had been taken. What followed was as close to a rebellion that the individualistic scientists at Harwell could possibly mount. Letters flew between them and Schonland's office. The experimentalists, whose complaints had initially been the most vocal, were now joined by the theoreticians, led by W B Thompson, who claimed (presumably because they would be separated from their computer* and Harwell's extensive library) that working at Winfrith would be practically impossible. Even some of the administrators joined in by condemning the travel and accommodation arrangements associated with a move to the wilds of Dorset.

The situation was in turmoil and to try and resolve it Schonland, late in October, brought the matter to the Research Group's Board of Management, of which he was chairman.† After a protracted discussion the Board approved the provision of new buildings for the fusion project at Winfrith, but only by a small majority, with both Spence (Head of

---

* The AERE's first computer, the IBM model 704, was installed in 1957.
† Its members were: B F J Schonland (chairman), D W Fry (deputy chairman), J R V Dolphin, R Spence, H M Finniston, J V Dunworth, D R Willson, T Le Cren (all from Harwell), J C C Stewart (Industrial Group), S C Curran (Weapons Group) and J A Jukes (London Office).

Chemistry) and Finniston (Head of Metallurgy) voicing their support for their fusion colleagues at Harwell. Soon after, at a special meeting of the Harwell Council called to thrash out the affair, the ranks of the dissenters increased significantly, with solid support coming from all the leading fusion scientists as well as from other Division Heads, most notably Dr Egon Bretscher of Nuclear Physics. A new case against the move was brought to bear by Dr W G Marley, Head of the Health Physics Division, who pointed out that the very remoteness of the Winfrith site was no accident. It had been selected because of the stringent safety regulations that would naturally apply to the fission reactor research to be undertaken there. Since no such safety restrictions applied to the fusion work, all those involved in it would be unnecessarily hamstrung by them.

Poor Schonland had a miserable time. He was caught very much on the horns of a dilemma not of his own making. Though sympathetic to all the arguments against moving the fusion project to Winfrith, his position of executive responsibility to the Board required that he had to defend it. Those who watched him at this most trying time could not fail to be impressed by the quiet dignity with which he conducted himself in the face of concerted opposition and considerable dissatisfaction at the way in which the views of Harwell were being trammelled almost by ministerial edict [19]. For a time it was clear that he saw no way of resolving it. Then, in November, there were two developments that enabled him to cut what was very much a Gordian knot.

Harwell's Chief Engineer and stout defender of the need to move ZETA II to Winfrith provided Schonland with a way of resolving the problem. J R V Dolphin wrote him two letters during November in which he suggested that the solution surely lay in locating the fusion work on an entirely new site close to both Harwell and the new Rutherford Laboratory then under construction next door. This initially seemed preposterous in the light of the Treasury-driven requirement to restrict any growth within the AEA. However, when Schonland tentatively floated the idea, it was soon apparent that the major issue of concern in London was not simply the cost but also the sheer size of Harwell and the management problems this was causing. Any mechanism for reducing the number of people at Harwell would be welcomed, and the case for an adjacent site was strengthened when further investigation even suggested that it would be no more expensive to build this new site alongside Harwell than to accommodate the fusion team at Winfrith.

Dolphin's second proposed route out of the morass was provided by the ever-increasing complexity into which the ZETA II design had descended. The diameter of the torus had been increased to 6 m, the current pulse lengthened to 0.5 or even 1 s (a 500-fold increase from that employed in the original machine), while the toroidal magnetic field was to be increased more than 30 times to 0.5 T. Cost estimates jumped to £5 000 000, while the number of additional professional staff required rose by a hundred. Even more radical changes were to follow. To meet the needs for a very rapid current rise-time, 200 MJ of stored energy and a total current capacity of 7 MA would require the design of a rather special power supply, the details of which had already been patented by Harwell. It was soon clear to everyone that ZETA II was by no means a theoretical *fait accompli* just awaiting engineering development: it was still very much an object of scientific investigation. The whole basis of the argument in favour of the move to Winfrith had been the perceived industrial applications that might follow from the development of this new toroidal system. Clearly such applications were now far less certain and demanded that the issue be reconsidered, so Schonland convened a meeting of the Board of Management on 18 November. With the array of technical facts displayed in front of them, and Dolphin's proposal on the table, the members soon agreed that the future of the fusion programme depended on the closest possible contact between the fusion team, other divisions at Harwell and even with university departments. All of these ruled out any move to Winfrith and Sir Edwin Plowden had little option but to agree, however reluctantly, to accept the situation now presented to him. Fusion research would therefore move, not to Winfrith, but just to Culham.

Culham was a former Royal Naval Air Station a mere 10 km from the Harwell site. The Admiralty agreed to its transfer and by May 1959 Culham was an AEA property about to be developed. Basil Schonland thus found himself in the very unaccustomed position of being a landlord with major construction projects requiring his considerable attention, and the irony was not lost on him. As he described in a letter to Anton Hales, his successor at the BPI in Johannesburg, he now found himself looking after 'another Harwell—at Winfrith Heath, an Isotope establishment at Wantage and another new establishment is mooted. Seeing that I always swore that I would have nothing to do with buildings and avoided getting tied up with the CSIR building programme I have only got what I deserve!' [20].

It had been a most demanding year and, in his Christmas letter to Felix, written just after the last of three functions that the Schonlands hosted at their home for Harwell staff and local dignitaries, Basil provided a résumé of the year's events. 'In addition to normal work and responsibilities and quite an astounding amount of abnormal work caused by 'flaps' and panics and reorganisation ... Cockcroft has been away to Russia and Tokyo and I have had to do his work a lot.' Such euphemisms hardly described the aftermath of Windscale and the drama of ZETA but Basil rarely let on to his younger brother quite how demanding his duties and responsibilities were at Harwell and Felix had much reading to do between the lines, while receiving another dig about the fallibility of engineers. 'It has been quite a year with fusion going wrong and fission giving rise to a great deal of high priority research work to solve the problems which you engineers always ignore until they raise their ugly heads. I expect next year will be worse' [21]. Certainly the next year would see many changes and the smoke from some of them was already on the horizon.

One momentous change would be the final departure from the British atomic energy scene of Sir John Cockcroft. Though not yet announced officially, those closest to him knew that Cockcroft would soon be leaving them to become Master of the recently founded Churchill College at Cambridge. Schonland kept the details under his hat but he did give Felix a hint: 'Cockcroft', he said, 'gets higher and higher into the stratosphere as you will learn in about a month, when the news gets out'. Also in the air was yet another change within the AEA itself. Sir William Penney, who would replace Cockcroft as the AEA's Member for Research, would assume executive responsibility for the Research Group, while other changes would also take place within the Development and Engineering Group and, ultimately, in the Weapons Group. This was all precipitated by the reversal of the decision, taken just a year before, to separate the functional from the executive responsibilities of the Board Members. The AEA's annual report explained it all rather obliquely as being due to 'experience' that had shown that 'the separation ... does not suit the particular circumstances ... as well as the previous arrangement which will accordingly be restored on 1 July 1959' [22]. Such instabilities at the highest levels of the AEA's management did little to inspire confidence amongst those in the laboratories who were wrestling with their own problems, though rather different in both detail and complexity.

*   *   *

ZETA II was abandoned officially in February 1959. It was obvious to all who were able to be dispassionate about it that the concept was well ahead of its time. CTR research still had some considerable way to go before anything like an industrial application could even be contemplated, let alone realized. Much more fundamental work was required and it was Harwell's recently patented large power supply, appropriately named Pandora, that would lead the way to what became known as the Intermediate Current Stability Experiment (ICSE[*]) — another toroidal system but one that should, at least in theory, be unconditionally stable.

Schonland was certainly under no illusions about the fundamental problems that still remained to be solved before controlled thermonuclear

---

[*] ICSE pronounced 'ice'.

fusion might be realized. Delivering the opening address at an International Convention on Thermonuclear Processes in London at the end of April 1959 he took great pains to point out that, whereas ZETA may have tamed the large-scale instabilities within the plasma, it was the small-scale effects, leading to extremely complex turbulence in the current-carrying skin of the plasma, that presented the ultimate challenge [23]:

> Whether this is due to subtle hydromagnetic disturbances, to the excitation of electrostatic plasma vibrations or to unstable oscillations in plasma magnetic fields we do not yet know. Until we do the achievement of stage one will resemble Mrs. Beaton's misquoted recipe for jugged hare — 'First catch your hare' — for it is an elusive one.

It was Harwell's objective, with the assistance of an Advisory Committee of eminent consultants from the universities and elsewhere, to 'understand the plasma and the forces that mould it. Until that is done it is idle to talk of a thermonuclear reactor'. The last thing Schonland wanted was any repetition of the painful events of the year before, when one unguarded answer to a journalist's question had caused a near-stampede and international repercussions.

Progress towards setting up ICSE was rapid. By May the project had received AEA approval to proceed and in July the Treasury gave its blessing for the expenditure of an estimated £1 500 000. The design group charged with the development was to be chaired by 'Bas' Pease, with engineering responsibility falling to Bob Carruthers, both of whom had been involved with the fusion programme from its inception. Hopes were once again high as work commenced.

Before following the saga of ICSE to its own somewhat ignominious conclusion it is important to understand how it evolved and also to gain an appreciation of the friction that soon developed between the various interest groups at Harwell and elsewhere. The scientific stimulus for this new approach had really come from the Geneva Conference in September the previous year. The remarkable atmosphere that existed there for free and open discussion between East and West was a significant milestone in the quest for controlled nuclear fusion, and the scientific debate was particularly lively. Theoretical work in the United States by Rosenbluth, who used magnetohydrodynamic techniques applied to the mathematically tractable case of cylindrical tubes, and a similar study at Harwell by R J Tayler, had shown that the instabilities in the plasma pinch that were seen by the experimentalists were very much to be expected. This was especially true if the current flowed uniformly through the plasma, but if it could be confined only to its surface then, in the presence of an external axial field within a conducting container, it could be stabilized. Further work had established an additional condition for stability, which required that the axial magnetic field within the plasma column should be oppositely-directed to that outside it. Though the theoreticians now maintained that ZETA's toroidal configuration was inherently unsuited to plasma stability, there was undeniable experimental evidence, from the very outset of the project at Harwell, that had shown that stability did indeed occur, if only for a brief period of time.

Of course, such a divergence between theory and practice was nothing new and its reconciliation depended on the backgrounds of the proponents. The experimentalists, such as Thonemann at Harwell, saw it as evidence of a deficiency in the theory, while the theorists maintained that ZETA's apparent stability was but a transient phenomenon which a more refined theory would probably identify. Where there was agreement was that an experiment had to be mounted to attempt to produce the stable conditions that were predicted when current flowed only within a skin on the surface of the plasma, whether toroidal or cylindrical. The mechanism for achieving this was well understood and would require the use either of extremely high frequencies in a system operating continuously or, as in the pulsed operation of these experimental devices, of pulses with very rapid transitions. Such current pulses would be produced by the very rapid rise-time of the Pandora power supply upon which so much would now depend.

The shape of the testing device had still to be agreed. Harwell's long experience with toroidal configurations naturally suggested that geometry in their eyes, but the somewhat later arrival on the fusion scene of the AWRE, following their need to diversify from purely military work in the face of a moratorium on the development of nuclear weapons, saw them plump for the recently favoured cylindrical system. Though Harwell's wishes ultimately prevailed, an AWRE undercurrent continued to flow and produced eddies of its own, as we will see.

ICSE was seen by some as the natural successor to ZETA, but Peter Thonemann, who more than anyone else could rightfully lay claim to being the pioneer of all thermonuclear research at Harwell, was not one of them [24]. Unconvinced by the theoretical arguments, he soon found himself at odds with some of his own colleagues and so decided to request a year's leave of absence, which he intended to spend at Princeton familiarizing himself with American fusion research and thinking about the subject's very future. Thonemann's loss to the project was a heavy blow but it was Cockcroft who sanctioned his leave of absence. Rather lose him temporarily, he argued, than risk losing him altogether if his disenchantment became total. Schonland was most unhappy at this decision but Cockcroft was still the AEA's Member for Scientific Research and had the ear of the Board, which Schonland did not. Such differences of opinion were not confined within Harwell's walls. Tensions had also arisen with Allibone at the AEI laboratory over the continuation of Harwell's funding of SCEPTRE, now into Marks III and IV. There were some at Harwell, most notably Flowers and Pease, who saw this work as a waste of effort, essentially mirroring what Harwell was doing anyway. Others, led by Fry who was now officially Schonland's deputy, countered that the AEI programme had been approved by Harwell, which surely implied support for it. When Penney took over from Cockcroft at the beginning of July he supported those ranged against AEI and indicated that financial support from Harwell would cease within a year. But in the general climate of confusion that was now so apparent, the matter rumbled on with no clear outcome on the immediate horizon.

Reconciliation of all these differences inevitably required a managerial directive and Schonland had frequently to take a firm stand one way or the other. At a time of such intense scientific debate, when emotion as much as reason played its part, it was inevitable that there would be those whose found themselves very much at odds with one decision or another. Camps formed and prejudices abounded. To some Schonland was aloof, remote and, because of his background, even ill-equipped to grasp the subtleties of the arguments; to others he was readily approachable, straight-forward and much respected as a scientist who was fully *au fait* with all the details. Comparisons with Cockcroft, and particularly with the Cockcroft-style, could not be avoided. Cockcroft's almost boyish enthusiasm for a new idea could be contrasted with Schonland's usually studied reaction and carefully considered opinions which some thought were merely obstructionist. Whereas Cockcroft spoke in monosyllables, which left considerable room for interpretation, Schonland was much more coherent and his yes and no were absolutely unambiguous. But where they differed even more markedly was Cockcroft's almost bluff disregard for administrative detail, while Schonland was precise to the point of perfection.

But perfection at a time of great change is usually unattainable and the strains at Harwell were much in evidence. Schonland, particularly, was handicapped by not being the responsible member on the AEA board and was thus powerless to intervene at the highest level on behalf of Harwell's interests as a whole. He had to implement policy decisions that were themselves often confused and contradictory, while at the same time he had to placate his troops who were decidedly restive. Plowden's decree that expansion of staff numbers within the authority was to be limited to eight per cent per annum caused great problems. Harwell's own expansion was checked while that of the new establishments at Winfrith and Culham went ahead. Most seriously affected was recruitment of theoretical physicists, the very people who were most sought after at this critical stage in the fusion programme [25]. If Harwell was to continue to have a role, then it was vital that it should be planning for the future by bringing in promising young researchers, but the authority's policy made that well-nigh impossible. Further complicating things was the delay in appointing someone to take over the fusion programme from Don Fry, who was now fully involved with developments at Winfrith. Whereas Thonemann might have seemed like the logical choice, administration was anathema to him, while his recently expressed views about ICSE would have further weakened any chance he may have had. The appointment went to J B Adams, an engineer of extraordinary flair and great competence, who had previously been on the Harwell staff and was then seconded to CERN, the European particle accelerator centre in Geneva. John Adams was offered the post of Director of the new establishment at Culham in April, and he indicated his acceptance in June, but made it clear that he would only be in a position to leave CERN twelve months later. Such delays were understandable but they did not help Schonland, who was juggling the many balls then in the air. Since Fry had served as his Deputy until leaving for Winfrith, Schonland was faced with finding his successor. After discussing the matter with Cockcroft he wrote personally to Dr F A Vick, then acting principal of the University College of North Staffordshire, and offered him the appointment [26]. To many at Harwell, Vick's appointment came as a great surprise because few there had ever heard of him and some were decidedly put out by it.

Arthur Vick[*] had known both Schonland and Cockcroft during the war when he was Assistant Director of Scientific Research at the Ministry of Supply with particular responsibilities for the fusing of anti-aircraft shells. His research experience was limited, and his background in the atomic field even more so, but he was a sound administrator and it would seem that both Cockcroft and Schonland were of the view that that was precisely what Harwell needed when, as *The Times* obituary of Vick put it many years later, 'Britain's nuclear programme was in danger of falling into the hands of warring factions'.

*　*　*

Close observers of the world of nuclear science and engineering in Britain in the closing years of the 1950s could not have missed seeing the occasional wisps of smoke, which signalled that all was not well at Harwell. The rapid developments taking place at Winfrith, the emergence of the new facility at Culham and the well-publicized setting up of the

[*] Sir Arthur Vick subsequently became Vice-Chancellor of Queen's University in Belfast in 1966. He retired in 1976 and then served as Pro-Chancellor and Chairman of Council of Warwick University until 1992.

National Institute under the name of the Rutherford Laboratory were public knowledge, but what was the future of Harwell itself? It was allegedly too large, its vision of the future was blurred and uncertain, and it seemed to be at war with itself. Certainly there had been something of an exodus of senior staff. Flowers, Finniston and Seligman had all left within the space of a few months. Thus, theoretical physics and metallurgy, two vitally important areas that underpinned almost everything that Harwell did, had lost their heads, while Henry Seligman was the energetic leader of the Isotopes Division which was one of the great success stories of the Establishment. Replacing them with men of equal calibre would not be easy. Was this just the normal movement of men of ambition and ability to more challenging positions offering more power and authority, or was there something else behind their departures?

One of the first to inquire was the journal *Nuclear Engineering*, which devoted its editorial in November 1959 to the subject [27]. 'Whither Harwell?' it asked. In attempting to answer his own question, the editor suggested that the Atomic Energy Research Establishment had become more and more isolated from 'the central themes of atomic energy development', while its contribution to the overall programme had become 'more and more ephemeral'. The departure of senior staff, the editorial suggested, was surely a contributory factor and salaries that were less attractive than those in other sectors did not help to retain people of potential. But the real reasons for the 'flux of people from Harwell', it claimed, were an 'inadequacy of purpose in the establishment, a feeling of frustration with somewhat overbearing administration and a consciousness of being divorced from the main stream of atomic energy development'. From being the most glamorous research establishment in Europe, involved in the most dynamic development programme of this century, Harwell had ossified and its research workers were 'hampered by an administrative system which would appear more suited to a government armament factory'!

Such grave charges were bound to raise eyebrows and amongst them were those of Lord Hailsham, the recently appointed Lord Privy Seal and Minister for Science and Technology, under whose wing the AEA now found itself. Hailsham was due to visit Harwell as the guest of Schonland and the retiring Chairman of the authority, Sir Edwin Plowden, whose extended term of office was soon to come to an end. The tone of that editorial and the events that evidently gave rise to it would undoubtedly be among the many issues to be discussed over lunch that day and indeed they were.

At about the same time as that editorial appeared, Schonland received a letter from D E H Peirson, Secretary of the Authority, enclosing what was referred to as the 'Finniston–Glyn' correspondence. Lord Glyn, as Sir Ralph Glyn, was previously the Member of Parliament for Abingdon, the district which included Harwell, and he was a long-time friend of Cockcroft who took a fatherly interest in all that went on behind Harwell's security fence [28]. Glyn and Finniston also had more than a nodding acquaintanceship. It would appear that in the course of a recent social visit to Glyn's home Finniston had some harsh words to say about Harwell and its future, and he repeated these in a subsequent letter to Glyn. Seeing it as his duty to communicate matters falling within his constituency to the relevant authorities, Glyn wrote to Peirson at the AEA and enclosed extracts from Finniston's letter. Peirson then deemed it only right that Schonland should be aware of the charges that had been levelled against the establishment and, evidently, against Schonland himself.

Schonland's response to those extracts that were relayed to him was immediate and unequivocal. He wrote a personal letter to Peirson on 16 December, with a copy to Sir William Penney, and informed them both that Lord Hailsham had been reassured by Plowden and himself over lunch. 'It was evident', said Schonland, 'that he had not taken [the matters raised by Glyn] very seriously'. But Schonland himself did and he then addressed the Finniston diatribe [29]:

> It is of course desirable to stop nonsense at an early stage, otherwise it can build up to damaging proportions. It arises to a large extent from individuals who have an axe to grind and have been disappointed in their ambitions. Two of them have used Lord Glyn as their sounding board, Seligman and Finniston, and some quite libellous statements have been made in the course of their correspondence which I have been shown but have no right to quote. Seligman's purpose in discrediting everyone from the Chairman downwards is well-known, and he has been helped by one or two of his henchmen. Monty [Finniston], who was anxious to be Deputy Director here, has now found himself out on a limb … and is looking for a new job.

Schonland then turned his attention to Harwell in order to counter the charge that it was 'lying idle', as implied by the article and suggested by the innuendoes in this now notorious Finniston–Glyn correspondence.

> When Culham and the National Institute get going, Harwell will, we all hope, diminish in size, but it will continue to be the Authority's centre for basic pure and applied work in Nuclear Physics, particularly Neutron Physics, Chemistry, Metallurgy, Electronics, Chemical Engineering etc. It has three large materials testing reactors and another (LIDO) for shielding work, and a programme which will continue for the next ten years without any falling off of pressure.

543

His own position too had seemingly been impugned by what had been said and he responded with vigour against the charge that the establishment lacked direction and that morale had been destroyed.

> My chief preoccupation during the last two years has been to ensure a satisfactory future for the daughter establishments to which we have given birth—Winfrith Heath, the National Institute and Culham. It is not very long since the morale amongst the Winfrith transferees and the C.T.R. boys was extremely low because they did not want to leave the maternal home. That all this is changed and great enthusiasm prevails is a measure of my success and that of Adams and D.W. Fry. The morale of those who will have to go to the National Institute is at present not high, but it will not be long before they too are full of beans.

As regards the future of Harwell itself he was in no doubt that it was in good hands.

> 'Nuclear Engineering' is worried about the future health of the Mother. I have no doubt that there will be no lack of stimulants to restore her to health but Penney and I, with Vick, need a little time to examine which of many such is the best. There is little 'reorganisation' needed at Harwell, but need for some forward planning which is indeed helped by the moves which are in progress. The solution to Harwell's problem is quite easy and straightforward; a few people will get hurt in the process but there is no danger that the Establishment will decline in importance— quite the contrary.
> Have you ever heard of the Wailing Wall?
>
> Yours
>
> B.F.J. Schonland

But all was not as well as Schonland made out. Personal ambition and the tensions that arise from it were not peculiar to Harwell; they smoulder almost continually in any organization of such size and complexity. But there were also deeper underlying problems and ICSE brought them all to the surface.

It had been announced by Penney at a press conference, with an air of some confidence, and by the Prime Minister in the House on 28 July, that ICSE was being planned by the AEA [30]. On the surface its purpose was purely scientific—to provide answers to some fundamental questions of physics—but it was also hoped that ICSE would restore the public's confidence in the authority and its various establishments that had been somewhat dented following the imbroglio over ZETA. It would also be the centre-piece at the new site at Culham, whose very existence depended on it being the home of a large-scale experiment that would open up new avenues in fusion research. After all, had not Schonland said as much in his press conference in May 1958? So, whether they liked it or not, the scientists at Harwell would have to live with the fact that ICSE was a prestige project that would naturally have a higher profile than anything at Harwell itself. It would, therefore, be the subject of careful scrutiny by those sections of the press who took notice of such things, while Parliament itself would certainly call for progress reports from time to time. Behind it all lurked the financial stringency that hung like a cloud over the country that was facing imminent inflation, while Macmillan's government took every step to combat it. Penney and Schonland were therefore most concerned when the estimated costs associated with the design and construction of ICSE soon began to rise almost inexorably.

By October 1959 a figure of £2 100 000 seemed likely; then £2 200 000 was predicted and soon it reached £2 500 000: Schonland was incensed. All along he had insisted on strict budgetary control with careful costing of every phase of the work, while negotiations with outside industrial organizations were only to take place with their senior personnel, in order to emphasize the prestige nature of the project and its status within Harwell. But, in spite of that, what he now saw appeared to be an exercise in profligate spending by the ICSE programme, and he told them so. Amongst the scientists and engineers working on the various facets of the project, this reaction of Schonland's did not go down at all well. There was a feeling of much resentment and a belief that both Schonland and Penney underestimated the technical complexity of ICSE and the consequent problems it presented. Whether they did or not, they certainly felt the cold hand of the Treasury on both their collars and there was very little room left for manoeuvre.

The *coup de grâce*, when it came, actually came from Harwell itself. Within the theoretical physics division there had been an increasing degree of concern about ICSE and some were even questioning the very scientific basis upon which it was so finely balanced. Could the crucial Rosenbluth plasma condition even be established? If so, would the plasma really be stable? Without detailed calculations or some experimental evidence from a small pilot study, was the ICSE concept not just pure speculation? These were indeed worrying questions and they came to the attention of their erstwhile colleague Brian Flowers, now at Manchester, when he was on a visit to Harwell in January 1960. Flowers, who was no supporter of Schonland, took the matter directly to Cockcroft, now only a part-time board member of the AEA and very much ensconced in his new domain in Cambridge. Cockcroft duly relayed them to Penney who immediately took soundings of his own by consulting his colleagues at the Atomic Weapons Establishment.

The scientific culture in the AWE at Aldermaston had always differed from that at Harwell, in that they were steeped in a tradition that required detailed analysis, which would enable them to predict, with a very high degree of certainty, how their rather peculiar 'experiments' would behave. Bombs left little chance to repeat an experiment! By contrast, Harwell's approach had always relied very heavily on extensive experimental work of a more conventional kind, to lead iteratively to a final solution. That was, after all, the Cockcroft method. At Penney's elbow now was John Adams, the Director-elect at Culham, whose experience at CERN had also been honed on calculable near-certainties. Since ICSE had never been in favour at Aldermaston, and Adams had his own vision about what he wished to see installed at Culham, the odds were therefore stacked very much against a Harwell without its front-line fusion team: Fry was now in charge at Winfrith, Flowers had left them for academia while Thonemann was on sabbatical. Therefore it would not be inaccurate to suggest, notwithstanding all Schonland's protestations to the contrary, that Harwell was indeed in some disarray [31].

Financial stringency soon forced the new Chairman of the authority to bring down the curtain. Sir Edwin Plowden retired at the end of 1959 and was ennobled. His successor as Chairman was Sir Roger Makins, previously British Ambassador to Washington and a man of vast experience of operating in the international atomic energy arena. Makins immediately asked Penney to provide him with a full account of the capital expenditure of the AEA and, particularly, that predicted for the Research Group. The dominant item was, of course, ICSE. What had started in May 1959 as an approved figure of £1 500 000 had escalated as if driven by some sort of chain reaction. By July 1960, little more than a year later, it had doubled. By mid-August, Penney estimated the likely cost to be in the region of £4 000 000. That same month the authority took the decision to limit any further expansion of the fusion programme, and ICSE was immediately under threat. This was further compounded by a considerable overspend on production facilities associated with the fissile programme in the north, but entirely unrelated to any Research Group activity. Schonland's immediate reaction at Harwell when confronted by Penney was to keep ICSE going at all costs, even if that meant restricting the funding of other, less pressing projects. However, with the centre of gravity of fusion research about to shift from Harwell to Culham with the arrival of Adams, the future of the Establishment's prestige project began to look increasingly bleak. To Adams, as he reviewed the fusion scene, ICSE appeared only as a huge drain on what were already limited resources. Having no previous attachment either to it or to Harwell's fusion programme, he readily supported Penney's inclination to cancel the ICSE experiment on the grounds of cost alone.

To muddy the waters still further, Thonemann returned from Princeton in April. Nothing he had seen in the States had changed his mind about ICSE and he had become increasingly alarmed at the way it had begun to dominate Harwell's fusion research. Steadfast in his refusal to move to Culham, he had managed, even before taking his sabbatical, to persuade Cockcroft to allow him to set up a small fusion group of his own at Harwell. Though Schonland had opposed this most vigorously, both on principle and because it would serve only to dilute Harwell's available research effort still further, the combined forces of Penney and Adams persuaded him to accede, however reluctantly, to Thonemann's request. Peter Thonemann was thus accorded the status of 'senior distinguished physicist' and allowed to form a small team to work on a topic entirely of his choosing: the Cockcroft ideal for Harwell seemed able to survive even when his very creation was in crisis! However, Penney did manage to introduce an air of reality into the affair by insisting that, in due course, Thonemann must move to Culham where he would report to Adams.

Who, though, would now lead the Harwell fusion programme in the interim before Culham was fully on stream? Pease[*] was the obvious choice but, like Thonemann, he eschewed the administrative responsibility which that would entail in favour of continuing with his own research. Then Thonemann, realizing that this might be his last opportunity to influence the direction of fusion research before Adams arrived, changed his mind and indicated that he was willing to take it on! But Adams, seeing what was afoot and naturally wary that no major changes should be made before he took over, immediately stepped in and called on Penney to set up a committee under his (Penney's) chairmanship to hold the fusion reins in the interim.

Schonland can only have watched these goings-on with feelings of despair and, no doubt, with much annoyance. Penney's numerous responsibilities at Aldermaston meant that he was clearly unable to run the fusion research programme as well, so he took the obvious decision and asked Schonland to do so in the normal course of his duties as Director of Harwell, and Schonland agreed.

But ICSE did not have long to run. It was eventually cancelled late in August 1960 [32] and so ended a saga of division and discord. What had begun as the most exciting project in the nuclear firmament—the idea of releasing almost limitless quantities of energy by the process of nuclear fusion—had stalled. So too had Harwell. There were those in the Atomic Energy Authority who expressed considerable relief, even satisfaction, at the demise of ICSE but they would, of course, because it was threatening to become a great drain on an already limited purse. Strangely, given the almost visceral attachment to the project in some quarters, the decision to terminate it was no cue for great mourning,

---

[*] Dr R S (Bas) Pease FRS became Director of the Culham Laboratory in 1967 and continued to serve in that post until 1981.

even within the Authority's own research fold. At the AWE in Aldermaston, for example, it had long been seen as an unworthy competitor which had merely stood in the way of their own schemes for entering the fusion race. To others, like John Adams waiting in the wings, it was just a distraction.

But there were repercussions from farther afield. Apparently, the decision to terminate ICSE had been taken without informing the Americans, and this after Britain had spent so long trying to persuade its trans-Atlantic cousin to adopt an open policy in all matters to do with such research. It was, so it was said, an unfortunate oversight but it was also a blunder that should not have occurred. Of all people Roger Makins should have known that. Closer to home the Government expressed its own disquiet. The new Minister for Science, Lord Hailsham, had stood by the authority when the decision to move all fusion research to Winfrith had been rescinded so that the new laboratory at Culham could become its home. Now, the very focal point of that research had met a rather ignominious end. Hailsham was annoyed and required a convincing explanation. European sensitivities had also been bruised because CERN in Geneva had John Adams ear-marked to become its Director-General, but had backed down in the face of a British veto because he was required at Culham, presumably to take charge of ICSE. What, they may well have asked when word reached them, would he now turn his hand to there? It was all very messy.

For Basil Schonland the ICSE affair had certainly taken its toll. Following so closely on the heels of the abandonment of ZETA II, and coupled with the tensions that were so rife within certain sectors of Harwell, it had weighed him down enormously. What had been an otherwise robust constitution showed the first signs of strain [33]. He was now approaching his sixty-fifth birthday and retirement was not far off. In September 1960 the authority announced a number of senior staff appointments. Dr Arthur Vick would become Director of the establishment, while the new Deputy Director would be Dr Robert Spence, who had long been Harwell's Chief Chemist. Until now the post of director of the AERE at Harwell had carried with it the dual responsibility for the running of the Research Group with all its diverse components and scattered laboratories.* Now, with this series of new appointments, sole responsibility for the Research Group would be in the hands of Sir Basil Ferdinand Jamieson Schonland CBE, FRS—his knighthood having been announced in the Queen's Birthday Honours List of June 1960.

---

* The Research Group comprised the AERE at Harwell (including the Isotope Laboratory at Wantage), AEE at Winfrith, the Radiochemical Laboratory at Amersham, and the Culham Laboratory, as well as the AERE outstations at Woolwich, Chatham, Bracknell and Oxford.

CHAPTER 25

'FOR THE FUTURE LIES WITH YOU'

On 22 December 1959 the Schonlands left Harwell for Denchworth House, in the village of that name, on the edge of the Berkshire Downs. What was to become their new home was an old rectory a matter of a few kilometres to the west and half an hour's travelling time from Harwell. With just a year to go before Basil's retirement it was natural that he and Ismay should be thinking of the years that lay ahead, and a home of their own was now a necessity. Their move coincided within a month with the departure of John and Elizabeth Cockcroft from Harwell for Cambridge. To many the fact that numbers 1 and 2 South Drive fell vacant almost simultaneously was unremarkable. To the few who noticed these things it was yet another event in the almost uncanny catalogue of those that marked the intertwining of the Cockcroft and Schonland careers, and it was certainly a coincidence if nothing else. The Schonlands' move was routine whereas that of the Cockcrofts marked the end of one era and the start of another. The man who, more than anyone in Britain, had taken nuclear science from its birthplace in the Cavendish, through the trauma of war and into the pioneering years of power-generation and on to the very frontiers of science was due a fitting response from the 'Establishment'—his establishment. It was only fitting too, in so many ways, that the man who bade him farewell was Schonland.

Packed into the Cockcroft Hall at Harwell on an early summer's day in 1959 were 250 of its employees representing every group of that very diverse community, then numbering well in excess of 6000. Sir John and Lady Cockcroft were the guests of honour; Basil and Ismay Schonland the hosts. Schonland's address had taken him some time to prepare and contained, for him, an unusual number of alterations, deletions and additions: it was clearly one of great importance and he wished to hone it to perfection. After paying Cockcroft the usual tributes, he reminded all present that Her Majesty had recently bestowed on Sir John the Order of Merit—an award, he said, that had given all at Harwell the greatest pleasure for 'An honour to you is an honour to us as well'. Then followed a little story of the great organist who visited the village church for a grand performance, which turned out to be laced with misadventure, a situation some recognized only too well much closer to home. But Schonland was quick to remind them that it was due to the combined efforts of them all, under Cockcroft's inspired and inspiring leadership, that had indeed 'given Britain and the World [another] grand performance'. He then turned to talk of the man himself, for though all in that audience knew the Cockcroft character, only some knew his handwriting. Schonland enlightened them all as to its general form and the notes that issued forth in that incomparable hand. It was, he said, those missives, mostly illegible and always so terse, that delivered the 'rockets' that kept them all on their toes. Quite which way they should then turn once they had received them was probably another matter entirely. Then came the parting gift that Basil Schonland presented to his colleague of so many years, and none could possibly have been more fitting. It was a model of ZETA, made by the apprentices of the establishment's Engineering Division, and it was indeed the 'big machine' with which the name of Cockcroft would always be associated [1].

Honours also came Schonland's way that year. The British propensity for recognizing the great and the good amongst its citizens was now about to honour him as well. But before the nation did so the world of academia doffed its ceremonial hat when, in January, a letter arrived from the Vice-Chancellor of the University of Southampton informing Schonland that the University wished to confer on him, at a ceremony in July, the honorary degree of Doctor of Science. The following month his old college, Gonville and Caius, at Cambridge, elected him to an Honorary Fellowship. Both were in certain recognition of his contribution to the furtherance of nuclear science in Britain and the honour that he had brought to his British *alma mater* and, because of its proximity to Harwell, to Southampton as well. Few special privileges attended these awards but at least he would receive invitations 'to all feasts and similar occasions', so it was said [2]. In a year that had seen much turmoil at Harwell and which had taken a great toll of Schonland himself, these gracious devices offered, at least, some comfort. But most satisfying of all was a letter he received in May the following year. It came in May from the Office of the Prime Minister [3].

10 Downing Street
Whitehall
May 4, 1960

Personal and Confidential

Sir,

I am asked by the Prime Minster to inform you that he has it in mind, on the occasion of the forthcoming list of Birthday Honours, to submit your name to The Queen with a recommendation that Her Majesty may be graciously pleased to approve that the Honour of Knighthood be conferred upon you.

Before doing so, the Prime Minister would be glad to be
assured that this mark of Her Majesty's favour would be agreeable
to you. I should be obliged if you would let me know at your
earliest convenience.

I am, Sir,

Your obedient Servant

(Signed) D. Bligh

B.F.J. Schonland, Esq., C.B.E.

This mark of Her Majesty's favour was most acceptable to a man for
whom service to King or Queen and Country had, ever since his boyhood days in Grahamstown, seemed mere second nature to him.
Ismay, for all her adherence to the socialist creed, was no Roundhead,
and she too was thrilled at the honour. On Tuesday, 5 July 1960, Basil
Schonland knelt before The Queen at Buckingham Palace to be dubbed
'Sir Basil'. In attendance amongst the many who witnessed the grand
occasion were Ismay and her sister-in-law 'Dot', the wife of Basil's
former room-mate of 1919 at Caius, Don Craib. It was a decidedly
happy occasion and the Schonland party spent a few days in London
afterwards to celebrate.

*   *   *

Restoring American confidence in British security had been an on-going
struggle throughout Harold Macmillan's years as Prime Minister. No
subject was more sensitive than the nuclear programme, and Harwell
had much ground to make up after the treachery and betrayal of Fuchs
and Pontecorvo at the beginning of the decade. Early in 1960, Schonland
found himself confronted by a case that had had its origins just three years
before. It concerned the access to classified material by a consultant to
Harwell and the man concerned was none other than Rudolf Peierls,
who had so vigorously championed Schonland's cause when the
University of Birmingham was seeking a new Vice-Chancellor in 1952.
It was in Peierls' department, too, that David Schonland, Basil and
Ismay's son, had completed his PhD just a few years before.

Peierls had been a most valued consultant to the theoretical physics
division at Harwell and was still acting in that capacity in 1957 when he
received a letter from the establishment informing him that 'for reasons of
administrative convenience' he would no longer have access to classified
documents. Surprised, he inquired why this should be, and was informed
that it was the standard procedure applied to all senior consultants whose
contracts were coming up for renewal. 'Administrative convenience'
turned out, on further inquiry, to be Harwell's rather clumsy way of
omitting to mention that it was actually American security sources that
had requested that he should have no access to such material. Annoyed

that Harwell had chosen, for whatever reason, not to be entirely frank with him, Peierls immediately withdrew as a consultant [4].

By 1960, Harwell's theoretical physics division was, as we have seen, somewhat depleted of senior staff and Schonland was seeking to augment those who were holding the fort with some high-calibre consultants, and naturally his thoughts turned once again to Peierls. The matter of security was clearly no longer an issue, if it ever had been, but it was still an embarrassment which Schonland hoped to set on one side. In February he wrote to Peierls and sounded him out informally about spending six weeks at Harwell in the summer: 'Before issuing a formal invitation, I write to ask you to forgive what happened in 1957. Cockcroft and I were placed in a most difficult position and could not have acted other-wise at the time. My formal invitation is one without any limitations whatever of a security nature and has Penney's full support' [5]. Peierls replied that was unable to accept the invitation because of a prior commit-ment in Chalk River. However, he found Schonland's apology for the way in which Harwell had handled the matter three years before not overly reassuring and was greatly disturbed that, though Schonland seemed to be implying that all security restrictions had been removed, there appeared to him still to be a risk that 'some new developments in high policy might at a later stage put you again in a similar position'. A position, presumably, of once again having to impose such a ban. Schonland's hands were tied. He could do little more than indicate how valuable Peierls' contribution would have been and stressed that: 'The other questions in your letter are outside my competence, as indeed was the whole initial trouble, but I hope that if we repeat the invitation as I am sure we would like to, you will be able to accept' [6]. And there the matter rested, at least for the time being.[*]

On 7 February 1960 Igor Kurchatov died. The news jolted Schonland, who was in the midst of the Peierls affair when he heard it. It was Kurchatov's remarkable lecture at Harwell in 1956 that helped break through the previously impenetrable barrier that existed between the Soviet Union and the West, but it was actually this giant of a man's wholeheartedness and *bonhomie* that had so endeared him to many at Harwell, and he was to return their hospitality many times over when Cockcroft and others paid official visits to Russia in later years. His passing was therefore greeted at Harwell with genuine sadness. It was immediately felt that some suitable tribute should be paid to him in the pages of *Nature,* and Schonland collaborated in writing it with Bas

---

[*] In 1963 Peierls was approached again to become a consultant to Harwell. He accepted after entering into brief correspondence with Sir Roger Makins on the matter [5]. Later he wrote: 'I reconsidered my position. By now all the people involved in the offending episode had left, and there seemed no point in keeping up resentment against an abstract body' [4, p 324].

Pease, who had been so closely involved with Harwell's fusion work since its inception [7].

*   *   *

Ill-health dogged Schonland that year when he suffered the first of a series of strokes that would become a feature of his declining years [8]. It was now clear that Arthur Vick should begin to familiarize himself with all matters of consequence in the Establishment, while Robert Spence would prepare to take over from him as Deputy Director. Schonland was now looking towards his retirement. He and Ismay had planned to spend six months in South Africa as soon as he had relinquished the reins at Harwell. In addition to enjoying a well-earned holiday he had three important functions to fulfil—a graduation ceremony at Rhodes, at which he was to preside as Chancellor, his first since 1954; the laying of the foundation stone of the new science building at St Andrew's College in Grahamstown; and a review for the Chamber of Mines of its research laboratories, a task for which his services had been sought for some while. Before attending to any of these, however, there was the need to disengage from his many other duties and responsibilities within the wider scientific community in England.

Since 1956 he had been a Member of the General Board of the National Physical Laboratory, a position for which has was nominated by the Royal Society. Echoes of 1949: the flurry of telegrams to Pretoria, the hurried flight to London and the interviews and discussions that eventually came to naught undoubtedly crossed his mind whenever he entered the NPL's portals. Whatever his recollections about those events, the meetings of the Board were certainly occasions for renewing old wartime acquaintanceships, for there he encountered, once again, Sir Charles Darwin, Professor R V Jones, Professor N F Mott, J A Ratcliffe and Sir Henry Tizard, who had all served alongside him, and theirs were the names that had been carved into the Schonland memory [9]. Now he would quietly slip away and consign those memories to history.

There was also the Royal Society itself. As early as 1942, in between his other manifold duties, Schonland found himself serving on a committee under James Chadwick which selected new Fellows from amongst the ranks of the physics community. After returning to England in 1954, he was once again invited to perform various functions in the Society's service and one such request had come from Nevill Mott in 1956. The Copley Medal, the Royal Society's most prestigious, was to be awarded to Patrick Blackett and Mott asked Schonland to 'write 500 words in support of Blackett' which he duly did [10]. Schonland, of course, had had more than a little interest in Blackett's theory of the magnetic field of large rotating bodies and before that in the 'penetrating radiation', the cosmic rays that Blackett had revealed so vividly using the Wilson cloud chamber. But it was their involvement in operational research that had occasionally brought them together during the war: Blackett, its founding father; Schonland, its battlefield practitioner. Schonland's citation made much of the physics, for that was its purpose, and left the rest to their wartime memories.

That same year the Royal Society invited Schonland himself to present the Wilkins Lecture, one of a number of named lectures in its programme. He chose as his subject the life of Benjamin Franklin since that year, appropriately enough, was the 250th anniversary of Franklin's birth. It was fitting that the man whom many saw as Franklin's twentieth-century successor should use this occasion in London to present, not his own achievements in the field of lightning, but rather to remind his audience and all who subsequently read his superb text, that it was the genius of Franklin that had laid bare the electrification of the thunder-cloud [11]. One senses, on reading Schonland's account, that he revelled in the opportunity given him by the Royal Society to divert his attention, however briefly, away from the issues of fusion and to allow his thoughts to retrace the steps that Franklin had taken when uncovering the mysteries of the 'electrical fire'. What became known as the 'Philadelphia experiment' was intended not as a mere curiosity piece but as a way, in Franklin's own words, 'to determine whether clouds that contain lightning are electrified or not'. The result, of course, was decidedly electrifying and Franklin's name will forever more be associated with that hazardous venture. Schonland quoted not the awe-struck reaction of the public but the conclusions of no less a scientist than Joseph Priestley, who described what Franklin had shown as 'the greatest [experiment], perhaps, that has been made in the whole compass of philosophy since the time of Sir Isaac Newton'. It was, said Schonland, when viewing Franklin's audacity some two hundred years later, no less awe-inspiring and could be compared with the explosion of the first atom bomb! He allowed his own work to appear just briefly when telling of Franklin's discoveries:

> All these results are in agreement with observations made within the last thirty years in different parts of the world. Except for a small low-lying positively charged region, which is sometimes evident when the active portion of a thundercloud is directly overhead, the bases of thunderclouds in their mature stages are negatively charged. The negative charge extends to a considerable height, above which is the main positive charge, too far away to diminish appreciably the effect of the negative charge below it. Franklin's observation that the inactive rear of a thundercloud is positively charged has also been confirmed by myself and others.

Though a far cry from Franklin's most colourful 'effluent stream of electric effluvium and the affluent stream that replaced it', Schonland's more staid description paid apt tribute to the work of his great predecessor, while underlining just a little of his own part in the quest that followed.

1956 had also seen the publication of the prestigious German *Handbuch der Physik*, an encyclopædic multi-volume set that defined the state of knowledge in the field. Schonland's chapter (in English) had been completed while he was still at the BPI, but before publication he had added some material that had just come to hand from Johannesburg. Radar measurements of the inter-stroke streamers within the thundercloud had been made by Frank Hewitt at the TRL and these formed the basis of a thesis he had recently submitted to Wits. The results, obtained with a 50 cm radar especially developed for such observations, added more weight to Schonland and Malan's theory that the upward-moving junction, or J-streamers as they called them, travelled progressively from the base of a cloud in a predominantly vertical channel [12]. Although this view was soon to be contested, it provided the springboard for what was to become a most fruitful area of research in South Africa, and elsewhere, using radar and radio techniques for the observation of lightning.* That Schonland's pioneering work on lightning and his visionary development of an elementary radar at the BPI at the very outbreak of war should converge was no great surprise, but its timing just as his career was drawing to its close was fitting indeed.

A somewhat unusual invitation to a scientist then saw Basil Schonland deliver the annual Stevenson Memorial Lecture at Chatham House in London in November 1958. It was unusual in that Sir Daniel Macauley Stevenson, the philanthropist in whose honour the lectures were held, had an interest in international affairs, not in science, and had founded university chairs in the subject at both Glasgow and the London School of Economics. Previous speakers had numbered amongst their ranks such dignitaries as Robert Schuman, a former French Prime Minister, and Mrs Vijaya Lakshmi Pandit, the sister of Nehru and Indian High Commissioner to London. Schonland, as the Director of Harwell, was therefore breaking the mould but he chose not to delve in any depth into physics; rather his subject would be 'The Invisible College' [13].

It was Robert Boyle, one of the original Fellows of the Royal Society and an experimenter of brilliance, who first used the title 'The Invisible College' in the seventeenth century to describe the enduring collection of academies, associations and societies into which natural philosophers had organized themselves. They had done so not only for the advancement of science but also for their own protection, for that was the age

* See, for example, D E Proctor 1981 'Radar observations of lightning' *J Geophysical Research* **86** 12109–12113, for an account of some of his work at the National Institute for Telecommunications Research, the successor of the TRL.

of Galileo and the Inquisition when heresy covered a multitude of sins. Three hundred years later, science and the society in which it functioned were still occasionally at odds and the Director of the Atomic Energy Research Establishment chose the occasion to touch on some of the issues. But he surprised his audience by looking, not at the effect that the scientific revolution had had on society, but rather at its impact on the scientist himself within the somewhat claustrophobic world of 'The Invisible College'. Stern traditions of trust, honour, integrity and many others had, he told them, enabled it to withstand the assaults of totalitarian regimes of many hues over the intervening years, and possibly none more so than in the half century that had just closed. But Schonland feared that a new threat lay in wait, not from without but from within due to the sheer size of 'The Invisible College' itself, and nowhere was this more evident than in the United States, where the number of scientists had almost doubled in six years. 'The *corps d'élite* has become an Army Corps' he said. More than three million original scientific papers were being published every year in the more than 40 000 journals that had sprung up to receive them. Conferences, symposia and colloquia abounded, with the recent 'Atoms for Peace' conference having attracted more than 5000 delegates. It was scientific saturation on a massive scale.

But there was even more, and here he echoed his words, spoken at Rhodes, just a while before. Technology itself was threatening to overwhelm science because of the huge demands it was making on the universities to produce increasingly large numbers of applied scientists and engineers at the expense of the fundamental subjects that were their very foundation. 'No increased amount of brilliant engineering or other applications of science will, in the long run, compensate for the absence of Faradays and Rutherfords and Flemings. The danger is that we may not maintain scientific scholarship in the state of vigour which drew Faraday to Davy and Rutherford to J.J. Thomson'. Where better, he suggested, to illustrate that very interdependence that was so vital to progress than at Harwell itself?

> Harwell has owed its success to the creation of an environment of freedom in which the pure scientist, the technologist, and even that rare bird, the inventor, can all feel at home.
>
> Contact between the man who discovers the causes of things and the man who applies these discoveries to practical ends is widespread and has long been going on. It is commonplace in the larger industries and is, or should be, the *raison d'être* of our national laboratories and research associations.

It was a question of balance and that required vigilance to ensure that things were never tipped too far one way or the other, and that introduced his next theme.

His audience were from a much broader church than was usual within any of the conclaves where scientific matters were discussed, and Schonland had sought to enlighten them about issues that should have been as pertinent to bankers and bishops as they were to the men whose daily lives seldom took them far from the reaches of 'The Invisible College'. Of course there were those for whom the very world of science and its underworld of technology, where gentlemen seldom ventured, were hardly subjects of polite conversation and they might have thought that that he was just apologizing for the shortcomings of science while, mercifully, sparing them the details. But he did not allow them any smugness in what some may have thought of as their higher calling. This was the peculiarly British disease that seemingly elevated those educated in the arts above their colleagues within the scientific disciplines, and he chose now to touch on this malady.

> I am very doubtful, however, whether the scientist in general
> deserves the sweeping criticisms I have quoted. The boot may well
> be on the other foot. It is easy to find out how much science the
> majority of Arts graduates know, for they know none. In the world
> of today this happy ignorance is tragic, for it inhibits proper public
> discussion of important issues. A knowledge of the meaning of a
> radio-isotope or a chromosome is at least as important as that of an
> irregular verb.

Whether he convinced them or not of the merits of his case, Schonland enjoyed the experience of venturing beyond the walls of science into areas of debate where few from what he always called the world of humanistic studies would ever go. He had ventured from one side to the other along a sensitive boundary between two cultures and he hoped that he might encourage some who heard him to step across as well. For his closing words he chose a line from Petrarch that probably said it all: 'Here stand I on the frontier between two peoples, looking both to the past and to the future'.

Such glimpses of the Schonland psyche were rare for anyone outside his immediate circle. There were some at Harwell who saw him as a rather austere, remote and even authoritarian figure whose purpose seemed only to place barriers in the way of perfectly reasonable avenues of scientific research. That was harsh. Politically-imposed imperatives, and a history of Cockcroftian indulgence, required a firm hand on a tiller that for too long had hardly been touched. Schonland's initial reserve, his strict adherence to formality and the high expectations he had of all who worked for him could appear to those who had not penetrated his natural shyness as examples of rigidity and inflexibility. In reality, he was neither. It is true that he felt most comfortable in a hierarchical system where both authority and responsibility were clearly defined. After all, he had flourished as a soldier for that very reason, but he was no martinet. To those who came to know him more intimately he was warm-hearted, good-humoured, enthusiastic and intensely supportive of ideas and proposals that were well-founded and had promise.

He also possessed an almost boyish sense of humour, disclosed by an ever-present twinkle in his eye, and it lurked not far below the surface of an outwardly rather formal demeanour. Occasionally it could be tempted to take the bait. Soon after taking over as Director of Harwell in 1958 he was informed that a glider had had to make a forced landing on the Harwell cricket field. This followed quite soon after a rather more serious incident when an American military aircraft had similarly to force-land on the Harwell runway that had played so important a part during the D-Day landings of 1944 when Harwell had an altogether different function. On that occasion in April 1957, to the concern of all the scientific staff, who watched from every possible vantage point, the US Air Force pilot and his ground-crew who had arrived to recover the jet fighter decided to make a rocket-assisted take-off the following day. After fitting rockets to both wings the aircraft accelerated down the old runway, but at the appointed moment when the rockets were to fire only that on the starboard wing did so. The aeroplane slewed off course and was heading for the building housing the GLEEP reactor, when its headlong rush was arrested by a hole in the ground. Fortunately for all concerned the pilot made a speedy if somewhat sheepish escape from the cockpit and soon after the runway was hastily covered up, in order to discourage others from landing on it. Therefore, when told of the glider's unannounced landing, Schonland's immediate reaction was to issue strict instructions that it should not be allowed to try and take off again! Both stories enlivened many a convivial gathering at Harwell for months afterwards [14].

*   *   *

Basil Schonland left Harwell for the last time on 9 January 1961. There was no fanfare and no grand farewell. Neither suited his style and any suggestion of ceremony would to him have seemed most inappropriate at such a time. Sir Roger Makins, Chairman of the Authority, wrote formally expressing his and the Board's most sincere appreciation of the work he had done for Harwell and for the furtherance of the atomic energy programme in Britain. His period of tenure as Director had coincided with that when the establishment was changing dramatically. From what had started as the AERE at Harwell had expanded in many directions to Wantage, Winfrith, Amersham and Culham, while right next door was the fledgling National Institute for Research in Nuclear Sciences, its high energy facility that had been so appropriately named after Lord Rutherford by Schonland himself. Makins continued: [15]

None of this would have been possible without your constant spur
to good scientific work and your wise and firm handling of the
administrative problems which these changes have entailed.
Without your leadership and guidance, the difficult task of setting
up a number of establishments outside Harwell but part of a single
Research Group would have gone very much less smoothly. When
a full account is written of the Authority's part in atomic energy
research your formative and abiding influence will clearly be seen.

At a gathering of senior staff the incoming Director, Dr Arthur Vick,
presented Schonland with a painting of the Harwell complex. It was, of
course, 'The Establishment' as Schonland always knew it and what
better tribute could there be to the man who so identified with it.

Within just two weeks Basil and Ismay were in South Africa, where
they would stay until the end of June. No matter how much he may have
wished to withdraw from the limelight, that was but a faint hope in a
country rather starved of heroes, as the years of Nationalist rule and
the enshrining of apartheid had taken a heavy toll. South Africa was
already developing an unpleasant stench in the international arena. The
local press fêted him with interviews and photographs. 'Harwell Chief
to do mystery job for S.A. mines' ran the headline of the *Pretoria News*,
while *The Star*, Johannesburg's main evening newspaper, which had
included him just a few months before in its 'Union Hall of Fame' as
one of South Africa's most famous sons since 1910, now picked his
brains about the possibilities for atomic power in South Africa [16].

South Africa's scientists wanted to hear him too and so there were
the inevitable invitations to speak at one function or another. Ismay
was insistent that these be restricted to no more than one or two, since
he had a major address to prepare for the graduation ceremony at
Rhodes in April and his health was a matter of some concern. Inevitably
the BPI offered him its hospitality and he relished the opportunity to meet
old friends once again. One address he did give was in the Great Hall of
the University of the Witwatersrand when he spoke about nuclear power
in Great Britain. The vote of thanks was proposed by Professor Friedel
Sellschop, now firmly in place in the Chair of Nuclear Physics at Wits.
Sellschop's words were warm and full of admiration for the man who
had become the elder statesman on the South African scientific stage
[17]. But there were some in the audience, such as Frank Hewitt, who
had known Schonland for a very long time, who noticed that the
burden of the past few years had taken a serious toll on the man who
had been such an inspirational scientific leader. Schonland had aged
beyond his years and his grip was no longer firm [18].

In April, the Schonlands were in Grahamstown. Basil was once
again in his birthplace, the city that had at its heart a cluster of schools
the likes of which were second to none in the country and a university

as good as any. Grahamstown was also a monument to the British settlers in South Africa long before such an edifice actually rose from the slopes of Gunfire Hill above the city. It was also the City of Saints, if only for Michael and George, honoured in its cathedral, and a host of others commemorated within the plethora of churches around and about, while Andrew and Aidan adorned two of its oldest schools but a stone's throw apart.

The university and St Andrew's College had long vied with each other as to which had the greater part to play in the weaning of one of the city's most celebrated men of science. Both were now to be honoured by his presence at important occasions that month. St Andrew's had benefited from the generosity of one of South Africa's many mining houses and was to open its new science laboratories. Spencer Chapman, its headmaster, lost no time in inviting Sir Basil Schonland to perform the ceremony [19]. Few duties gave him more pleasure as science, St Andrew's and South Africa's most powerful industry, the mining of gold, combined in a mood of rare optimism at a time when events elsewhere in the country were the cause for much concern.

Just a year before, on 21 March 1960, 69 black people were shot dead by police at the town of Sharpeville to the south of Johannesburg. Accounts of the 'Sharpeville Massacre' had reverberated around the world, while South Africa's heart momentarily stopped beating. A crowd of several thousand Africans surrounding Sharpeville's police station, in protest against the so-called pass laws, which required only people of colour to carry identification papers or 'passes' at all times, had been fired on without warning by 75 nervous policemen. International condemnation was fierce, the Johannesburg Stock Exchange plummeted, many liberal-thinking South Africans planned to leave the country, while the government of Hendrik Verwoerd saw it as just the culmination of the Communist plot against the white peoples of South Africa. The nation faced the biggest crisis in its history. Away from such areas, in the cities and towns, the air was tense but life carried on very much as usual. The divided white nation reacted much as would be expected. The Nationalists called for punitive legislation to make gatherings, such as that at Sharpeville, illegal and soon there was a rash of laws to ban, to censor and to imprison on the merest pretext of a threat to 'national security'. On the other hand, those in opposition to the government were themselves divided. Calls for moderation, but little else, came from the United Party, at whose helm General Smuts had stood throughout the war, only to be so summarily dismissed from government just three years after it ended. Now the United Party was all but leaderless and certainly rudderless. The Liberal Party that had striven so hard for a political system based on the simple precepts of democracy found itself attacked with almost equal ferocity from both

Figure 32. The Chancellor, Sir Basil Schonland CBE FRS, addressing the congregation of Rhodes University in Grahamstown on 8 April 1961 when he delivered a message of hope tinged with the reality of the political situation in South Africa. Second from the left is the university's Vice-Chancellor, Dr Tom Alty. (Reproduced by kind permission of Dr Mary Davidson.)

wings of white political opinion, while they were largely ignored by the mass of black people who saw them, at best, as ineffectual or, at worst, as hypocrites.

South Africa's universities were divided as well. In 1959 the Nationalist government passed legislation under the decidedly disingenuous title of the 'Extension of University Education Act', which effectively imposed apartheid on the university system. Until then all South African universities had the right to admit students on the basis of academic merit alone. After the passage of the Act this was denied them, except in special cases which required ministerial consent if 'people of colour' were to be admitted to any of the 'white' universities. Opposition to this blatant interference with university autonomy, and the obvious infringement of human rights, was vociferous from those universities whose language of instruction was English (UCT, Wits, Natal and Rhodes), but was essentially non-existent on the Afrikaans-speaking campuses of Stellenbosch, Pretoria and Potchefstroom.

With the passing of this act, Rhodes had been dealt yet another blow. In 1957 the government, without prior consultation, saw fit for

561

ideological reasons alone to rescind an Act of Parliament of 1949 by which Rhodes University and the South African Native College at Fort Hare had been affiliated. Now, at the stroke of a ministerial pen, that most precious link was broken to the great distress of both academic establishments. Then, just a short time later, this most recent legislation went even further and drove an even more racially dogmatic wedge between them. Not only were they to be separately administered, but all contacts between them had been all but severed. The academic communities across the country responded almost in unison. In Grahamstown a solemn procession of university staff and students, headed by the Vice-Chancellor, walked in

Figure 33.　On the occasion of the official opening of the science laboratory at St Andrew's College, Grahamstown, on 19 April 1961. (Reproduced by kind permission of the Head-master, St Andrew's College, Grahamstown.)

562

protest down the High Street [20]. Similar protests and processions that took place in the other university cities were contemptuously brushed aside as the will of the majority in Parliament prevailed, and so South Africa entered a new phase in its history that was fuelled by fear and bigotry.

On 8 April 1961, Sir Basil Schonland CBE FRS, with a host of honorary degrees after his name, rose as Chancellor of Rhodes University to deliver his Graduation Address before an expectant audience. They would not be disappointed, for this was vintage Schonland. Though no orator in the grand style, Schonland was a compelling speaker, more for what he said than how he said it, for he was by nature a quiet man and his voice displayed the characteristics of his reserved, undemonstrative personality. But what he said was powerful, logical and eminently sound. As in 1951, when he addressed that congregation as their newly-installed Chancellor, the university's first, he spoke as a scientist but with the tutored fluency of a man of letters. His themes now were Thought, Truth and what he called the 'fellowship of learning', the process by which the Ionian Greeks under such giants as Pythagoras and Democritus had turned their minds to the problems and the issues of their day, and to philosophical concepts that still remain unresolved, even in ours. His Invisible College of the recent past was much in evidence. Reasoned discussion, constructive debate and an unceasing quest for the truth were, he said, the marks of the true scholar who saw no boundaries between nations and no limitations on knowledge. For examples to illustrate these concepts he pointed to the international collaboration then beginning to develop in the world of nuclear physics. Without some 'supra-national organization' that straddled all political boundaries, progress would have been stunted or even stifled completely. Now its home, the great laboratory in Geneva, contained the largest particle accelerator in the world, and access to it was a matter only of scientific priority, not national interest nor commercial gain. Lest anyone think that such ideals were new and such collaboration a recent phenomenon beyond their reach, their Chancellor told his audience of the order issued to the Congressional fleet during the American War of Independence. It came from none other than Benjamin Franklin and stated emphatically that the ships of the British scientific expedition under Captain Cook were to be allowed unhindered passage, despite the state of war that existed between their two nations. Likewise, the Napoleonic Wars had seen both Edward Jenner and Humphrey Davy given free access to France in the name of scientific collaboration. 'Insularity', said Schonland, was 'very much out of fashion'.

His message was simple and he spoke not just about science. Great ideas flourished in a world where barriers were few but such freedoms had to be cherished and defended. To ask how and why is part of man's birthright, he reminded them, and this was no less so in the physical or the biological sciences than in any other area of academic study far removed from both. 'The ideas of Copernicus on astronomy and of Darwin on evolution have been as effective in changing thought on spiritual and social problems as those of Newton and Rutherford in extending knowledge, and power over inanimate nature'.

It was an address that was meant to inspire and no doubt it did, but for all Schonland's sincerity and his encouragement of the highest of mankind's ideals, deep down he felt a void in the pit of his stomach. All he had said could have been said at any English university on a similar occasion and those just capped would then leave, to face the world outside, suitably inspired and heartened by these words, for they were truly universal. But South Africa stood on the edge of darkness, its future in the hands of men whose philosophy was underpinned by fear and who were driven by the urge of isolation. Most alarming of all was that they propagated their beliefs as the enlightened truth. The future looked bleak and Schonland saw it as his duty to broach the subject, even though it was customary on such occasions to steer well clear of anything remotely political. His closing words touched this most sensitive nerve [21]:

> I hold it to be improper, and indeed wrong, for a speaker on an occasion like this, to develop political ideas of any kind. I must however urge you as you leave to join the proud universal community of which I have spoken, to strive with all your might to seek and find, and defend—the Truth.
>
> If you do this you are free men or women. If you do not you are slaves.
>
> Examine with an open mind whether some of the ideas which you hold are readily based on truth, or whether they are founded on prejudice, which so often disguises itself as historical necessity; whether they are truly Christian or arise from self-interest which so often masquerades as patriotism.
>
> You cannot be blamed for the past; but as the clocks of South Africa approach the hour of midnight and a new day is about to dawn, *your* new day, remember Cromwell's angry outburst: "I beseech you by the Bowels of Christ, consider well lest ye be mistaken."
>
> For the future lies with you.

These were the words of a man who cared deeply for his university and for the country that had given *him* the opportunity to fly. None had the right to deny that to others.

The Schonlands were confronted by the harsh realities of South African politics during their extended stay in the country. What had been a land of seemingly immeasurable promise in those early, post-war years was now a country torn apart by nationalism, divided by language and trembling at the rise of rampant black consciousness. What to many white South Africans, over generations, had seemed a benign system of paternalism was now stark in its insistence on the total segregation of the races and the undisguised dominance of the white Afrikaner Nationalist over every aspect of life in the land. Apartheid was real, pernicious and cruel; it was a stake driven through the heart of the country and its peoples by a government of men whose minds were locked in the past. Basil and Ismay were shocked and saddened by it all, but intensely heartened by the many South Africans who stood firm against the tide of repression welling up around them.

Amongst the many were Quintin and Maida Whyte, who continued to struggle for justice and an improvement in the lot of South Africa's indigenous population through the Institute of Race Relations and the Bureau for Literacy and Literature, which they served so selflessly. Basil and Ismay, though far removed from the harsh realities of apartheid, had remained staunch supporters of everything the Whytes were trying to achieve and they were frequent correspondents. When word reached the Whytes of Schonland's knighthood Quintin Whyte wrote immediately to express his and his wife's delight [22].

> We cannot appreciate the range and depth of your knowledge and brilliance. We just hear and read, and from the record of your achievements which appear in public, we judge that in your field you are on the Everest of human achievement. What delights Maida and myself more than that is the recognition of the essential humility and quietness which you both possess. These qualities — of humility, quietness, awareness — together with the strength which comes from achievement are the marks of greatness. In this striving, aggressive, and so politically-minded world, it is so rare to meet this combination of qualities that one feels shy when one does. And it is of these qualities that leadership and real distinction are constituted, and Maida and I are happy and grateful that this has been so signally recognised in you both.

Had Schonland ever sought a political career, there would have been many a constituency in South Africa that would have welcomed him with open arms.

* * *

On a previous visit to South Africa, in 1957, Schonland had been approached by the Technical Adviser to the Chamber of Mines, Michael Falcon, for his advice and assistance in finding someone suitable as head of the chamber's small research laboratory in Johannesburg. Until now no such post had existed within the large and complex gold mining industry upon which the country depended so heavily for its wealth. At first glance, mining and science were not natural bedfellows. The rugged individuality of miners themselves, coupled with the vast scale of the engineering problems encountered when mines became progressively deeper and hotter, had seen engineers rather than scientists well to the fore in all aspects of mine design and operation. However, problems associated both with the strata-control aspects of mining and the well-being of the miners themselves, had made some work of an applied scientific nature especially important.

On the latter, Mickey Falcon had identified two areas of particular need. The first was some means of measuring and controlling the amount of dust produced in the course of mining, while the other was the acclimatization of the miners so that they could work efficiently and safely under conditions of elevated temperature and humidity. In correspondence between them, he and Schonland had also considered the type of man needed to lead this research effort and they had agreed that he should be 'an experimental physicist with a practical bent', but not necessarily with any mining experience, for his chief quality should be a 'flair for seeing what are the possibilities of the job'. With this as his starting point Schonland had written letters to various scientific colleagues in England in an attempt to find the right man, but none had elicited the names of any promising candidates [23]. It was then agreed that Falcon himself should visit England to consult his counterparts in the British mining industry and possibly interview anyone who might emerge during that process. This he duly did, but though his discussions were fruitful, his search was not.

Schonland was approached again late in 1959 with the request that he should visit South Africa to review what research activity there was and to advise on the appointment of someone to coordinate it. He agreed but made it clear that urgent matters at Harwell made it impossible for him to consider doing so in the immediate future. Of course, gold mining in South Africa was almost inextricably linked with uranium and there were certainly implications in all this for Harwell. Schonland had, therefore, to tread warily as any involvement of the Director of Harwell might easily be misconstrued in some quarters. He wrote immediately to Sir William Penney, who was now the AEA's Board Member for Research, and asked whether the board would have any objection to his undertaking such an assignment. He informed Penney that the greater part of the chamber's work was connected with safety in mining – 'rock bursts, silicosis and heat' – but there 'may be a small fraction of my investigation devoted to Uranium'. This, he felt, should present no particular problems to the AEA because any advice he might give the chamber was only likely to cause the price of the metal to come down 'to the advantage of the Authority' [24]. Penney agreed and, in January 1960, the Gold Producers' Committee of the Chamber asked Schonland formally to review all aspects of the research being conducted by the industry and to make recommendations.

The time for such a mission had now come and Basil Schonland found himself, once more, dealing with mining men whose previous indifference towards research was beginning to change, as the problems they confronted advanced on so many fronts. In April 1961, after he and Ismay had returned to Johannesburg from Grahamstown, he began what was to be seven weeks of detailed discussions with the various mining houses and numerous visits to their research facilities, to the Mining Engineering Department at Wits and to the Pneumoconiosis Research Unit of the SA Institute of Medical Research. There were also meetings with his former colleagues at both the BPI and the CSIR on matters in which they clearly had some expertise. A preliminary verbal report was given to the chamber at the beginning of May and the final document following a month later [25]. It revealed that research activity within the mining industry was rather like the curate's egg, but the hen that laid it eked out a somewhat neglected existence.

He was most impressed by the Applied Physiology Laboratory: 'an almost unique institution with a unique problem', he called it, and noted that it clearly drew all its strength from the 'energetic direction' of the man in charge. This laboratory had the task of assessing the physiological effects of demanding physical work under conditions of extreme heat and humidity, and of devising acclimatization methods that would allow the miner to work safely without fear of heat stroke. To assist in this the laboratory had developed a complex human calorimeter and Schonland saw this as work of the very highest order. By contrast, he found the Dust and Ventilation Laboratory 'apparently not qualified to assist ventilation engineers in the important question of heat flow from rock to air'. In addition, its efforts had to a large extent been misdirected in rather fruitless pursuits, such as measuring the surface area of quartz dust particles, when such information seemed to be of little practical value in combating problems of ventilation.

He expressed considerable dismay at another of its tasks, seemingly unrelated both to its purpose and its capabilities: the design of radio communication equipment for underground rescue operations. As this was an area in which Schonland himself had shown more than a little interest over many years it was not surprising that his eye fell on this rather obvious turkey. The fact that it 'has taken at least eight years since the basic specification was worked out from trials by the Telecommunications Research Laboratory of the C.S.I.R.', with little progress to show for it, rather irked him. But he was being unfair. That such a task should fall within the compass of dust and ventilation research was an accident of history because the electronics engineers who he thought were making rather heavy weather of the radio were also responsible for all the electronic equipment used elsewhere within the laboratories, and they just happened to be located within the Dust and Ventilation Laboratory.* He clearly thought the task was beyond them.

The more fundamental problems of mining gold had been explained to him as well. Rock-breaking by explosives and the characteristics of rock drills were both subjects under investigation by the Department of Mining Engineering at Wits under contract to the Chamber. The design of shafts to improve the efficiency of airflow and the investigation of improved methods to prevent the flooding of underground workings had recently become topics of investigation by the CSIR, while the very stability of the ground itself, when extensively excavated, and the fracture zones that formed around such excavations, were also matters of prime importance. Both had received the attention of the CSIR, and particularly of the BPI, where seismological investigation was now the dominant activity at Schonland's old laboratory. In addition, the thermodynamic issues involved in the take-up of heat from rock at great depth by cooling air was a matter of considerable importance, but what research had been done by one of the groups was sporadic and isolated. It was, he felt, by no means an intractable problem but, again, was probably beyond the capabilities of the staff entrusted to investigate it.

He then turned to the question of increased mechanization of the mines. Gold mining was a massively labour-intensive industry, but was that not because of the ready supply of labour on which the industry could call? Drilling and charging the face with explosives, coordinated blasting followed by the removal and hoisting to the surface of tonnes of broken rock for processing and treatment leading to the recovery of the gold-bearing ore was the daily pattern involving tens of thousands of men. Such work not only provided for their employment but it sustained the lives of millions both in South Africa and its neighbouring territories. The mining of gold was much more than the pursuit of riches, it was the very backbone of the region's economy and even the culture of its peoples had been influenced, even moulded, by it. Increased mechanization, even were it feasible, was therefore not an issue to be considered in isolation. The problem was as much to do with the resources of the country as with the engineering complexities of narrow seams, hard rock, punishing temperatures and phenomenal depth. It was not surprising, therefore, that the industry spawned in the face of these challenges

---

* Schonland had also accepted too readily the assertions of Trevor Wadley, who did the original work at the TRL, that the problem of radio communication underground had been solved and all that remained was the production of suitable equipment. In theory, Wadley's proposals were indeed sound, but they assumed the sort of ideal conditions of geology that are never obtained in practice. Even forty years later, and after considerable research effort on an international scale, no such radio system is in general use underground in mines.

was proud, resourceful and fiercely independent. If he had not really appreciated that before he readily understood it now. Not only was it the diversity of the geological strata across the length and breadth of the goldfields that made a uniform approach to increased mechanization quite impractical but many factors unconnected with machines militated against it. However, talk of ultra-deep mining to depths of 3.5 km and more, which he heard frequently, surely made some form of coordinated research into mechanization imperative even if it concentrated solely on ways of hauling more broken rock from the stopes to the surface while still breaking it underground by drill and explosives.

All of this raised the question of an Institute of Mining Research to serve the industry as a whole. It was an idea that had surfaced years before when the CSIR itself was in its formative years, but the Chamber's Technical Advisory Committee considered such a venture premature, though they did not exclude it entirely. What was almost self-evident, though, was the need for operational research into both mining methods and mechanization, and this should be carried out immediately by the Chamber's own experts, those of the various groups, and by the university. There was a need too for a similar study to be made of the issues that affected the productivity of the underground worker. Not only his health but the matter of his motivation were of prime importance since, in the end, nothing was more important to the industry than a stable labour force, but such stability depended upon a variety of factors requiring in-depth research. Valuable assistance had apparently already been provided by the National Institute of Personnel Research of the CSIR under the direction of the able Dr Simon Biesheuvel.

So much depended upon the well-being, the motivation, the training and the selection of people who made up the complete constituency of mining that the Schonland Report, if it could be categorized at all, was as much a scientific appreciation of the needs of a dynamic industry as an appraisal of its sociological complexities. Schonland's brief, of course, was to make recommendations as to the future direction that research should take and this he did explicitly in a number of areas, none of which is particularly germane here. However, the quality of that research would depend fundamentally upon the calibre of the people involved and, in this regard, he was somewhat alarmed at the relative inexperience of many of those charged with carrying it out. For this he blamed the employment policy of the mining industry itself. It was apparent that persons with postgraduate qualifications saw little future for themselves in mining: its only channel for advancement of its technical staff was the route through production rising from underground manager to mine manager. Outside of that only ventilation engineers were accorded anything like the status and salary likely to attract first class university graduates. This was a serious shortcoming

569

and one that could not be allowed to continue. It should surely be a matter of concern that the major industry in the country was getting far less than its rightful share of the university output.

The situation within the various laboratories that he had visited was even more alarming. Except in the highly impressive Applied Physiology Laboratory, the majority of staff, he asserted, were underqualified for the responsibilities they were carrying. Most possessed only bachelor's degrees. And his conclusion were forthright:

> I should emphasise that to staff laboratories with inadequately qualified people can be a sheer waste of money. The Chamber's best laboratory has a far higher proportion of men with a post-graduate university training then any other laboratory and it is getting value for money for that reason. In Europe it is now recognised that the possession of a B.Sc. degree does not qualify a man for appointment as a research scientist. All such people must enter with an Honours, an M.Sc. or a PL.D [sic]* degree obtained at a University *before*† appointment. Working as an extra-mural student for a post-graduate degree is no substitute for working in an atmosphere of the university itself. The supply of post-graduates in South Africa does not easily permit of this, but an effort could be made to improve matters by instituting a system of Chamber of Mines post-graduate scholarships for Honours and M.Sc. students at universities.

When writing his report Schonland might well have experienced a feeling of *déjà vu*, for the problems that beset South Africa's mining industry were so reminiscent of those that had confronted Britain's armed forces when war appeared imminent in 1939. The inclination to persist with the tried and tested, though patently out-moded, methods of a previous campaign could only be changed from within. That required the appointment of a scientific adviser with the personality and the flexibility to be able to win over men who had already won their spurs and who did not suffer ill-informed interlopers gladly. He knew the situation only too well. The question of a scientific adviser to the Chamber of Mines therefore exercised his mind and he recommended that the industry place the organization of its research 'under a man who is selected because of his training and experience in scientific research and its management. If he has no acquaintance with the outlook and special expertise of the gold mining industry this would be a disadvantage, but considerable less of a disadvantage than placing a large research organisation under a mining man with little experience of how to obtain good results from pure and applied scientists, including getting rid of them.'

---

* An obvious misprint for PhD.
† Schonland's emphasis.

The impact of the Schonland Report was almost immediate. The chamber resolved to appoint a suitable man to the position of Scientific Adviser, but chose to give him the title of Research Adviser, a name more in line with the thinking of the engineers in the groups. In view of its previous lack of success in attracting anyone suitable from England, the search concentrated on South Africa and Schonland was asked for his suggestions. He put forward two names: Dr W S Rapson, a chemist, and Dr N Stutterheim, a chemical engineer, who were then both serving as Vice-Presidents of the CSIR. Bill Rapson, whom Schonland himself had appointed as the first Director of the National Chemical Research Laboratory of the CSIR in 1946, became the chamber's first Research Adviser in 1962 and two years later the Chamber of Mines Research Organization was established.

*   *   *

Feeling physically much invigorated after their stay in South Africa, the Schonlands returned in the summer to England where Basil still had two unfulfilled ambitions: the first, to write another book, the other to re-establish some contact with the world of atmospheric phenomena. In 1962 the opportunity to do so presented itself when he become President of Section A: Physics and Mathematics of the British Association for the Advancement of Science, having declined a earlier invitation due to the demands of Harwell in 1959 [26]. This brought him into contact with Professor B J Mason, recently appointed to the Chair of Cloud Physics at Imperial College in London. Mason was then leading a substantial research group that was working on the mechanisms by which electric charges and fields were set up in thunderstorms and, in Schonland, he found a more than enthusiastic supporter. Their mutual interest in thunderstorm activity soon led to Schonland accepting an appointment as an honorary lecturer in Mason's department, a role which both amused and delighted him. Together, he and Mason combined their respective theories: Schonland's, on the cylindrical charge structure within a thundercloud that he had first formulated at the BPI with Dawie Malan and Mason's, which described how the charge distribution came about as a result of the interaction between hailstones and super-cooled water droplets.

In August 1962 the British Association held its annual meeting in Manchester and Schonland took the chair for Section A. For his presidential address he chose the topic that he and Ted Allibone had investigated in the Metro-Vick laboratory many years before—the relationship between lightning and the long electric spark. Now, with Mason's model of charge generation to bolster it, he departed from the usual lecture by showing a film specially made for him by Messrs Ferranti Ltd of Hollinwood in Lancashire. He described it in a letter to Felix as a cartoon without the 'funny little people in it'; but it was a bit more than that. The Ferranti company had developed the technique of animation to the point where it had been used very successfully in the making of training films and, at their suggestion, it was used to illustrate the processes involved in the lightning discharge by slowing them down 100 000 times in a colourful, cartoon-like sequence lasting all of ten minutes. Schonland wrote the explanatory commentary. This animated combination of the Schonland and Mason models was well-received by the audience, but their very simplicity prompted the question whether Schonland thought that thunderstorms really do behave as the Mason theory of charge formation propounded. 'If they have any darned sense they will. The physics is so simple and elegant, it has to be right', was his reply [27]. Almost an encapsulation, one might say, of the Schonland and the Cavendish view of physics itself.

Basil Schonland's last contribution to the physics of lightning was actually made during the closing paragraphs of his presidential address in Manchester. In the months immediately preceding that meeting he had drawn up a questionnaire, which he had circulated in the villages of the Austrian Alps. His interest had been sparked by many reports, emanating from educated people frequenting those parts, of the existence of ball lightning. Many years before, at the Cavendish, Schonland had been most sceptical whenever this strange phenomenon had come up in conversation. It was the mercurial Peter Kapitza whose theories on the subject had sounded most bizarre, and Schonland had dismissed them. Now he began to have second thoughts. The response he received from Austria was by no means overwhelming (five replies in all, of which he rejected two) but provided him with much food for thought. To the assembled gathering in Manchester he put forward an explanation for the 'luminous soap bubble' that had been observed, more than once, in areas of dry, weathered rock [28].

> It is possible to explain these reports if the luminous ball was, in fact, a slowly burning bubble of reactive gas, such as hydrogen, nitrous oxide, active nitrogen, oxygen or methane. Combustion for a few seconds may have been maintained by catalysts of hot pulverised material, carbon or rock dust, created by the same agency as the gases. This agency could only be the exceptionally heavy current of the return stroke flowing some distance along the earth in wet fissures in the ground selected for its path in preference to resistive rock. Such a selective flow of the heavy ground current is well established and some stream beds retain evidence of it in the magnetism of their rocky sides.

In the almost forty years since he offered this explanation for the generation of ball lightning no definitive theory nor unequivocal experiment has yet been described in the scientific literature to explain, once and for all, the mechanism at work. Schonland's suggestion that the lightning return stroke could be both the cause of the burning bubble of reactive gas, as well as that of the catalysts of hot, pulverized material, carbon or rock dust that maintained it was echoed in a letter to *Nature** published in February 2000 and hailed as the most plausible explanation yet presented. Schonland never wrote nor spoke on the subject again. However, his description of a possible mechanism to explain these infrequent phenomena may well have identified it many years before, but its publication was buried within the proceedings of the British Association for the Advancement of Science. Ironically, just a while before, when delivering his lecture at Chatham House, he had voiced his own concerns about the avalanche of published material that nowadays threatens almost to overwhelm science itself. It is just likely that his own elucidation of ball lightning may have suffered that very fate.

A final and most significant honour came Basil Schonland's way in 1962 when the Institution of Electrical Engineers in London awarded him the Faraday Medal. Instituted in 1922 to commemorate the fiftieth anniversary of the founding of the Society of Telegraph Engineers, the forerunner of the IEE, it is the Institution's most prestigious medal that is awarded to the person, whether a member of the Institution or not, adjudged to be most worthy of it. The citation read: 'For the outstanding part he has played in the development of electrical science and engineering, in particular in the field of nuclear power'. He was certainly in good company. Previous recipients had been Lord Rutherford himself, awarded the medal in 1930, Oliver Heaviside, Sir J J Thomson, Sebastian de Ferranti and Ambrose Fleming, amongst many distinguished others. In keeping with the tradition that all recipients of the medal should be recorded on film for posterity he was invited by the IEE to appear before the cameras when he delivered a short address of acceptance.

Significantly, he used the occasion to pay tribute to his many colleagues at Harwell who had worked alongside him in securing for Britain its pre-eminent position in the field of nuclear power. But he singled out by name just one man to whom he owed so much: Bernard Price, 'a great Englishman and a great engineer who helped to finance the institute for geophysical research which bears his name, at a time when such things were rare'. Price, he said, had done this because of his 'deep understanding of the value of basic research as a foundation for engineering and technology' and he continued: 'He would be happy

---

*J Abrahamson and J Dinniss 2000 'Ball lightning caused by oxidation of nanoparticle networks from normal lightning strikes on soil' *Nature* **403** 519–521.

to know that in return we did something for the power and the communication industries by laying bare the unexpected fundamental processes
which govern the development of the lightning flash and the long electric
spark and the radio waves they send out'. Certainly Bernard Price was no
ordinary engineer—a species for whom Schonland had, on occasion,
reserved the odd barbed remark. Now, when reflecting on his own
career he summed up that special relationship rather well [29]:

> In all this collaborative work I have come to realise that the
> engineer is the salt of the earth. Perhaps the engineers will concede
> that the scientists are the pepper, and that judicious mixture of both
> ingredients is essential for success.

Schonland himself was very much a scientist but, as Robert Spence
recorded after his death, he was also a man of 'good judgement, administrative skill and common sense' who, during some of the most intense
and bitter controversies, 'bore himself with a simple dignity which
derived naturally from the patent integrity of his character' [30].

*   *   *

In June 1962 a double blow fell upon the Council of Rhodes University
when it was announced at their meeting on the 22nd that the Chancellor
of the university, Sir Basil Schonland CBE FRS, had tended his resignation. Members of the council were stunned, for they were still pondering
the implications of the previous item on their agenda, in which Dr
Thomas Alty who, since 1951, had been Rhodes's first Principal and
Vice-Chancellor had announced his intention to retire the following
year. The council minutes recording the Chancellor's resignation offered,
by way of explanation, the fact that Schonland had 'taken up permanent
residence in England' [31]. He had indeed, but had done so all of eight
years before! There was no mention of ill-health or other factors beyond
his control that would have led him to this decision, so there must have
been much more to it, and indeed there was.

The reason was never publicly disclosed but it was the odour of
political interference that had begun to pervade all South Africa's universities at that time that had everything to do with it. Grahamstown was
celebrating its centenary in 1962 and the festivities were to be attended
by the first State President of the recently created Republic of South
Africa, Mr C R Swart. Since Grahamstown was at the very heart of British
settler country, the annual commemoration of Settlers Day that year was
to culminate in the unveiling of a Precinct Stone on Gunfire Hill, where
plans were already far advanced for the construction of a monument to
honour those Settlers of 1820 [32]. The council of the university thought
it fitting that an honorary degree should be conferred upon the State
President on that important occasion in the life of the city, but this

raised a storm of protest in some quarters of the university constituency who well remembered that it was 'Blackie' Swart who, as Minister of Justice, had piloted through the infamous Extension of Universities Act that had forced apartheid upon the campuses across the land just three years before. Feelings ran high. A petition calling on the University Council to rescind its decision was drawn up and circulated, but it was rebuffed and its originator, a senior lecturer in the Department of Philosophy, and all who signed it were rebuked by Council.[*]

The Chancellor heard of the council's decision from Tom Alty and wrote back immediately expressing his dismay that Rhodes could even contemplate awarding an honorary degree to the man who had so blatantly defaced all that a university held dear. The reply he received from Alty setting out the university's stance evidently fell far short of satisfying him. The council seemingly felt it only proper that Rhodes should honour the office of State President in this way and the fact that the incumbent happened to be C R Swart was an accident of timing. But Schonland rejected this and immediately submitted his resignation, but left it to Alty to disclose the reasons that lay behind it. Alty chose not to make Schonland's views known and the university's official chronicler only recorded that the chancellor had resigned 'for personal reasons' [34].

It was a bitter decision for Basil Schonland to have to take, but it was taken on principle, and throughout his life he had never shirked from such action if he deemed it necessary. Ismay stood four-square behind him—on matters of fundamental freedoms and the rights of all to enjoy them she was unyielding. Sadly, the university that meant so much to both of them, and that Basil Schonland cherished more than any other because of the part it had played in moulding his life and in setting him on his path to greater things, had lost one of its greatest sons for reasons, presumably, of fear and a lack of courage. But then, it could be argued, he no longer had to stand firm against the Nationalists' policies nor had he to expose himself to personal vilification or worse. One feels, though, that he would not have wilted even had he been in the front line.

*   *   *

Basil Schonland had long maintained an interest in gardening and evidence of this was to be seen wherever the Schonlands lived, whether in South Africa or in England. However, it was only in retirement that he had the time to pursue his hobby with any dedication and Denchworth House provided him with that opportunity. He had become something of

[*] Only 26 academics and just a single member of the university's Senate actually put their signatures to the petition [33].

an expert, especially on shrubs, and was soon called upon by colleagues for his advice. Lord Plowden also nurtured an interest, but was less familiar with the details. Could Schonland suggest something suitable for his garden? '*Escallonia*, a handsome, hardy evergreen' was Basil's considered opinion. It might well have been his view of Harwell itself. The AERE Horticultural Society, when word reached them of their Director's interest and expertise, invited to him to become their President and he accepted with enthusiasm.

Then there was also the matter of writing another book. Though he had told John Cockcroft some time before that writing popular science was something to be shunned, Schonland had clearly undergone a change of heart and now wished to prepare a second edition of *The Flight of Thunderbolts* that had appeared first in 1950. The spur for this was actually the Gregynog Lecture that he had given at the University College of Aberystwyth in 1960. This special commemorative lecture had been an annual event at the college since 1932 and the choice of subject was as catholic as the speakers were eminent. Schonland followed such personages as A L Rowse on 'The Elizabethan State and the Celtic Borderlands'; Sir Solly Zuckerman, displaying admirable breadth himself, on 'The Seasons'; and Sir Ifor Williams, much closer to home, who addressed the subject of 'Llyfr Taliesin'. Schonland himself spoke, not surprisingly, on 'Lightning'. His audience that night reflected the complete spectrum of university life, as well as interested people from the wider community, and so the story he had recounted in his first venture on to the popular shelves provided an excellent foundation on which to lay some of the more recent research findings. It apparently succeeded so well in public that a second edition of the book was called for and it was duly published in 1964 [35].

But his major venture in retirement was the writing of *The Atomists*, an account of the science and the circumstances behind the discoveries in atomic physics between 1805 and 1933, and of the men and women who were central to them. It was, according to Ismay, his way of repaying the great debt of gratitude he owed the Cavendish, just as *The Flight of Thunderbolts* before it had set out to do for the BPI. But it became a major task and, in the end, the research and the long hours of writing were too much for him. Inevitably, as the effects of his arteriosclerosis took their toll, Ismay began to play an increasingly active part in the writing of the final chapters.

The Atomists was published in 1968 to mixed reviews. Though regarded generally as a most readable account of a remarkable period in scientific history, it was criticized by the more discerning reviewers for the surprisingly large number of errors it contained. Some, of fact, revealed a too ready reliance on secondary sources, but most were matters of nomenclature and the occasional misspelling of proper names. None would have escaped the scrutiny of the Schonland of old but, sadly, as the years advanced so that razor-sharp faculty went into decline. One reviewer, after describing the book as 'the best popular account of the subject to date' warned historians of science that if they wanted accurate popular histories they would have to write them themselves! This was harsh, but then Schonland's own meticulous work over so many years was well-known and such unfortunate solecisms that had slipped through the editorial net did not go unnoticed. Nor had some forgotten his earlier venture into the realms of popular science. 'The Atomists', wrote a reviewer in the *American Historical Review*, 'requires more physics of its readers than did Schonland's earlier, elegant vulgarization of his own special field, *The Flight of Thunderbolts* (1950). Those properly prepared will probably also appreciate his brief, lively, faulty atomic story' [36]. It was indeed sad that Schonland, the scourge of the imprecise and the inaccurate, should himself fall victim to these demons at what was almost the final hurdle.

But the Schonland memory should not be scarred by a final pursuit that went on too long. We should remember him not for some relatively minor inaccuracies that blotted an otherwise fascinating book but rather by something he wrote just a year or two before. In January 1963 Lord Hankey, the 'Prince of Secretaries', died and Basil Schonland was invited by the Royal Society to write his biographical memoir.

Maurice Pascal Alers Hankey, First Baron Hankey of the Chart, often referred to as the greatest background figure in British history, was the confidante of Prime Ministers and the power behind the great machinery of State through two World Wars, and the often stagnant shallows in between. It was, therefore, a great tribute to Schonland to be called upon to celebrate the life and contributions of a man whose very presence, though almost unobtrusive, had galvanized great men into action. Schonland accepted the commission both willingly and enthusiastically and set out to pay due honour to a man for whom he had always had the very highest regard.

There had been more than a little affinity between them — seeded, one must presume, by the fact that Hankey's wife had been born in the Cape, but it was really kindled because both men had at various times been close associates of General Smuts. It was when he was Secretary to Lloyd George's War Cabinet of 1917, of which Smuts was the only colonial member and its intellectual colossus, that Hankey and the great South African first met. Each undoubtedly impressed the other, for neither could be ignored. At the outbreak of the Second World War Hankey himself was a member of the War Cabinet and, under Churchill, became Paymaster General in which capacity he ranged far and wide, and it was in one of those many roles that he and Schonland had first encountered each other. Hankey's imaginative schemes to increase the numbers of technically trained men and women to operate and maintain the radars in Britain and overseas, had drawn Schonland in to advise how the SSS might assist. Soon afterwards Schonland's own services had been required at the ADRDE(ORG), and it was Hankey who informed Smuts of the significant contribution that this South African scientist was now making towards the war effort. That communication sowed a very important seed in Smuts' mind and South Africa's scientific future then took root.

What Schonland eventually wrote was a superb memoir of the life of one of Britain's finest sons [37]. It captured the very essence of the man, self-effacing yet with a will of steel and a determination that knew no bounds. He traced Hankey's career from youthful Subaltern of Marines in 1898, serving aboard venerable battleships of an era soon to close, to intelligence officer in the Admiralty, where the young Hankey's quite precocious ability soon brought him to the attention of the formidable Admiral Sir John Fisher. As a linguist with an innate grasp of the scientific method, Hankey made quite startling progress. By 1908, he was a secretary to the Committee of Imperial Defence; seven years later, at the height of the First World War, he was *the* Secretary. From then on he became the undoubted *éminence grise* of the civil service, with unparalleled influence in determining the part that science would play in warfare. The concept of the tank, so enthusiastically embraced by Churchill and so precipitately used by Haig, was his, as was the idea of the convoy to defeat the submarine and the flame-thrower for use in the event of an invasion. Schonland readily identified the elements of operational research that lay behind all Hankey's proposals, for Hankey had laid its very foundations. Well before the next war threatened, Hankey had seen the need to mobilize scientific effort. With the support of the Royal Society in 1940, he became the chairman of the Cabinet's Scientific Advisory Committee that played such an important part in ensuring that senior men were available for crucial scientific work while, in parallel, he chaired its Engineering Advisory Committee with similar aims and objectives. By the war's end his contributions towards ensuring the Allied victory were legion.

For all his involvement with the most bellicose activities of mankind, Maurice Hankey had the most gentle self-deprecatory manner. He was the quintessential public servant and Schonland captured the real quality of the man admirably when he wrote:

> He had indefatigable energy and an incredible memory. Above all
> he had complete integrity. But he also had the tact and discretion
> which so often made it possible for him to reconcile discordant
> elements, a combination of modesty and friendliness which
> disarmed hostility towards new ideas.

Many will have said that Basil Schonland himself had been endowed with such qualities. Fitting it was that he should have been called upon to record for posterity the life and the work of such a man.[*]

*   *   *

Basil Schonland died peacefully in a nursing home near 'The Down House' in Shawford on Friday 24 November 1972. He was 76 and his last years were spent quietly at the home that he and Ismay had bought to be passed on to the family. The Schonland mind remained alert almost to the end but his ability to walk began to leave him, following a number of small strokes. He was eventually persuaded to give up smoking but the damage had, by then, been done. The was no memorial service and his ashes were scattered on the Downs where he loved to walk with Ismay.

In 1973 the Royal Society published T E Allibone's excellent biographical memoir, which set out in vivid detail the life and work of one of South Africa's greatest scientists and one of Great Britain's most loyal adopted sons [38]. It was a fitting tribute from one who had been associated with him longer almost than anyone else, except for one man—John Cockcroft—who died five years earlier, still in harness, as Master of Churchill College, Cambridge.

The Times published an obituary just a day after Schonland's death. It was followed soon after by a letter from Robert Spence, soon to become Director of Harwell himself. Spence wrote of the man who had given his all to Harwell, at a time when it was undergoing great change. It was service, he said, of the highest order during moments of often intense and bitter controversy [39]. In South Africa, Schonland's passing was marked by a fulsome tribute from the CSIR and by short pieces in many newspapers. In 1985, at a ceremony at the University of the Witwatersrand, the President of the CSIR, Dr C F Garbers, unveiled a plaque renaming the University's Nuclear Physics Research Unit the 'Schonland Research Centre for Nuclear Sciences'. The driving force behind that was Friedel Sellschop, by now a Deputy Vice-Chancellor of the University and the director of that laboratory which had been recognized as a Centre of Research Excellence in South Africa [40]. In December 1999 the *Financial Mail*, a South African newspaper, published a special millennium issue in which was recorded the 'South African Achievers of the Century'. The panel of judges selected Sir Basil Schonland CBE FRS as South Africa's 'Scientist of the Century'.

---

[*] In 1969 Schonland contributed something to the Royal Society's Biographical Memoir of P J du Toit (1969 *Biog. Mems. Fellows of the Royal Society* **15** 247–266). Due to his failing health the burden of writing it fell to Denys Kingwill of the CSIR with considerable input from Ismay Schonland.

NOTES AND SOURCES

Throughout this listing of sources the abbrevations BFJS and IMS appear frequently. They refer to Basil and Ismay Schonland respectively.

Chapter 1 EARLY YEARS

[1] *Standard Encyclopædia of Southern Africa* 1971 Selmar Schönland p 519; 21 Sept 1940 *Nature* p 395; *Memories of my father,* unpublished notes prepared by Dr Mary Davidson, daughter of Basil Schonland, February 1996.

[2] Brown A C (ed) 1977 *A History of Scientific Endeavour in South Africa* (Cape Town: The Royal Society of South Africa) p 481.

[3] Currey R F 1970 *Rhodes University 1904–1970* (Grahamstown: Rhodes University) p 7.

[4] Pakenham T 1979 *The Boer War* (London: Macdonald & Co) p 2.

[5] Ibid p 22.

[6] *Memories* ref [1].

[7] Notes (undated) on the life of B F J Schonland by Ismay Schonland supplied to the author by Dr Mary Davidson.

[8] Rhodes University Cory Library MS 11008 St Andrew's Preparatory School quarterly report B Schönland.

[9] Rhodes University Cory Library PR 2148 Newspaper cutting.

[10] Currey *op cit* p 9.

[11] Rhodes University Cory Library PR 2148 Ibid.

[12] Rhodes University Cory Library MS 11017 Letter to Felix (undated) from 46 Stationary Hospital BEF France.

[13] *St Andrew's College Magazine* vol XXV No 119 Pt 3 Sept 1908 pp 106–107.

[14] Rhodes University Cory Library MS 11009 St Andrew's College quarterly report Schönland B F J.

[15] Rhodes University Cory Library MS 11010 St Andrew's College quarterly report.

[16] *St Andrew's College Magazine* vol XXXL Pt 3 No 123 Sept 1909 p 120.

[17] Ibid vol XXXII Pt 2 No 126 June 1910 p 49.

[18] Ibid Pt 3 No 127 Oct 1910 p 86.

[19] Smuts J C 1952 *Jan Christian Smuts* (London: Cassell & Co. Limited) p 119.

[20] Brown *op cit* p 246.

[21] Rhodes University Cory Library PR 2148 'Old Boys' Successes' Newspaper cutting.

[22] Rhodes University Cory Library MSS 11012 & 11013.

[23] Pakenham *op cit* p 549.

[24] *St Andrew's College Magazine* vol XXXII No 128 Pt 4 Dec 1910 pp 129–143.

Chapter 2 RHODES TO CAMBRIDGE

[1] Currey R F 1970 *Rhodes University 1904–1970* (Grahamstown: Rhodes University).

[2] Rhodes University Cory Library MS 17 260/1 Register of Students; Senate Minutes MS 17 504/2.

[3] Packenham T 1979 *The Boer War* (London: Futura) p 572.

[4] Austin B A 1995 'Wireless in the Boer War' *The Royal Engineers Journal* **109** 3 232–238.

[5] Ismay Schonland (undated) notes on BFJS supplied to the author by Dr Mary Davidson.

[6] Dowsett H M 1923 *Wireless Telegraphy and Broadcasting* (London: The Gresham Publishing Co) pp 7–8.

[7] Rhodes University Cory Library MS 11018 Letter: Schonland to Felix from 'Somewhere in France' 7.3.1916.

[8] Rhodes University Cory Library MS 11031 Letter from Sir Thomas Graham to Schonland's parents 13.3.1938.

[9] Ismay Schonland *op cit* p 1.

[10] *St Andrew's College Magazine* Oct. 1910 vol XXXII Pt 3 No 127 p 81.

[11] Currey *op cit* p 2.

[12] Rhodes University Cory Library Newspaper cutting Cape Town 28.1.1912.

[13] Supplied to the author by The Warden, Rhodes House, Oxford, File No 1378 St Andrew's College *Regulations for the Rhodes Scholarships* 1903.

[14] Ismay Schonland p 1.

[15] Currey *op cit* p 29.

[16] Brown A C (ed) 1977 *A History of Scientific Endeavour in South Africa* (Cape Town: Royal Society of South Africa) p 375.

[17] CSIR Archives Pretoria Record of Military Service B F J Schonland transcribed from UDF Records AG(1)P1/11036/1.

[18] University of Cape Town Archive (Schonland personal file) Letter from Selmar Schönland to the Registrar 19.12.1921.

[19] Rhodes University Cory Library MS 17504 Senate Minutes 24.2.1915 p 354.

[20] Ibid Senate Minutes 12 May 1915 Bk III p 111.

[21] Currey *op cit* p 38.

[22] National Archives of South Africa, CES 106 ES70/2081/14 Letter from Selmar Schönland to Commissioner for Enemy Subjects 6.3.1915.

[23] Ibid ES70/2081/14 Reply from Commissioner to Schönland 8.3.

[24] Rhodes University Cory Library MS 10742 Letter from W P Schreiner to Selmar Schönland 7.3.1915.

[25] Brown *op cit* p 384.

[26] Rhodes University Cory Library PR 2148. At this stage of Schonland's career there appears to have been no consistency in the use or otherwise of the umlaut. It is used as often as it is omitted.

[27] Schonland B F J Diary 1915–, a diary kept by 2nd Lt Schonland RE (Signals) between 6 July 1915 and 11 December 1915. Kindly lent to the author by Dr Robert Davidson, Schonland's grandson.

Chapter 3 FOR KING AND COUNTRY

[1] Schonland's Diary: July–December 1915. This remarkable little book is the basis for almost everything in this chapter and should be taken as the source of all material quoted here unless otherwise stated.

[2] Austin B A 1995 'Wireless in the Boer War' *The Royal Engineers Journal* **109** 232–238.

[3] Adams R M 1970 *Through to 1970 – Royal Signals Golden Jubilee* (London: Royal Signals Institution) ch 5.

[4] Rhodes University Cory Library MS 11016 Letter BFJS to his mother 12.8.1915.

[5] Terraine J 1990 *Douglas Haig – The Educated Soldier* (London: Leo Cooper) p 106 et seq.

[6] Nalder R F H (1958) *The Royal Corps of Signals* (Royal Signals Institution) p 90.

[7] Rhodes University Cory Library PR 2148 Newspaper cutting from Grahamstown undated but headlined 'Letters from Overseas' and undoubtedly referring to events in December 1915.

Chapter 4 IN COMMAND

[1] Rhodes University Cory Library MS 11015: Field Service Post Card 24.1.1915 (as marked but actually 1916).

[2] Priestley R E *The Signal Service in the European War of 1914 to 1918* (Chatham: W & J Mackay and Co Ltd).

[3] Rhodes University Cory Library MS 1018 Letter: BFJS to Felix 7.3.1916 from 'Somewhere in France'.

[4] Phillips D 1979 'William Lawrence Bragg' *Biographical Memoirs of Fellows of the Royal Society* **25** (London: The Royal Society) p 94.

[5] Rhodes University Cory Library MS 11020 Letter: BFJS to Felix 14.4.1916.

[6] Rhodes University Cory Library MS 11023 Letter: BFJS to Felix 8.10.1916 from W/T C.C.

[7] Nalder R F H 1958 *The Royal Corps of Signals* (London: Royal Signals Institution) p 105.

[8] Adams R M 1970 *Through to 1970* (London: Royal Signals Institution) p 39.

[9] Rhodes University Cory Library MS 11022 Letter: BFJS to his mother 1 October 1916.

[10] Terraine J 1963 *Douglas Haig – The Educated Soldier* (London: Leo Cooper) pp 218–219.

[11] Rhodes University Cory Library MS 11017 Letters: undated and MS1025 22.10.1917 from BFJS to Dick.

[12] Rhodes University Cory Library MS 11024 Letter: BFJS to Felix 29.10.1916.

[13] Uys I 1983 *Delville Wood* (Johannesburg: Uys Publishers) p 274; Amended figures were supplied more recently to the author by the Director, SA National Museum of Military History, Johannesburg.

[14] Schonland B F J 1919 'W/T RE' *The Wireless World* **7** pp 174–178, 261–267, 394–397, 452–445.

[15] 'ARMY BOOK 3' Schonland's notebook of wireless circuit diagrams was kindly lent to the author by Schonland's grandson, Dr Robert Davidson.

[16] UCT Archives BFJS's application of 29.12.1921 for the senior lectureship in Physics at the University and the accompanying letter from himself to 'The Office of the SA High Commissioner in London' 28.12.1921. The amount of detailed personal information provided by Schonland in this application form is remarkable, particularly as regards his war service. One notes that, at the end of the war he was in command of about one-third of the total complement of officers and men in the wireless branch of the Royal Engineers Signal Service.

[17] UCT Archives: (Undated) copy of the testimonial letter from Maj Rupert Stanley submitted by BFJS when applying for the UCT post.

[18] Rhodes University Cory Library MS 11025 Letter: BFJS to Dick 22.10.1917; Dowsett HM 1924 *Wireless Telephony and Broadcasting* vol II (London: The Gresham Publishing Co Ltd) p 142.

[19] Humby SR and Schonland B F J 1919 'The wavelengths radiated from oscillating valve circuits' *The Electrician* 17 October pp 443–444.

[20] Nalder *op cit* p 123.

[21] Appleton E V L 1918 'Note on the production of continuous electrical oscillations by the three-electrode valve' *The Electrician* 27 December pp 743–744.

[22] Ref [16] *op cit.*

[23] Terraine *op cit* p 378.

[23] Rhodes University Cory Library MS 11026 Letter: BFJS to Felix 31.3.1918.

[24] Rhodes University Cory Library MS 11028 Letter: BFJS to Felix 3.6.1918.

[25] Rhodes University Cory Library MS 11029 Letter: BFJS to Felix 31.7.1918.

[26] Rhodes University Cory Library MS 11027 Letter: BFJS to Felix 2.8.1918.

[27] Rhodes University Cory Library MS 11030 Letter: BFJS to Felix 16.11.1918.

Chapter 5 FROM THE CAVENDISH TO CAPETOWN

[1] SANDF Archive File AG (1)P1/11036/1 BFJS's Record of Service.

[2] UCT Archive: Extract from Army Book 439 Officer's Record of Service written by Colonel H T G Moore DSO RE Chief Signal Officer submitted by BFJS amongst his references to UCT in 1921.

[3] Letter from Schonland's daughter, Dr Mary Davidson, to the author, December 1996; Rhodes University Cory Library MS 11046 Felix Schonland's letter to the Cory Librarian 11.3.1963.

[4] See Smuts J C 1952 *Jan Christian Smuts* (London: Cassell & Co Ltd). This biography of Field Marshal Smuts was written by his son who bore the same names as his father. In the interests of historical accuracy it should be noted that Smuts's second name is as given here and not in the form *Christiaan* in which it so often appears, quite erroneously, especially in South Africa.

[5] South African National Archive Smuts Papers Vol 206 No 119 Letter from Selmar Schönland to J C Smuts 26.8.1919.

[6] Supplement to the *London Gazette* 3.6.1919 No 6793; CSIR Archive File B26 54 58: MID confirmed in letters from the Ministry of Defence to the South African Scientific Counsellor London 22.12.1972 and 22.1.1973. In T E Allibone's Royal Society Biographical Memoir of Schonland (vol 19, 1973) he states the Schonland was mentioned twice in dispatches. This was not so.

[7] CSIR Archive Doc B26 54 35 *Professor William Hofmeyr Craib*. This document was written by Craib himself as part of the contribution towards the history of the CSIR.

[8] Letter to the author from the archivist at Gonville and Caius College, Cambridge 24.6.1998.

[9] Notes prepared by Ismay Schonland and supplied to the author by Dr Mary Davidson 1996.

[10] UCT Manuscripts and Archives. Copy of a reference for Schonland written by E V Appleton (undated *ca* Dec 1921).

[11] Lord Blake and Nichols C S 1986 *Dictionary of National Biography* Sir Basil Ferdinand Schonland (1896–1972) p 761.

[12] Eve A S 1939 *Rutherford* (London: Macmillan) p 199.

[13] Rutherford E, Chadwick J and Ellis C D 1930 *Radiations from Radioactive Substances* (Cambridge University Press) p 201.

[14] Fleming C A 1971 'Ernest Marsden (1889–1970)' *Biographical Memoirs of Fellows of the Royal Society* **17** 463–495.

[15] Crowther J G 1974 *The Cavendish Laboratory 1874–1974* (London: Macmillan) p 187.

[16] Thomson G P 1955 'George Frederick Charles Searle (1864–1954)' *Biographical Memoirs of Fellows of the Royal Society* **1** 247–252.

[17] UCT Manuscripts Archive. Copy of a reference written for Schonland by G F C Searle (undated *ca* Dec 1921).

[18] Halliday E C 1970 *Some memories of Prof C T R Wilson*. An unpublished recollection kindly given to the author by Dr D E Proctor of Johannesburg.

[19] Crowther *op cit* p 167.

[20] Ibid p 164.

[21] Wood A 1946 *The Cavendish Laboratory* (Cambridge University Press) p 28.

[22] Crowther *op cit* p 201.

[23] Letter to the author from Ellie Clewlow, archivist at Gonville and Caius College 5.11.1997.

[24] Rhodes University Cory Library, MS 14814: Letter (undated but from its contents it must be early 1921) from BFJS to Sherwood Watson.

[25] Crowther *op cit* p 291.

[26] Ref [24] *op cit*.

[27] Crowther J A and Schonland B 1922 'On the scattering of β-rays' *Proceedings of the Royal Society* **A100** 526–550.

[28] Schonland B F J 1922 'On the scattering of β-particles' *Proceedings of the Royal Society* **A101** 299–311.

[29] Brown A C (ed) 1977 *A History of Scientific Endeavour in South Africa* (Cape Town: The Royal Society of South Africa) p 378.

[30] UCT Main Archive: McMillan Collection 'Schonland' papers.

Chapter 6 AND THEN THERE WAS LIGHTNING

[1] Brown A C 1977 *A History of Scientific Endeavour in South Africa* (Cape Town Royal Society of South Africa) p 65.

[2] Ibid p 375.

[3] Ibid p 67.

[4] Sir Carruthers Beattie (1866–1946) read physics at the University of Edinburgh before taking up the *1851 Exhibition* in the years 1894–1897 at the Universities of Vienna, Berlin and Glasgow where he studied the behaviour of conductors in a magnetic field. He retired as vice-chancellor of the University of Cape Town in 1937.

[5] Grindley E N 1975 *The UCT Physics Department in the 1920s* Physics Research Bulletin 19–24 kindly provided to the author by Dr John Juritz, Dept of Physics, UCT.

[6] Cory Library MS 14814 Letter (undated) from BFJS at 'Chartfield' , Kalk Bay, near Cape Town to Sherwood Watson in Grahamstown.

[7] Hargreaves J K 1995 *The solar–terrestrial environment* (Cambridge University Press) p 335.

[8] Halliday E 1986 'Reminiscences of physics before the nuclear age' *Meson* (a newsletter of the CSIR) p 7.

[9] McCrea W 1988 'Sir Richard van der Riet Woolley' *Biographical Memoirs of Fellows of the Royal Society* **34** 923–982.

[10] Rutherford E, Chadwick J and Ellis CD 1930 *Radiations from Radioactive Substances* (Cambridge University Press) p 417.

[11] Schonland B F J 1923 'The passage of cathode rays through matter' *Proceedings of the Royal Society* **A104** 235–247. Schonland's brother Felix was later to claim that Basil Schonland had 'devised the Pentode valve which the Philips Company of Eindhoven took up from his published work and commercialised' : Cory Library MS 16046, letter to Librarian of 11.3.1963. There is no evidence to support this but certainly Schonland's use of his suppresser grid pre-dated the Dutch work since it appeared both in this paper and in the next, which he submitted to *Nature* **115** of 4 April 1925 p 497. British Patent 287,958 was issued to Holst and Tellegen of Philips on 26 March 1928 following their application of 24.12.1926.

[12] Brown *op cit* p 74.

[13] Schonland B F J 1925 'The passage of cathode rays through matter' *Proceedings of the Royal Society* **A113** 210.

[14] Newnham College Cambridge Register vol 1 1871–1923 entry for Isabel Marian Craib.

[15] Author's interview with Dr Mary Davidson, Schonland's daughter, 18.8.1997. Shortly after the First World War, Schonland had been diagnosed with a medical condition that, at the time, was considered serious enough to prevent him from obtaining life insurance. He was therefore always reluctant to incur any debts and so rented rather than bought the houses the family occupied.

[16] Cory Library MS 14 814: Letter from BFJS to Sherwood Watson 19.6.1924.

[17] Cambridge University Archives Registry of personal records *B F J Schonland*. Information supplied to the author by the Deputy Keeper of the Archives Dr E Leedham-Green 3.8.1998.

[18] Phillips H 1993 *The University of Cape Town 1918-1948. The Formative Years* (Cape Town: University of Cape Town Press) p 47.

[19] At a ceremony in London in 1962 when Schonland was awarded the Institution of Electrical Engineers Faraday Medal it was stated that he 'became the first man to utilize an electron beam of 100,000 volts'. 1962 'Presentation of Institution Honours and Prizes for 1962' *Journal of the Institution of Electrical Engineers* 291–292.

[20] UCT Archive 1924 *University of Cape Town Quarterly* vol 7 nos 3 & 4.

[21] Ibid ref [16].

[22] Schonland B F J 1925 'The absorption of cathode rays in aluminium' *Nature* **115** 497.

[23] Eve A S 1939 *Rutherford* (New York: Macmillan) p 307.

[24] Archives of the *Royal Commission of 1851* letter from Professor A Ogg 13.4.1927 in which he quotes Rutherford's remarks on Schonland.

[25] Rutherford E, Chadwick J and Ellis CD 1930 *Radiations from Radioactive Substances* (Cambridge University Press) p 417-423.

[26] Schonland B F J 1926 'The scattering of cathode rays' *Proceedings of the Royal Society* **A113** 87–106.

[27] Ismay Schonland's recollections of Basil Schonland, written soon after his death for T E Allibone, his Royal Society biographer, and supplied to the author by Dr Mary Davidson, their daughter.

[28] 1925 'A discussion on the ionization in the atmosphere and its influence on the propagation of wireless signals' held at the Imperial College of Science on 28 November 1924 *Proceedings of the Physical Society* **37** 2D–50D.

[29] Ibid p 32D.

[30] Ibid p 46D.

[31] Cambridge University Manuscript Library ADD 8702 Box 2. Schonland and C T R Wilson continued this correspondence from early 1925 until 1951. It makes fascinating reading. Wilson's letters were full of old-world charm, usually apologetic for the lateness of his reply to Schonland's previous one and almost all opening with the phrase 'I wonder what you will think of me! I never meant to be so long in writing.' They also are remarkable for the depth of physical insight they contain as Wilson encouraged Schonland in his experimental programme designed to test Wilson's own theories of thundercloud electrification and then the phenomenon of 'the penetrating radiation' he believed was generated within the cloud itself. His last letter to Schonland, written in 1951, expressed his pleasure at having the second edition of Schonland's book '*Atmospheric Electricity*' dedicated to himself.

[32] Schonland B F J and Craib J 1927 'The electric fields of South African thunderstorms' *Proceedings of the Royal Society* **A114** 229–243.

[33] Boys C V 1926 'Progressive lightning' *Nature* **118** 20 Nov p 749.

[34] Strangeways H J 1996 'Lightning, Trimpis and Sprites' *Review of Radio Science 1993–1996* (W Ross Stone ed) pp 741–780.

[35] Simpson G C 1927 The mechanism of a thunderstorm' *Proceedings of the Royal Society* **A114** 376–401.

[36] Schonland B F J 1928 'The polarity of thunderclouds' *Proceedings of the Royal Society* **A118** 233–251.

[37] Schonland B F J ibid 'The interchange of electricity between thunderclouds and the earth' *Proceedings of the Royal Society* **A118** 252–262.

Chapter 7 ENTER COCKCROFT

[1] Brown A C (ed) 1977 *A History of Scientific Endeavour in South Africa* (Cape Town: Royal Society of South Africa) p 474.

[2] UCT Archive Minutes of Council 26.4.1927.

[3] Archives of the *Royal Commission for the Exhibition of 1851*: Letter from Professor A Ogg to Mr Evelyn Shaw, Secretary of the Commission 13.4.1927.

[4] Ibid.

[5] Ibid: Letter from Sir W H Bragg of 17.6.1927 to the Secretary of the Royal Commission of 1851.

[6] Schonland B F J 1929 'A new electroscope' *Proceedings of the Cambridge Philosophical Society* **25** 340–343.

[7] Schonland B F J 1928 'The scattering of cathode rays' *Proceedings of the Royal Society* **A119** 673–680.

[8] Rutherford E, Chadwick J and Ellis C D 1930 *Radiations from Radioactive Substances* (Cambridge: The University Press) pp 227–234.

[9] Ref [7] *op cit* p 680.

[10] Crowther J G 1974 *The Cavendish Laboratory 1874–1974* (London: Macmillan) p 235.

[11] Brown A 1998 'Patrick Blackett: sailor, scientist, socialist' *Physics World* **11** 35–38.

[12] Hartcup G and Allibone T E 1984 *Cockcroft and the Atom* (Bristol: Adam Hilger).

[13] Cockburn S and Ellyard D 1981 *Oliphant* (Adelaide: Axiom Books) p 36.

[14] Archive of the Royal Commission for the Exibition of 1851: Report of Sir Richard Glazebrooke 1930 *Opinion on the Report of work done during the tenure of an Overseas Scholarship for one year (1927–28) by Basil Ferdinand Jamieson Schonland OBE BA PhD nominated by the University of Cape Town.*

[15] Proceedings of the Meeting of the *British Association for the Advancement of Science* held in Cape Town July 1929 kindly supplied to the author by Christina Tyree of the British Association on 3.2.1997.

[16] Eve A S 1939 *Rutherford* (New York: Macmillan) p 330.

[17] Tobias P V 1977 *A Century of Research in Human Biology and Palaeo-Anthropology in Southern Africa* ch 9 in [1] pp 228–232.

[18] Schonland B F J 1930 'Thunder-storms and penetrating radiation' *Proceedings of the Royal Society* **A130** 37–63.

[19] Letters and notes to the author from Dr Mary Davidson, included amongst which are Ismay Schonland's personal recollections of life in Cape Town.

[20] Phillips H 1993 *The University of Cape Town 1918–1948: The Formative Years* (Cape Town: University of Cape Town) p 47.

[21] Burke B F and Graham-Smith F 1997 *An Introduction to Radio Astronomy* (Cambridge: Cambridge University Press) p 105.

[22] Letter to the author from Professor Anton Hales 27.9.1998.

[23] Schonland B F J and Viljoen J P T 1933 'On a Penetrating Radiation from Thunderstorms' *Proceedings of the Royal Society* **A** 314–333.

Chapter 8 THE WINGS BEGIN TO SPREAD

[1] University of Cape Town Archive: Council Minutes 24.3.1930.

[2] Pask T P 1930 'An approach to the study of lightning and allied phenomena' *Transactions of the South African Institution of Electrical Engineers* **21** 72–79.

[3] Quoted in Hewitt F J 1976 'Symposium on High Voltage Engineering in South Africa — Opening Address' *Transactions of the South African Institution of Electrical Engineers* **67** 2–4.

[4] Bozzoli G R 1997 'Dr Bernard Price' in *Forging Ahead – South Africa's Pioneering Engineers* (Johannesburg: Witwatersrand University Press) p 189.

[5] Monk C J 1933 'Lightning Investigation Committee' *Transactions of the South African Institution of Electrical Engineers* **24** 14.

[6] University of Cape Town Archive: Letter from Sir Carruthers Beattie 9.2.1933.

[7] Appleton E V 1932 *Thermionic Vacuum Tubes* (London: Methuen).

[8] Author's interview with Dr Mary Davidson 18.8.1997.

[9] Smuts J C 1952 *Jan Christian Smuts* (London: Cassell & Co) pp 314–322.

[10] Eve A S 1939 *Rutherford* (New York: Macmillan) p 346.

[11] Wilson C T R 1925 'The electric field of a thundercloud and some of its effects' *Proceedings of the Physical Society of London* **37** 32D–37D with a contribution from C G Simpson p 46D.

[12] Schonland B F J 1931 Trans (Section A) *Lightning* British Association Meeting London p 356.

[13] Letter from BFJS to Ismay 27.9.1931 from the collection of Dr Mary Davidson kindly lent to the author.

[14] Simpson G C 1927 'The mechanism of a thunderstorm' *Proceedings of the Royal Society* **A** 376–401.

[15] Allibone T E 1973 'Basil Ferdinand Jamieson Schonland' *Biographical Memoirs of Fellows of the Royal Society* **19** 629–653.

[16] Ref [13] *op cit.*

[17] Simpson G C and Scrase F J 1937 'The distribution of electricity in thunderclouds' *Proceedings of the Royal Society* **A** 309–352.

[18] Schonland B F J 1943 'Thunderstorms and their electrical effects' *Proceedings of the Physical Society* **55** 445–458.

[19] Cambridge University Manuscript Library ADD 8702: Letter of 20.2.1932 from R A Watson Watt to BFJS.

[20] Ref [13] *op cit.*

[21] Schonland B F J and Allibone T E 1931 'Branching of lightning' *Nature* **128** 794–795.

[22] Rhodes House Oxford: Correspondence between the Trustees of the Rhodes Scholarships and the Dominion Secretary of the Trust in South Africa of 25.9.1925 and 2.10.1925 kindly supplied to the author by archivist 29.10.1998.

[23] Lord Rayleigh 1945 'Charles Vernon Boys' *Obituary Notices of Fellows of the Royal Society* pp 771–788.

[24] Buderi R 1996 *The Invention that Changed the World* (New York: Simon & Schuster) p 38.

[25] Boys C V 1928 'Progressive lightning' *Nature* **122** 310–311.

[26] Boys C V 1931 Letter (untitled) *Nature* **127** 125.

[27] Cambridge University Manuscript Library ADD 8702 Box 2: Correspondence between C V Boys and BFJS 17.12.1931, 14.7.1932, 6.7.1933; G C Simpson to BFJS (undated) but from its contents is apparently 1951.

[28] Halliday E C 1933 'On the propagation of a lightning discharge through the atmosphere' Philosophical Magazine **S7** 409–420.

[29] Schonland B F J 1932 *Atmospheric Electricity* (London: Methuen).

[30] Ref [3] p 3 *op cit*.

[31] Schonland B F J 1933 'The development of the lightning discharge' *Transactions of the South African Institution of Electrical Engineers* **24** 145–153.

[32] Halliday [28] *op cit*.

[33] Boys C V 1933 'Progressive lightning: a new stereoscope' *Nature* **131** 492–494.

[34] Schonland B F J 1950 *The Flight of Thunderbolts* (Oxford: Clarendon Press) p 70.

[35] Schonland B F J and Collens H 1934 'Progressive lightning' *Proceedings of the Royal Society* **A** 654–674.

[36] Schonland B F J and Collens H 1933 'Development of the lightning discharge' *Nature* **132** 407–408.

[37] Cambridge University Manuscript Library ADD8702: Letter from N Ernest Dorsey US National Bureau of Standards to BFJS 27.9.1933.

[38] University of Cape Town Archive Council Minutes (Memorandum) p 2074 25.6.1935.

[39] Brown A C 1977 *A History of Scientific Endeavour in South Africa* (Cape Town: The Royal Society of South Africa) p 74.

[40] University of Cape Town Archive Council Minutes p 1947 26.9.1933.

[41] Schonland B F J 1948 'Scientific and industrial research' *Journal of the South African Institution of Engineers* **47** 19–32.

[42] Cambridge University Manuscript Library ADD 8702 Box 2: BFJS to H Raikes 23.7.1934.

[43] University of Edinburgh (Special Collections) Appleton Papers (MS2300/C24): BFJS to E V Appleton 29.7.1934 kindly supplied to the author by Mr John Bradley.

[44] Halliday E C 1934 'Thunderstorms and the penetrating radiation' *Proceedings of the Cambridge Philosophical Society* **30** 206–215.

[45] *Proc of the General Assembly of the International Scientific Radio Union* (URSI) London September 1934 pp 60–70; and Schonland B F J 1935 Atmospherics and lightning' *South African Journal of Science* **XXXII** 31.

[46] Allibone T E and Schonland B F J 1934 'Development of the spark discharge' *Nature* **134** 736–737.

Chapter 9 TWO CHAIRS AND THE BPI

[1] University of Cape Town Archive: Council Minutes 30.6.1935.

[2] Ibid *Principal's Letter Book* vol 1 9.2.1933 p 369.

[3] Op cit Council Minutes 25.6.1935.

[4] Op cit Memorandum to the UCT Council 23.6.1935.

[5] Schonland B F J 1935 'Atmospherics and lightning' *South African Journal of Science* **32** 24–31.

[6] Watson Watt R A, Herd J F and Bainbridge-Bell L H 1933 *Applications of the Cathode Ray Oscillograph in Radio Research* (London: HMSO).

[7] Letter to Schonland from the Research Grant Board Pretoria 28.3.1936 supplied to the author by Dr Allon Poole, Department of Physics Rhodes University Grahamstown.

[8] Hodges D B 1935 'Radio direction finding with the cathode ray oscillograph' *South African Journal of Science* **32** 113–117.

[9] Malan D J, Schonland B F J and Collens H 1935 'Intensity variations in the channel of the return lightning stroke *Nature* **136** 831.

[10] Cairns J E J 1927 'Atmospherics at Watheroo, Western Australia' *Proceedings of the Institution of Radio Engineers* **15** 985–997.

[11] Op cit ref [5] p 31 referring to Jansky K G 1932 'Directional studies of atmospherics at high frequencies' *Proceedings of the Institution of Radio Engineers* **20** 1920–1932.

[12] University of Cape Town Archive: Council Minutes 30.7.1935.

[13] Munro G H and Huxley L G H 1932 'Shipboard observations with cathode-ray tube direction-finder' *Journal of the Institution of Electrical Engineers* **71** 488–496.

[14] Schonland B F J and Hodges D B 1936 'The relation between thunderstorms and atmospherics in Southern Africa' *Transactions of the Royal Society of South Africa* **24** 81–93.

[15] Schonland B F J 1936 'Report on work of Lightning Research Committee' *Transactions of the South African Institution of Electrical Engineers* **27** 26–29.

[16] University of the Witwatersrand Archive: Letter to B J Schönland from the Assistant Registrar 31.12.1935.

[17] Murray B K 1997 *Wits – the Open Years* (Johannesburg: Witwatersrand University Press) p 4.

[18] University of the Witwatersrand Archive Council Minutes File 25/a : Letter to Members of Council from the Principal 14.4.1936 with attached memoranda Misc.C/9/36 and C/10/36.

[19] University of the Witwatersrand Archive File S3/25: Letter from H R Raikes to Schonland 24.4.1936.

[20] Adams E B 1981 *Discussions with William Hofmeyr Craib*—a transcription of a tape recording made with Craib and housed in the library of the Department of Medicine, University of the Witwatersrand.

[21] University of Cape Town Archive Council Minutes 23.6.1936.

[22] Personal communication to the author from Dr Mary Davidson 12.12.1998.

[23] Op cit ref [18].

[24] University of the Witwatersrand Archive File 3/8 Geo BPI Board of Control Minutes vol 1 (Geo.C/3/37).

[25] University of Cape Town Archive: Minutes of the Board of the Faculty of Science 29.7.1935 4(b); University of the Witwatersrand Archive BPI Minutes (Geo.C/19/38) 18.3.1938; letter to the author from Dr A L Hales 21.10.1998.

[26] *Statuta et Decreta Universitatis Oxoniensis* (1912) Appx C.14 *The Halley Lecture* pp 502–503.

[27] Op cit ref [22].

[28] Oxford Historical Register Supplement 1931–1950: Endowed Lectureships p 20.

[29] Viljoen J P T and Schonland B F J 1933 'The distribution of the ionizing particles of the penetrating radiation in relation to the magnetic meridian' *Philosophical Magazine* **16** 449–456.

[30] University of the Witwatersrand Archive File GEO C/25/37: BPI Director's Report 4.5.1937.

[31] Crowther J G 1974 *The Cavendish Laboratory 1874–1974* (London: Macmillan) p 226-230.

[32] Oliphant M L E and Lord Penney 1968 'John Douglas Cockcroft (1897–1967)' *Biographical Memoirs of Fellows of the Royal Society* p 142.

[33] University of the Witwatersrand Archive (File GEO C/9/37) 3b 12.2.1937; and (GEO C/16/37) 3d and 7 2.4.1937.

[34] Schonland B F J 1937 'The diameter of the lightning channel' *Philosophical Magazine* **23** 503–508.

[35] Schonland B F J 1938 *The Lightning Discharge being The Halley Lecture delivered on 28 May 1937* (Oxford: Clarendon Press).

[36] Interview with Professor Frank Nabarro FRS in Johannesburg 7.5.1997 and in Cambridge 27.9.2000.

[37] Schonland B F J 1937 'The lightning discharge' *Transactions of the South African Institution of Electrical Engineers* **28** 204–212.

[38] University of the Witwatersrand Archive (File GEO C/44/370) 23.7.1937.

[39] Ibid (File GEO C/84/37) 20.12.1937.

[40] Op cit ref [20].

[41] Rhodes University Cory Library: MS 10 740: Letter BFJS to Selmar Schönland 30.10.1937.

[42] Brown A 1997 *The Neutron and the Bomb – A Biography of Sir James Chadwick* (Oxford: Oxford University Press) p 162.

[43] Hartcup G and Allibone T E 1984 *Cockcroft and the Atom* (Bristol: Adam Hilger) p 83.

Chapter 10 BERNARD PRICE'S INSTITUTE

[1] University of the Witwatersrand Archives File GEO C/19/38: Minutes of the 8th Meeting of the BPI Board of Control 18 March 1938.

[2] Murray B K 1982 *WITS. The Early Years* (Johannesburg: University of the Witwatersrand Press) p 292.

[3] Rhodes University Cory Library MS 11031, MS 11032 and MS 11033.

[4] Archives of *The Royal Commission for the Exhibition of 1851* London Schonland file: BFJS letter to E Shaw Esq 2.4.1938.

[5] Schonland B F J 1938 'Progressive lightning IV – the discharge mechanism' *Proceedings of the Royal Society* **A164** 132–150.

[6] Schonland B F J, Collens H and Malan D J 1934 'Development of the lightning discharge' *Nature* **134** 177–178.

[7] Watson Watt R A and Appleton E V 1923 'On the nature of atmospherics – I' *Proceedings of the Royal Society* **A103** 84–102.

[8] Cairns J E I 1927 'Atmospherics at Watheroo, Western Australia' *Proceedings of the Institution of Radio Engineers* pp 985–997.

[9] Malan D J, Schonland B F J and Collens H 1935 'Intensity variations in the channel of the return lightning stroke' *Nature* **136** 831.

[10] Collens H 1937 'Cameras for the investigation of lightning strokes' *Transactions of the South African Institution of Electrical Engineers* **28** 214–215.

[11] Ratcliffe J A 1975 'Robert Alexander Watson-Watt' *Biographical Memoirs of Fellows of the Royal Society* **21** 549–568.

[12] Appleton E V and Chapman F W 1937 'On the nature of atmospherics—IV' *Proceedings of the Royal Society* **A158** 1–22.

[13] Schonland B F J, Hodges D B and Collens H 1938 'Progressive lightning. V— A comparison of photographic and electrical studies of the discharge process' *Proceedings of the Royal Society* **A166** p 56-75.

[14] University of the Witwatersrand Archive GEO. C/19/38 Minutes of the 8th Meeting of the BPI Board of Control section 4(a) 'Relations with other Organisations'.

[15] Anon *South Africa—some facts and figures* 1938 (London: South Africa House) pp 12, 54.

[16] Churchill Archive Cambridge Schonland papers SCHO7: BFJS letter to A L Hales 29.1.1958.

[17] University of the Witwatersrand Archive: BPI Board of Control GEO.C/23/ 29 para 2b 17.3.1939.

[18] Op cit ref [14] para 7 18.3.1938.

[19] Ibid GEO.C/12/39 para 4.

[20] Wits Archive (BPI File): Brochure to mark the occasion of the official opening of the BPI on 21 October 1938. There is some confusion about the date on which this opening actually took place. This brochure states 21 October whereas the Annual Report of the BPI of 1938 (GEO. C/12/39), item 4, cited above states 22 October. Amongst the invitations issued was one personally from Schonland to C V Boys, then aged 83, 'but getting younger every year' as he told Schonland in his reply. However, he felt the journey to South Africa would be beyond him and sent 'best wishes' instead (source: Cambridge University Manuscripts Library ADD 8702, Box 2).

[21] Wits Archive GEO C/12/39 para 6(g) and C/23/39 para 2(E).

[22] Brown A C (ed) 1977 *A History of Scientific Endeavour in South Africa* (Cape Town: Royal Society of South Africa) p 362.

[23] Letter to the author from Dr Mary Davidson 19.2.1996.

[24] Letter to the author from Dr A L Hales 21.10.1998.

[25] Wits Archive GEO. C/51/39 *Report to the Chamber of Mines on the work done by the Consultative Panel of the BPI during the year 1938*.

[26] Bozzoli G R 1995 *A Vice-Chancellor Remembers* (Randburg: Alphaprint) p 42.

[27] Wits Archive GEO C/51/39 Minutes of 11th meeting of the BPI Board of Control item 9 13.6.1939.

[28] Wits Archive: GEO C/108/39 *Annual Report to Council 1939*. This list of names shows the spread of collaborating institutions and specialities: D B Hodges and W E Phillips (Natal); D J Malan (UCT); from Wits there were E C Halliday (Physics), G R Bozzoli (Electrical Engineering); A L Hales and H O Oliver (Applied Mathematics), C G Wiles and W H Aarts (Physics), M A Cooper; G D Walker (Wits Tech. Coll.); E C Bullard (Cambridge) as well as N Sellick, J Peake and R A Jubb (Rhodesian Meteorological Service).

[29] Schonland B F J and Hodges D B 1939 'Direction-finding of sources of atmospherics and South African meteorology' *Quarterly Journal of the Royal Meteorological Society* **66** 23–46.

[30] Schonland B F J 1938 *Thunder-clouds, Shower-clouds and their Electrical Effects* ch 12 in *Terrestrial Magnetism and Atmospheric Electricity* (New York: McGraw Hill) pp 657–678.

[31] Op cit ref [28] part 4.

[32] Schonland B F J, Elder J S, van Wyk J W and Cruickshank G A 1939 'Reflections of atmospherics from the ionosphere' *Nature* **143** 893–894.

[33] Ibid p 893.

[34] Schonland B F J, Elder J S, Hodges D B, Phillips W E and van Wyk J W 1940 'The wave form of atmospherics at night' *Proceedings of the Royal Society* **A176** 180–202.

Chapter 11 THE SSS AT WAR

[1] Wits Archive GEO C/82/39 Minutes of the BPI Board of Control 6 October 1939 item 4: 'Position of the Institute in the Present Emergency'.

[2] Schonland B F J 1951 'The work of the Bernard Price Institute of Geophysical Research 1938–1951' *Transactions of the South African Institution of Electrical Engineers* **42** 241–258.

[3] Letter to the author from Dr Mary Davidson 19.2.1996.

[4] Smuts J C 1952 *Jan Christian Smuts* (London: Cassell & Co) p 376.

[5] Gilbert M 1983 *Finest Hour* (London: Heinemann) p 18.

[6] Public Record Office AIR 2/4487 JDC 45th Minutes 20.5.1938.

[7] Ibid Minute from R A Watson Watt to D.C.A.S. 17.8.1938.

[8] Ibid Draft Paragraph for CAS Dominion Liaison Letter 73A.

[9] Ibid AIR 41/12 Radar in Raid Reporting *Disclosure of R.D.F. Information to the Dominions* p 72.

[10] Jacobs F J, Bouch, R J du Preez S and Cornwell R 1975 *South African Corps of Signals* (Pretoria: SADF Documentation Service) p 51.

[11] *The Times* 5 August 1993: *Obituary of Major-General H G Willmott SAAF.*

[12] Oliphant M L E and Lord Penney 1968 'John Douglas Cockcroft 1897–1967' *Biographical Memoirs of Fellows of the Royal Society* **14** 155.

[13] Clark R W 1962 *The Rise of the Boffins* (London: Phoenix House) p 62.

[14] Turner L C F, Gordon-Cummings H R and Betzler J E 1961 *War in the Southern Oceans* (Oxford: The University Press) p 4.

[15] Public Record Office AIR 2/4487 S.40952 and [9] *op cit* p 75.

[16] Ref [9] *op cit* p 74.

[17] Fleming C A 1971 'Ernest Marsden 1889–1971' *Biographical Memoirs of Fellows of the Royal Society* **17** 480; Unwin R S 1992 'The development of radar in New Zealand in World War II' *The Radioscientist* **3** 8.

[18] Phillips W E 1979 Vote of thanks after the 27th Bernard Price Memorial Lecture by Major General T G E Cockbain 'Radar development in South Africa with special reference to air defence' *Transactions of the South African Institution of Electrical Engineers* **70** 95.

[19] Ref [10] *op cit* p 51.

[20] Letter to the author from Professor A L Hales 21.10.1998.

[21] Ref [2] *op cit* p 253.

[22] SANDF Archive Pretoria folder 125 report EA7 *South African Units in East Africa – South African Corps of Signals.*

[23] Author's interview with Professor Emeritus G R Bozzoli 3.4.1997.

[24] Neale B T *CH – the first operational radar* ch 8 in Burns R W (ed) *Radar Development to 1945* (London: Peter Peregrinus) pp 132–150.

[25] Swords S S 1986 *Technical History of the Beginnings of Radar* (London: Peter Peregrinus) p 239.

[26] Schonland B F J *War Diary – Special Signals Services UDF*. This exceptionally useful document was written initially by Schonland and then continued by Bozzoli. Whereas it commences with Schonland's meeting with Marsden it is apparent that much of the earlier material was probably written some while after the events actually occurred because it contains some inconsistencies and is at odds with official records in certain cases. It changed to a more formal and daily catalogue of activities on 8 February 1941.

[27] Bozzoli G R 1995 *A Vice-Chancellor Remembers* (Randburg: Alphaprint) p 48.

[28] Ref [26] *op cit*: hereinafter referred to as the *War Diary*; November 1939.

[29] Ref [27] *op cit* p 48.

[30] Brown L 1999 *A Radar History of World War II: Technical and Military Imperatives* (Bristol: Institute of Physics Publishing) p 83.

[31] Hanbury Brown R 1991 *Boffin* (Bristol: Adam Hilger) pp 11–12.

[32] Katz L and Phillips W E 1938 'Two station direction finding with the cathode ray oscillograph in South Africa' *South African Journal of Science* **25** 184–192.

[33] Ref [23] *op cit*.

[34] SANDF Archive AG 736/7/1 Secret letter to the Secretary for Defence from the Deputy Chief of the General Staff, signed by Lt Col F Collins 21 December 1939: *Establishment Table for a Special Wireless Section (Home Defence)*.

[35] Hewitt F J 1975 Military History Journal (South Africa) *South Africa's role in the development and use of Radar in World War II.*

[36] Letter to the author from Dr F J Hewitt 19.11.1991.

[37] Ref [26] *op cit War Diary*, February 1940.

[38] Brown J A 1990 *The War of a Hundred Days* (Rivonia: Ashanti Publications) p 49.

[39] Ref [35] *op cit.*

[40] SANDF Archive AG/736/7/1 28.5.1940.

[41] Imperial War Museum London: Schonland papers 86/63/1.

[42] Ref [35] *op cit.*

[43] Ref [38] *op cit* p 51.

[44] SANDF Archive Officer's Record of Service: Brigadier B F J Schonland CBE.

[45] Letter to the author from Dr F J Hewitt 15.6.1998.

[46] Hewitt F J 1990 'SA Radar in World War II – the outside story' *Elektron* **7** 8.

[47] Public Record Office AIR2/4465 S.5734 Pt 1.

[48] Ibid S.47922/Signals 'Report on RDF Sites – Mombasa' 29.7.1940.

[49] Ibid S.5734 para 8 'Mombasa' 4.9.1940.

[50] Ibid AIR 2/4465 S.5734/Signals Secret Cypher Message X.7128 3.10.1940.

[51] Ref [26] *op cit War Diary*: 'Major Schonland returned to the Union September 22nd 1940. Arrangements were then made for the construction of more sets for use in E. Africa'.

[52] Ref [50] *op cit* Air Ministry MINUTE SHEET to ACAS(R). No 98a 9.10.1940.

[53] Ibid Message from ACAS(R) to HQ RAF ME No 99a 10.10.1940.

[54] Ref [38] *op cit* p 81.

[55] Ref [4] *op cit* p 403.

[56] Furlong P J 1991 *Between Crown and Swastika* (Johannesburg: Witwatersrand University Press) p 149.

[57] Ref [26] *op cit War Diary*.

[58] Public Record Office AIR 2/4465 Report of 21.11.1940 from S.Ldr Yuill to Sig.4 (Air Min.); also 'Summary of present position in Middle East' 7.3.1941 (v) Kenya (a) In Operation MRU (No 218 MRU) at Mombasa.

[59] Ibid HQ RAF ME to A.M.(R) HQ Mediterranean 22.1.1941.

[60] Ref [26] *op cit War Diary* September 1940.

[61] Brain P 1993 *South African Radar in World War II* (Cape Town: The SSS Radar Book Group) p 27; and letter to the author from Dr F J Hewitt 12.2.2000.

[62] Ref [26] *op cit War Diary*.

[63] SANDF Archive AG213/6/2 Special Signal Services South African Corps of Signals 26.12.1941.

[64] Ibid AG(1)736/7 Letter from Col F Collins D Sigs to A G: Appointments to Special Signals Services.

[65] Ref [18] *op cit* p 96.

[66] Public Record Office AVIA/887 Radio Branch HQ RAF ME letter to A P Rowe 30.12.1940.

[67] Ref [26] *op cit War Diary* February 1941.

[68] Ref [14] *op cit* p 47.

[69] Author's telephonic interview with one of those volunteers D R Forte Johannesburg 4.4.1997.

[70] Ref [26] *op cit War Diary* 25 February 1941.

Chapter 12 'COULD I BE OF SERVICE?'

[1] Author's telephonic conversation with D R Forte 4.4.1997.

[2] SANDF Archive Pretoria AG(1)P1/11036/1 Officers Record of Service B F J Schonland.

[3] Letter B F J Schonland to Ismay Schonland (BFJS to IMS) written 'At Sea' 31.3.1941. This most valuable collection of Schonland's letters to his wife was very kindly made available to the author by Schonland's daughter, Dr Mary Davidson.

[4] Letter BFJS to IMS 10.4.1941.

[5] Brown A C 1977 *A History of Scientific Endeavour in South Africa* (Cape Town: The Royal Society of South Africa) p 92.

[6] CSIR Archives—Ministry of Defence ref. A/68/Gen/9874/MS3; Adams E B 1990 *In Search of Truth—A Portrait of Don Craib* (London: Royal Society of Medicine Services Limited).

[7] Standard Encyclopædia of Southern Africa 1971 vol 3 *Maj-Gen I P de Villiers* (Cape Town: Nasou Ltd).

[8] Ref [4] *op cit*.

[9] Author's interview with Mrs Ann Oosthuizen (BFJS's daughter) Johannesburg 18.4.1997.

[10] Oliphant M L E and Lord Penney 1968 'John Douglas Cockcroft 1897–1967' *Biographical Memoirs of Fellows of the Royal Society* pp 139–188; Interview with Sir Maurice Wilkes, Cambridge 14.7.1997.

[11] Swords S S 1986 *Technical History of the Beginnings of Radar* (London: Peter Peregrinus) p 266.

[12] Burns R W 1994 'Impact of technology on the defeat on the U-boat September 1939–May 1943' IEE Proc Sci Meas Technol **141** 343–355.

[13] Bowen E G 1987 *Radar Days* (Bristol: Institute of Physics Publishing) ch 10, p 11.

[14] Wilkes M V 1985 *Memoirs of a Computer Pioneer* (Cambridge, MA: The MIT Press) p 62.

[15] Pile F A 1947 'The Anti-Aircraft Defence of the United Kingdom 28 July 1939 to 15 April 1945' *The London Gazette* 18 December pp 5973–5994.

[16] Clark R W 1962 *The Rise of the Boffins* (London: Phoenix House) p 142.

[17] Sayer A P 1950 *Army Radar* (London: The War Office) p 50.

[18] Public Record Office WO 291/597 Schonland B F J 28 February 1944 Army Operational Research Group Memorandum No 263 *The Operational Performance of Army Radar Equipment 1939–1943*.

[19] Crowther J G and Whiddington R 1947 *Science at War* (London: HMSO) p 96; Imperial War Museum Schonland papers 86/63/1 Cockcroft J D 1945 *General Account of Army Radar* an unpublished report.

[20] Ref [14] *op cit* p 65.

[21] Imperial War Museum 86/63/1 A50/LEB *History of the Origins of Operational Research in AA Command* by L E Bayliss 1945.

[22] Royal Military College of Science Reports Section R/81/517 Working paper OR/WP/8 Reading K section 4 p 103 R W Shephard; author's interview with Professor Emeritus F R N Nabarro, Johannesburg 7.5.1997.

[23] Ref [17] *op cit* p 228.

[24] Ref [21] *op cit* p 4.

[25] Van Der Vat D 1988 *The Atlantic Campaign* (London: Grafton Books) p 297.

[26] Jones R V 1978 *Most Secret War* (London: Hamish Hamilton) p 92-105.

[27] Gilbert M 1983 *Finest Hour* (London: Heinemann) p 668.

[28] Ref [14] *op cit* p 66.

[29] Pile F A 1949 *Ack-Ack* (London: George G Harrap & Co) p 114.

[30] Watson-Watt R A 1957 *Three Steps to Victory* (London: Odhams).

[31] Ref [21] *op cit* p 6.

[32] Ref [1] *op cit*.

[33] Adams K A H 1975 *Transactions of the South African Institution of Electrical Engineers* **66**. Vote of thanks following the paper 'Radar—past, present and future' by T V Peter pp 236–240.

[34] Letter BFJS to IMS 3.5.1941.

[35] Ref [14] *op cit* p 70.

[36] Letter BFJS to IMS 12.5.1941.

[37] Schonland B F J 1964 'Maurice Pascal Alers Hankey 1877–1963' *Biographical Memoirs of Fellows of the Royal Society* **10** 137–146.

[38] Letter BFJS to IMS 25.5.1941.

[39] Letter BFJS to IMS 3.6.1941.

[40] Brain P 1993 *South African Radar in World War II* (Cape Town: The SSS Radar Book Group) p 28.

[41] Public Record Office AIR 2/4465 AM(R)HQ Med from HQ RAF ME 22.1.1941.

[42] Public Record Office AIR 2/4466 Minute Sheets 196 Grp Capt C S Lang DD of S4 to SAT 18.3.1941.

[43] Letters BFJS to IMS 7 and 14.6.1941.

[44] CSIR Archive Pretoria File B26; 54; 58 Union Defence Force Record of Service B F J Schonland.

[45] Rowe A P 1948 *One Story of Radar* (Cambridge: Cambridge University Press) p 96.

[46] Letter BFJS to IMS 20.7.1941.

[47] Letter BFJS to IMS 25.7.1941.

[48] Telegram BFJS to IMS 26.7.1941.

[49] SSS War Diary 23.7.1941.

[50] Ref [40] *op cit* p 30.

[51] Public Record Office AIR 41/12 *Radar in Raid Reporting* Appx No 14 p 564.

[52] Imperial War Museum 86/63/1 Schonland papers 'Army Operational Research' a single hand-written page seemingly never completed by Schonland.

Chapter 13 SOLDIERS AND CIVILIANS

[1] Imperial War Museum 86/63/1 Schonland's hand-written notes 'Some notes on the history & organisation of AORG' and 'Army Operational Research'; Letter BFJS to IMS 8.8.1941.

[2] Pile F A 1949 *Ack-Ack* (London: George G Harrap & Co) p 114.

[3] See [1] BFJS to IMS.

[4] Mott N F 1986 *A Life in Science* (London: Taylor & Francis) p 63.

[5] Wilkes M V 1985 *Memoirs of a Computer Pioneer* (Cambridge, MA: The MIT Press) p 70.

[6] Letter BFJS to IMS 10.8.1941.

[7] Swords S S 1986 *Technical History of the* beginnings of Radar (London: Peter Peregrinus) p 87.

[8] Clark R W 1962 *The Rise of the Boffins* (London: Phoenix House Ltd) pp 126–151.

[9] Author's interview with Professor Emeritus F R N Nabarro FRS Johannesburg 7.5.1997; also SA/AC paper C67/3/4/48 'Operational Research in the British Army' Royal College of Military Science (RMCS) R/79/1702 (Reading U) 'Presentation of Results' p 241; and Imperial War Museum, Schonland papers 86/63/1 'Further notes on the art of writing a scientific paper for military consumption, being comments, hitherto unpublished, on "How to write a Report—by a Staff Officer"'.

[10] Letter to the author from Dr J S Hey FRS 30.7.1997.

[11] Wilkes [5] *op cit* p 71; Dr F J Hewitt letter to the author 15.6.1998.

[12] Wilkes M V 1997 *Schonland in relation to Cockcroft* notes prepared for the author 6.8.1997.

[13] Author's interview with Professor Sir Maurice Wilkes FREng FRS Cambridge 14.7.1997.

[14] Letter BFJS to IMS 12.1.1942.

[15] Letter BFJS to IMS 25.1.1942.

[16] Hartcup G and Allibone T E 1984 *Cockcroft and the Atom* (Bristol: Adam Hilger) p 118.

[17] Letters BFJS to IMS [6] *op cit.*

[18] Letters BFJS to IMS 8 & 14.2.1942.

[19] Wilkes [5] *op cit* p 78.

[20] Sayer A P 1950 *Army Radar* (London: The War Office) p 208.

[21] Jones R V 1978 *Most Secret War* (London: Hamish Hamilton) p 236.

[22] Correspondence with the author from D H Preist 26.10.1996 and 22.11.1996.

[23] Correspondence with the author from Professor R V Jones 4.11.1996; also Sayer [20] *op cit* pp 207–208.

[24] The Bruneval raid has been described in great detail in a number of books, some of which are: (a) Millar G 1974 *The Bruneval Raid* (London: The Bodley Head); (b) Clark R W 1962 *The Rise of the Boffins* (London: Phoenix House) pp 176–180; (c) Johnson B 1978 *The Secret War* (London: Arrow Books) pp 123–128 and many more. The part that Basil Schonland played in its planning is not referred to in (a) or (c) while T E Allibone in his Royal Society Biographical Memoir of Schonland, written in 1973, somewhat overstates it by maintaining that Schonland 'planned the Bruneval raid' (p 640). When Schonland and Don Preist met it was with the intention of guessing what electronic devices might be inside the various cubicles attached to the Würzburg and how best to remove them. Schonland showed Preist how to use a form of plastic explosive to shear a metal bar though quite for what purpose is not clear since it was not intended that Preist should be involved in the dismembering of the radar. Preist's name is spelt incorrectly in most of the published accounts of the raid.

[25] Churchill Archive Cambridge: Schonland papers SCHO 2. Letter from Schonland to Professor Leo Brandt, Director, Arbeitsgemeinschaft für Forschung des Landes Nordheim-Westfalen 6.4.1960.

[26] Public Record Office AIR 20/1631 'Intelligence aspect of the Bruneval Raid'.

[27] Van Der Vat D 1988 *The Atlantic Campaign* (London: Grafton Books) p- 352–362.

[28] Cockburn R 1988 *The Radio War* ch 24 in *Radar Development to 1945* Burns R W (ed) (London: Peter Peregrinus) p 344.

[29] Wilkes [5] *op cit* p 75.

[30] Hey [10] *op cit*; and Royal Military College of Science R/81/517 Reading K Army Operational Research p 101.

[31] Jansky K G 1932 'Directional studies of atmospherics at high frequencies' *Proceedings of the Institution of Radio Engineers* **20** 1920–1932; *op cit* 1933 vol **21** 'Electrical disturbances apparently of extraterrestrial origin' pp 1387–1398. Schonland was well aware of Jansky's work because, as mentioned in ch 9, he had studied these results with interest and had commented upon them in his Presidential Address to Section A of the South African Association for the Advancement of Science at Paarl in 1935.

[32] Wilkes [13] *op cit*; Letter to the author from Sir Bernard Lovell OBE FRS 19.8.1992. Appleton, though, soon changed his mind and wasted no time in publishing his ideas on the phenomenon as soon as wartime restrictions were lifted. His letter to *Nature* of 3.11.1945 actually preceded that from Hey, which was still in the form of an AORG Report (see [33] below) and only appeared in *Nature* some months later (vol 157, 12.2.1946, pp 47-48). Then, in September the following year, Appleton and Hey collaborated by

writing a joint letter (*Nature* vol 158 7.9.1946, p 339) in which they reported on an experiment made in July which established that the VHF radiation from the sun was circularly polarized. In that same issue M Ryle and D D Vonberg, at the Cavendish, reported on their observations of solar radiation at 175 MHz while, in Australia, similar observations had been made at 200 MHz by a group under the direction of J L Pawsey, another former Cavendish man, and their letter appeared even earlier (*Nature* vol 157 9.2.1946, pp 158-159). The science of radio astronomy was conceived in America by K G Jansky in 1932 and then fostered by his fellow countryman G Reber in 1940, but the real impetus came from the wartime work in England and in Australia with its protagonists, including A C B Lovell, all having been intimately connected with various aspects of wartime radar.

[33] PRO WO 291/255 Hey J S AORG Report No 275 (13.6.45) 'Solar Radiations in the 4 to 6m wavelength band on 17th and 28th February 1942'.
[34] Letter BFJS to IMS [14] *op cit.*
[35] Murray B K 1982 *Wits: The Early Years* (Johannesburg: Witwatersrand University Press) p 65.
[36] Zuckerman S 1978 *From Apes to Warlords* (London: Hamish Hamilton) p 403.
[37] Letters BFJS to IMS. This series of letters of 12.1.1942, 17.1.1942, 25.1.1942, 31.1.1942, 8.2.1942 and 14.2.1942 contains a remarkable personal record of Schonland's life in the early days of the ADRDE(ORG) and, indeed, of the organisation itself.
[38] Nabarro [9] *op cit.*
[39] Zuckerman's post-war writing on this subject is a little confusing. In his autobiography [36] p 123 he states that the limiting over-pressure (i.e. pressure above atmospheric) is 'nearly 500 lb a square inch' (psi) whereas the prevailing doctrine was that 'a man could not tolerate even a tenth of that pressure'; therefore 50 psi, presumably. However, in *Proceedings of the Royal Society* **A342** (1975) 465–480 he states that the 'critical level of over-pressure was thought to be 7 psi' whereas the actual value lay 'between 300 and 500 psi' : a difference of the order of one hundred times between his findings and the prevailing view at the time.
[40] Hastings M 1979 *Bomber Command* (London: Michael Joseph) p 112.
[41] Nabarro [9] *op cit.*
[42] Imperial War Museum: Schonland papers 86/63/1 AORG Christmas Memorandum 1943.
[43] Letter IMS to BFJS 19.4.1942.
[44] South African National Defence Force (SANDF) Archive Pretoria AG 736/7/1 Cipher Telegram P.1460 3.6.1942 to DECHIEF from OPPOSITELY; and reply from D.Sigs of 5.6.1942.
[45] Letter IMS to Dot, her sister-in-law, Doris Craib, written from England 31.10.1942.
[46] National Archives of South Africa: Smuts Private Papers 252/184.
[47] Smuts J C 1952 *Jan Christian Smuts* (London: Cassell) p 421-423.
[48] Nabarro [9] *op cit.*
[49] Sayer [20] *op cit* p 289.
[50] Royal Military College of Science: War Office and Ministry of Supply Joint paper 287/Gen/752 Jan 1943.

[51] Obituary of Spencer Robert Humby (1892–1959), The Wykehamist Society, Winchester.

[52] Royal Military College of Science: Reports Section R/81/517 Working paper OR/WP/8: *Work on Signals and Jamming* by Major E W B Gill Nov 1944 who, in 1924, had assisted Appleton in setting up his receiver in Oxford for the experiment from which he determined the existence of the ionosphere. Gill's remarkably candid report was evidently written after Schonland had left the AORG in response to a request from Schonland's successor as super-intendent, Colonel O M Solandt, who asked each section to provide a short history of its work including that during its previous existence within ADRDE(ORG). Unfortunately only some were produced or possibly survived the post-war 'weeding' that took place and so none exists for AORS 4, 7, 9 and 10.

[53] Jones [21] *op cit* pp 287–299.

[54] Private communication to the author from Sir Maurice Wilkes FREng FRS.

[55] Lovell A C B 1991 *Echoes of War* (Bristol: Adam Hilger) p 80.

[56] Gill [52] *op cit* pp 174–176.

[57] Jones [20] *op cit* p 302 and Hastings [40] *op cit* p 208.

[58] Gill [52] *op cit* p 179 and Nabarro [9] *op cit.*

[59] Pubic Record Office WO291/492 AORG Memorandum 144 'The relative merits of HF (3-30 Mc/s) and VHF (30–300 Mc/s) for short distance communications' Aug 1943.

[60] Gill [52] *op cit* p 180.

[61] Imperial War Museum 86/63/1 Schonland papers 'General Account of Army Radar' by Cockcroft J D (undated) but this same document also appears in *IEE Proceedings Pt A* **132** (1985) pp 327–339 under the title of *Memories of Radar Research* by J D Cockcroft with the quotation in question on p 339.

[62] Public Record Office WO291/117 AORG Report No 126 'Simple Sky-Wave Aerials for Wireless Communication Over Short and Medium Distances' 27.9.1943. This report is typical of the series of reports and memoranda issued by AORG on this subject and represents some of the earliest research anywhere on a subject of much importance fifty years later which now goes by the name of NVIS—' Near Vertical Incidence Skywave'.

[63] Institute of Physics (successor to the Physical Society)—list of winners of the Charles Chree Medal and Prize: www.iop.org/IOP/Awards/chree.html.

[64] Schonland B F J 1943 'Thunderstorms and the electrical effects' *Proceedings of the Physical Society* **55** 445–458.

[65] Wits Archive GEO C/30/43 Minutes of a special meeting of the Board of the BPI held at the Country Club on 1.6.1943.

[66] Smuts [47] *op cit* pp 429–436.

[67] Letter to the author from Professor D K Hill FRS 22.11.1997.

[68] Malherbe E G 1981 *Never a Dull Moment* (Cape Town: Howard Timmins) p 278. Malherbe was Smuts's Director of Military Intelligence during the war and later became Principal of Natal University College and then, as the University of Natal, its first vice-chancellor.

[69] National Archives of South Africa *Smuts Private Papers* 1943—260/172 no 1145.

[70] Pile [2] *op cit* p 359.

[71] Churchill Archive Cambridge: Schonland papers SCHO 1 Letter from BFJS to Sir Keith Hancock 2.1.1957; Interview with Dr Margaret Nabarro who, as Miss Dalziel, served as Schonland's personal assistant at the AORG and took minutes at the meetings between Smuts and Schonland.

[72] National Archives of South Africa: Smuts papers [69] *op cit* 263 no 185: Letter from BFJS to the Prime Minister 17.12.1943.

Chapter 14 MONTGOMERY'S SCIENTIST

[1] Crowther J G and Whiddington R 1947 *Science at War* (London: HMSO) p 89.

[2] Clark R W 1962 *The Rise of the Boffins* (London: Phoenix House) p 3.

[3] Royal Military College of Science (R/79/1702) *Operational Research in the British Army* 1939–1945 SAAC paper C67/3/4/48.

[4] Imperial War Museum: Schonland papers 86/63/1 Schonland's undated hand-written recollections of 'Operational Research with the 21st Army Group, British Liberation Army before and after the invasion of the Continent, June 6th 1944'.

[5] Fraser D 1982 *Alanbrooke* (London: Collins) p 315.

[6] Schonland *op cit* [4].

[7] Sayer A P 1951 *Army Radar* (London: The War Office) p 206.

[8] Public Record Office WO291/614 AORG Memorandum No. 281 (19.2.1944) *Investigation of mutual interference between radar and communications equipment under the congested conditions of an opposed landing*—final report on Exercise 'Feeler'.

[9] Public Record Office WO291/597 AORG Memorandum No 263 (28.2.1944) *The Operational Performance of Army Radar Equipment. 1939–1943.*

[10] Sayer *op cit* [7].

[11] Fraser [5] *op cit* p 374.

[12] Hastings M 1984 *Overlord* (London: Michael Joseph) p 29.

[13] Chalfont A 1977 *Montgomery of Alamein* (London: Magnum Books) p 262.

[14] Hastings [12] *op cit* p 51.

[15] Churchill Archive Cambridge: Cockcroft papers CKFT 20/30 undated letter from Schonland to Cockcroft.

[16] Imperial War Museum: Schonland papers 86/63/1 undated type-written and hand-written notes (a) *Some Recollections of my time with 21st Army Group* and (b) *Operational Research with the 21st Army Group, British Liberation Army before and after the invasion of the Continent, June 6th 1944.*

[17] Schonland *op cit* [4].

[18] Notes of an interview conducted with Lord Swann in 1989 by Professor T Copp (Wilfrid Laurier University, Toronto) and kindly supplied to the author.

[19] Public Record Office WO291/616 AORG Memorandum No. 284: *Minutes of a conference on 'Window' held at AORG on March 4th 1944.*

[20] Schonland [16] *op cit* (a); Twinn J E 1985 'The use of " Window" to simulate the approach of a convoy of ships towards a coastline' in *Radar Development to 1945* ed R W Burns (London: Peter Peregrinus) pp 416–422.

[21] Obituary of Sir Edgar Williams CB, CBE, DSO *The Times* 29.6.1995.

[22] Schonland B F J 1951 *On being a Scientific Adviser to a Commander-in-Chief* type-script supplied to the author by Professor D F S Fourie University of South Africa; also in Imperial War Museum: Montgomery Papers BLM 140/6.

[23] Letter to the author from Professor D K Hill FRS 22.11.1997.

[24] Schonland [22] *op cit* p 12.

[25] SANDF Archive Pretoria: AG(1)P1/11036/1 Service Record Schonland Basil Ferdinand Jamieson.

[26] Ellis L F 1962 *Victory in the West* (London: HMSO) vol 1 p 134.

[27] Chalfont [13] *op cit* p 272.

[28] Imperial War Museum: Schonland papers 86/63/1 *Operational Research in North West Europe – The work of No 2 ORS with 21 Army Group: Jun 44–Jul 45*. This report, written soon after the end of the war by Lt Col M M Swann of No 2 ORS, is the most complete assessment of the part played by Operational Research within the Army from D-day until the end of the war in Europe.

[29] Ellis [26] *op cit* pp 311–316.

[30] Swann [28] *op cit* p 282.

[31] Author's interview with Professor D K Hill 3.7.1999.

[32] Zuckerman S 1978 *From Apes to Warlords* (London: Hamish Hamilton) pp 266– 277.

[33] See for example [12] *op cit* p 262 for a dispassionate view written long after the dust had settled; Montgomery B L 1958 *The Memoirs of Field Marshal Montgomery* (London: Collins) ch 14 for those of the Commander himself and [26] *op cit* for the record of the official historian.

[34] Ellis [26] *op cit* p 351.

[35] Schonland [22] *op cit* and Swann [28] *op cit.*

[36] SANDF Archive, Pretoria: RVH1 letter from Schonland to Omond Solandt 7.5.1951. That Blackett was 'no longer in good odour with his naval collea-gues' is also recounted by Zuckerman [32] *op cit* p 266, who also relates how J D Bernal, for a time joint scientific adviser with Zuckerman himself to Mountbatten's Combined Operations HQ, was also 'somewhat unpopu-lar' for raising with his masters 'unnecessary hares about obstacles on the beaches'.

[37] Copp T 1991 'Scientists and the Art of War: Operational Research in 21 Army Group' *Royal United Services Institute Journal* **136** 65–69 and Swann [28] *op cit* pp 284–287.

[38] Montgomery B L 1946 *Normandy to the Baltic* (London: Hutchinson) p 118.

[39] Letter from BFJS to O M Solandt 19.8.1944 kindly supplied to the author by Professor T Copp (Wilfrid Laurier University, Toronto).

[40] Letter BFJS to IMS 30.7.1944.

[41] Sayer [7] *op cit* p 289.

[42] Copies of correspondence between Schonland and Solandt supplied to the author by Professor T Copp, Wilfrid Laurier University, Toronto. In a series of fourteen D.O. (demi-official) letters written between early August and late October 1944 Schonland and Solandt exchanged views, sometimes dys-peptic, on all manner of subjects to do with the research and aspirations of the personnel at AORG and the rapidly evolving state of Operational Research within 21 Army Group. Solandt, often piqued by the administrative

malaise that seemed to have submerged the War Office, and always apologetic for his poor use of the typewriter, paid considerable attention to small details while trying to persuade Schonland to authorize the transfer to the battlefront of various members of the AORG. In reply Schonland, his exasperation occasionally betraying itself in bursts of spleen, had repeatedly to remind all at Roehampton that 'people have no right to push themselves on me merely because I have asked if they would be willing to come, if asked for, or that I hoped to get them over'. To Solandt he wrote with some sarcasm: 'I keep getting letters from Nab [Nabarro] about plans & ideas. What actually am I expected to do about them? What would you do in my place? The study of and reporting on our operations is the duty of BGS(Ops), not of the S.A. BGS(Ops) consults me a lot but always as an adviser, since I am not at the present time holding the post of Chief of Staff or C in C'.

[43] Schonland [22] *op cit* p 18.
[44] Hibbert C 1962 *The Battle of Arnhem* (London: Batsford); Montgomery B L 1958 *The Memoirs of Field-Marshal Montgomery* (London: Collins) ch 16; Ellis [26] *op cit* vol II, ch 2.
[45] Golden L 1984 *Echoes from Arnhem* (London: William Kimber) ch 8.
[46] Public Record Office WO 291/117 AORG Report No 126 (27 September 1943) *Simple Sky-wave Aerials for Wireless Communications over Short and Medium Distances*. This was the first of two reports on this subject issued by the AORG. It described, in detail, the requirements to be met by simple wire antennas necessary to radiate energy directly upwards towards the ionosphere and provided elementary design rules for the choice of appropriate length for use between 2 and 8 MHz with the inclusion, if necessary, of a simple switched capacitor network to ensure that all common radio sets in use in the Army could easily be tuned into the antenna. As such it expanded upon the information provided in the Corps of Signals own document *Signal Training Part 2, Pamphlet No IX* issued earlier that year. The second, WO 291/149, AORG Report No 160 (25 November 1943) *Sky-wave Communication over short and medium distances, with special reference to the Far Eastern theatre of the war* took the problem further by including the effects of both ionospheric absorption and atmospheric noise, produced mainly by lightning activity over global distances, on the transmitter power required and the antennas to be deployed for near vertical incidence propagation over distances 'up to about 150 miles'.
[47] Public Record Office WO 291/689 AORG Memorandum No 360 (13 July 1944) *Frequency Prediction for Short and Medium Distance Sky-Wave Working during September 1944*.
[48] Golden [45] *op cit* p 159.
[49] Ibid p 160.
[50] Ryan C 1974 *A Bridge Too Far* (London: Hamish Hamilton) p 126; Warner P 1982 *Phantom* (London: Kimber) p 21.
[51] Schonland B F J 1948 'Scientific and industrial research' *Journal of the South African Institution of Engineers* **47** 19–32.
[52] Swann [28] *op cit* p 291.
[53] Churchill Archive Cambridge: Cockcroft papers CKFT 20/30.
[54] Copp [42] *op cit* Letters BFJS to O M Solandt 7.10.1944 and 28.10.1944.

[55] Imperial War Museum London: Schonland papers 86/63/1.

Chapter 15 WHEN SMUTS CALLED
[1] CSIR Archive A-0/1 vol 1 Notes by H R Burrows for the meeting of the National Research Council of 25.11.1942.
[2] Bozzoli G R 1997 *Forging Ahead – South Africa's Pioneering Engineers* (Johannesburg: Witwatersrand University Press) p 181.
[3] Kingwill D G 1990 *The CSIR – the first 40 years* (Pretoria: CSIR) p 6.
[4] National Archives of South Africa: Smuts papers vol 263 no 185 letter BFJS to J C Smuts 17.11.1943.
[5] CSIR Archive A-0/1 vol 1 letter S H Haughton to Brigadier B F J Schonland Main HQ 21 Army Group, British Liberation Army 13.10.1944.
[6] CSIR Archive: *Minutes of 20th General Meeting of the National Research Board* 14.2.1944 para 3.
[7] CSIR Archive: *Memorandum on the proposed reorganization of research in the Union of South Africa* by J Smeath Thomas 23.2.1944.
[8] Churchill Archive Cambridge: SCHO1 letter BFJS to Sir Keith Hancock 2.1.1957 when Hancock was writing his biography of Smuts and had asked Schonland for his comments about Smuts's attitude towards science and scientists in South Africa.
[9] CSIR Archive: *Minutes of 23rd General Meeting of the National Research Board* 25.1.1945 para 2.
[10] Letter BFJS to O Solandt 4.1.1945 supplied to the author by Professor T Copp Wilfrid Laurier University Toronto.
[11] Churchill Archive Cambridge: [8] *op cit* p 2.
[12] CSIR Archive: S3/25 letter to Humphrey Raikes from the Secretary to the Prime Minister 2.1.1945.
[13] Op cit Letter BFJS to Raikes 1.2.1945.
[14] Friedman B 1975 *Smuts – A Reappraisal* (Johannesburg: Hugh Keartland) p 161.
[15] Author's interview with Dr Mary Davidson – Schonland's daughter 12.8.1997.
[16] Churchill Archive Cambridge: CKFT 20/30 Letter BFJS to Cockcroft 25.2.1945; and Imperial War Museum: Schonland papers 86/63/1 letter Cockcroft to BFJS 29.3.1945.
[17] Murray B K 1997 *Wits – The 'Open' Years* (Johannesburg: Witwatersrand University Press) p 276.
[18] Sparks A 1990 *The Mind of South Africa* (London: Heinemann) p 162; and Malherbe E G 1981 *Never a Dull Moment* (Cape Town: Howard Timmins) p 308.
[19] CSIR Archive: A-0/1 Vol.1 Schonland's notes after his meeting with Dr Malan, Dr Stals and Mr Eric Louw 23.4.1945.
[20] Op cit Letter BFJS to Sir Edward Appleton 23.4.1945.
[21] Lang J 1986 *Bullion Johannesburg* (Johannesburg: Jonathan Ball) p 371.
[22] Ibid p 372.
[23] Gowing M 1974 *Independence and Deterrence* vol 1 (London: Macmillan) p 378.
[24] Ref [20] *op cit*.
[25] Malherbe *op cit* p 278.

[26] CSIR Archive: B26; 54; 56 Personal recollections of the early years of the CSIR written under the title 'Initial Contribution by Dr S. Meiring Naudé'.

[27] Ibid A-0/1 vol 2 Minutes of the Meeting of the Exploratory Committee of the CSIR (27 Aug 1945) 'Location of the CSIR'.

[28] Ref [26] *op cit* p 2.

[29] Author's interview with Dr W S Rapson, Johannesburg, 5.5.1997.

[30] University of the Witwatersrand Archive: GEO C/11/45 Minutes of the BPI Board of Control 27.3.1945.

[31] Lovell A C B 1990 *Astronomer by Chance* (London: Macmillan) pp 108–120.

[32] Ref [27] *op cit* Letter from the Secretary for External Affairs to the Scientific Adviser to the Prime Minister 31.7.1945.

[33] Kingwill [3] *op cit* p 106.

[34] Wits Archive: Geo C/62/45 Minutes of the 14th Ordinary Meeting of the Board of the Bernard Price Institute of Geophysical Research 13.7.1945. A subsequent set of minutes Geo C/62A/45 was issued nearly a year later (8.5.1946) in which Dr Price's introductory remarks were quite extensively changed, possibly at the request of Schonland, because the later set contains far less hyperbole about the role Schonland played in England at the AORG and within 21 Army Group, though Price's opening paragraph more than made up for that!

[35] CSIR Archive: Radar Museum file FJH 1/58 Letter from Ismay Schonland to Dr Frank Hewitt undated but *ca* October 1975; and newspaper cutting from the *Sunday Times* (Johannesburg) 28.9.1947; See also Sayer A P 1950 *Army Radar* (London: The War Office) p 188: 'Perhaps the most important contribution made by South Africa was in the field of Operational Research, where B N J [sic] Schonland – rising to the rank of Brigadier – was loaned by the South African Government to take charge of this work'.

[36] Wits Archive: File 25/a Misc C/10/36 Memorandum on the establishment of an Institute of Geophysical Research at the University of the Witwatersrand 14.4.1936.

[37] Wits Archive: GEO C/19/38 Minutes of the 8th Meeting of the BPI Board of Control item 4a 18.3.1938.

[38] CSIR Archive: A-0/1 vol 2 Minutes of the meeting of the Exploratory Committee on the CSIR held at the Union Buildings, Pretoria 27.8.1945.

[39] Kingwill ref [3] *op cit* p 23.

[40] South African National Archives: Smuts papers URU 2267 232 Document from the Prime Minister's Office.

[41] CSIR Archive: A-0/1 vol 2 Summary of activities for the period 5.10.1945 to 2.2.1946.

[42] Letter to the author from Dr F J Hewitt 29.10.1993; and Wadley T L 1954 'Variable-frequency crystal-controlled generators and receivers' *Transactions of the South African Institution of Electrical Engineers* **45** 77–99; also Thrower K R 1993 'Racal and the RA17 HF Communications Receiver (part 1)' *Radio Bygones* **25** 4–19; and Austin B A 1994 'The Racal RA17 (Letter correcting an error of fact)' ibid **30** 28–29.

[43] Nature **156** Untitled 8.12.1945 p 696.

[44] Rhodes University Cory Library for Historical Research: BFJS honorary degrees MS 14685.

[45] Imperial War Museum Schonland papers 86/63/1 Letters to BFJS from: Pile

15.1.1945; de Guingand 16.2.1945; Hill 7.2.1945; Solandt 4.3.1945; Johnson 14.5.1945; Bowles 14.6.1945; Humby 23.7.1945.

[46] CSIR Archives: ref [38] *op cit* Tabled paper X2 para 3.

[47] Letters: BFJS to O. Solandt 10.5.45 and Solandt to BFJS 6.6.45 supplied to the author by Professor T Copp, Wilfrid Laurier University, Toronto.

[48] Brown A C (ed) 1977 *A History of Scientific Endeavour in South Africa* (Cape Town: Royal Society of South Africa) p 194.

[49] Kingwill [3] *op cit* p 26.

[50] Wits Archive: GEO C/34/46 Minutes of the 15th Ordinary Meeting of the Board of the BPI 8.5.1946.

Chapter 16 NATIONALISTS AND INTERNATIONALISTS

[1] The Royal Society Report on the Empire Scientific Conference June–July 1946 vol 1 (London: The Royal Society 1948) p 11.

[2] SANDF Archive: PM 102/3 Commonwealth Scientific Conference. Letter BFJS to Secretary for Defence 17.11.1945.

[3] Letter BFJS to IMS 17.6.1946.

[4] Cambridge University Manuscripts Library: ADD3702 (Box 2) Congregation held at Cambridge on Monday 24.6.1946.

[5] Letter BFJS to IMS 9.6.1946.

[6] Gowing M 1974 *Independence and Deterrence* vol 1 (London: Macmillan) p 8.

[7] Smuts J C 1952 *Jan Christian Smuts* (London: Cassell & Co) p 518.

[8] South African Press Association 29.6.1946 'Schonland Wants Research Body for Africa'.

[9] Smuts ref [7] *op cit* p 497.

[10] Ref [3] *op cit*.

[11] Wits Archives: Misc GEO C/11/47 BPI Director's Report February 1947.

[12] Malan D J and Schonland B F J 1947 'Progressive Lightning VII. Directly-correlated photographic and electrical studies of lightning from near thunderstorms' *Proceedings of the Royal Society* **A191** 485–503.

[13] MacGorman D R and Rust W D 1998 *Red Sprites and Blue Jets* in ch 5 of *The Electrical Nature of Storms* pp 116–117.

[14] Schonland B F J and Gane P G 1947 Trans SAIEE **38** *A Lightning Warning Device* pp 119–125.

[15] Smuts ref [7] *op cit* p 487.

[16] Gane P G and Schonland B F J 1948 'The Ceraunometer' *Weather* **3** 174–178.

[17] CSIR Archive: Minutes of the 9th meeting of the Executive Committee 11.2.1947; Vermeulen D J and Blignaut P J 1961 'Underground radio communications and its use in mine emergencies' *Transactions of the South African Institution of Electrical Engineers* **52** 94–109.

[18] SANDF Archive Pretoria: DC 974/70 Designation of ACF Units 'South African Corps of Scientists' 15.8.1947; Government Gazette No 3865 5.9.1947; AG(1)(C)P1/11036/1 Appointment as Director S.A. Corps of Scientists, A.C.F. 18.10.1947.

[19] Ibid: British Government to SA Secretary for External Affairs (DG 17850/325) 4.1.1946; Reply from CGS (CGS 65/3) 2.2.1946; Secretary of Defence to CGS (DG 17850/325) 12.2.46; CGS minute (CGS 65/2) 19.3.1946.

[20] *Natal Mercury* 29.10.46 'Hush-hush Empire Scientific Talks'.

[21] SANDF Archive Pretoria: 'Post War Scientific Work' 65/3 CGS Houer 309 3.9.1945.

[22] Ibid: (DSD/SEC/7) Extracts from official correspondence dealing with the establishment of the South African Corps of Scientists. Personal communication (May 1997) to the author from Professor D F S Fourie of the University of South Africa.

[23] *Pretoria News* 31.10.47 Editorial 'Patience for the Scientists'.

[24] Gowing ref [6] ibid p 379.

[25] Hartcup G and Allibone T E 1984 *Cockcroft and the Atom* (Bristol: Adam Hilger) p 136.

[26] Public Record Office: Harwell Papers AB 6/139 'Liaison with South Africa': Letter BFJS to J D Cockcroft 10.9.1945.

[27] Gowing ibid ref [6] p 378.

[28] Public Record Office: *op cit* Letter Cockcroft to Makins 17.6.1947; Cockcroft to Schonland 15.7.1947.

[29] Ibid: Cockcroft to Tizard 15.7.1947 and Tizard's reply 17.7.1947.

[30] *Sunday Times* (South Africa) 2.5.48 'Schonland for Overseas Talks on Atomic Energy?'.

[31] Malherbe E G (1981) *Never a Dull Moment* (Cape Town: Howard Timmins) p 366.

[32] Letter BFJS to IMS 5.6.48 written from Washington.

[33] Cambridge University Library: Manuscripts ADD 8702 Box 2 Letter 16.8.1948 Sir John Cockcroft to BFJS.

[34] Wits Archive: Misc GEO C/15/48 BPI Director's Report February 1948.

[35] Ibid: GEO C/85/51 Anonymous memorandum of 1951 but almost certainly written by Schonland on the establishment of a section on Precambrian geology, geochemistry and tectonics at the Bernard Price Institute.

[36] *Die Transvaler* 9.10.58 'Instrument van twee Suid-Afrikaners in Spoetnik III'. The scientific paper describing the field mill appeared in D J Malan and B F J Schonland 1950 'An electronic fluxmeter of short response-time for use in studies of transient field-changes' *Proceedings of the Physical Society* (*London*) **63** 402–408.

[37] Schonland B F J 1950/1 'Hendrik Johannes van der Bijl' *Obituary Notices of Fellows of the Royal Society* **VII** 27–33.

[38] Bozzoli G R 1997 *Forging Ahead* (Johannesburg: Witwatersrand University Press) pp 172–198. It is worth clarifying the names of Price's two institutes at Wits since the preposition in each is important and they are different. The earth and its atmosphere are the domain of the Bernard Price Institute *of* Geophysical Research while early man and his ancestors are studied within the Bernard Price Institute *for* Palaeontological Research.

[39] National Archives of South Africa: URU 2611 12 27.10.1948.

Chapter 17 'COME OVER TO MACEDONIA AND HELP US!'

[1] Cambridge University Manuscripts Library: ADD 8702 Box 2 Letter E V Appleton to BFJS 10.1.1949.

[2] Churchill Archives Cambridge: CKFT 20/30 Letter BFJS to John Cockcroft undated but *ca* mid-January 1949.

[3] Hartcup G and Allibone T E 1984 *Cockcroft and the Atom* (Bristol: Adam Hilger) p 156.

[4] Ref [1] *op cit* Letter Appleton to BFJS 28.1.1949.

[5] Ibid. Letter Cockcroft to BFJS 2.2.1949.

[6] Ibid. Letter Darwin to BFJS 6.2.1949.

[7] Ibid. Letter Cockcroft to BFJS 23.2.1949.

[8] Ibid. Letter Tizard to BFJS 8.3.1949.

[9] Ibid. Letter Cockcroft to BFJS 9.3.1949.

[10] Ibid. Letter Cockcroft to BFJS 19.3.1949 and another of 28.3.1949.

[11] Ibid. Secret message BFJS to Rowlands undated but presumably 14.3.1949 in the context of those that follow.

[12] Ibid. Letter BFJS from Lt Col G W H Peters Staff Officer UKSLO Pretoria 9.4.1949.

[13] Ibid. Letter BFJS to Dr D F Malan 4.4.1949.

[14] Ref [1] *op cit* ADD 8702 Box 1 Letter from Dept of External Affairs in Cape Town to BFJS 12.4.1949.

[15] National Physical Laboratory Archives: NPL Executive Committee Minutes 27.9.1949 p 7.

[16] Royal Society Archives: Minutes of Council (item 14) 16.12.1948 p 3.

[17] Ref [14] *op cit* BFJS's (undated) hand-written notes of his meetings at Shell-Mex House, Ministry of Supply, London with Sir Ben Lockspeiser, Secretary of the DSIR, the Lord President of the Privy Council, Herbert Morrison; Sir Archibald Rowlands, Permanent Secretary of the Ministry and various officials.

[18] Ibid. Letter 17.1.1950 from Sir Charles Darwin to BFJS written many months after these meetings in London. Darwin wrote: 'The HQ, DSIR have long been celebrated for being wonderfully 'ham-handed' , but in this case it was further complicated by the change-over from Appleton to Lockspeiser'.

[19] Ref [14] *op cit* Letters 7.6.1949 and 14.6.1949 to BFJS from Lockspeiser and Morrison.

[20] Ibid. Letter from Sir Henry Tizard to BFJS 2.8.1949.

Chapter 18 COCKCROFT'S MAN

[1] Public Record Office: Harwell Papers AB6/139 Press Release from Shell-Mex House 11.6.1949.

[2] *Memories of my Father* Dr Mary Davidson's unpublished notes written for the author February 1996.

[3] Personal communication to the author from Sir Maurice Wilkes 6.8.1997.

[4] Cambridge University Manuscripts Library: ADD 8702 Letter BFJS to Sir Henry Tizard 21.8.1949.

[5] University of the Witwatersrand The Gubbins Library: QE 500 *Geophysical Conference* 27, 28 and 29 July 1949.

[6] *Pretoria News* 28.7.1949 Editorial 'Science and the Nation'.

[7] University of the Witwatersrand Archive: Misc GEO C/77/50 BPI Director's Report for 1949.

[8] D G Kingwill 1990 *The CSIR – the First 40 Years* (Pretoria: CSIR) p 17.

[9] Public Record Office: Harwell papers AB6/139 Letter S M Naudé to J D Cockcroft 19.8.1948.

[10] M Gowing 1974 *Independence and Deterrence: Britain and Atomic Energy 1945–1952* vol 1 (London: Macmillan) p 381.

[11] CSIR Archive: S M Naudé file B26; 54; 66 'Nuclear Physics' (undated).

[12] Public Record Office: *op cit* Letter J D Cockcroft to BFJS 24.4.1950.

[13] CSIR *loc cit* 'Initial Contribution by Dr S Meiring Naudé' – a record of his time at the CSIR (39 pp). Naudé's recollections reflect a degree of rodomontade in his make-up, especially with reference to his own contributions to the CSIR while tending, as well, to portray Schonland in a rather less than positive light.

[14] National Archives of South Africa: URU 2684 3266 Document in Afrikaans 5.10.49 signed by D F Malan.

[15] J C Smuts 1952 *Jan Christian Smuts* (London: Cassell & Company Ltd) p 519.

[16] Headlines in the *Sunday Times*, the *Star* and the *Pretoria News* in the days immediately before and during the Conference.

[17] *Pretoria News* 17.10.1949 Editorial 'The Backward Continent'.

[18] Kingwill [ref 8] *op cit* p 62.

[19] Rhodes University Cory Library MS 14 668 LL.D Laudation B F J Schonland University of Natal 19.3.1949.

[20] Cambridge Manuscript Library: *op cit* Letter Sidney Smith to BFJS 28.12.1949.

[21] Ibid Letter BFJS to D D Forsyth Secretary to the Prime Minister 26.1.1950.

[22] Ibid Letter Cockcroft to BFJS 22.1.1950.

[23] Ibid Letter BFJS to Cockcroft 31.1.1950.

[24] Ibid Letter BFJS to the Hon D F Malan 17.2.1950.

[25] Ibid Letter Department of External Affairs to BFJS 21.2.1950.

[26] *Rand Daily Mail* (Johannesburg) Commenting on the Marshall Clark affair 7, 24, 26 February 1950.

[27] *Hansard* of the Parliament of the Union of South Africa 14.4.1950 p 4334.

Chapter 19 SCIENTIA ET LABORE

[1] Wits Archive; Misc GEO C/80/50 Minutes of the Board of the BPI 1.3.1950.

[2] Personal correspondence to the author from Dr F J Hewitt 29.10.1993.

[3] Public Record Office Harwell papers AB6/139 Correspondence between Cockcroft and Schonland 19.6.1950; 7.7.1950 and 11.7.1950.

[4] Personal communication to the author from T V Peter 2.12.1999.

[5] CSIR Archive: A-0/1 vol 2 Minutes of the 14th meeting of the CSIR Council in Cape Town 17–19 July 1950 (TRL Report E.43).

[6] Ibid Minutes of the CSIR Council of 20.9.1950.

[7] Kingwill D G 1990 *The CSIR – the First 40 Years* (Pretoria: CSIR) p 16.

[8] Oliphant M L E and Lord Penney 1968 'John Douglas Cockcroft' *Biographical Memoirs of Fellows of the Royal Society* **14** 153.

[9] CSIR Archive: B26; 54; 56 Document (undated) entitled 'Initial Contribution by Dr S Meiring Naudé'.

[10] Personal communication to the author from Mrs Sheila Lloyd, a former CSIR employee, Cape Town 16.5.1997.

[11] Malan D J and Schonland B F J 1950 'An electrostatic fluxmeter of short response-time for use in studies of transient field-changes' Proceedings of the Physical Society **63** 402–408; and Schonland B F J and Malan D J 1950 'The distribution of electricity in thunderclouds' *Archives Met Geophys Bioklim* **3** 64–69.

[12] Schonland B F J 1950 *The Flight of Thunderbolts* (Oxford: Clarendon Press); with reviews in *Nature* of 19.5.1950; the *American Scientist* of July 1951 and the *Journal of the Franklin Institute* of May 1952.

[13] Cambridge University Manuscript Library: ADD 8702 Box 1 Letter BFJS to J D Cockcroft 31.1.1950; Imperial War Museum: Schonland papers 86/63/1 'General Account of Army Radar' (J D Cockcroft) typescript.

[14] CSIR Archive: S3/25 Letter BFJS to H R Raikes 5.5.1950.

[15] Wits Archive: GEO C/255/48 and Misc C84/49; Murray B K 1982 *Wits – The Early Years* (Johannesburg: Witwatersrand University Press) p 183.

[16] Wits Archive: GEO C/83/50 Minutes of the BPI Board 1.3.1950.

[17] Lang J 1986 *Bullion Johannesburg* (Johannesburg: Jonathan Ball) p 389.

[18] Murray ref [15] *op cit* p 64.

[19] Wits Archive: GEO C/85/51 Minutes of the BPI Board 21.3.1951.

[20] Ibid Paper A *Report of the Director for the Year 1950* 21.2.1951.

[21] Malan D J and Schonland B F J 1951 'The electrical processes in the intervals between the strokes of a lightning discharge' *Proceedings of the Royal Society* **A206** 145–163; 1951 'The distribution of electricity in thunderclouds' Ibid **A209** 158–177; Ogawa T and Brook M 1969 'Charge distribution in thunderstorm clouds' *Quarterly Journal of the Royal Meteorological Society* **95** 513–525; Personal communication to the author from Dr D E Proctor 2.12.1999.

[22] Schonland B F J 1951 'The work of the Bernard Price Institute of Geophysical Research 1938–1951' *Transactions of the South African Institution of Electrical Engineers* **42** 241–258.

[23] Rhodes University Cory Library for Historical Research: Addresses by Dr B F J Schonland CBE LL.D DSc FRS on his installation as the first Chancellor of Rhodes University 25.10.1951.

[24] Currey R F 1970 *Rhodes University 1904 – 1970* (Grahamstown: Rhodes University) p 118.

[25] Rhodes University Cory Library for Historical Research: Address to the Congregation of Rhodes University at the South African Native College Fort Hare 26.10.1951.

[26] Sparks A 1990 *The Mind of South Africa* (London: Heinemann) p 32.

[27] Mandela N R 1994 *Long Walk to Freedom* (London: Little, Brown and Company) p 69 *et seq*.

[28] Author's interview with Schonland's daughter Dr Mary Davidson, Winchester 18.8.1997.

[29] *Hansard* (South Africa) – Debates of the House of Assembly 3rd Session 10th Parliament **71** 17.4.1950 p 4335.

[30] Schonland B F J 1951 *On being a Scientific Adviser to a Commander-in-Chief* Copy supplied to the author by Professor D F S Fourie of the University of South Africa May 1997; also in Imperial War Museum: Montgomery papers BLM 140/6.

[31] South African National Defence Force (SANDF) Archive: RVN1 Letters BFJS to de Guingand 19.2.1951 and 2.3.1951; from Tizard 17.4.1951; to Solandt 7.5.1951 and to du Toit 24.7.1951.

[32] Imperial War Museum: Schonland Papers 86/63/1 Letter de Guingand to Schonland 21.3.1951.

[33] SANDF Archive: DC2623 (S) Minutes of the 1st Meeting of the Defence Science Policy Committee held at the University of the Witwatersrand 23.4.1951.

[34] SANDF Archive: ref [31] *op cit* Letter BFJS from H R Raikes 11.10.1951.

[35] Hartcup G and Allibone T E 1984 *Cockcroft and the Atom* (Bristol: Adam Hilger Ltd) pp 168 and 190.

[36] Murray B K 1997 *Wits – The Open Years* (Johannesburg: Witwatersrand University Press) p 139.

[37] Wits Archive: S3/25 Letter BFJS to Humphrey Raikes 12.12.1951.

[38] SANDF Archive: ref [31] *op cit* HGS 65/1 Letter BFJS from CGS 22.1.1952.

[39] Cambridge University Manuscript Library: ADD8702 Letter C T R Wilson to BFJS 4.7.1951.

[40] Wits Archive: 3/8 GEO vol II BPI Director's Report for 1952.

[41] Schonland B F J 1952 'The work of Benjamin Franklin on thunder-storms and the development of the lightning rod' *Journal of the Franklin Institute* **253** 375–392.

[42] The Nuffield Foundation, London, Sixth Annual Report 1952 *University of the Witwatersrand, Bernard Price Institute of Geophysical Research: Nuffield Research Unit* pp 93–94.

[43] Public Record Office AB6/915 Letter J D Cockcroft to Mr F C How 28.2.1952 and How's reply 29.2.1952.

[44] Schonland B F J 1952 'The South African Association for the Advancement of Science, its past and its future' *South African Journal of Science* **49** 61–68.

[45] Cambridge University Manuscript Library: ADD 8702 Box 2 Letter Zuckerman to BFJS 13.11.1952.

[46] Schonland B F J 1953 'The pilot streamer and the long spark' *Proceedings of the Royal Society* **A220** 25–38. The suggestion of a pilot streamer first appeared in Schonland's Halley Lecture 1937 (Oxford: The Clarendon Press) p 11 and in detail in Schonland B F J 1938 'Progressive lightning IV – the discharge mechanism' *Proceedings of the Royal Society* **A164** 137 et seq.

Chapter 20 'COMETH THE HOUR'

[1] Murray B K 1997 *Wits: The Open Years* (Johannesburg: Witwatersrand University Press) pp 65, 138, 140 and 147.

[2] Imperial War Museum: Schonland papers 86/63/1 Letter J D Cockcroft to BFJS 29.3.1945.

[3] Author's interview with Professor F R N Nabarro FRS Johannesburg 7.5.1997.

[4] Murray *op cit* p 277.

[5] Cambridge University Manuscript Library: ADD 8702 Letter BFJS to F R N Nabarro 30.11.1952.

[6] Ibid. Letter M L Oliphant to BFJS 20.1.1953.

[7] Brown A 1997 *The Neutron and the Bomb* (Oxford: Oxford University Press) p 189.

[8] Gowing M M 1974 *Independence and Deterrence* vol 1 (London: Macmillan) p 335.

[9] Cambridge University Manuscript Library: *op cit* Telegram BFJS from Oliphant 7.4.1953.

[10] Forster S G and Varghese M M 1996 *The Making of The Australian National University 1946–1996* (Sydney: Allen & Unwin) pp 116–118.

[11] Sparks A 1990 *The Mind of South Africa* (London: Heinemann) p 183.

[12] Esterhuyse W and Nel P 1990 *The ANC and its Leaders* (Cape Town: Tafelberg) p 9.

[13] Luthuli A 1962 *Let my People Go* (Glasgow: Collins) p 117.

[14] Sparks *op cit* p 194.

[15] Mandela N R 1994 *Long Walk to Freedom* (London: Little, Brown and Co) pp 119–120.

[16] Luthuli *op cit* p 115.

[17] Letter BFJS to IMS 10.8.41.

[18] Paton A 1964 *Hofmeyr* (London: Oxford University Press) pp 293–294.

[19] Public Record Office Harwell papers AB6/915 Letter BFJS to J D Cockcroft 10.8.1953.

[20] Op cit.

[21] Hartcup G and Allibone T E 1984 *Cockcroft and the Atom* (Bristol: Adam Hilger) p 176.

[22] Rhodes University Cory Library: MS 11037 Letter BFJS to Felix Schonland 20.5.1954.

[23] Ibid. MS 10746 BFJS to Felix 18.6.1954.

[24] Schonland B F J and Malan D J 1954 'Upward stepped leaders from the Empire State Building' *Journal of the Franklin Institute* **258** 271–275.

[25] University of Natal Pietermaritzburg: GOV Address by the Chancellor Dr B F J Schonland CBE, LL.D, DSc, FRS at the ceremony to commemorate the fiftieth anniversary of the founding of Rhodes University College, 24th September 1954.

Chapter 21 HARWELL

[1] Allibone T E 1973 'Basil Ferdinand Jamieson Schonland' *Biographical Memoirs of Fellows of the Royal Society* **19** 629–653.

[2] Gowing M 1974 *Independence and Deterrence – Britain and Atomic Energy, 1945–1952* vol 1 (London: Macmillan) p 39.

[3] Oliphant M L E and Lord Penney 1968 'John Douglas Cockcroft' *Biographical Memoirs of Fellows of the Royal Society* **14** 139–188.

[4] Gowing M 1974 *Independence and Deterrence* vol 2 (London: Macmillan) p 21.

[5] Ibid p 17.

[6] Public Record Office: AB6/139 Letter G W Raby to J D Cockcroft 19.1.1948.

[7] Hartcup G and Allibone T E 1984 *Cockcroft and the Atom* (Bristol: Adam Hilger) p 195.

[8] Oliphant and Penney *op cit* p 206.

[9] Jay K E B 1955 *Atomic Research at Harwell* (London: Butterworth).

[10] Gowing vol 2 *op cit* p 15.

[11] Jay K E B 1956 *Calder Hall* (London: Methuen) p 1.

[12] Brown A 1998 'Patrick Blackett: sailor, scientist, socialist' *Physics World* **11** 35-38; Hartcup and Allibone *op cit* p 138.

[13] Brown A 1997 *The Neutron and the Bomb – A Biography of Sir James Chadwick* (Oxford University Press) p 205.

[14] Gowing vol 2 *op cit* ch 16.

[15] Gowing vol *op cit* pp 104–112.

[16] Ibid pp 210–212.

[17] Anon 1947 'Peaceful applications of nuclear energy' *Nature* **160** 451–453.

[18] Gowing vol 2 *op cit* ch 22 for a most detailed discussion of the *modus operandi* of the individual factories under Hinton and of the design and performance details of the various piles or reactors that contributed enormously to the development of the British atomic programme.

[19] Ibid p 392.

[20] Gowing vol 1 *op cit* p 231.

[21] Hartcup and Allibone *op cit* p 159.

[22] Spence R 1967 'Twenty-one years at Harwell' *Nature* **214** 343–344, 436–438.

[23] Gowing vol 1 *op cit* pp 425–433.

[24] The Earl of Birkenhead 1961 *The Prof in Two Worlds – the official life of Professor F A Lindemann, Viscount Cherwell* (London: Collins) ch 11.

Chapter 22 FISSION BUT NO FUSION

[1] Vick F A 1965 *The Atomic Energy Research Establishment, Harwell* in Cockcroft J D (ed) *The Organization of Research Establishments* (Cambridge University Press) pp 55–77.

[2] Spence R 1967 'Twenty-one years at Harwell' *Nature* **214** 343–344, 436–438.

[3] Gowing M 1974 *Independence and Deterrence* vol 2 (London: Macmillan) pp 255–256.

[4] Hartcup G and Allibone T E 1984 *Cockcroft and the Atom* (Bristol: Adam Hilger) p 185.

[5] Churchill Archive Cambridge: SCHO 7 Letter BFJS to A Hales 4.11.1954.

[6] *Harlequin* vol 5 no 2 Christmas 1954 Foreword written by B F J Schonland. The *Harlequin* was the social magazine of Harwell that contained news of a general nature of interest to its community of scientists and their families.

[7] Letter to the author from Dr J E Johnston 14.2.2000.

[8] Johnston *op cit.*

[9] Cmnd 9389 1955 *A Programme of Nuclear Power* (London: HMSO).

[10] Arnold L 1995 *Atomic Energy in Britain 1939 – 1995* (Didcot: UKAEA) pp 12, 13 and Appx III; Personal communication to the author from Mrs Lorna Arnold 5.3.2000.

[11] Goodlet B L 1953 'The outlook for economic nuclear power' *Engineering* **176** 345–347.

[12] Hartcup and Allibone *op cit* p 186.

[13] Glueckauf E 1977 'Robert Spence' *Biographical Memoirs of Fellows of the Royal Society* **23** 521.

[14] Keith S T 1993 *The Fundamental Nucleus: A study of the impact of the British atomic energy project on basic research* (London: HMSO) p 34.

[15] Hartcup and Allibone *op cit* p 199.

[16] United Kingdom Atomic Energy Authority *Second Annual Report* 1955–1956 (London: HMSO) p 2.

[17] Hendry J and Lawson J D 1993 *Fusion Research in the UK 1945–1960* (London: HMSO) p 29.

[18] Gowing *op cit* p 139; also Public Record Office (HO 334) Register R1/12792 for Schonland's registration as a citizen of the United Kingdom and Colonies on 3 March 1955. In an interview with the *Sunday Express* on 11 June 1960 he disclosed that he had given up his South African nationality six years before but refused to discuss why he had done so.

[19] AWE Rowley File 0019 Pt 1B (ES1/15) Letter from Penney to BFJS 6.8.1955.

[20] Ibid Minutes of 1st Ψ Meeting Section A 20.12.1955. The Ψ Committee was to meet every six months. Section A was made up of those fully cleared to discuss weapons reactions while Section B was restricted to physics details. Both were under Schonland's chairmanship.

[21] A caption to the photograph showing the party about to board their aircraft appeared in the atomic industry magazine *Atom Industry* December 1955 p 1.

[22] Letter to the author from Dr Mary Davidson, Schonland's daughter 19.2.1996.

[23] Hartcup and Allibone *op cit* p 203.

[24] Hendry and Lawson *op cit* p 38.

[25] Pease R S and Schonland B F J 1960 'Obituary: Academician I V Kurchatov' *Nature* **185** 887.

[26] Hendry and Lawson *op cit* p 30.

[27] Personal communications to the author from Dr R S Pease 16.2.1998 and 17.3.2000.

[28] *The Times* (London) 26.4.1956 'Soviet Scientist's Lecture: Questions answered at Harwell' p 10; and 27.4.1956 'Heavy Hydrogen Experiments: Russia's remarkable contribution' p 6.

[29] Churchill Archive Cambridge: SCHO1 11/DD/1 Correspondence between F R N Nabarro, J P F Sellschop and BFJS between December 1955 and September 1956.

[30] Letter to the author from Emeritus Professor J P F Sellschop 12.10.1998.

[31] Churchill Archive Cambridge: SCHO3 11/DD/2 Letter BFJS to I D Macrone 29.5.1956.

[32] CSIR Archive: B26; 54; 35 'History of the CSIR' a contribution by the late Dr W H Craib; also 'Discussions with William Hofmeyr Craib' transcription of a tape recording made of discussions with Craib in 1982 by E R and S R Adams in the library of the Department of Medicine, University of the Witwatersrand, Johannesburg.

[33] University of Cape Town, The Harry Oppenheimer Institute of African Studies: Minutes of a Special Meeting of Council 6.6.1956.

[34] Churchill Archive Cambridge: SCHO1 Letter F C Robb to BFJS 10.7.1956.

[35] Jay K E B 1956 *Calder Hall* (Norwich: Jarrold & Sons) p iii.

[36] Churchill Archive Cambridge: SCHO8 11/DD/9(1) Visit to America September 1957 and 11/DD/9(2) visit to India and Pakistan January 1957.

[37] *The Star* (Johannesburg) 26.6.1957, 'Wits honour is his greatest—Future does not frighten atom scientist' : an interview with Schonland conducted by one of the newspaper's columnists.

[38] Public Record Office: AB6/1693 Letters BFJS to S M Naudé 28.3.1956 and to Sir Edwin Plowden 13.3.1957.

[39] Cambridge University Manuscript Library: ADD 3702 Box 2 Letter from S B Hobson to BFJS 15.9.1957.

Chapter 23 ZETA AND WINDSCALE

[1] See Hendry J and Lawson J D 1993 *Fusion Research in the UK* 1945–1990 (London: HMSO). This account of the work at Harwell, its associated sites and its various collaborating organizations presents by far the most accurate and detailed review of the underlying scientific theory, the engineering problems and the contributions of all the persons most closely associated with the British fusion programme that the author has yet seen. It therefore served as the primary source for much of the material in this chapter.

[2] Brown A P 1997 *The Neutron and the Bomb – A biography of Sir James Chadwick* (Oxford University Press) p 189.

[3] Hartcup G and Allibone T E 1984 *Cockcroft and the Atom* (Bristol: Adam Hilger) p 202.

[4] Churchill Archive Cambridge SCHO7 Letter BFJS to F R N Nabarro 20.11.1958.

[5] Personal communication to the author from Dr R S Pease *Some random recollections of Basil Schonland* 12.2.1998.

[6] Lawson J D 1957 'The industrial applications of thermonuclear reactions' *Nature* **180** 780–782.

[7] Hansard 11.11.1957 p 611.

[8] Hartcup and Allibone *op cit* p 205.

[9] Ibid p 206; Hendry and Lawson *op cit* p 51.

[10] *The Times* 24.1.1958 'New Harwell Director Named; Changes in Atomic Authority' p 5; South African Press Association 24.1.1958.

[11] UKAEA Fourth Annual Report 1 April 1957–31 March 1958 (London: HMSO) para 23.

[12] Grant P J 1959 'Wigner energy storage – its relation to reactor design' *Nuclear Engineering* **4** 69–72.

[13] Gowing M 1974 *Independence and Deterrence – Britain and Atomic Energy, 1945–1952* vol 2 (London: Macmillan) p 234.

[14] Ibid p 32.

[15] See Arnold L 1995 *Windscale 1957 – Anatomy of a Nuclear Accident* 2nd edition (London: Macmillan) for an extensive and most readable account of the fire in Pile No 1 at Windscale; the events that led up to it, the subsequent inquiry and the official reports that followed. This excellent book was the major source of relevant material in this chapter.

[16] Hartcup and Allibone *op cit* p 211.

[17] Arnold *op cit* ch 5 *Damage Assessment and Damage Control* pp 60–75.

[18] Ibid p 67.

[19] Horne A 1989 *Macmillan 1957–1986* (London: Macmillan) pp 51–59.

[20] Rhodes University Cory Library: MS 11042 Letter BFJS to Felix 29.12.1957.

[21] Letter to the author from E H Cooke-Yarborough 26.1.2000.

[22] *Report on the Accident at Windscale No 1 Pile on 10th October 1957* reproduced as Appendix XI in Arnold's *Windscale 1957 – Anatomy of a Nuclear Accident op cit* pp 189–204.

[23] Horne *op cit* p 54.

[24] Arnold *op cit* pp 66 and 81.

[25] Cmnd 302 8.11.57 *Accident at Windscale No 1 Pile on 10th October 1957* (London: HMSO).

[26] Arnold *op cit* p 93.

[27] Gowing M 1990 'Lord Hinton of Bankside OM' *Biographical Memoirs of Fellows of the Royal Society* **36** 229.

[28] Ref [20] *op cit.*

[29] Arnold *op cit* p 91.

[30] Letter to the author from R F Jackson 5.2.1998; see also Cottrell A 1981 'Annealing a nuclear reactor: an adventure in solid-state engineering' *Journal of Nuclear Materials* **100** 64–66.

[31] Rhodes University Cory Library: MS 11043 Letter BFJS to Felix 6.4.1958.

[32] Arnold *op cit* p 95 and ref [20] *op cit.*

[33] Ref [20] *op cit.*

Chapter 24 'CAPTAIN AND NOT THE FIRST OFFICER'

[1] *Nature* 25 January 1958 'Harnessing nuclear energy' (editorial) **181** 213.

[2] *New Scientist* 16 January 1958 'News held back' (editorial) **3** 7.

[3] Thonemann P C, Butt E P, Carruthers R, Dellis A N, Fry D W, Gibson A, Harding G N, Lees D J, McWhirter R W P, Pease R S, Ramsden S A and Ward S 'Production of high temperatures and nuclear reactions in a gas discharge' 25 January 1958 *Nature* **181** 217–220.

[4] Cockcroft J D 1958 'The next stages with Zeta' *The New Scientist* **3** 14.

[5] Gibson A 1958 Nature **181** *Controlled Thermonuclear Reactions* – a summary of the meeting to discuss recent research results held on 5 February 1958 pp 803–806.

[6] Rhodes University Cory Library: MS 11043 Letter BFJS to Felix 6.6.1958 though in the text he refers to it being Easter Monday which means it was 6 April.

[7] Churchill Archive Cambridge: SCHO2 Letters BFJS to Frank Hewitt 10.2.1958; Cyrus Smith, Information Attaché 12.2.1958; J P Jordi South African Press Association 17.2.1958; de Guingand and Solandt 11.2.1958; Wrightson 24.1.1958; Lance-Bombardier H L Smith 9.2.1958.

[8] Cambridge University Manuscript Library: ADD 8702 Box 1 Letter from Field Marshal Montgomery to BFJS 10.4.1958.

[9] *Harlequin* **22** 1958 'A note From Dr B F J Schonland, Director A.E.R.E.' p 13.

[10] Cmnd 338 December 1957 *Report of the Committee appointed by the Prime Minister to examine the Organisation of certain parts of the UKAEA* (London: HMSO).

[11] Hartcup G and Allibone T E 1984 *Cockcroft and the Atom* (Bristol: Adam Hilger) p 200.

[12] Fry D W 1960 Atom **42** *The Atomic Energy Establishment, Winfrith* pp 10–12.

[13] Schonland B F J 1958 Atom **21** *Experimental Results with Zeta*. A statement made by the Director of the Atomic Energy Research Establishment, Harwell at a Press Conference held in London on Friday 16th May, 1958 p 2.

[14] Hendry J and Lawson J D 1993 *Fusion Research in the UK 1945–1960* (London: HMSO) p 55. This most comprehensive document is the definitive account of this vital period in Harwell's history and, unless stated otherwise, is the source of much of the material in this chapter.

[15] Ibid p 72.

[16] Willis J A V 1959 'The National Institute' *Harlequin* **26** 22.

[17] UKAEA Fifth Annual Report 1 April 1958–31 March 1959 p 60.

[18] Hendry and Lawson *op cit* p 73.

[19] Letter to the author 31.12.1997 from R M Fishenden who was Head of the Scientific Administration Office at Harwell at that time.

[20] Churchill Archive: SCHO7 Letter BFJS to Anton Hales 31.12.1958.

[21] Rhodes University Cory Library: MS 11044 Letter BFJS to Felix 21.12.1958.

[22] UKAEA ref [17] *op cit* para 18.

[23] Schonland B F J 1959 *Atom* **32**. Opening address at a Convention on Thermo-nuclear Processes organized by the Institution of Electrical Engineers and the Nuclear Engineering Conference held in London on 29 and 30 April pp 10–12.

[24] Author's telephonic interview with Professor P C Thonemann 26.4.2000.

[25] Vick F A 1965 *The Atomic Energy Research Establishment, Harwell* ch4 in *The Organization of Research Establishments* edited by Sir John Cockcroft (Cambridge University Press) p 58.

[26] Vick F A 1998 *Sir Basil Schonland – Some Reminiscences* prepared for the author by Sir Arthur Vick shortly before his death in September that year. Vick's account of receiving a hand-written letter from Schonland inviting him to become Deputy Director with a view ultimately to succeeding Schonland as Director conflicts with that in [11] and in M C Crowley-Milling's *John Bertram Adams – Engineer Extraordinary* 1993 (Gordon and Breach), both of which ascribe Vick's appointment to Cockcroft without any prior consultation. Though now seen to be erroneous, such an eventuality was by no means unlikely!

[27] *Nuclear Engineering* 1959 **4** November editorial *Whither Harwell?* pp 379–380.

[28] Hartcup and Allibone *op cit* p 215.

[29] Churchill Archive Cambridge: SCHO2 Letter from BFJS to D E H Peirson 16.12.1959.

[30] Hansard col 38 para 56 28.7.1959.

[31] Hendry and Lawson *op cit* p 78 and 85.

[32] Atom 1960 **49** November *UKAEA Press Releases* p 4.

[33] Spence R 1972 *The Times* 5 December p 18 *Letter to the Editor* following the publication of Schonland's obituary on 25 November. Spence related how Schonland's health had begun to fail during 1960 and that had prompted him to decide to retire.

Chapter 25 'FOR THE FUTURE LIES WITH YOU'

[1] Churchill Archive Cambridge: SCHO4 11/DD/3 (undated) Miscellaneous lectures, addresses etc delivered by Sir Basil Schonland.

[2] *op cit* SCHO 2 Letter to BFJS from D G James Vice-Chancellor of Southampton University 1.1.1959 and one from N Kurti 25.2.1959 congratulating him on his election to the Fellowship of Caius.

[3] Cambridge University Manuscript Library: ADD3702 Box 2 Letter from the Office of the Prime Minister to BFJS 4.5.1960.

[4] Peierls R 1985 *Bird of Passage* (Princeton: Princeton University Press) p 324.

[5] I am indebted to Professor R H Dalitz FRS of the Department of Theoretical Physics at the University of Oxford for supplying me with this extract from Schonland's letter to Peierls of 3.2.1960 and for permission to quote from it.

[6] Churchill Archive Cambridge: SCHO2 Letter from Peierls to BFJS 6.3.1960. The matter did not rest there. In 1999 the columns of *The Spectator* (29 May, 10 July, 17 July and 24 July) reverberated to charge and counter charge about Peierls allegedly being the Russian spy who went under the crypto-nyms of *Vogel* and *Pers* as described in the so-called *Venona* decrypts of Soviet intelligence communications just recently made public. See West N 1999 *Venona* (London: Harper Collins).

[7] Pease R S and Schonland B F J 1960 'Academician I V Kurchatov' *Nature* **185** 887.

[8] Author's interview with Dr Mary Davidson (BFJS's daughter) 18.8.1997.

[9] Lists of Members of the General Board of the NPL for the years 1956–1961, National Physical Laboratory, Queens Road, Teddington.

[10] Churchill Archive Cambridge: SCHO3 Letter to BFJS from N Mott 13.7.1956 and reply 28.8.1956. Mott actually asked Schonland to write in support of Blackett for the Royal Medal even though Blackett had already been awarded it in 1940. Schonland corrected this in his reply. I am indebted to Sir Bernard Lovell OBE FRS for his correspondence on this subject.

[11] Schonland B F J 1956 'Benjamin Franklin: natural philosopher' *Proceedings of the Royal Society* **A235** 433–444.

[12] Schonland B F J 1956 *Handbuch der Physik* S Flügge (ed) *The Lightning Discharge* (Berlin: Springer-Verlag) pp 576-628.

[13] Schonland B F J 1959 'The Invisible College' *International Affairs* **35** 141–150.

[14] *ECHO*, a publication of UKAEA Corporate Communications: *Harwell 1946-1996 Special Anniversary Edition*. A US Air Force type T33 'Starfire' made an unscheduled landing at Harwell on 22 April 1957. A personal account of the incident was given to the author on 16.5.1997 by Professor Frank Brooks, Emeritus Professor of Nuclear Physics at UCT, who was a Harwell Fellow in 1955 and member of staff of the Nuclear Physics section at Harwell until 1963.

[15] Cambridge University Manuscript Library: ADD8702 Box 1 Letter from Sir Roger Makins to BFJS 9.1.1961.

[16] Cartwright A P *The Star* Johannesburg 'The " Boffin" whom Britain wanted at Harwell' , Monday 18 April 1960 pp 13–14; 'Schonland sees delay in atomic electricity here' , Ibid 23 January 1961.

[17] Letter to the author from Professor J P F Sellschop 12.10.1998.

[18] Letter to the author from Dr F J Hewitt 16.3.2000.

[19] Churchill Archive Cambridge: SCHO2 Correspondence between BFJS and Frederick Spencer Chapman, Headmaster of St Andrews's College, Grahamstown 12.4.1960, 28.4.1960 and 15.5.1960.

[20] Currey R F 1970 *Rhodes University 1904–1970* (Grahamstown: Rhodes University) pp 137–139.

[21] Rhodes University, Grahamstown 1961 *Graduation Address by the Chancellor Sir Basil F J Schonland CBE DSc (Cape Town and Southampton), LLD (Rhodes and Natal) FRS*. Missing from these post-nominal letters was the honorary DSc awarded to him by Wits in 1957.

[22] University of the Witwatersrand Manuscripts Library: AD1502 Q A Whyte papers. The correspondence in this collection between Quintin Whyte and both the Schonlands spans a period from May 1954, when Schonland was still at the BPI, until 1963 when he and Ismay had retired to Denchworth House.

[23] Churchill Archive Cambridge: SCHO1 Letter from BFJS to Sir Owen Wansborough-Jones Chief Scientist at the Ministry of Supply as well as similar correspondence (in SCHO3) to the National Coal Board and to the Ministry of Power 31.7.1957.

[24] Ibid SCHO2 Letter BFJS to Sir William Penney 1.2.1960.

[25] Schonland B F J 1961 Report commissioned by the Transvaal and Orange Free State Chamber of Mines: *A Review of the Research Activities supported by the Gold Producers' Committee of the Chamber of Mines*. I am indebted to Dr J M Stewart, Mining Consultant to the Chamber of Mines of South Africa, for sending me this report and for permission to quote from it.

[26] Churchill Archive Cambridge: SCHO2 Invitation to BFJS become President of Section A of the British Association for the Advancement of Science 20.10.1959 and Schonland's reply 30.10.1959.

[27] Letter to the author from Sir John Mason FRS 9.8.1999.

[28] Schonland B F J 1962 Presidential Address—Section A British Association Manchester 31 August *Lightning and the Long Electric Spark* pp 306–313.

[29] The Institution of Electrical Engineers 13.3.1962 Faraday Medallist Talking Film [sic] Sir Basil Schonland CBE FRS Hon.M(SA)IEE.

[30] Spence R 1972 *The Times* (Letters) 5.12.1972.

[31] Rhodes University Cory Library: Minutes of the meeting of the Council held on Friday 22.6.1962.

[32] Neville T (undated) *More Lasting than Bronze – A story of the 1820 Settlers National Monument* (Pietermaritzburg: The Natal Witness Printing and Publishing Company) p 27.

[33] I am indebted for this information to Professor Emeritus Terence Beard, former head of the Department of Political Studies at Rhodes, who was responsible for drawing up and circulating the petition; and also to Mrs Ann Oosthuizen, Schonland's younger daughter who, as the wife of the late Professor 'Daantjie' Oosthuizen, Professor of Philosophy at Rhodes at the time, for her recollections of the incident (interview 18.4.1997). Professor Oosthuizen was away from Rhodes on sabbatical leave at the time these events occurred.

[34] Currey R F 1970 *op cit* p 155.

[35] Schonland B F J 1964 *The Flight of Thunderbolts* 2nd edition (Oxford: Clarendon Press).

[36] Schonland B F J 1968 *The Atomists (1805–1933)* (Oxford: Clarendon Press) Review by J L Heilbron *American Historical Review* December 1968.

[37] Schonland B F J 1964 'Maurice Pascal Alers Hankey, First Baron Hankey of the Chart 1877–1963' *Biographical Memoirs of Fellows of the Royal Society* **10** 137–146.

[38] Allibone T E 1973 'Basil Ferdinand Jamieson Schonland 1896–1972' *Biographical Memoirs of Fellows of the Royal Society* **19** 629–653.

[39] *The Times* 25.11.1972; Spence R *op cit* 5.12.1972.

[40] Personal communication from Professor Emeritus J P F Sellschop 28.4.2000.

INDEX

Defence) (SACS)185; Superintendent
ADRDE(ORG) 229; Superintendent
AORG 252; Scientific Adviser to 21
Army Group 271 *et seq*; South African
Corps of Scientists 417; promoted
Lieutenant 48; Captain 51; Major 185;
Lt Colonel 199; Colonel 235; Brigadier
252
nationality 9, 474
offers made to, Australian National
University (ANU) 436–8, British
scientific establishments 364–79,
University of Toronto 390–1
personal characteristics: an inspirational
leader 231, 254; 'a slightly forbidding
figure'184; 'austere, remote and
authoritarian' 557; beleaguered 542;
charades 184; enthusiasm 180; formal
demeanour 558; humour 249, 298,
558; ill health 553; innate friendliness
and charm 273; inspired confidence
233; self-doubt 225, 234, 271, 288;
shyness; 184, 557; 'the odd precipitate
decision' 233; urbane and benign
exterior 278; personal research: as a
young soldier 49; at The Cavendish
Laboratory 58; on β *scattering* 58 *et seq*;
on *atmospheric electricity* 83 *et seq*; on
the penetrating radiation 99; on
lightning 110 *et seq*; at the *BPI* (pre-
war) 151 *et seq*; at the *BPI* (post-war)
400 *et seq*;
relationship with Cockcroft 98, 218, 230,
233–4, 272, 298, 308, 358, 367, 371, 374,
380 et seq, 413, 419, 422, 436, 443, 449,
446, 514, 523, 540, 549
relations with the mining industry 324,
336, 350, 565
research, direction of: BPI 142 *et seq*, 395
et seq; CSIR 302 *et seq*;
Harwell 449 *et seq*
schools: Victoria Infants 3; St Andrew's
Preparatory 3; St Andrew's College 6
'South Africa's outstanding individual
contribution to our war effort' 321
'soldier and scientist' 252
speculates on uses of atomic energy 490
university education: Rhodes University
College 11; Cambridge 19
visits: East Africa 488; Canada 422; India
487; Pakistan 487; Rhodesia 488;
Russia 474; Sweden 128; South Africa
488, 559; United Kingdom 338, 364;
USA 127, 421
writes: *Atmospheric Electricity* 112, 421;
The Flight of Thunderbolts 398; *On being
a Scientific Adviser to a Commander-in-
Chief* 413; *The Atomists* 577
Windscale board of enquiry 512
Winfrith controversy 530
ZETA 528
ZETA II 530
'Schonland Park' *fn* 398
Schonland Research Centre for Nuclear
Sciences 579
Schonland, Ann (daughter) 77, 118, 250, 365
Schonland, David (son) 77, 104, 163, 235, 250,
342, 520, 551
Schonland, Ismay (Isabel Marion) (wife) 205,
217–8, 261, 280, 315, 321, 361, 417, 422,
485, 553, *fn* 579
Basil's confidante 205–6, 225, 246
custodian of official SSS documents 202
encouragement for SAIRR 442
Head of Womens' Residence at UCT 136
a 'home of their own' 549
hosted tennis parties for the SSS 184
issues of morale within the SSS 298
marriage to Schonland 77
South Africa's racial policies 441–2
promoted Captain in Women's Auxiliary
Army Service 288
returns to South Africa 488
sails to the USA with Basil in 1934 127
'Smuts too occupied with world
problems' 298
Smuts's visit to the Schonland home
141–2
socialist ideals 309, 411, 551
spends summer of 1929 at Wits 102
joins Basil in England 245
undertakes RDF and signal liaison duties
in London 250
visits Fort Hare 408
Schonland, Mary (daughter) 77, 104, 141–2,
144, 235, 250
Schreiner, Olive 406
Schreiner, W P 21, 24
Schuster, P 359
'Scientia ' fn 398
Scientific Adviser to the Army Council
(SAAC) 235, 252, 265
Scientific Adviser to General (later Field
Marshal) Montgomery 231, 382, 447